Symmetries in Science XV 2011

Journal of Physics: Conference Series Volume 380

Bregenz, Austria
31 July – 5 August 2011

ISBN: 978-1-62276-847-9
ISSN: 1742-6588

Printed from e-media with permission by:

Curran Associates, Inc.
57 Morehouse Lane
Red Hook, NY 12571

Some format issues inherent in the e-media version may also appear in this print version.

Printed by Curran Associates, Inc. (2013)

For permission requests, please contact the Institute of Physics
at the address below.

Institute of Physics
Dirac House, Temple Back
Bristol BS1 6BE UK

Phone: 44 1 17 929 7481
Fax: 44 1 17 920 0979

techtracking@iop.org

Additional copies of this publication are available from:

Curran Associates, Inc.
57 Morehouse Lane
Red Hook, NY 12571 USA
Phone: 845-758-0400
Fax: 845-758-2634
Email: curran@proceedings.com
Web: www.proceedings.com

Papers

Symmetries in Science XV

Dieter Schuch and Michael Ramek

Abstract Metrics

PREFACE

FREUNDE DER UNIVERSITÄT

Bregenz, the peaceful monastery of Mehrerau and the Opera on the Floating Stage again provided the setting for the international symposium 'Symmetries in Science'. The series which has been running for more than 30 years brings together leading theoreticians whose area of research is, in one way or another, related to symmetry. Since 1992 the meeting took place biannually in Brengez until 2003. In 2009, with the endorsement of the founder, Professor Bruno Gruber, we succeeded in re-establishing the series without external funding. The resounding success of that meeting encouraged us to continue in 2011 and, following on the enthusiasm and positive feedback of the participants, we expect to continue in 2013.

Yet again, our meeting in 2011 was very international in flavour and brought together some 30 participants representing 12 nationalities, half of them from countries outside the European Union (from New Zealand to Mexico, Russia to Israel). The broad spectrum, a mixture of experienced experts and highly-motivated newcomers, the intensive exchange of ideas in a harmonious and relaxed atmosphere and the resulting joint projects are probably the secrets of why this meeting is considered to be so special to its participants.

At the resumption in 2009 some leading experts and younger scientists from economically weak countries were unable to attend due to the lack of financial resources. This time, with the very worthy and unbureaucratic support of the 'Vereinigung von Freunden und Förderern der J W Goethe-Universität Frankfurt am Main' (in short: 'Friends and Supporters of the Frankfurt University'), it was possible for all candidates to participate. In particular some young, inspired scientists had the chance of presenting their work to a very competent, but also friendly, audience. We wish to thank the 'Freunde und Förderer' for supporting Symmetries in Science XV.

Almost all participants contributed to the publication of this Conference Proceedings. There were also two colleagues who would have liked to participate but were unable to do so (M Horn and A Solomon); we agreed to incorporate their manuscripts into the Proceedings.

We would like to thank the staff of Collegium Mehrerau for their hospitality. Once more special thanks to the Schenk Family for their continuing friendship and generous support. The informal evening meetings in these stimulating surroundings probably contributed as much to the scientific success as the lectures during the day and many personal contacts were made.

Last, but not least, we would like to thank Yvette again for her unremitting support.

Dieter Schuch and Michael Ramek
Frankfurt am Main and Graz, July 2012

Peer review statement

Abstract Metrics

All papers published in this volume of *Journal of Physics: Conference Series* have been peer reviewed through processes administered by the proceedings Editors. Reviews were conducted by expert referees to the professional and scientific standards expected of a proceedings journal published by IOP Publishing.

Dual algebraic structures for pairing models

F. Iachello

Center for Theoretical Physics, Sloane Physics Laboratory,
Yale University, New Haven, CT 06520, USA

E-mail: francesco.iachello@yale.edu

Abstract. Duality relations for pairing models are discussed in an unified framework for bosons and fermions. Quantum phase transitions for pairing systems are studied and analogies and differences between bosonic and fermionic systems are highlighted.

1. Introduction

Pairing models acquired an important role in physics when the theory of superconductivity was introduced (1957) [1]. In 1958, they were introduced in the description of finite systems: atomic nuclei [2]. In recent years they have become again of interest for applications to mesoscopic systems: atomic condensates, atomic clusters, metallic grains, etc. For finite systems, the Hamiltonian operator for pairing models can be written as

$$H = \sum_{km} \varepsilon_k a^\dagger_{km} a_{km} + \frac{1}{4} \sum_{kk'mm'} G_{kk'} (-)^{j_{k'}-m'} a^\dagger_{k'm'} a^\dagger_{k',-m'} (-)^{j_k - m} a_{k,-m} a_{km}, \tag{1}$$

where $k = 1, 2, ...,$ j_k=integer (bosons) and j_k=half-integer (fermions). It is of importance to provide a classification scheme for pairing models, and to find the eigenvalues and eigenvectors of H. Duality relations provide a great simplification in solving this problem.

2. Dual algebraic structures

The fundamental algebra, G, of many-boson and many-fermion systems is composed of bilinear products of creation and annihilation operators

$$a^\dagger_{m'} a^\dagger_m, a^\dagger_{m'} a_m, a_{m'} a_m \;\; ; \;\; (m', m) = 1, ..., n \;\; ; \;\; \mathrm{G} \equiv \left\{ \begin{array}{c} \mathrm{Sp}(2n,\mathcal{R}) \\ \mathrm{SO}(n) \end{array} \right\} \begin{array}{c} \text{bosons} \\ \text{fermions} \end{array} . \tag{2}$$

If bosons transform as irreps of SO(3) and fermions as irreps of SU(2) labelled by j, (rotationally invariant systems), it is convenient to use $m = j, j-1, ..., -j$, and time-reversed creation and annihilation operators $a^\dagger_m \to a^\dagger_{j,m}$ and $a^\dagger_{\bar{m}} \to (-)^{j-m} a^\dagger_{j,-m}$. For a single value of j, the number of allowed states is $n = 2j+1$.

The algebra G has two commuting subalgebras:

(1) Number-conserving, G_1. This is the unitary algebra U(n) composed of elements $G_{m'm} = a^\dagger_{m'} a_m$ and classification scheme

$$\left\{ \begin{array}{c} \mathrm{Sp}(2n,\mathcal{R}) \\ \mathrm{SO}(2n) \end{array} \right\} \supset \mathrm{U}(n) \supset \left\{ \begin{array}{c} \mathrm{SO}(n) \\ \mathrm{Sp}(n) \end{array} \right\} \begin{array}{c} \text{bosons} \\ \text{fermions} \end{array} . \tag{3}$$

Here SO(n) and Sp(n) leave invariant $\sum_m a^\dagger_m a^\dagger_{\bar{m}} |0\rangle$. The representations of U(n), characterized by the total number of particles N, are totally symmetric for bosons and totally antisymmetric for fermions, with Young tableaux

$$\begin{array}{cc} \square\square...\square & \begin{array}{c}\square\\ \square\\ \cdots\\ \square\end{array} \\ N_B(\text{bosons}) & N_F(\text{fermions}) \end{array} \tag{4}$$

The representations of SO(n) and Sp(n) contained in U(n) are also totally symmetric (bosons) and antisymmetric (fermions). They are characterized by the quantum number v, called seniority, introduced by Racah in 1949 [3]. The branching of U(n) into SO(n) and Sp(n) was worked out years ago and is given by

$$\begin{array}{cc} \mathrm{U}(n) \supset \mathrm{SO}(n) & v = (N \mathrm{mod} 2), ..., N-2, N \\ \mathrm{U}(n) \supset \mathrm{Sp}(n) & \begin{array}{c} v = (N' \mathrm{mod} 2), ..., N'-2, N' \\ N' \equiv \min(N, n-N) \end{array} \end{array} . \tag{5}$$

(2) Non-numberconserving, G_2. This is the so-called quasi-spin algebra composed of three elements

$$S_+ = \frac{1}{2}\sum_m a^\dagger_m a^\dagger_{\bar{m}}, \quad S_- = \frac{1}{2}\sum_m a_{\bar{m}} a_m, \quad S_0 = \frac{1}{4}\sum_m \left(a^\dagger_m a_m + \vartheta a_{\bar{m}} a^\dagger_{\bar{m}} \right), \tag{6}$$

where $\vartheta = +1$ for bosons, SU(1,1), and $\vartheta = -1$ for fermions, SU(2). The representations of SU(1,1)$\supset$U(1) and SU(2)$\supset$U(1) are characterized by S (quasi-spin) and M (quasi-spin projection). This gives the alternate decomposition

$$\begin{array}{cc} \mathrm{Sp}(2n,\mathcal{R}) \supset \mathrm{SU}(1,1) \supset \mathrm{U}(1) & \text{bosons} \\ \mathrm{SO}(2n) \supset \mathrm{SU}(2) \supset \mathrm{U}(1) & \text{fermions} \end{array} . \tag{7}$$

States of pairing models can thus be classified in terms of two commuting algebras G_1 and G_2 with labels in one-to-one correspondence (duality relations)

$$S = \frac{1}{2}(\Omega + \vartheta v), \quad M = \frac{1}{2}(N + \vartheta\Omega), \tag{8}$$

where $\Omega = (j + \frac{1}{2}) = \frac{n}{2}$. For a complete classification scheme of many-body bosonic and fermionic systems, the algebras SO(n) and Sp(n) need to be further reduced

$$\begin{array}{c} \mathrm{Sp}(2n,\mathcal{R}) \supset \left\{ \begin{array}{c} \mathrm{U}(n) \supset \mathrm{SO}(n) \supset ... \supset \mathrm{SO}(3) \supset \mathrm{SO}(2) \\ [\mathrm{SU}(1,1) \supset \mathrm{U}(1)] \otimes [\mathrm{SO}(n) \supset ... \supset \mathrm{SO}(3) \supset \mathrm{SO}(2)] \end{array} \right\} \\ \mathrm{SO}(2n) \supset \left\{ \begin{array}{c} \mathrm{U}(n) \supset \mathrm{Sp}(n) \supset ... \supset \mathrm{SO}(3) \supset \mathrm{SO}(2) \\ [\mathrm{SU}(2) \supset \mathrm{U}(1)] \otimes [\mathrm{Sp}(n) \supset ... \supset \mathrm{Spin}(3) \supset \mathrm{Spin}(2)] \end{array} \right\} \end{array} . \tag{9}$$

3. Two-level systems

Two-level systems have a variety of applications in physics, especially those in which $j_1 = 0$ (scalar boson) and $j_2 = \ell$ ($2\ell + 1$-fold degenerate boson), called s-b boson systems [4]. The group theory of generic two-level systems has now been completely worked out [5]. Duality relations provide a great simplification in the study of two-level systems. The fundamental algebra for these systems is still G=$\left\{ \begin{array}{c} \mathrm{Sp}(2n,\mathcal{R}) \\ \mathrm{SO}(2n) \end{array} \right\}$ but with $n = n_1 + n_2$.

3.1. Dual algebras for two-level models

The dual algebras for two-level models are:

(1) The number-conserving algebra $U(n_1+n_2)$ with elements $G^{(g)}_{a_1a_1}=\left(a_1^\dagger\times\tilde{a}_1\right)^{(g)}$, $G^{(g)}_{a_1a_2}=\left(a_1^\dagger\times\tilde{a}_2\right)^{(g)}$, $G^{(g)}_{a_2a_1}=\left(a_2^\dagger\times\tilde{a}_1\right)^{(g)}$, $G^{(g)}_{a_2a_2}=\left(a_2^\dagger\times\tilde{a}_2\right)^{(g)}$. The algebra $U(n_1+n_2)$ has in turn several (non-dual) algebras, two of which are of importance for the pairing problem. [The two subalgebras correspond to the two physically relevant cases of $\varepsilon=\varepsilon_2-\varepsilon_1=0$ (strong coupling limit) and $\varepsilon=$ large (weak coupling limit)]. This can be summarized in the schemes

$$\begin{array}{c}U(n_1+n_2)\\ [N]\end{array}\supset\left\{\begin{array}{c}SO(n_1+n_2)\\ v\\ U_1(n_1)\otimes U_2(n_2)\\ [N_1]\otimes[N_2]\end{array}\right\}\supset\begin{array}{c}SO_1(n_1)\\ v_1\end{array}\otimes\begin{array}{c}SO_2(n_2)\\ v_2\end{array}\supset\ldots \tag{10}$$

with branching $U(n_1+n_2)\supset SO(n_1+n_2)$ given by $v=(N\mathrm{mod}2),...,N-2,N$ and branching $SO(n_1+n_2)\supset SO(n_1)\otimes SO(n_2)$ given by $v=v_1+v_2+2n_v$, $(n_v=0,1,...[v/2])$, for bosons, and

$$\begin{array}{c}U(n_1+n_2)\\ \{N\}\end{array}\supset\left\{\begin{array}{c}Sp(n_1+n_2)\\ v\\ U_1(n_1)\otimes U_2(n_2)\\ \{N_1\}\otimes\{N_2\}\end{array}\right\}\supset\begin{array}{c}Sp_1(n_1)\\ v_1\end{array}\otimes\begin{array}{c}Sp_2(n_2)\\ v_2\end{array}\supset\ldots \tag{11}$$

with branching $U(n_1+n_2)\supset Sp(n_1+n_2)$ given by $v=(N'\mathrm{mod}2),...,N'-2,N'$, where $N'=\min(N,n-N)$, and branching $Sp(n_1+n_2)\supset Sp(n_1)\otimes Sp(n_2)$ given by $v=v_1+v_2+2n_v$, $(n_v=0,1,...,[v/2])$, $\left|(v_1-v_2)-\frac{1}{2}(n_1-n_2)\right|\le\frac{1}{2}(n_1+n_2)-v$, for fermions.

Eigenvalues of the quadratic Casimir operators of algebras appearing in the chains above in the symmetric representations of $SO(n)$ and antisymmetric representations of $Sp(n)$ can be obtained from the basic relations

$$\left\langle\left\{\begin{array}{c}C_2[SO(n)]\\ C_2[Sp(n)]\end{array}\right\}\right\rangle=2v\left(\vartheta v+n-2\vartheta\right). \tag{12}$$

For $U(n)$ the linear Casimir operator has trivial eigenvalues $\langle C_1[U(n)]\rangle=N$.

(2) Number-nonconserving subalgebras $SU_1(1,1)\otimes SU_2(1,1)$ for bosons and $SU_1(2)\otimes SU_2(2)$ for fermions. These are constructed from the quasi-spin algebra

$$[S_0,S_+]=+S_+,\quad [S_0,S_-]=-S_-,\quad [S_+,S_-]=-2\vartheta S_0, \tag{13}$$

with quadratic Casimir operator $\mathbf{S}^2$ and eigenvalues $\langle\mathbf{S}^2\rangle=S(S-\vartheta)$, specifically

$$\begin{aligned}\langle\mathbf{S}^2\rangle &= \left\{\begin{array}{c}S(S-1)\\ S(S+1)\end{array}\right\}\text{with}\left\{\begin{array}{c}S=\frac{1}{4},\frac{3}{4},\frac{5}{4},...\\ S=0,\frac{1}{2},1,...\end{array}\right\}\\ &\quad\text{and}\left\{\begin{array}{c}S_0=S,S+1,S+2,...\\ S_0=-S,...,+S\end{array}\right\},\end{aligned} \tag{14}$$

for bosons ($\vartheta=1$) and fermions ($\vartheta=-1$) respectively. [The boson representations are projective representations of $SU(1,1)$ [6], [7]]. Introducing a quasi-spin algebra for each of the two systems, $k=1,2$, satisfying duality relations

$$\begin{aligned}S_k &= \frac{1}{2}\left(\Omega_k+\vartheta v_k\right)\\ M_k &= \frac{1}{2}\left(N_k+\vartheta\Omega_k\right)\end{aligned} \tag{15}$$

where $M_k = \langle S_{k0}\rangle$, $\Omega_k = \frac{1}{2}(2j_k+1)$, and $v_k = 0,1,2,...$ [$v_k \leq N_k$ for bosons and $v_k \leq \min(N_k, 2\Omega_k - N_k)$ for fermions], one can obtain the elements of the combined algebra by simple addition

$$S_+ = S_{1+} + \sigma S_{2+},\quad S_- = S_{1-} + \sigma S_{2-},\quad S_0 = S_{10} + S_{20}, \tag{16}$$

where $\sigma = \pm 1$ signifies an inner automorphism. The label S characterizing the representations of the combined algebras is given by the rules

$$\left\{ \begin{array}{c} S \geq S_1 + S_2 \\ |S_1 - S_2| \leq S \leq |S_1 + S_2| \end{array} \right\} \text{for} \left\{ \begin{array}{cc} \mathrm{SU}_{12}(1,1) & \text{bosons} \\ \mathrm{SU}_{12}(2) & \text{fermions} \end{array} \right\}. \tag{17}$$

Thus, in the dual space we have

$$\begin{array}{c} \mathrm{SU}_1(1,1)\otimes \mathrm{SU}_2(1,1) \\ S_1 \otimes S_2 \end{array} \supset \left\{ \begin{array}{c} \mathrm{SU}_{12}(1,1) \\ S \\ \mathrm{U}_1(1)\otimes \mathrm{U}_2(1) \\ M_1 \otimes M_2 \end{array} \right\} \supset \begin{array}{c} \mathrm{U}_{12}(1) \\ M \end{array} \tag{18}$$

for bosons and

$$\begin{array}{c} \mathrm{SU}_1(2)\otimes \mathrm{SU}_2(2) \\ S_1 \otimes S_2 \end{array} \supset \left\{ \begin{array}{c} \mathrm{SU}_{12}(2) \\ S \\ \mathrm{U}_1(1)\otimes \mathrm{U}_2(1) \\ M_1 \otimes M_2 \end{array} \right\} \supset \begin{array}{c} \mathrm{U}_{12}(1) \\ M \end{array} \tag{19}$$

for fermions.

3.2. Hamiltonian duality relations

The generic pairing Hamiltonian for multi-level models is given in (1). This Hamiltonian can be easily rewritten in terms of the dual non-number conserving quasi-spin algebra

$$H = \sum_k \varepsilon_k \left(2S_{k0} - \vartheta\Omega_k\right) + \sum_{k'k} G_{k'k} S_{k',+} S_{k,-}. \tag{20}$$

H conserves quasispin S_k (or seniority v_k) and $S_0 = \frac{1}{2}(N + \vartheta\Omega)$. It can be easily diagonalized in the basis $|S_1M_1, S_2M_2, ...\rangle$.

3.3. Dynamic symmetries of the two-level pairing Hamiltonian

(1) Number-conserving subalgebra $\mathrm{U}(n_1+n_2)$. The generic two-level pairing Hamiltonian H can be re-written in terms of all the invariants in (10)

$$H = aC_1[\mathrm{U}(n_1+n_2)] + b_1C_1[\mathrm{U}_1(n_1)] + b_2C_1[\mathrm{U}_2(n_2)] + b\left\{ \begin{array}{c} C_2[\mathrm{SO}(n_1+n_2)] \\ C_2[\mathrm{Sp}(n_1+n_2)] \end{array} \right\} \tag{21}$$

for bosons and fermions respectively. Dynamical symmetries are situations in which H contains only invariants of a chain [8, Ch.11]. We have two possible dynamical symmetries:

(I) $b = 0$ (weak coupling limit) with chain $\mathrm{U}(n_1+n_2)\supset\mathrm{U}(n_1)\otimes\mathrm{U}(n_2)$ and eigenvalues

$$E^{(I)} = aN + b_1N_1 + b_2N_2. \tag{22}$$

(II) $b_1 = b_2 = 0$ (strong coupling limit) with chains $\mathrm{U}(n_1+n_2)\supset\mathrm{SO}(n_1+n_2)$ for bosons and $\mathrm{U}(n_1+n_2)\supset\mathrm{Sp}(n_1+n_2)$ for fermions, and eigenvalues

$$E^{(II)} = aN + 2bv\left(\vartheta v + n_1 + n_2 - 2\vartheta\right). \tag{23}$$

Journal of Physics: Conference Series **380** (2012) 012001 doi:10.1088/1742-6596/380/1/012001

To the Hamiltonian in (21), also invariants of the common subalgebras can be added, still retaining the dynamical symmetry

$$H' = H + c_1 \left\{ \begin{array}{c} C_2[\mathrm{SO}(n_1)] \\ C_2[\mathrm{Sp}(n_1)] \end{array} \right\} + c_2 \left\{ \begin{array}{c} C_2[\mathrm{SO}(n_2)] \\ C_2[\mathrm{Sp}(n_2)] \end{array} \right\}. \tag{24}$$

The energy formulas are modified to:

(I) $b = 0$

$$E^{(I)} = aN + b_1 N_1 + b_2 N_2 + 2c_1 v_1 (\vartheta v_1 + n_1 - 2\vartheta) + 2c_2 v_2 (\vartheta v_2 + n_2 - 2\vartheta), \tag{25}$$

(II) $b_1 = b_2 = 0$

$$\begin{aligned} E^{(II)} &= aN + 2bv (\vartheta v + n_1 + n_2 - 2\vartheta) + 2c_1 v_1 (\vartheta v_1 + n_1 - 2\vartheta) \\ &\quad + 2c_2 v_2 (\vartheta v_2 + n_2 - 2\vartheta). \end{aligned} \tag{26}$$

Spectra of pairing systems in the strong coupling limit are shown in Fig.1, illustrating the difference between bosons and fermions.

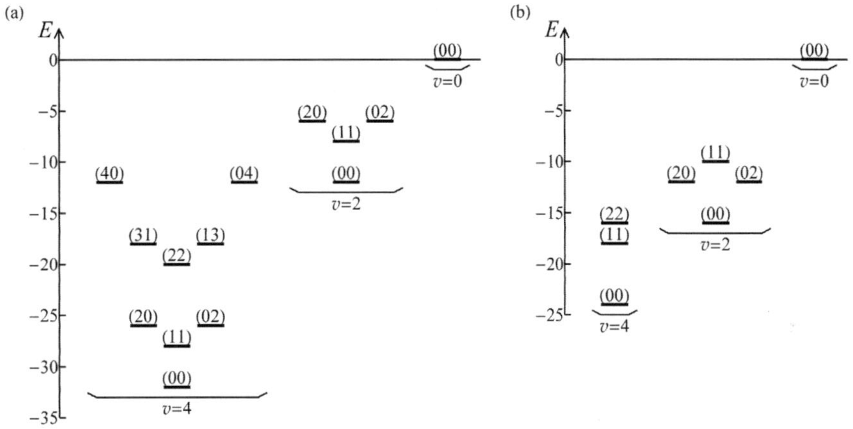

Figure 1. Energy diagram for the bosonic $\mathrm{SO}(n_1 + n_2)$ and fermionic $\mathrm{Sp}(n_1 + n_2)$ dynamical symmetries of the two-level pairing model. The degeneracy within irreps of the two-level algebra (brackets labeled by v) is split according to the single-level algebra irrep $(v_1 v_2)$. (a) Energy levels for the bosonic SO(6)⊃SO(3)⊗SO(3) dynamical symmetry Hamiltonian ($j_1 = j_2 = 1$). (b) Energy levels for the fermionic Sp(8)⊃Sp(4)⊗Sp(4) dynamical symmetry Hamiltonian ($j_1 = j_2 = \frac{3}{2}$). The Hamiltonain in each case is chosen as in (24) with $b = -\frac{1}{2}$ and $c_1 = c_2 = +\frac{1}{2}$. The total occupation in both cases is $N = 4$, which for the fermionic example gives half-filling.

(2) Number-nonconserving subalgebras $\mathrm{SU}_1(1,1)\otimes\mathrm{SU}_2(1,1)$ for bosons and $\mathrm{SU}_1(2)\otimes\mathrm{SU}_2(2)$ for fermions. The generic two-level pairing Hamiltonian in the dual space is

$$\begin{aligned} H &= \varepsilon_1 (2S_{10} - \vartheta\Omega_1) + \varepsilon_2 (2S_{20} - \vartheta\Omega_2) + G_{11} S_{1+} S_{1-} + G_{22} S_{2+} S_{2-} \\ &\quad + G_{12} (S_{1+} S_{2-} + S_{2+} S_{1-}). \end{aligned} \tag{27}$$

Dynamical symmetries can be easily constructed in this space.

(I) Weak coupling limit, $G_{11} = G_{22} = G_{12} = 0$, with algebra chains $\mathrm{SU}_1(1,1)\otimes\mathrm{SU}_2(1,1)\supset\mathrm{U}_1(1)\otimes\mathrm{U}_2(1)$ (bosons) and $\mathrm{SU}_1(2)\otimes\mathrm{SU}_2(2)\supset\mathrm{U}_1(1)\otimes\mathrm{U}_2(1)$ (fermions), and eigenvalues

$$E^{(I)} = \varepsilon_1 (2M_1 - \vartheta\Omega_1) + \varepsilon_2 (2M_2 - \vartheta\Omega_2). \tag{28}$$

By duality this can be rewritten as in (22) (with $a = 0$)

$$E^{(I)} = \varepsilon_1 N_1 + \varepsilon_2 N_2. \tag{29}$$

(II) Strong coupling limit, $\varepsilon_1 = \varepsilon_2 = 0$, $G_{11} = G_{22} = G_{12} = G$, with chains $SU_1(1,1)\otimes SU_2(1,1)\supset SU_{12}(1,1)$ (bosons) and $SU_1(2)\otimes SU_2(2)\supset SU_{12}(2)$ (fermions), Hamiltonian $H = GS_+S_-$, and eigenvalues

$$E^{(II)} = G\left[S(S-\vartheta) - S_0(S_o - \vartheta)\right]. \tag{30}$$

By duality the Hamiltonian can be rewritten as

$$H = G\frac{1}{4}\left[-\vartheta C_1[\mathrm{U}(n_1+n_2)] + C_2[\mathrm{U}(n_1+n_2)] + \frac{1}{2}\left\{\begin{array}{c} C_2[\mathrm{SO}(n_1+n_2)] \\ C_2[\mathrm{Sp}(n_1+n_2)] \end{array}\right\}\right] \tag{31}$$

with eigenvalues

$$E^{(II)} = G\frac{1}{4}\vartheta\left[N\left(N+2\vartheta\Omega-2\right) - v\left(v+2\vartheta\Omega-2\right)\right], \tag{32}$$

a variation of (25). [Cases when $G_{11} \neq G_{22} \neq G_{12}$ can also be solved]. The eigenvalues of H in the strong coupling limit (symmetry II) are shown in Fig.2, again emphasizing the difference between bosons and fermions.

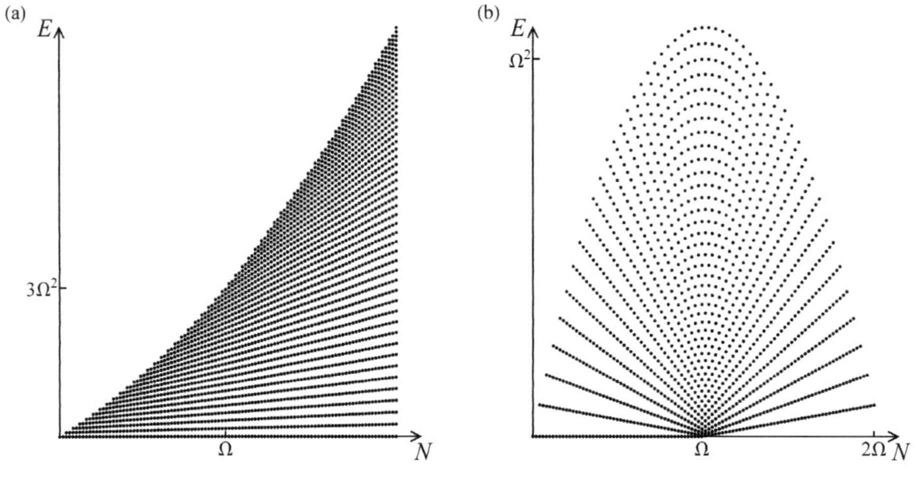

Figure 2. Eigenvalues of the pairing interaction term $4S_+S_-$, which determines the energy spectrum of the two-level pairing model in the strong-coupling limit, shown for (a) bosonic and (b) fermionic systems, as a function of filling N. The eigenvalues are given by (30). The axes are labeled generically, to indicate the asymptotic (large-Ω) dependence, but the specific points shown for illustration are calculated for $\Omega = 50$.

4. Pairing phase transitions

Quantum phase transitions (QPT) are qualitative changes in the ground state of a system induced by a change in one (or more) parameters appearing in the quantum Hamiltonian describing the system. For two phases, α and β, it is convenient to write

$$H = (1-\xi)\,H_\alpha + \xi H_\beta, \tag{33}$$

where $\xi = [0,1]$ is the control parameter. For pairing models the two phases are the weak coupling phase, α, and the strong coupling phase, β. QPTs in pairing models are induced by a change in the level splitting, ε/G. They can be studied both classically and quantum mechanically [9]. Duality relations are useful also in the study of QPTs in pairing models.

4.1. Quantal analysis

A convenient way to analyze simultaneously pairing phase transitions in both bosonic and fermionic systems is to introduce the Hamiltonian

$$H = \frac{(1-\xi)}{N} N_2 \pm \vartheta \frac{\xi}{N^2} 4 S_+ S_-, \tag{34}$$

and diagonalize it as a function of ξ, thus constructing the so-called correlation diagram. The $\pm$ signs in (34) correspond to repulsive or attractive interactions. For purposes of presentation it is convenient to consider, for repulsive pairing, the Hamiltonian

$$H_+ = \frac{(1-\xi)}{N} N_2 + \frac{\xi}{N^2} \left[4 S_+ S_- - \left\{ \begin{array}{c} N(N+2\Omega-2) \\ N(2\Omega-N+2) \end{array} \right\} \right], \tag{35}$$

and, for attractive pairing,

$$H_- = \frac{(1-\xi)}{N} N_2 - \frac{\xi}{N^2} 4 S_+ S_-. \tag{36}$$

The correlation diagram of two-level pairing models is shown in Fig. 3.

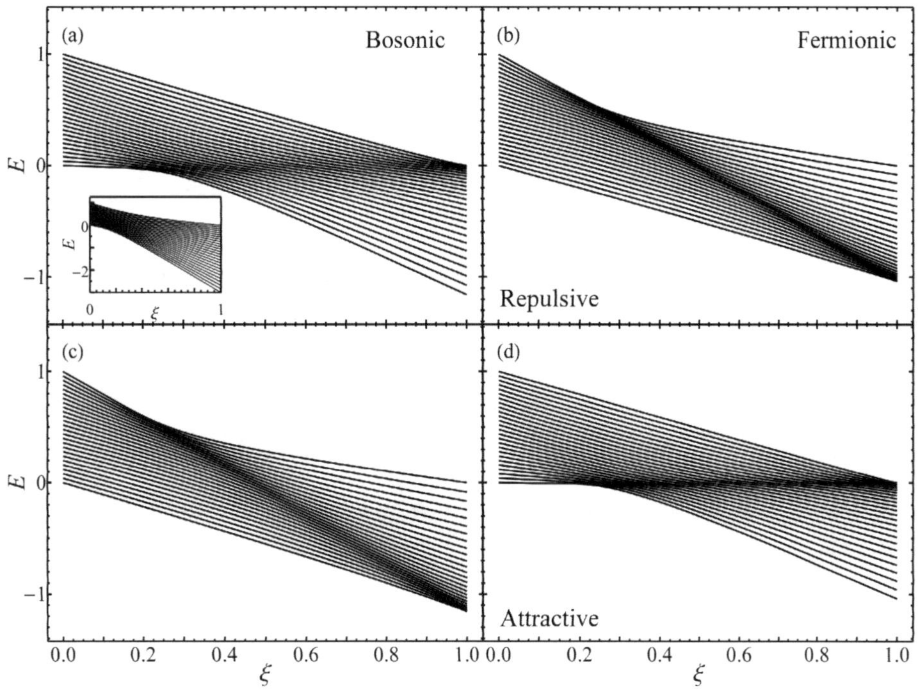

Figure 3. Eigenvalues of the bosonic two-level pairing model, with level degeneracies $n_1 = n_2 = 5$ (at left), and fermionic two-level pairing model, with level degeneracies $n_1 = n_2 = 50$ (at right), for repulsive (at top) and attractive (at bottom) pairing interactions, shown for the $(v_1 v_2)$ subspace, as functions of the control parameter ξ between the weak- and strong-coupling limits. All calculations are for $N = 50$, thus for $\Omega \ll N$ in the boson case and $\Omega = N$ (half-filling) in the fermionic case. The alternative regime in which the bosonic system also has $\Omega \approx N$ is shown (specifically, for $n_1 = n_2 = 51$ and $N = 50$) in the inset to panel (a). The Hamiltonians $H_\pm$ of (35) and (36) are used in the calculations.

4.2. Classical analysis

In order to do the classical analysis, one needs to introduce the geometry of pairing models. In s-b models this has been done by introducing the coset $U(n_2+1)/U(n_2) \otimes U(1)$ with dimension dim = $2n_2$. For general two-level pairing models, one needs to introduce cosets $U(n_1+n_2)/U(n_1) \otimes U(n_2)$ with dimension dim $= 2n_2n_1$. The classical analysis in this multidimensional space is rather difficult. Instead, in the dual space, the classical analysis can be done easily by introducing cosets SU(1,1)/SO(1,1) for bosons or SU2)/SO(2) for fermions. This analysis will not be discussed here.

4.3. Differences and similarities between bosonic and fermionic systems

The onset of criticality in pairing systems depends on the number of particles, N, and the available states, Ω. The ratio N/Ω is calling filling. For bosonic systems, one can take the limit $N \to \infty$. The onset of critical phenomena occurs when $N \gg \Omega$. For fermionic systems the occupancy is limited to $n_1 + n_2$. The limit $N \to \infty$ must be taken simultaneously as $\Omega \to \infty$. The case of $N = 2\Omega$ (complete filling) is trivial, since all states are occupied. Phase transitions in fermionic systems are best studied at half-filling, $N = \Omega$. Also, as one can see from Fig. 3, a fermionic system with attractive interactions has similar critical properties of a bosonic system with repulsive interactions. Finally, even for bosonic and fermionic systems with the same number of available states, fixed $n_1 + n_2$, there are differences, as shown in Fig. 4. While for bosonic systems the critical behavior does not depend on the individual degeneracies, n_1 and n_2, for fermionic systems there is a dramatic dependence.

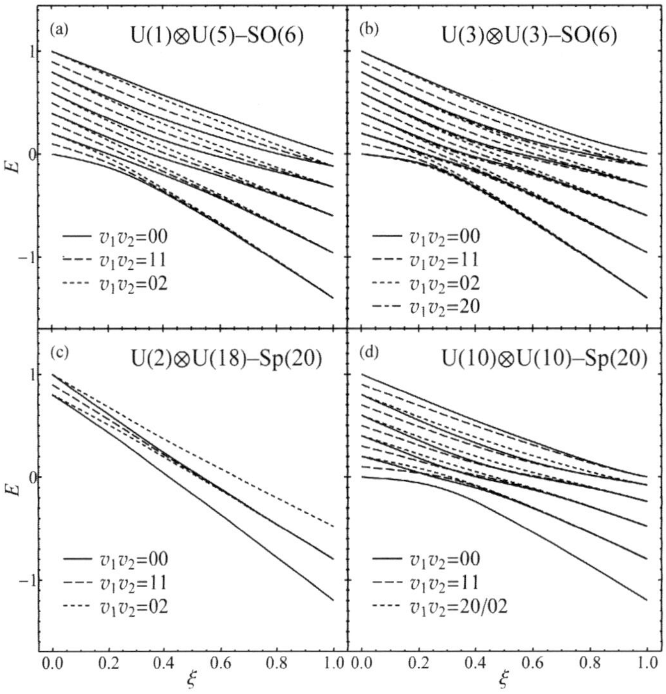

Figure 4. Eigenvalues for the bosonic two-level pairing model, with level degeneracies (a) $n_1 = 1$ and $n_2 = 5$ or (b) $n_1 = 3$ and $n_2 = 3$, and for the fermionic two-level pairing model, with level degeneracies (c) $n_1 = 2$ and $n_2 = 18$ or (d) $n_1 = 10$ and $n_2 = 10$, as a function of the control parameter ξ, between the weak- and strong-coupling limits. All calculations are for $N = 10$. Eigenvalues are shown only for the lowest seniority subspaces (v_1v_2), specifically those with $v_1 + v_2 \leq 2$. The Hamiltonians $H_\pm$ of (35) and (36) are used in the calculations.

5. Conclusions

The analysis of two-level pairing models has been completed. This analysis can be easily generalized in the dual space to multi-level systems by combining quasi-spin. The only Wigner-Racah algebra needed in this space is that of SU(2) for fermions or SU(1,1) for bosons. Multilevel systems are of interest in a variety of fields. Two examples are: (a) Nuclear physics, where usually one has to deal with ∼5-6 levels with $j = 1/2, 3/2, 5/2, 7/2, ...$ and (b) metallic grains where one has to deal with ∼100 levels but with $j = 1/2$.

For many-body finite systems with interactions other that pairing, one needs further explicit breaking of SO(n) and Sp(n). Duality relations are also of help in these cases, since they allow a simple construction of a basis for the diagonalization of H, as for example, in the Interacting Boson Model of nuclei [10].

Acknowledgments

This work was supported by the US DOE under grant DE-FG02-91ER-40608. This presentation is based on [5]. I wish to thank M.A. Caprio for providing the figures, and J. Kotila for formatting the paper.

References

[1] Bardeen J, Cooper L N and Schrieffer J R 1957 *Phys. Rev.* **106** 162 ; **108** 1175
[2] Bohr A, Mottelson B R and Pines D 1958 *Phys. Rev.* **110** 936
[3] Racah G 1949 *Phys. Rev.* **76** 1352
[4] Cejnar P and Iachello F 2007 *J. Phys.* A: Math. Theor. **40** 581
[5] Caprio M A, Skrabacz J H and Iachello F 2011 *J. Phys.* A: Math. Theor. **44** 075303
[6] Ui H 1968 *Ann. Phys.* **49** 69
[7] Arima A and Iachello F 1976 *Ann. Phys.* (N.Y.) **99** 253
[8] Iachello F 2006 *Lie Algebras and Applications*, Lecture Notes in Physics, Vol. 708 (Berlin: Springer)
[9] Iachello F 2011 *Rivista del Nuovo Cimento* **34** 617
[10] Iachello F and Arima A 1987 *The Interacting Boson Model* (Cambridge: Cambridge University Press)

Partial dynamical symmetries in quantum systems

A Leviatan

Racah Institute of Physics, The Hebrew University, Jerusalem 91904, Israel

E-mail: ami@phys.huji.ac.il

Abstract. We discuss the the notion of a partial dynamical symmetry (PDS), for which a prescribed symmetry is obeyed by only a subset of solvable eigenstates, while other eigenstates are strongly mixed. We present an explicit construction of Hamiltonians with this property, including higher-order terms, and portray their significance for spectroscopy and shape-phase transitions in nuclei. The occurrence of both a single PDS, relevant to stable structures, and of several PDSs, relevant to coexistence phenomena, are considered.

1. Introduction

Models based on spectrum generating algebras form a convenient framework to examine underlying symmetries in dynamical systems, and have been used extensively in diverse areas of physics [1]. Notable examples in nuclear physics are Wigner's spin-isospin SU(4) supermultiplets [2], SU(2) single-j pairing [3], Elliott's SU(3) model [4], symplectic model [5], Ginocchio's monopole and quadrupole pairing models [6], interacting boson models (IBM) for even-even nuclei [7] and boson-fermion models (IBFM) for odd-mass nuclei [8]. Similar algebraic techniques have proven to be useful in the structure of molecules [9, 10] and of hadrons [11]. In such models the Hamiltonian is expanded in elements of a Lie algebra, (G_0), called the spectrum generating algebra. A dynamical symmetry occurs if the Hamiltonian can be written in terms of the Casimir operators of a chain of nested algebras, $G_0 \supset G_1 \supset \ldots \supset G_n$ [12]. The following properties are then observed. (i) All states are solvable and analytic expressions are available for energies and other observables. (ii) All states are classified by quantum numbers, $|\alpha_0, \alpha_1, \ldots, \alpha_n\rangle$, which are the labels of the irreducible representations (irreps) of the algebras in the chain. (iii) The structure of wave functions is completely dictated by symmetry and is independent of the Hamiltonian's parameters.

A dynamical symmetry provides clarifying insights into complex dynamics and its merits are self-evident. However, in most applications to realistic systems, the predictions of an exact dynamical symmetry are rarely fulfilled and one is compelled to break it. The breaking of the symmetry is required for a number of reasons. First, one often finds that the assumed symmetry is not obeyed uniformly, *i.e.*, is fulfilled by only some of the states but not by others. Certain degeneracies implied by the assumed symmetry are not always realized, (*e.g.*, axially deformed nuclei rarely fulfill the IBM SU(3) requirement of degenerate β and γ bands [7]). Secondly, forcing the Hamiltonian to be invariant under a symmetry group may impose constraints which are too severe and incompatible with well-known features of the dynamics (*e.g.*, the models of [6] require degenerate single-nucleon energies). Thirdly, in describing systems in-between two different structural phases, *e.g.*, spherical and deformed nuclei, the Hamiltonian by necessity mixes terms with different symmetry character. In the models mentioned above, the required

Table 1. Generators, Casimir operators, $\hat{C}_k(G)$, of order $k = 1, 2, 3$ and their eigenvalues for algebras G in the IBM. Here $\hat{n}_s = s^\dagger s$, $\hat{n}_d = \sqrt{5}U^{(0)}$, $\hat{N} = \hat{n}_s + \hat{n}_d$, $\hat{L}_m = \sqrt{10}U^{(1)}_m$, $\hat{Q}_m = \Pi^{(2)}_m - \frac{\sqrt{7}}{2}U^{(2)}_m$, $\Pi^{(2)}_m = d^\dagger_m s + s^\dagger \tilde{d}_m$, $\bar{\Pi}^{(2)}_m = i(d^\dagger_m s - s^\dagger \tilde{d}_m)$, $U^{(\ell)}_m = (d^\dagger\,\tilde{d})^{(\ell)}_m$, where $\tilde{d}_m = (-1)^m d_{-m}$.

Algebra	Generators	Casimir operator $\hat{C}_k(G)$	Eigenvalues $\langle\hat{C}_k(G)\rangle$
O(3)	$U^{(1)}$	$\hat{L}\cdot\hat{L}$	L(L+1)
O(5)	$U^{(1)}, U^{(3)}$	$2(U^{(1)}\cdot U^{(1)} + U^{(3)}\cdot U^{(3)})$	$\tau(\tau+3)$
O(6)	$U^{(1)}, U^{(3)}, \Pi^{(2)}$	$\hat{C}_2(\mathrm{O}(5)) + \Pi^{(2)}\cdot\Pi^{(2)}$	$\Sigma(\Sigma+4)$
SU(3)	$U^{(1)}, \hat{Q}$	$2\hat{Q}\cdot\hat{Q} + \frac{3}{4}\hat{L}\cdot\hat{L}$	$\lambda^2 + (\lambda+\mu)(\mu+3)$
		$-4\sqrt{7}\hat{Q}\cdot(\hat{Q}\times\hat{Q})^{(2)} - \frac{9}{2}\sqrt{3}\hat{Q}\cdot(\hat{L}\times\hat{L})^{(2)}$	$(\lambda-\mu)(2\lambda+\mu+3)$ $\times(\lambda+2\mu+3)$
U(5)	$U^{(\ell)}$ $\ell = 0,\dots,4$	$\hat{n}_d$, $\hat{n}_d(\hat{n}_d+4)$	n_d, $n_d(n_d+4)$
U(6)	$U^{(\ell)}$ $\ell = 0,\dots,4$ $\Pi^{(2)}$, $\bar{\Pi}^{(2)}$, $\hat{n}_s$	$\hat{N}$, $\hat{N}(\hat{N}+5)$	N, $N(N+5)$

symmetry breaking is achieved by including in the Hamiltonian terms associated with (two or more) different sub-algebra chains of the parent spectrum generating algebra. In general, under such circumstances, solvability is lost, there are no remaining non-trivial conserved quantum numbers and all eigenstates are expected to be mixed. A partial dynamical symmetry (PDS) [13] corresponds to a particular symmetry breaking for which some (but not all) of the virtues of a dynamical symmetry are retained. The essential idea is to relax the stringent conditions of *complete* solvability so that the properties (i)–(iii) are only partially satisfied. It is then possible to identify several types of partial dynamical symmetries. PDS of type I corresponds to a situation where *some* of the states have *all* the dynamical symmetry. In this case, properties (i)-(iii) are fulfilled exactly, but by only a subset of states. PDS of type II corresponds to a situation for which *all* the states preserve *part* of the dynamical symmetry. In this case, there are no analytic solutions, yet selected quantum numbers (of the conserved symmetries) are retained. PDS of type III has a hybrid character, for which *some* of the states preserve *part* of the dynamical symmetry.

In what follows we discuss algorithms for constructing Hamiltonians with partial dynamical symmetries and demonstrate their relevance to quantum systems. For that purpose, we employ the interacting boson model (IBM) [7], widely used in the description of low-lying quadrupole collective states in nuclei in terms of N interacting monopole (s) and quadrupole (d) bosons representing valence nucleon pairs. The bilinear combinations $\{s^\dagger s,\, s^\dagger d_m,\, d^\dagger_m s,\, d^\dagger_m d_{m'}\}$ span a U(6) algebra, which serves as the spectrum generating algebra. The IBM Hamiltonian is expanded in terms of these generators and consists of Hermitian, rotational-scalar interactions which conserve the total number of s- and d- bosons, $\hat{N} = \hat{n}_s + \hat{n}_d = s^\dagger s + \sum_m d^\dagger_m d_m$. Three dynamical symmetry limits occur in the model with leading subalgebras U(5), SU(3), and O(6), corresponding to typical collective spectra observed in nuclei, vibrational, rotational, and γ-unstable, respectively. Relevant information on these algebras is collected in Table 1.

A geometric visualization of the model is obtained by an energy surface

$$E_N(\beta,\gamma) = \langle \beta,\gamma;N|\hat{H}|\beta,\gamma;N\rangle \ , \tag{1}$$

defined by the expectation value of the Hamiltonian in the coherent (intrinsic) state [14, 15]

$$|\beta,\gamma;N\rangle = (N!)^{-1/2}(b^{\dagger}_{c})^{N}\,|0\,\rangle \ , \tag{2a}$$
$$b^{\dagger}_{c} = (1+\beta^2)^{-1/2}[\beta\cos\gamma d^{\dagger}_{0} + \beta\sin\gamma(d^{\dagger}_{2}+d^{\dagger}_{-2})/\sqrt{2} + s^{\dagger}] \ . \tag{2b}$$

Here (β,γ) are quadrupole shape parameters whose values, (β_0,γ_0), at the global minimum of $E_N(\beta,\gamma)$ define the equilibrium shape for a given Hamiltonian. The shape can be spherical $(\beta = 0)$ or deformed $(\beta > 0)$ with $\gamma = 0$ (prolate), $\gamma = \pi/3$ (oblate), $0 < \gamma < \pi/3$ (triaxial), or γ-independent. The equilibrium deformations associated with the dynamical symmetry limits are $\beta_0 = 0$ for U(5), $(\beta_0 = \sqrt{2}, \gamma_0 = 0)$ for SU(3) and $(\beta_0 = 1, \gamma_0$ arbitrary) for O(6).

2. Construction of Hamiltonians with partial dynamical symmetries

PDS of type I corresponds to a situation for which the defining properties of a dynamical symmetry (DS), namely, solvability, good quantum numbers, and symmetry-dictated structure are fulfilled exactly, but by only a subset of states. An algorithm for constructing Hamiltonians with PDS has been developed in [16] and further elaborated in [17]. The analysis starts from the chain of nested algebras

$$\begin{array}{ccccccc} G_{\rm dyn} & \supset & G & \supset & \cdots & \supset & G_{\rm sym} \\ \downarrow & & \downarrow & & & & \downarrow \\ [h] & & \langle\Sigma\rangle & & & & \Lambda \end{array} \tag{3}$$

where, below each algebra, its associated labels of irreps are given. Eq. (3) implies that $G_{\rm dyn}$ is the dynamical (spectrum generating) algebra of the system such that operators of all physical observables can be written in terms of its generators; a single irrep of $G_{\rm dyn}$ contains all states of relevance in the problem. In contrast, $G_{\rm sym}$ is the symmetry algebra and a single of its irreps contains states that are degenerate in energy. Assuming, for simplicity, that particle number is conserved, then all states, and hence the representation $[h]$, can then be assigned a definite particle number N. For N identical particles the representation $[h]$ of the dynamical algebra $G_{\rm dyn}$ is either symmetric $[N]$ (bosons) or antisymmetric $[1^N]$ (fermions) and will be denoted, in both cases, as $[h_N]$. The occurrence of a DS of the type (3) signifies that the Hamiltonian is written in terms of the Casimir operators of the algebras in the chain,

$$\hat{H}_{DS} = \sum_{G} a_G\,\hat{C}(G) \ , \tag{4}$$

and its eigenstates can be labeled as $|[h_N]\langle\Sigma\rangle\ldots\Lambda\rangle$; additional labels (indicated by ...) are suppressed in the following. The eigenvalues of the Casimir operators in these basis states determine the eigenenergies $E_{DS}([h_N]\langle\Sigma\rangle\Lambda)$ of $\hat{H}_{DS}$. Likewise, operators can be classified according to their tensor character under (3) as $\hat{T}_{[h_n]\langle\sigma\rangle\lambda}$.

Of specific interest in the construction of a PDS associated with the reduction (3), are the n-particle annihilation operators $\hat{T}$ which satisfy the property

$$\hat{T}_{[h_n]\langle\sigma\rangle\lambda}|[h_N]\langle\Sigma_0\rangle\Lambda\rangle = 0 \ , \tag{5}$$

for all possible values of Λ contained in a given irrep $\langle\Sigma_0\rangle$ of G. Equivalently, this condition can be phrased in terms of the action on a lowest weight (LW) state of the G-irrep $\langle\Sigma_0\rangle$,

$$\hat{T}_{[h_n]\langle\sigma\rangle\lambda}|LW;\,[h_N]\langle\Sigma_0\rangle\rangle = 0 \ , \tag{6}$$

from which states of good Λ can be obtained by projection. Any n-body, number-conserving normal-ordered interaction written in terms of these annihilation operators and their Hermitian conjugates (which transform as the corresponding conjugate irreps),

$$\hat{H}' = \sum_{\alpha,\beta} A_{\alpha\beta}\, \hat{T}^{\dagger}_{\alpha}\hat{T}_{\beta} \ , \tag{7}$$

has a partial G-symmetry. This comes about since for arbitrary coefficients, $A_{\alpha\beta}$, $\hat{H}'$ is not a G-scalar, hence most of its eigenstates will be a mixture of irreps of G, yet relation (5) ensures that a subset of its eigenstates $|[h_N]\langle\Sigma_0\rangle\Lambda\rangle$, are solvable and have good quantum numbers under the chain (3). An Hamiltonian with partial dynamical symmetry is obtained by adding to $\hat{H}'$ the dynamical symmetry Hamiltonian, $\hat{H}_{DS}$ (4), still preserving the solvability of states with $\langle\Sigma\rangle = \langle\Sigma_0\rangle$,

$$\hat{H}_{PDS} \;=\; \hat{H}_{DS} + \hat{H}' \ . \tag{8}$$

If the operators $\hat{T}_{[h_n]\langle\sigma\rangle\lambda}$ span the entire irrep $\langle\sigma\rangle$ of G, then the annihilation condition (5) is satisfied for all Λ-states in $\langle\Sigma_0\rangle$, if none of the G irreps $\langle\Sigma\rangle$ contained in the $G_{\rm dyn}$ irrep $[h_{N-n}]$ belongs to the G Kronecker product $\langle\sigma\rangle \times \langle\Sigma_0\rangle$. So the problem of finding interactions that preserve solvability for part of the states (3) is reduced to carrying out a Kronecker product. The arguments for choosing the special irrep $\langle\Sigma\rangle = \langle\Sigma_0\rangle$ in Eq. (5), which contains the solvable states, are based on physical grounds. A frequently encountered choice is the irrep which contains the ground state of the system. The above algorithm is applicable to any semisimple group.

PDS of type II corresponds to a situation for which *all* the states of the system preserve *part* of the dynamical symmetry, $G_0 \supset G_1 \supset G_2 \supset \ldots \supset G_n$. In this case, there are no analytic solutions, yet selected quantum numbers (of the conserved symmetries) are retained. This occurs, for example, when the Hamiltonian contains interaction terms from two different chains with a common symmetry subalgebra [18], *e.g.*,

$$G_0 \supset \left\{ \begin{array}{c} G_1 \\ G'_1 \end{array} \right\} \supset G_2 \supset \ldots \supset G_n \ . \tag{9}$$

If G_1 and G'_1 are incompatible, *i.e.*, do not commute, then their irreps are mixed in the eigenstates of the Hamiltonian. On the other hand, since G_2 and its subalgebras are common to both chains, then the labels of their irreps remain as good quantum numbers.

An alternative situation where PDS of type II occurs is when the Hamiltonian preserves only some of the symmetries G_i in the DS chain and only their irreps are unmixed. A systematic procedure for identifying interactions with such property was proposed in [19]. Let $G_1 \supset G_2 \supset G_3$ be a set of nested algebras which may occur anywhere in the chain, in-between the spectrum generating algebra G_0 and the invariant symmetry algebra G_n. The procedure is based on writing the Hamiltonian in terms of generators, g_i, of G_1, which do not belong to its subalgebra G_2. By construction, such Hamiltonian preserves the G_1 symmetry but, in general, not the G_2 symmetry, and hence will have the G_1 labels as good quantum numbers but will mix different irreps of G_2. The Hamiltonians can still conserve the G_3 labels *e.g.*, by choosing it to be a scalar of G_3. The procedure involves the identification of the tensor character under G_2 and G_3 of the operators g_i and their products, $g_i g_j \ldots g_k$. The Hamiltonians obtained in this manner belong to the integrity basis of G_3-scalar operators in the enveloping algebra of G_1 and, hence, their existence is correlated with their order.

PDS of type III combines properties of both PDS of type I and II. Such a generalized PDS [20] has a hybrid character, for which *part* of the states of the system under study preserve *part* of the dynamical symmetry. In relation to the dynamical symmetry chain of Eq. (3), with associated basis, $|[h_N]\langle\Sigma\rangle\Lambda\rangle$, this can be accomplished by relaxing the condition of Eq. (5), so that it holds only for *selected* states Λ contained in a given irrep $\langle\Sigma_0\rangle$ of G and/or selected (combinations of) components λ of the tensor $\hat{T}_{[h_n]\langle\sigma\rangle\lambda}$. Under such circumstances, let $G' \neq G_{sym}$ be a subalgebra of G in the aforementioned chain, $G \supset G'$. In general, the Hamiltonians, constructed from these tensors, in the manner shown in Eq. (7), are not invariant under G nor G'. Nevertheless, they do possess the subset of solvable states, $|[h_N]\langle\Sigma_0\rangle\Lambda\rangle$, with good G-symmetry $\langle\Sigma_0\rangle$ (which now span only part of the corresponding G-irrep), while other states are mixed. At the same time, the symmetry associated with the subalgebra G', is broken in all states (including the solvable ones). Thus, part of the eigenstates preserve part of the symmetry. These are precisely the requirements of PDS of type III.

3. SU(3) partial dynamical symmetry

The SU(3) DS chain of the IBM and related quantum numbers are given by [7]

$$\begin{array}{ccccc} \mathrm{U}(6) & \supset & \mathrm{SU}(3) & \supset & \mathrm{O}(3) \\ \downarrow & & \downarrow & & \downarrow \\ [N] & & (\lambda,\mu) & K & L \end{array} \quad . \tag{10}$$

For a given U(6) irrep $[N]$, the allowed SU(3) irreps are $(\lambda,\mu) = (2N-4k-6m, 2k)$ with k, m non-negative integers, such that, $\lambda, \mu \geq 0$. The multiplicity label K is needed for complete classification and corresponds geometrically to the projection of the angular momentum on the symmetry axis. The values of L contained in a given SU(3) irrep (λ,μ), are obtained from the known $SU(3) \supset O(3)$ reduction. The states $|[N](\lambda,\mu)KL\rangle$ form the (non-orthogonal) Elliott basis [4] and the Vergados basis $|[N](\lambda,\mu)\tilde{\chi}L\rangle$ [7] is obtained from it by a standard orthogonalization procedure. The two bases coincide in the large-N limit and both are eigenstates of a Hamiltonian with SU(3) DS. The latter, for one- and two-body interactions, can be transcribed in the form

$$\hat{H}_{DS} = h_2\left[-\hat{C}_2(\mathrm{SU}(3)) + 2\hat{\mathrm{N}}(2\hat{\mathrm{N}}+3)\right] + C\,\hat{C}_2(\mathrm{O}(3)) \;, \tag{11}$$

where $\hat{C}_2(G)$ is the quadratic Casimir operator of G, as defined in Table 1. The spectrum of $\hat{H}_{DS}$ is completely solvable with eigenenergies

$$E_{\mathrm{DS}} = h_2\, 6\left[2N(k+2m) - k(2k-1) - 3m(2m-1) - 6km\right] + CL(L+1) \;, \tag{12}$$

and $(\lambda,\mu) = (2N-4k-6m, 2k)$. The spectrum resembles that of an axially-deformed rotovibrator and the corresponding eigenstates are arranged in SU(3) multiplets. In a given SU(3) irrep (λ,μ), each K-value is associated with a rotational band and states with the same L, in different K-bands, are degenerate. The lowest SU(3) irrep is $(2N,0)$, which describes the ground band $g(K=0)$ of a prolate deformed nucleus. The first excited SU(3) irrep $(2N-4,2)$ contains both the $\beta(K=0)$ and $\gamma(K=2)$ bands. States in these bands with the same angular momentum are degenerate. This β-γ degeneracy is a characteristic feature of the SU(3) limit of the IBM which, however, is not commonly observed. In most deformed nuclei the β band lies above the γ band. In the IBM framework, with at most two-body interactions, one is therefore compelled to break SU(3) in order to conform with the experimental data.

The construction of Hamiltonians with SU(3)-PDS of type I is based on identification of n-boson operators which annihilate all states in a given SU(3) irrep (λ,μ), chosen here to be the ground band irrep $(2N,0)$. For that purpose, we consider the following two-boson SU(3) tensors, $B^{\dagger}_{[n](\lambda,\mu)\tilde{\chi};\ell m}$, with $n=2$, $(\lambda,\mu)=(0,2)$ and angular momentum $\ell = 0, 2$

$$B^{\dagger}_{[2](0,2)0;00} \propto P^{\dagger}_{0} = d^{\dagger}\cdot d^{\dagger} - 2(s^{\dagger})^2 \;, \tag{13a}$$

$$B^{\dagger}_{[2](0,2)0;2m} \propto P^{\dagger}_{2m} = 2d^{\dagger}_{m}s^{\dagger} + \sqrt{7}\,(d^{\dagger}\,d^{\dagger})^{(2)}_{m} \;. \tag{13b}$$

The corresponding Hermitian conjugate boson-pair annihilation operators, P_0 and P_{2m}, transform as $(2,0)$ under SU(3), and satisfy

$$\begin{aligned} P_0\,|[N](2N,0)K=0,L\rangle &= 0\ , \\ P_{2m}\,|[N](2N,0)K=0,L\rangle &= 0\ , \qquad L=0,2,4,\ldots,2N\ . \end{aligned} \tag{14}$$

Equivalently, these operators satisfy

$$\begin{aligned} P_0|\beta=\sqrt{2},\gamma=0;N\rangle &= 0\ , \\ P_{2m}|\beta=\sqrt{2},\gamma=0;N\rangle &= 0\ , \end{aligned} \tag{15}$$

where $|\beta=\sqrt{2},\gamma=0;N\rangle$, is the condensate of Eq. (2) with the SU(3) equilibrium deformations. It is the lowest-weight state in the SU(3) irrep $(\lambda,\mu)=(2N,0)$ and serves as an intrinsic state for the SU(3) ground band. The rotational members of the band $|[N](2N,0)K=0,L\rangle$, Eq. (14), are obtained from it by O(3) projection, and span the entire SU(3) irrep $(\lambda,\mu)=(2N,0)$. The relations in Eqs. (14)-(15) follow from the fact that the action of the operators $P_{\ell m}$ leads to a state with $N-2$ bosons in the U(6) irrep $[N-2]$, which does not contain the SU(3) irreps obtained from the product $(2,0)\times(2N,0)=(2N+2,0)\oplus(2N,1)\oplus(2N-2,2)$. In addition, P_0 satisfies

$$P_0\,|[N](2N-4k,2k)K=2k,L\rangle = 0\ , \quad L=K,K+1,\ldots,(2N-2k)\ . \tag{16}$$

For $k>0$ the indicated L-states span only part of the SU(3) irreps $(\lambda,\mu)=(2N-4k,2k)$ and form the rotational members of excited $\gamma^k(K=2k)$ bands. This result follows from the fact that P_0 annihilates the intrinsic states of these bands, $|\gamma^k(K=2k)\rangle \propto (P^{\dagger}_{2,2})^k|\beta=\sqrt{2},\gamma=0;N-2k\rangle$.

Following the general algorithm, a two-body Hamiltonian with partial SU(3) symmetry can now be constructed as in Eq. (7), $\hat{H}'=h_0P^{\dagger}_0P_0+h_2P^{\dagger}_2\cdot\tilde{P}_2$, where $\tilde{P}_{2m}=(-)^mP_{2,-m}$. For $h_2=h_0$, this Hamiltonian is an SU(3) scalar, while for $h_0=-5h_2$, it transforms as a $(2,2)$ SU(3) tensor component. The scalar part is related to the quadratic Casimir operator of SU(3)

$$\hat{\theta}_2 \equiv P^{\dagger}_0P_0+P^{\dagger}_2\cdot\tilde{P}_2 = -\hat{C}_2(\mathrm{SU(3)})+2\hat{\mathrm{N}}(2\hat{\mathrm{N}}+3)\ , \tag{17}$$

and is simply the first term in $\hat{H}_{DS}$, Eq. (11). In accord with Eq. (8), the two-body SU(3)-PDS Hamiltonian is thus given by

$$\hat{H}_{PDS} \;=\; \hat{H}_{DS}+\eta\,P^{\dagger}_0P_0\ . \tag{18}$$

The $P^{\dagger}_0P_0$ term is not diagonal in the SU(3) chain, however, Eqs. (14)-(16) ensure that $\hat{H}_{PDS}$ retains selected solvable states with good SU(3) symmetry. Specifically, the solvable states are members of the ground $g(K=0)$

$$|N,(2N,0)K=0,L\rangle \quad L=0,2,4,\ldots,2N \tag{19a}$$
$$E_{PDS}=CL(L+1) \tag{19b}$$

and $\gamma^k(K=2k)$ bands

$$|N,(2N-4k,2k)K=2k,L\rangle \quad L=K,K+1,K+2,\ldots,(2N-2k) \tag{20a}$$
$$E_{PDS}=h_2\,6k(N-2)(2N-2k+1)+CL(L+1) \qquad k>0\ . \tag{20b}$$

The remaining eigenstates of $\hat{H}_{PDS}$ do not preserve SU(3) and, therefore, get mixed.

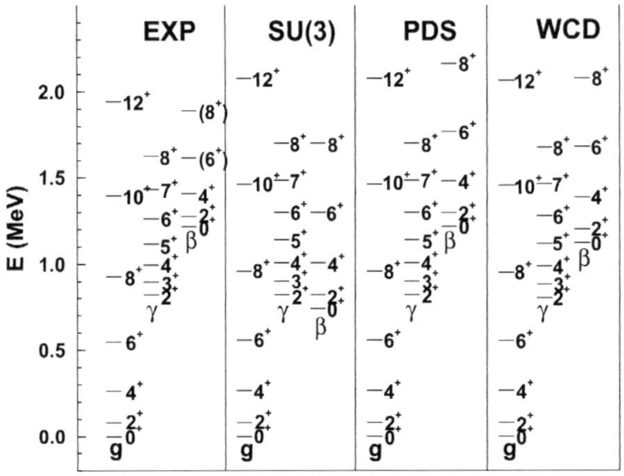

Figure 1. Spectra of ^{168}Er ($N = 16$). Experimental energies (EXP) are compared with IBM calculations in an exact SU(3) dynamical symmetry [SU(3)], in a broken SU(3) symmetry (WCD) and in a partial dynamical SU(3) symmetry (PDS). The latter employs the Hamiltonian of Eq. (18), with $h_2 = 4$, $\eta = 4$, $C = 13$ keV [21].

The empirical spectrum of ^{168}Er is shown in Fig. 1 and compared with SU(3)-DS, SU(3)-PDS and broken SU(3) calculations [21]. The SU(3)-PDS spectrum shows an improvement over the schematic, exact SU(3) dynamical symmetry description, since the β-γ degeneracy is lifted.

Table 2. B(E2) branching ratios from states in the γ band in ^{168}Er. The column EXP lists the experimental ratios, PDS is the SU(3) partial dynamical symmetry calculation and WCD is a broken SU(3) calculation [21].

L_i^π	L_f^π	EXP	PDS	WCD	L_i^π	L_f^π	EXP	PDS	WCD
2_γ^+	0_g^+	54.0	64.27	66.0	6_γ^+	4_g^+	0.44	0.89	0.97
	2_g^+	100.0	100.0	100.0		6_g^+	3.8	4.38	4.3
	4_g^+	6.8	6.26	6.0		8_g^+	1.4	0.79	0.73
3_γ^+	2_g^+	2.6	2.70	2.7		4_γ^+	100.0	100.0	100.0
	4_g^+	1.7	1.33	1.3		5_γ^+	69.0	58.61	59.0
	2_γ^+	100.0	100.0	100.0	7_γ^+	6_g^+	0.74	2.62	2.7
4_γ^+	2_g^+	1.6	2.39	2.5		5_γ^+	100.0	100.0	100.0
	4_g^+	8.1	8.52	8.3		6_γ^+	59.0	39.22	39.0
	6_g^+	1.1	1.07	1.0	8_γ^+	6_g^+	1.8	0.59	0.67
	2_γ^+	100.0	100.0	100.0		8_g^+	5.1	3.57	3.5
5_γ^+	4_g^+	2.91	4.15	4.3		6_γ^+	100.0	100.0	100.0
	6_g^+	3.6	3.31	3.1		7_γ^+	135.0	28.64	29.0
	3_γ^+	100.0	100.0	100.0					
	4_γ^+	122.0	98.22	98.5					

The quality of the calculated PDS spectrum is similar to that obtained in the broken-SU(3) calculation, however, in the former the ground $g(K=0_1)$ and $\gamma(K=2_1)$ bands remain solvable with good SU(3) symmetry, $(\lambda,\mu)=(2N,0)$ and $(2N-4,2)$ respectively. At the same time, the excited $K=0_2^+$ band involves about 13% SU(3) admixtures into the dominant $(2N-4,2)$ irrep. Since the wave functions of the solvable states (19)-(20) are known, one can obtain *analytic* expressions for matrix elements of observables between them. For example, the most general one-body E(2) operator reads $T(E2)=\alpha\hat{Q}+\rho\Pi^{(2)}$, in the notation of Table 1. Since $\hat{Q}$ is an SU(3) generator, it cannot connect different SU(3) irreps, hence only $\Pi^{(2)}$, which is a (2,2) SU(3) tensor, contributes to $\gamma\to g$ transitions. Accordingly, the calculated B(E2) ratios for $\gamma\to g$ transitions involve ratios of known SU(3) isoscalar factors and lead to parameter-free predictions. The latter, as shown in Table 2, are in excellent agreement with experiment, thus confirming the relevance of SU(3)-PDS to the spectroscopy of ^{168}Er [21].

The construction of SU(3)-PDS Hamiltonians with higher-order terms follows the general algorithm and is based on identification of n-boson operators which annihilate all states in the irrep $(2N,0)$. For $n=3$, we consider the following SU(3) tensors, $\hat{B}^\dagger_{[n](\lambda,\mu)\tilde{\chi};\ell m}$,

$$
\begin{aligned}
\hat{B}^\dagger_{[3](2,2)0;00} &\propto W_0^\dagger = 5P_0^\dagger s^\dagger - P_2^\dagger\cdot d^\dagger \quad , \quad \hat{B}^\dagger_{[3](2,2)2;2m} \propto W_{2m}^\dagger = P_0^\dagger d_m^\dagger + 2P_{2m}^\dagger s^\dagger \quad , \\
\hat{B}^\dagger_{[3](2,2)0;2m} &\propto V_{2m}^\dagger = 6P_0^\dagger d_m^\dagger - P_{2m}^\dagger s^\dagger \quad , \quad \hat{B}^\dagger_{[3](2,2)2;\ell m} \propto W_{\ell m}^\dagger = (P_2^\dagger d^\dagger)^{(\ell)}_m \quad \ell=3,4 \qquad \text{(21a)}\\
\hat{B}^\dagger_{[3](0,0)0;00} &\propto \Lambda^\dagger = P_0^\dagger s^\dagger + P_2^\dagger\cdot d^\dagger \ . \qquad \text{(21b)}
\end{aligned}
$$

The operators $W_0^\dagger$, $W_{2m}^\dagger$, $V_{2m}^\dagger$, $W_{3m}^\dagger$, $W_{4m}^\dagger$ in Eq. (21a) span the irrep $(\lambda,\mu)=(2,2)$, while $\Lambda^\dagger$ of Eq. (21b) transforms as $(\lambda,\mu)=(0,0)$. The latter SU(3)-scalar operator is related to the cubic and quadratic Casimir operators of SU(3), defined in Table 1,

$$
2\Lambda^\dagger\Lambda = \hat{C}_3(SU(3)) - 2\hat{N}(4\hat{N}+3)(2\hat{N}+3) + 3(2\hat{N}+3)\hat{\theta}_2 \ , \qquad (22)
$$

where $\hat{\theta}_2$ is given in Eq. (17). In the presence of two- and three-body terms, the dynamical-symmetry Hamiltonian and eigenenergies for states with $(\lambda,\mu)=(2N-4k-6m,2k)$, read

$$
\begin{aligned}
\hat{H}_{DS} &= h_1\,\Lambda^\dagger\Lambda + h_2\,\hat{\theta}_2 + C\,\hat{L}\cdot\hat{L} \ , \qquad \text{(23a)}\\
E_{DS} &= +h_1\,54m\,[\,N(2k+2m+1) - k(2k-1) - (2m-1)(2m+1) - 6km\,]\\
&\quad +h_2\,6\,[\,2N(k+2m) - k(2k-1) - 3m(2m-1) - 6km\,] + C\,L(L+1) \ . \qquad \text{(23b)}
\end{aligned}
$$

The SU(3)-PDS Hamiltonian has the following SU(3)-tensor expansion

$$
\begin{aligned}
\hat{H}_{PDS} &= \hat{H}_{DS} + \eta\,P_0^\dagger P_0 + a_1\,W_0^\dagger W_0 + a_2\left(W_0^\dagger\Lambda + \Lambda^\dagger W_0\right) + a_3\,W_2^\dagger\cdot\tilde{W}_2\\
&\quad + a_4\,V_2^\dagger\cdot\tilde{V}_2 + a_5\left(W_2^\dagger\cdot\tilde{V}_2 + V_2^\dagger\cdot\tilde{W}_2\right) + a_6\,W_3^\dagger\cdot\tilde{W}_3 + a_7\,W_4^\dagger\cdot\tilde{W}_4 \ . \qquad (24)
\end{aligned}
$$

The relations of Eq. (14) ensure that the operators $W_{\ell m}$, V_{2m} and Λ of Eq. (21) annihilate the states of the SU(3) ground band $g(K=0)$. The solvable eigenstates and eigenenergies of $\hat{H}_{PDS}$ are those shown in Eq. (19). In addition, the operator Λ (21b) annihilates *all* states in the irreps $(2N-4k,2k)$

$$
\Lambda\,|[N](2N-4k,2k)K;L\rangle = 0 \ . \qquad (25)
$$

This property follows from the fact that the U(6) irrep $[N-3]$ does not contain SU(3) irreps obtained from the product $(0,0)\times(2N-4k,2k)$. Using Eqs. (16) and (25), we can identify the following sub-class of SU(3)-PDS Hamiltonians

$$
\hat{H}_{PDS-1} = \hat{H}_{DS} + h_3\,P_0^\dagger P_0 + h_4\,P_0^\dagger s^\dagger s P_0 + h_5\left(\Lambda^\dagger s P_0 + P_0^\dagger s^\dagger\Lambda\right) \ , \qquad (26)
$$

with additional solvable states which are the members of the $g(K=0)$ and $\gamma^k(K=2k)$ bands listed in Eqs. (19)-(20).

A second sub-class of SU(3)-PDS Hamiltonian corresponds to the choice

$$\hat{H}_{PDS-2} \;=\; \hat{H}_{DS} + h_6\, W_2^\dagger \cdot \tilde{W}_2 + h_7\, W_3^\dagger \cdot \tilde{W}_3 \;. \tag{27}$$

$\hat{H}_{PDS-2}$ has a solvable ground band $g(K=0)$, Eq. (19), and a solvable $\beta(K=0)$ band

$$|N,(2N-4,2)K=0,L\rangle \quad L=0,2,4,\ldots,(2N-4) \tag{28a}$$

$$E_{PDS-2} = h_2\, 6(N-2)(2N-1) + CL(L+1)\;. \tag{28b}$$

This result follows from the fact that W_{2m} and W_{3m} annihilate the intrinsic state of this band, $|\beta(K=0)\rangle \propto (\sqrt{2}P_0^\dagger - P_{2,0}^\dagger)|\beta=\sqrt{2},\gamma=0;N-2\rangle$, as well as the projected Elliott basis states

$$W_{\ell m}|[N](2N-4,2),K=0,L\rangle \;=\; 0 \qquad \ell=2,3\;. \tag{29}$$

Three-body terms allow an additional solvable symmetry-conserving operator

$$\hat{\Omega} \;=\; -4\sqrt{3}\,\hat{Q}\cdot(\hat{L}\times\hat{L})^{(2)}\;. \tag{30}$$

This operator is constructed from SU(3) generators, hence is diagonal in (λ,μ). It breaks, however, the aforementioned K-degeneracy. A well defined procedure exists for obtaining the eigenstates of $\hat{\Omega}$ and corresponding eigenvalues $\langle\hat{\Omega}\rangle$ [22, 23]. For example, for the irreps $(\lambda,0)$ and $(\lambda,2)$ with λ even, we have

$$(\lambda,0)\; K=0, \quad L=0,2,4,\ldots,\lambda: \qquad \langle\hat{\Omega}\rangle = (2\lambda+3)L(L+1)\;, \tag{31a}$$

$$(\lambda,2)\; K=2, \quad L=3,5,7,\ldots,\lambda+1,\lambda+2: \qquad \langle\hat{\Omega}\rangle = (2\lambda+5)[L(L+1)-12]\;, \tag{31b}$$

$$(\lambda,2)\; K=0, \quad L=0: \qquad \langle\hat{\Omega}\rangle = 0\;, \tag{31c}$$

$$(\lambda,2)\; K=0,2, \quad L=2,4,6,\ldots,\lambda: \quad \langle\hat{\Omega}\rangle = (2\lambda+5)[L(L+1)-6] \pm 6\sqrt{(2\lambda+5)^2 + L(L+1)(L-1)(L+2)}\;. \tag{31d}$$

Several works have examined the influence of the symmetry-conserving operator $\hat{\Omega}$ (30) on nuclear spectra, within the IBM [23–25] and the symplectic shell model [26, 27]. It is interesting to note that the operator $\hat{\Omega}$ can be expressed in terms of $\hat{H}_{PDS-1}$ (26) and $\hat{H}_{PDS-2}$ (27) as

$$\begin{aligned}\hat{\Omega} \;=\;& -2(2\hat{N}-1)\hat{\theta}_2 + 2\Lambda^\dagger\Lambda + (4\hat{N}+3)\hat{L}\cdot\hat{L}\\ &+6(\hat{N}-1)P_0^\dagger P_0 - 2\left(\Lambda^\dagger s P_0 + P_0^\dagger s^\dagger \Lambda\right) + 2\,W_2^\dagger\cdot\tilde{W}_2 + 4\,W_3^\dagger\cdot\tilde{W}_3\;.\end{aligned} \tag{32}$$

4. O(6) partial dynamical symmetry

The O(6) DS chain of the IBM and related quantum numbers are given by [7]

$$\begin{array}{ccccccc} U(6) & \supset & O(6) & \supset & O(5) & \supset & O(3) \\ \downarrow & & \downarrow & & \downarrow & & \downarrow \\ [N] & & \langle\Sigma\rangle & & (\tau) & n_\Delta & L \end{array}\;, \tag{33}$$

For a given U(6) irrep $[N]$, the allowed O(6) and O(5) irreps are $\Sigma = N,\ N-2,\ldots 0$ or 1, and $\tau = 0,\ 1,\ \ldots\Sigma$, respectively. The values of L contained in the O(5) irrep (τ) are obtained from the known O(5) $\supset$ O(3) reduction and n_Δ is a multiplicity label. The eigenstates $|[N]\langle\Sigma\rangle(\tau)n_\Delta L\rangle$

are obtained with a Hamiltonian with O(6) DS which, for one- and two-body interactions, can be transcribed in the form

$$\hat{H}_{\rm DS} \;=\; h_0\left[-\hat{C}_2({\rm O(6)}) + \hat{N}(\hat{N}+4)\right] + B\,\hat{C}_2({\rm O(5)}) + C\,\hat{C}_2({\rm O(3)}) \;. \tag{34}$$

Here the quadratic Casimir operators, $\hat{C}_2(G)$, are defined in Table 1. The spectrum of $\hat{H}_{\rm DS}$ is completely solvable with eigenenergies

$$E_{\rm DS} \;=\; 4h_0\,(N-v+2)v + B\,\tau(\tau+3) + C\,L(L+1) \;. \tag{35}$$

The spectrum resembles that of a γ-unstable deformed rotovibrator, where states are arranged in bands with O(6) quantum number $\Sigma = N-2v$, $(v=0,1,2,\ldots)$. The ground band $(v=0)$ corresponds to the O(6) irrep with $\Sigma = N$. The O(5) and O(3) terms in $\hat{H}_{\rm DS}$ (34), govern the in-band rotational splitting. The lowest members in each band have quantum numbers $(\tau=0,\,L=0)$, $(\tau=1,\,L=2)$ and $(\tau=2,\,L=2,4)$.

The construction of Hamiltonians with O(6)-PDS of type I is based on identification of n-boson operators which annihilate all states in a given O(6) irrep, $\langle\Sigma\rangle$, chosen here to be the ground band irrep $\langle\Sigma\rangle = \langle N\rangle$. For that purpose, a relevant operator to consider is

$$\hat{B}^{\dagger}_{[2]\langle 0\rangle(0)0;00} \;\propto\; P^{\dagger}_0 = d^{\dagger}\cdot d^{\dagger} - (s^{\dagger})^2 \;. \tag{36}$$

The corresponding Hermitian conjugate boson-pair annihilation operator, P_0, transforms also as $\langle\Sigma\rangle = \langle 0\rangle$ under O(6) and satisfies

$$P_0\,|[N]\langle N\rangle(\tau)n_{\Delta}L\rangle = 0 \;, \qquad \tau = 0,1,2,\ldots,N \;. \tag{37}$$

Equivalently, this operator satisfies

$$P_0|\beta=1,\gamma;N\rangle \;=\; 0 \;, \tag{38}$$

where $|\beta=1,\gamma;N\rangle$, is the condensate of Eq. (2) with the O(6) equilibrium deformations. It is the lowest-weight state in the O(6) irrep $\langle\Sigma\rangle = \langle N\rangle$ and serves as an intrinsic state for the O(6) ground band. The rotational members of the band, $|[N]\langle N\rangle(\tau)n_{\Delta}L\rangle$, Eq. (37), are obtained from it by O(5) projection, and span the entire O(6) irrep $\langle\Sigma\rangle = \langle N\rangle$. The relations in Eqs. (37)-(38) follow from the fact that the action of the operator P_0 leads to a state with $N-2$ bosons in the U(6) irrep $[N-2]$, which does not contain the O(6) irrep $\langle N\rangle$ obtained from the product of $\langle 0\rangle \times \langle N\rangle$.

Since both $P^{\dagger}_0$ and P_0 (36) are O(6) scalars, they give rise to the following interaction

$$P^{\dagger}_0 P_0 \;=\; -\hat{C}_{\rm O(6)} + \hat{N}(\hat{N}+4) \;, \tag{39}$$

which is simply the O(6) term in $\hat{H}_{\rm DS}$, Eq. (34), with an exact O(6) symmetry. Thus, in this case, unlike the situation encountered with SU(3)-PDS, the algorithm does not yield an O(6)-PDS of type I with two-body interactions. In the IBM framework, an Hamiltonian with a genuine O(6)-PDS of this class, requires higher-order terms.

Focusing on three-body interactions with O(6)-PDS, one considers the following two three-boson O(6) tensors, $\hat{B}^{\dagger}_{[n]\langle\sigma\rangle(\tau)n_{\Delta}\ell m}$, with $n=3$, $\sigma=1$ and $\ell=0,2$,

$$\hat{B}^{\dagger}_{[3]\langle 1\rangle(0)0;00} \;\propto\; P^{\dagger}_0 s^{\dagger} \;, \tag{40a}$$

$$\hat{B}^{\dagger}_{[3]\langle 1\rangle(1)0;2m} \;\propto\; P^{\dagger}_0 d^{\dagger}_m \;. \tag{40b}$$

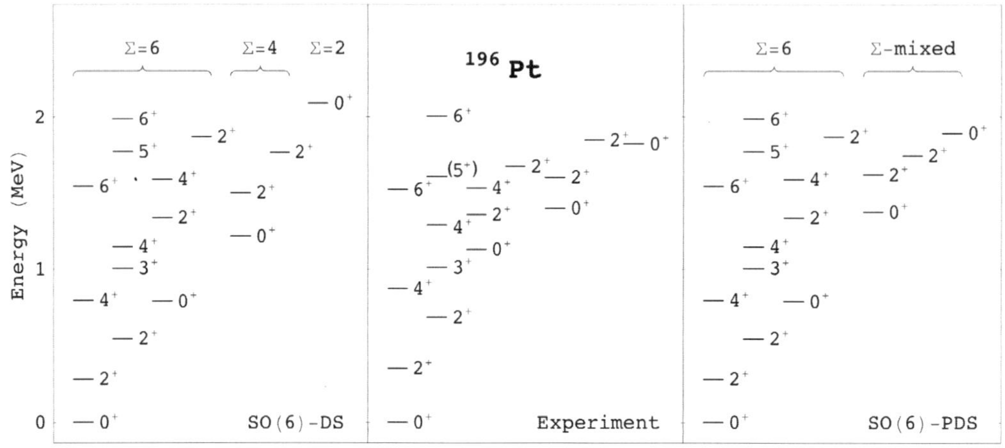

Figure 2. Observed spectrum of ^{196}Pt compared with the calculated spectra of $\hat{H}_{\rm DS}$ (34), with O(6) dynamical symmetry (DS), and of $\hat{H}_{\rm PDS}$ (41) with O(6) partial dynamical symmetry (PDS). The parameters in $\hat{H}_{\rm DS}$ ($\hat{H}_{\rm PDS}$) are $h_0 = 43.6\,(30.7)$, $B = 44.0\,(44.0)$, $C = 17.9\,(17.9)$, and $\eta = 0\,(8.7)$ keV. The boson number is $N = 6$ and Σ is an O(6) label [17].

The relation of Eq. (37) ensures that sP_0 and d_mP_0 annihilate the states of the O(6) ground band. The only three-body interactions that are partially solvable in O(6) are thus $P_0^\dagger \hat{n}_s P_0$ and $P_0^\dagger \hat{n}_d P_0$. Since the combination $P_0^\dagger(\hat{n}_s + \hat{n}_d)P_0 = (\hat{N}-2)P_0^\dagger P_0$ is completely solvable in O(6), we can transcribe the O(6)-PDS Hamiltonian in the form

$$\hat{H}_{\rm PDS} = \hat{H}_{\rm DS} + \eta\, P_0^\dagger \hat{n}_s P_0 \;, \tag{41}$$

Table 3. Observed (EXP) and calculated B(E2) values (in e^2b^2) for ^{196}Pt. For both the exact (DS) and partial (PDS) O(6) dynamical symmetry calculations, the E2 operator is $T(E2) = e_B[\,\Pi^{(2)} + \chi\, U^{(2)}\,]$ with $e_B = 0.151$ eb and $\chi = 0.29$. Only the state 0_3^+ has a mixed O(6) character [17].

Transition	EXP	DS	PDS	Transition	EXP	DS	PDS
$2_1^+ \to 0_1^+$	0.274 (1)	0.274	0.274	$2_3^+ \to 0_2^+$	0.034 (34)	0.119	0.119
$2_2^+ \to 2_1^+$	0.368 (9)	0.358	0.358	$2_3^+ \to 4_1^+$	0.0009 (8)	0.0004	0.0004
$2_2^+ \to 0_1^+$	3.10^{-8}(3)	0.0018	0.0018	$2_3^+ \to 2_2^+$	0.0018 (16)	0.0013	0.0013
$4_1^+ \to 2_1^+$	0.405 (6)	0.358	0.358	$2_3^+ \to 0_1^+$	0.00002 (2)	0	0
$0_2^+ \to 2_2^+$	0.121 (67)	0.365	0.365	$6_2^+ \to 6_1^+$	0.108 (34)	0.103	0.103
$0_2^+ \to 2_1^+$	0.019 (10)	0.003	0.003	$6_2^+ \to 4_2^+$	0.331 (88)	0.221	0.221
$4_2^+ \to 4_1^+$	0.115 (40)	0.174	0.174	$6_2^+ \to 4_1^+$	0.0032 (9)	0.0008	0.0008
$4_2^+ \to 2_2^+$	0.196 (42)	0.191	0.191	$0_3^+ \to 2_2^+$	< 0.0028	0.0037	0.0028
$4_2^+ \to 2_1^+$	0.004 (1)	0.001	0.001	$0_3^+ \to 2_1^+$	< 0.034	0	0
$6_1^+ \to 4_1^+$	0.493 (32)	0.365	0.365				

Here the dynamical symmetry Hamiltonian, $\hat{H}_{DS}$, is that of Eq. (34), since no new terms are added to it at the level of three-body interactions. The solvable states are members of the γ-unstable deformed ground band

$$|[N]\langle N\rangle(\tau)n_{\Delta}L\rangle\ , \qquad \tau = 0, 1, 2, \ldots, N \tag{42a}$$
$$E_{PDS} = B\tau(\tau+3) + CL(L+1)\ . \tag{42b}$$

The experimental spectrum and E2 rates of ^{196}Pt are shown in Fig. 2 and Table 3. The O(6)-DS limit is seen to provide a good description for properties of states in the ground band ($\Sigma = N$). This observation was the basis of the claim [28] that the O(6)-DS is manifested empirically in ^{196}Pt. However, the resulting fit to energies of excited bands is quite poor. The 0_1^+, 0_3^+, and 0_4^+ levels of ^{196}Pt at excitation energies 0, 1403, 1823 keV, respectively, are identified as the bandhead states of the ground ($v = 0$), first- ($v = 1$) and second- ($v = 2$) excited vibrational bands [28]. Their empirical anharmonicity, defined by the ratio $R = E(v = 2)/E(v = 1) - 2$, is found to be $R = -0.70$. In the O(6)-DS limit these bandhead states have $\tau = L = 0$ and $\Sigma = N, N-2, N-4$, respectively. The anharmonicity $R = -2/(N + 1)$, as calculated from Eq. (35), is fixed by N. For $N = 6$, which is the appropriate boson number for ^{196}Pt, the O(6)-DS value is $R = -0.29$, which is in marked disagreement with the empirical value. A detailed study of double-phonon excitations within the IBM, has concluded that large anharmonicities can be incorporated only by the inclusion of at least cubic terms in the Hamiltonian [29]. In the IBM there are 17 possible three-body interactions [7]. One is thus confronted with the need to select suitable higher-order terms that can break the DS in excited bands but preserve it in the ground band. On the basis of the preceding discussion this can be accomplished by the O(6)-PDS Hamiltonian of Eq. (41). The spectrum of $\hat{H}_{\rm PDS}$ is shown in Fig. 2. The states belonging to the $\Sigma = N = 6$ multiplet remain solvable with energies (42b), which obey the same DS expression, Eq. (35). States with $\Sigma < 6$ are generally admixed but agree better with the data than in the DS calculation. For example, the bandhead states of the first- (second-) excited bands have the O(6) decomposition $\Sigma = 4$: 76.5% (19.6%), $\Sigma = 2$: 16.1% (18.4%), and $\Sigma = 0$: 7.4% (62.0%). Thus, although the ground band is pure, the excited bands exhibit strong O(6) breaking. The calculated O(6)-PDS anharmonicity for these bands is $R = -0.63$, much closer to the empirical value, $R = -0.70$. It should be emphasized that not only the energies but also the wave functions of the $\Sigma = N$ states remain unchanged when the Hamiltonian is generalized from DS to PDS. Consequently, the E2 rates for transitions among this class of states are the same in the DS and PDS calculations. Thus, the additional three-body term in the Hamiltonian (41), does not spoil the good O(6)-DS description for this segment of the spectrum. This is evident in Table 3 where most of the E2 data concern transitions between $\Sigma = N = 6$ states.

5. Coexistence of partial dynamical symmetries

The examples considered in previous sections involved Hamiltonians with a single PDS, describing stable structures, *e.g.*, well-deformed nuclei. Multiple partial dynamical symmetries can occur in systems undergoing quantum phase transitions (QPTs) [30]. The latter are structural changes induced by a variation of parameters in the Hamiltonian. Such ground-state phase transitions are a pervasive phenomenon observed in many branches of physics [31], and are realized empirically in nuclei as transitions between different shapes. In the IBM, the dynamical symmetry limits correspond to possible phases of the system and the relevant Hamiltonians for studying shape-phase transitions involve terms with from different DS chains [15]. The nature of the phase transition is governed by the topology of the surface $E_N(\beta,\gamma)$, Eq. (1), which serves as a Landau's potential with the equilibrium deformations as order parameters. The surface at the critical-point of a first-order transition is required to have two degenerate minima, corresponding to the two coexisting phases. Specifically, the first-order critical surface is

$$E_N(\beta,\gamma=0) \;\; = \;\; 2h_2N(N-1)(1+\beta^2)^{-2}\beta^2\,(\beta-\beta_0)^2\ , \tag{43}$$

 IOP Publishing
doi:10.1088/1742-6596/380/1/012002

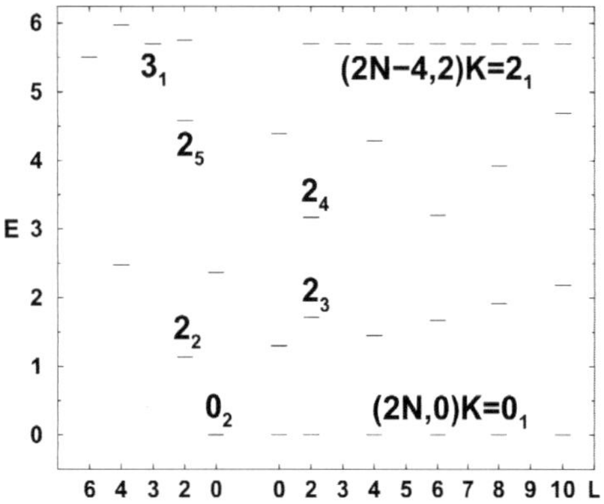

Figure 3. Spectrum of $\hat{H}_{cri}(\beta_0 = \sqrt{2})$, Eq. (44), with $h_2 = 0.05$, $N = 10$. $L(K = 0_1)$ and $L(K = 2_1)$ are the solvable SU(3) states of Eqs. (19)-(20) with $k = 0$ and $k = 1$, respectively. $L = 0_2, 3_1$ are the solvable U(5) states of Eq. (49) [30].

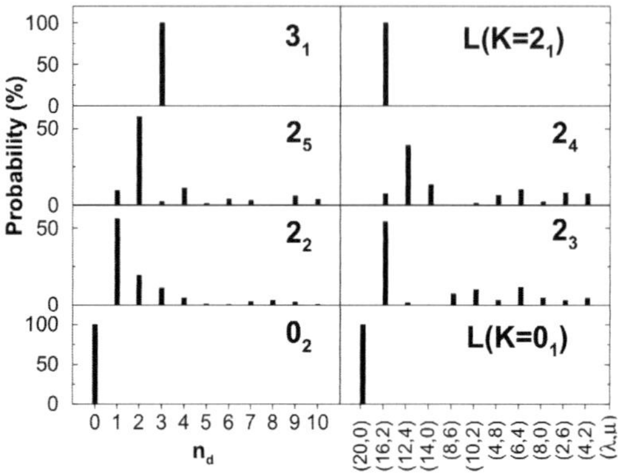

Figure 4. U(5) (n_d) and SU(3) $[(\lambda, \mu)]$ decomposition for selected spherical and deformed states in Fig. 3 [30].

and has degenerate spherical and deformed minima at $\beta = 0$ and $(\beta = \beta_0, \gamma = 0)$, corresponding to spherical and axially-deformed shapes. A barrier of height $h = h_2 N(N-1)(1 - \sqrt{1+\beta_0^2})^2/2$ separates the two minima. Such a surface can be obtained from Eq. (1) with the following critical-point Hamiltonian

$$\hat{H}_{cri}(\beta_0) \;=\; h_2\, P_2^\dagger(\beta_0) \cdot \tilde{P}_2(\beta_0)\;, \quad P_{2m}^\dagger(\beta_0) = \beta_0 \sqrt{2} d_m^\dagger s^\dagger + \sqrt{7}\,(d^\dagger\, d^\dagger)^{(2)}_m\;. \tag{44}$$

$P_{2m}(\beta_0)$ annihilates the condensate of Eq. (2) with $(\beta = \beta_0, \gamma = 0)$ as well as the states of good L, $|\,\beta_0; N, L\rangle$, projected from it

$$P_{2m}(\beta_0)\,|\,\beta_0, \gamma = 0; N\rangle \;=\; 0\;, \tag{45a}$$
$$P_{2m}(\beta_0)\,|\,\beta_0; N, L\rangle \;=\; 0\;, \qquad L = 0, 2, 4, \ldots, 2N\;. \tag{45b}$$

Consequently, $\hat{H}_{cri}(\beta_0)$ has a solvable zero-energy prolate-deformed ground band, composed of the L-projected states mentioned above

$$|\beta_0; N, L\rangle \quad E = 0 \qquad L = 0, 2, 4, \ldots, 2N\;. \tag{46}$$

The multipole form of $\hat{H}_{cri}(\beta_0)$ (44) is given by

$$\hat{H}_{cri}(\beta_0) = h_2\left[2(\beta_0^2\hat{N} - 2)\hat{n}_d + 2(1-\beta_0^2)\hat{n}_d^2 + 2\hat{C}_2(\mathrm{O}(5)) - \hat{C}_2(\mathrm{O}(3)) + \sqrt{14}\beta_0\Pi^{(2)}\cdot U^{(2)}\right], \tag{47}$$

where the various operators are defined in Table 1. The $\hat{n}_d$, $\hat{n}_d^2$, $\hat{C}_2(O(5))$ and $\hat{C}_2(O(3))$ terms in Eq. (47) belong to the dynamical symmetry Hamiltonian of the U(5) chain

$$\begin{array}{ccccccc} \mathrm{U}(6) & \supset & \mathrm{U}(5) & \supset & \mathrm{O}(5) & \supset & \mathrm{O}(3) \\ \downarrow & & \downarrow & & \downarrow & & \downarrow \\ [N] & & \langle n_d\rangle & & (\tau) & n_\Delta & L \end{array}\;, \tag{48}$$

and describe the dynamics of an anharmonic spherical vibrator. The $\Pi^{(2)}\cdot U^{(2)}$ term can connect states with $\Delta n_d = \pm1$ and $\Delta\tau = \pm1, \pm3$, hence breaks the U(5) DS. Nevertheless, $\hat{H}_{cri}(\beta_0)$ has selected solvable eigenstates with good U(5) symmetry,

$$|N, n_d = \tau = L = 0\rangle \qquad E = 0\;, \tag{49a}$$
$$|N, n_d = \tau = L = 3\rangle \qquad E = 6h_2[\beta_0^2(N-3) + 5]\;, \tag{49b}$$

Journal of Physics: Conference Series **380** (2012) 012002 doi:10.1088/1742-6596/380/1/012002

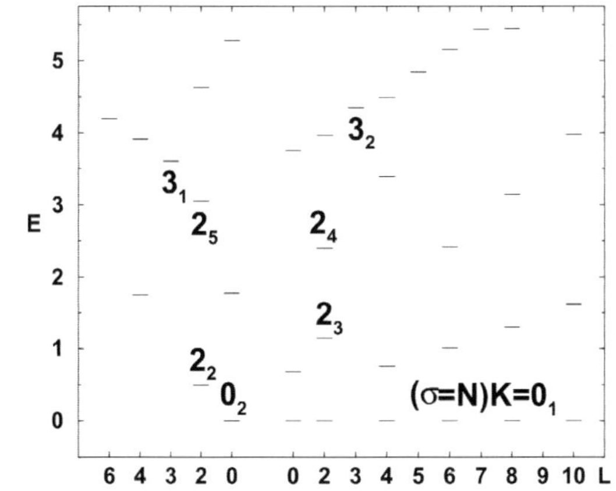

Figure 5. Spectrum of $\hat{H}_{cri}(\beta_0 = 1)$, Eq. (44), with $h_2 = 0.05$ and $N = 10$. $L(K = 0_1)$ are the solvable states of Eq. (46) with $\beta_0 = 1$, which have good O(6) symmetry $\sigma = N$, but broken O(5) symmetry. $L = 0_2, 3_1$ are the solvable U(5) states of Eq. (49) [30].

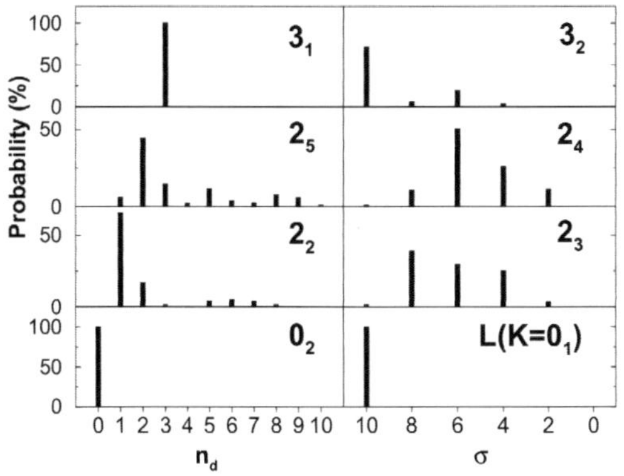

Figure 6. U(5) (n_d) and O(6) (σ) decomposition for selected spherical and deformed states in Fig. 5 [30].

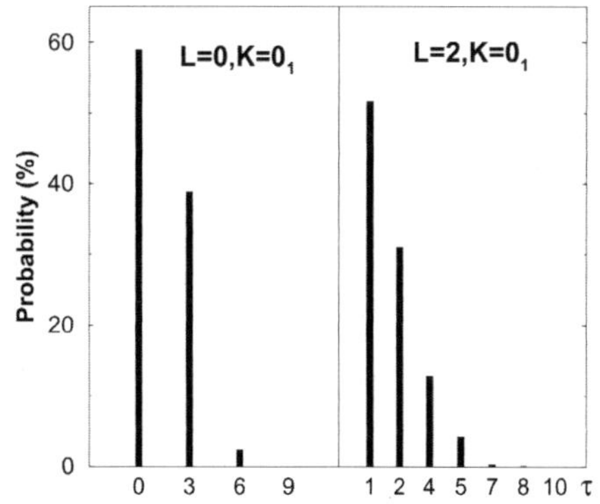

Figure 7. O(5) (τ) decomposition for the $L = 0, 2$ states, Eq. (46) with $\beta_0 = 1$, members of the ground band ($K = 0_1$) of $\hat{H}_{cri}(\beta_0 = 1)$. Both states have O(6) symmetry $\sigma = N$ [30].

and therefore, by construction, $\hat{H}_{cri}(\beta_0)$ exhibits U(5)-PDS of type I.

For $\beta_0 = \sqrt{2}$, the critical Hamiltonian of Eq. (44) is recognized to be a special case of the Hamiltonian of Eq. (18), shown to have SU(3)-PDS of type I. As such, it has a subset of solvable eigenstates, Eqs. (19)-(20), which are members of deformed ground $g(K = 0)$ and $\gamma^k(K = 2k)$ bands with good SU(3) symmetry, $(\lambda, \mu) = (2N - 4k, 2k)$. In addition, $\hat{H}_{cri}(\beta_0 = \sqrt{2})$ has the spherical states of Eq. (49), with good U(5) symmetry, as eigenstates. The spherical $L = 0$ state, Eq. (49a), is exactly degenerate with the SU(3) ground band, Eq. (19), and the spherical $L = 3$ state, Eq. (49b), is degenerate with the SU(3) γ-band, Eq. (20) with $k = 1$. The remaining levels of $\hat{H}_{cri}(\beta_0 = \sqrt{2})$, shown in Fig. 3, are calculated numerically and their wave functions are spread over many U(5) and SU(3) irreps, as is evident from Fig. 4. This situation, where some states are solvable with good U(5) symmetry, some are solvable with good SU(3) symmetry and all other states are mixed with respect to both U(5) and SU(3), defines a U(5) PDS of type I coexisting with a SU(3) PDS of type I.

For $\beta_0 = 1$, the critical Hamiltonian of Eq. (47) involves the Casimir operators of O(5) and O(3) which are diagonal in the corresponding quantum numbers, σ and τ, of the O(6)-DS chain, Eq. (33). It also contains a term involving $\hat{n}_d$ which is a scalar under O(5) but can connect states differing by $\Delta\sigma = 0, \pm 2$ and a $\Pi^{(2)} \cdot U^{(2)}$ term which induces both O(6) and O(5) mixing subject to $\Delta\sigma = 0, \pm 2$ and $\Delta\tau = \pm 1, \pm 3$. Although $\hat{H}_{cri}(\beta_0 = 1)$ is not invariant under O(6), it has a solvable prolate-deformed ground band, Eq. (46) with $\beta_0 = 1$, which has good O(6) symmetry, $\langle\sigma\rangle = \langle N\rangle$, but broken O(5) symmetry. In addition, $\hat{H}_{cri}(\beta_0 = 1)$ has the spherical states of Eq. (49), with good U(5) symmetry, as eigenstates. The remaining eigenstates in Fig. 5 are mixed with respect to both U(5) and O(6), as is evident from the decomposition shown in Fig. 6. Apart from the solvable U(5) states of Eq. (49), all eigenstates of $\hat{H}_{cri}(\beta_0 = 1)$ are mixed with respect to O(5) [including the solvable O(6) states of Eq. (46) with $\beta_0 = 1$, as shown in Fig. 7]. It follows that the Hamiltonian has a subset of states with good U(5) symmetry and a subset of states with good O(6) but broken O(5) symmetry, and all other states are mixed with respect to both U(5) and O(6). These are precisely the required features of U(5) PDS of type I coexisting with O(6) PDS of type III.

Second-order quantum phase transitions between spherical and deformed γ-unstable nuclei, can be accommodated in the IBM, by mixing the DS Hamiltonians of the U(5) chain, Eq. (48), and the O(6) chain, Eq. (33). Since U(5) and O(6) are incompatible, yet both chains have a common $O(5) \supset O(3)$ segment, we encounter a situation similar to that described in Eq. (9), which gives rise to O(5)-PDS of type II [18].

6. Concluding remarks

The notion of partial dynamical symmetry generalizes the concepts of exact and dynamical symmetries. In making the transition from an exact to a dynamical symmetry, states which are degenerate in the former scheme are split but not mixed in the latter, and the block structure of the Hamiltonian is retained. Proceeding further to partial symmetry, some blocks or selected states in a block remain pure, while other states mix and lose the symmetry character. A partial dynamical symmetry lifts the remaining degeneracies, but preserves the symmetry-purity of the selected states.

Having at hand concrete algorithms for identifying and constructing Hamiltonians with PDS, is a valuable asset. It provides selection criteria for the a priori huge number of possible symmetry-breaking terms, accompanied by a rapid proliferation of free-parameters. This is particularly important in complicated environments when many degrees of freedom take part in the dynamics and upon inclusion of higher-order terms in the Hamiltonian. Futhermore, Hamiltonians with PDS break the dynamical symmetry (DS) but retain selected solvable eigenstates with good symmetry. The advantage of using interactions with a PDS is that they can be introduced, in a controlled manner, without destroying results previously obtained with a DS for a segment of the spectrum. These virtues greatly enhance the scope of applications of algebraic modeling of quantum many-body systems.

PDSs appear to be a common feature in algebraic descriptions of dynamical systems. They are not restricted to a specific model but can be applied to any quantal systems of interacting particles, bosons, as demonstrated in the present contribution, and fermions [13, 32, 33]. They are also relevant to the study of mixed systems with coexisting regularity and chaos, where they lead to a suppression of chaos [34, 35].

Acknowledgments

This work is supported by the Israel Science Foundation.

[1] Bohm A, Néeman Y and Barut A O eds 1988 *Dynamical Groups and Spectrum Generating Algebras* (Singapore: World Scientific)
[2] Wigner E 1937 *Phys. Rev.* **51** 106
[3] Kerman A 1961 *Ann. Phys.* **12** 300
[4] J.P. Elliott 1958 *Proc. Roy. Soc.* A **245** 128; 562
[5] D.J. Rowe 1985 *Rep. Prog. Phys.* **48** 1419
[6] J.N. Ginocchio 1980 *Ann. Phys.* **126** 234
[7] Iachello F and Arima A 1987 *The Interacting Boson Model* (Cambridge: Cambridge Univ. Press)
[8] Iachello F and Van Isacker P 1991 *The Interacting Boson-Fermion Model* (Cambridge: Cambridge Univ. Press)

[9] Iachello F and Levine R D 1994 *Algebraic Theory of Molecules* (Oxford: Oxford Univ. Press)
[10] A. Frank and P. Van Isacker 1994 *Algebraic Methods in Molecular and Nuclear Physics*, (New York: Wiley)
[11] Bijker R, Iachello F and Leviatan A 1984 *Ann. Phys.* **236** 69; *ibid* 2000 **284** (2000) 89
[12] Iachello F 2006 *Lie Algebras and Applications* (Berlin: Springer-Verlag)
[13] Leviatan A 2011, Prog. Part. Nucl. Phys. **66** 93
[14] Ginocchio J N and Kirson M W 1980 *Phys. Rev. Lett.* **44** 1744
[15] Dieperink A E L, Scholten O and Iachello F 1980 *Phys. Rev. Lett.* **44** 1747
[16] Alhassid Y and Leviatan A 1992 *J. Phys. A* **25** L1265
[17] García-Ramos J E, Leviatan A and Van Isacker P 2009 *Phys. Rev. Lett.* **102** 112502
[18] Leviatan A, Novoselsky A and Talmi I 1986 *Phys. Lett. B* **172** 144
[19] Van Isacker P 1999 *Phys. Rev. Lett.* **83** 4269
[20] Leviatan A and Van Isacker P 2002 *Phys. Rev. Lett.* **89** 222501
[21] Leviatan A 1996 *Phys. Rev. Lett.* **77** 818
[22] De Meyer H, Vanden Berghe G and Van der Jeugt J 1985 *J. Math. Phys.* **26** 3109
[23] Vanden Berghe G, De Meyer H E and Van Isacker P 1985 *Phys. Rev. C* **32** 1049
[24] Bonatsos D 1985 *Phys. Lett. B* **200** 1
[25] Vanthournout J 1990 *Phys. Rev. C* **41** 2380
[26] Rosensteel G, Draayer J P and Weeks K J 1984 *Nucl. Phys. A* **419** 1
[27] Draayer J P and Rosensteel G 1985 *Nucl. Phys. A* **439** 61
[28] Cizewski J A *et al.* 1978 *Phys. Rev. Lett.* **40** 167
[29] García-Ramos J E, Arias J M and Van Isacker P 2000 *Phys. Rev. C* **62** 064309
[30] Leviatan A 2007 *Phys. Rev. Lett.* **98** 242502
[31] Carr L Ed. 2010 *Understanding Quantum Phase Transitions* (CRC press)
[32] Escher J and Leviatan A 2000 *Phys. Rev. Lett.* **84** 1866; *ibid* 2002 *Phys. Rev. C* **65** 054309
[33] Van Isacker P and Heinze S 2008 *Phys. Rev. Lett.* **100** 052501
[34] Whelan N , Alhassid Y and Leviatan A 1993 *Phys. Rev. Lett.* **71** 2208
[35] Leviatan A and Whelan N D 1996 *Phys. Rev. Lett.* **77** 5202

Spectrum generating algebras for few-body systems

Roelof Bijker

Instituto de Ciencias Nucleares, Universidad Nacional Autónoma de México, A.P. 70-543, 04510 México, D.F., México

E-mail: bijker@nucleares.unam.mx

Abstract. I discuss an algebraic treatment of few-body systems based on spectrum generating algebras for the relative motion of the clusters. Particular attention is paid to the formalism for identical clusters including a discussion of the permutation symmetry and a study of special solutions which are shown to correspond to the harmonic oscillator, the deformed oscillator, the oblate symmetric top and the spherical top with tetrahedral symmetry.

1. Introduction

Algebraic models and spectrum generating algebras have played an important role in the study of both many-body and few-body systems. Generally speaking, in algebraic models energy eigenvalues and eigenvectors are obtained by diagonalizing a finite-dimensional matrix, rather than by solving a set of coupled differential equations in coordinate space. The benefits and merits of this type of models are illustrated beautifully by the interacting boson model (IBM), which has been very successful in the description of collective states in nuclei [1]. Its dynamical symmetries correspond to the quadrupole vibrator, the axially symmetric rotor and the γ-unstable rotor in a geometric description. In addition to these special solutions, the IBM can describe intermediate cases between any of them equally well. The first extension of the algebraic approach to few-body systems was the vibron model [2], which was introduced to describe vibrational and rotational excitations in diatomic molecules. The dynamical symmetries of the vibron model correspond to the (an)harmonic oscillator and the Morse oscillator. In spectroscopic studies these algebraic methods provide a powerful tool to study symmetries and selection rules, to classify the basis states, and to calculate matrix elements of physical observables.

The general procedure is to introduce a $U(\nu+1)$ spectrum generating algebra for a bound-state problem with ν degrees of freedom in which all states are assigned to the symmetric representation $[N]$ of $U(\nu+1)$. For the $\nu=5$ quadrupole degrees of freedom in collective nuclei this led to the introduction of the $U(6)$ interacting boson model [1]. Similarly, the $U(4)$ vibron model was proposed to describe the dynamics of the $\nu=3$ dipole degrees of freedom of the relative motion of two objects, *e.g.* two atoms in a diatomic molecule [2], two clusters in a nuclear cluster model [3], or a quark and antiquark in a meson [4]. An application to the three-body system involves the six degrees of freedom of two relative vectors which in the algebraic approach leads to a $U(7)$ spectrum generating algebra [5] as an extension of the vibron model. The $U(7)$ model was developed originally to describe the relative motion of the three constituent quarks in baryons [5], but has also found applications in molecular physics [6, 7] and nuclear physics (^{12}C as a cluster of three α particles) [8].

The aim of this contribution is to introduce the so-called Algebraic Cluster Model (ACM) to the n-body system in which the relative motion of the clusters is described in terms of a $U(3n-2)$ spectrum-generating algebra, to discuss the permutation symmetry for the case of identical objects, to study some special solutions and to establish the connection between algebraic interactions and geometric models. In the ACM, the full rotational-vibrational structure is present from the outset. There is no need for Born-Oppenheimer type approximations.

2. The Algebraic Cluster Model

In this section, I introduce the Algebraic Cluster Model for n-body systems. The eigenvalues and eigenfunctions are obtained by matrix diagonalization instead of by solving a set of differential equations. The ACM is an interacting boson model to describe the relative motion of the n clusters based on the spectrum generating algebra of $U(\nu+1)$ with $\nu = 3(n-1)$. As special cases the ACM contains the vibron model [2] for two-body problems ($n=2$) and the $U(7)$ model [5, 7, 8] for three-body clusters ($n=3$).

First, I introduce relative Jacobi coordinates

$$\vec{\rho}_k = \frac{1}{\sqrt{k(k+1)}} \left(\sum_{i=1}^{k} \vec{r}_i - k\vec{r}_{k+1} \right) , \qquad k = 1,\ldots,n-1 , \tag{1}$$

and their conjugate momenta. Here $\vec{r}_i$ are the coordinates of the i-th cluster ($i = 1,\ldots,n$). Instead of a formulation in terms of coordinates and momenta the method of bosonic quantization is used which consists of introducing a dipole boson with $L^P = 1^-$ for each independent relative coordinate and an auxiliary scalar boson with $L^P = 0^+$

$$s^\dagger , \; b^\dagger_{k,m} , \tag{2}$$

with $k = 1,\ldots,n-1$ and $m = -1,0,1$. The scalar boson does not represent an independent degree of freedom, but is added under the restriction that the Hamiltonian commutes with the number operator

$$\hat{N} = s^\dagger s + \sum_k \sum_m b^\dagger_{k,m} b_{k,m} , \tag{3}$$

i.e. the total number of bosons $N = n_s + \sum_k n_k$ is conserved. The set of $(3n-2)^2$ bilinear products of creation and annihilation operators spans the Lie algebra of $U(3n-2)$. All operators, such as the Hamiltonian and electromagnetic transition operators, are expanded into elements of this algebra. This procedure leads to a compact spectrum generating algebra of $U(3n-2)$ whose model space is spanned by the symmetric irreducible representation $[N]$ which contains the oscillator shells with $n_b = \sum_k n_k = 0,1,2,\ldots,N$. The introduction of the scalar boson makes it possible to investigate the dynamics of three vector degrees of freedom including situations in which there is a mixing of oscillator shells.

For n identical objects, the Hamiltonian has to be invariant under the permuation group S_n. The permutation symmetry of n identical objects is determined by the transposition $P(12)$ and the cyclic permutation $P(12\cdots n)$ [9]. All other permutations can be expressed in terms of these two elementary ones. The transformation properties under S_n of all operators in the model follow from those of the building blocks. The scalar boson, $s^\dagger$, transform as the symmetric representation $[n]$, whereas the dipole Jacobi bosons, $b^\dagger_k$ with $k = 1,\ldots,n-1$ transform as the $n-1$ components of the mixed symmetry representation $[n-1,1]$.

Next, one can use the multiplication rules for S_n to construct construct physical operators with the appropriate symmetry properties. As a result, the most general one- and two-body

Hamiltonian that describes the relative motion of a system of n identical clusters, is a scalar under the permutation group S_n, is rotationally invariant, conserves parity as well as the total number of bosons, is given by

$$\begin{aligned} H &= \epsilon_0\, s^\dagger \tilde{s} - \epsilon_1 \sum_k b_k^\dagger \cdot \tilde{b}_k + u_0\, s^\dagger s^\dagger \tilde{s}\tilde{s} - u_1 \sum_k s^\dagger b_k^\dagger \cdot \tilde{b}_k \tilde{s} \\ & + v_0 \left[\sum_k b_k^\dagger \cdot b_k^\dagger\, \tilde{s}\tilde{s} + \text{h.c.} \right] + \sum_L \sum_{ijkl} v_{ijkl}^{(L)}\, [b_i^\dagger \times b_j^\dagger]^{(L)} \cdot [\tilde{b}_k \times \tilde{b}_l]^{(L)} , \end{aligned} \tag{4}$$

with $\tilde{b}_{k,m} = (-1)^{1-m} b_{k,-m}$ with $k = 1, \ldots, n-1$ and $\tilde{s} = s$. The invariance under the permutation symmetry imposes restrictions on the coefficients $v_{ijkl}^{(L)}$. By construction, the first five terms in Eq. (4) are invariant under the permutation group S_n.

In general, the eigenvalues and corresponding eigenvectors are obtained numerically by diagonalization in a coupled harmonic oscillator basis. By construction, the wave functions are characterized by the total number of bosons N, angular momentum and parity L^P, and their transformation property under t the permutation group S_n. The relation between harmonic oscillators and permutation symmetry was studied by Kramer and Moshinsky [9]. In particular, it was shown that for the n-body problem, the reduction $U(3n-3) \supset U(3) \otimes U(n-1)$ with $U(3) \supset SO(3)$ and $U(n-1) \supset O(n-1) \supset S_n$ provides a basis for an explicit construction of states with good permutation symmetry. Here $U(3)$ denotes the symmetry group for the coupled harmonic oscillator and $U(n-1)$ is the symmetry group associated with the transformation of the indices of the $n-1$ relative Jacobi vectors. In the application to the ACM, it is preferred to generate the wave functions with good permutation symmetry $|\psi_t\rangle$ numerically by diagonalization of a S_n invariant Hamilltonian. The permutation symmetry t of a given wave function can then be determined from the transformation properties under $P(12)$ and $P(12\cdots n)$ [9] with applications to systems of three and four identical clusters [5, 10].

3. Geometric analysis

The geometric properties of the algebraic Hamiltonian of Eq. (4) can be studied by using a coherent state as a variational wave function for the n-body system

$$|N; \vec{\alpha}_1, \ldots, \vec{\alpha}_{n-1}\rangle = \frac{1}{\sqrt{N!}} (b_c^\dagger)^N |0\rangle , \tag{5}$$

with

$$b_c^\dagger = \sqrt{1 - \sum_k \vec{\alpha}_k \cdot \vec{\alpha}_k^*}\; s^\dagger + \sum_k \vec{\alpha}_k \cdot \vec{b}_k^\dagger . \tag{6}$$

The condensate boson $b_c^\dagger$ is parametrized in terms of $3(n-1)$ complex variables corresponding to the $n-1$ relative Jacobi coordinates and their conjugate momenta For the geometric analysis of the ACM Hamiltonian it is convenient to use spherical rather than cartesian coordinates and momenta. We transform the variables $\vec{\alpha}_k$ to intrinsic coordinates (q_k, θ_k, ϕ_k) and their conjugate momenta [11, 12]

$$\alpha_{k,\mu} = \frac{1}{\sqrt{2}} \sum_\nu \mathcal{D}_{\mu\nu}^{(1)}(\phi_k, \theta_k, 0)\, \beta_{k,\nu} , \tag{7}$$

with

$$\begin{pmatrix} \beta_{k,1} \\ \beta_{k,0} \\ \beta_{k,-1} \end{pmatrix} = \begin{pmatrix} [-p_{\phi_k}/\sin\theta_k - i p_{\theta_k}]/q_k\sqrt{2} \\ q_k + i p_k \\ [-p_{\phi_k}/\sin\theta_k + i p_{\theta_k}]/q_k\sqrt{2} \end{pmatrix} , \tag{8}$$

for $k = 1, \ldots, n-1$. The classical limit of the ACM Hamiltonian is then given by the coherent state expectation value of the normal ordered Hamiltonian divided by the number of bosons N

$$H_{\rm cl} \;=\; \frac{1}{N}\langle N;\vec{\alpha}_1,\ldots,\vec{\alpha}_{n-1} \mid : H : \mid N;\vec{\alpha}_1,\ldots,\vec{\alpha}_{n-1}\rangle \;. \tag{9}$$

4. Special solutions

The Algebraic Cluster Model has a rich algebraic structure, which includes both continuous and discrete symmetries. It is of general interest to study limiting cases of the Hamiltonian of Eq. (4), in which the energy spectra can be obtained in closed form. We first consider two dynamical symmetries of the S_n invariant ACM Hamiltonian. These dynamical symmetries can be studied in general for the n-body problem for any number of clusters and are shown to correspond to the $3(n-1)$-dimensional spherical and deformed oscillators.

In addition, there are some other cases of S_n invariant Hamiltonians for which approximate solutions can be obtained in a semiclassical mean-field analysis. For the case of three-body clusters I discuss the oblate symmetric top and for four-body clusters the spherical top with tetrahedral symmetry.

4.1. Harmonic oscillator

For $v_0 = 0$ in Eq. (4), there is no coupling between different harmonic oscillator shells. The oscillator is harmonic if all terms, except ϵ_0 and ϵ_1, are set to zero; otherwise it is anharmonic. This dynamical symmetry corresponds to the reduction

$$\left| \begin{array}{ccccc} U(3n-2) & \supset & U(3n-3) & \supset & SO(3n-3) \\ N & , & n_b & , & \sigma \end{array} \right\rangle \;. \tag{10}$$

We consider the one-body Hamiltonian

$$H_1 \;=\; \epsilon \sum_{km} b^{\dagger}_{k,m} b_{k,m} \;, \tag{11}$$

whose eigenvalues are those of a $3(n-1)$-dimensional spherical oscillator

$$E_1 \;=\; \epsilon\, n_b \;. \tag{12}$$

The label n_b represents the total number of oscillator quanta $n_b = \sum_k n_k = 0, 1, \ldots, N$. The energy levels are grouped into oscillator shells characterized by n_b and parity $P = (-1)^{n_b}$. The levels belonging to an oscillator shell are further classified by the symmetric irreducible representation σ of $SO(3n-3)$ with $\sigma = n_b, n_b-2, \ldots, 1$ or 0 for n_b odd or even (see Fig. 1). This special case is called the $U(3n-3)$ limit.

The classical limit of the Hamiltonian in Eq. (11) is given by

$$H_{1,\rm cl} \;=\; \epsilon \sum_k \vec{\alpha}_k \cdot \vec{\alpha}^*_k \;=\; \epsilon \frac{1}{2} \sum_k \left(p_k^2 + q_k^2 + \frac{L_k^2}{q_k^2} \right) \;, \tag{13}$$

where L_k^2 is the angular momentum in polar coordinates of the k-th oscillator

$$L_k^2 \;=\; p_{\theta_k}^2 + \frac{p_{\phi_k}^2}{\sin^2\theta_k} \;. \tag{14}$$

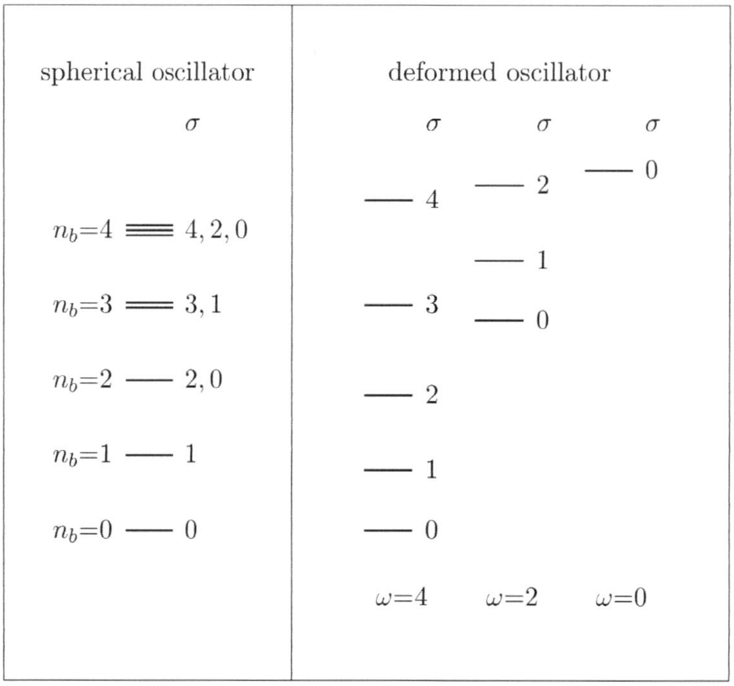

Figure 1. Comparison of the spectrum of σ multiplets in the $U(3n-3)$ limit (left) and the $SO(3n-2)$ limit (right) for the four-body problem ($n=4$). The number of bosons is $N=4$.

A change of variables of the coordinates q_k to hyperspherical coordinates in index space, the hyperradius q and the angles $\chi_1, \ldots, \chi_{n-2}$

$$\begin{aligned} q_1 &= q \sin\chi_1 \cdots \sin\chi_{n-3} \sin\chi_{n-2} \;, \\ q_2 &= q \sin\chi_1 \cdots \sin\chi_{n-3} \cos\chi_{n-2} \;, \\ &\vdots \end{aligned} \tag{15}$$

$$q_{n-1} = q \cos\chi_1 \;, \tag{16}$$

and their conjugate momenta, p and p_{χ_k}, reduces the classical limit to

$$H_{1,\mathrm{cl}} = \epsilon \frac{1}{2} \left(p^2 + q^2 + \frac{\Lambda^2_{n-1}}{q^2} \right) \;. \tag{17}$$

Here Λ^2_{n-1} is the generalized angular momentum for rotations in $3(n-1)$ dimensions

$$\begin{aligned} \Lambda^2_{n-k} &= p^2_{\chi_k} + \frac{L^2_{n-k}}{\cos^2\chi_k} + \frac{1}{\sin^2\chi_k} \Lambda^2_{n-k-1} \;, \qquad k = 1, \ldots, n-2 \;, \\ \Lambda^2_1 &= L^2_1 \;. \end{aligned} \tag{18}$$

Eq. (17) confirms the interpretation of the $U(3n-3)$ limit of the ACM in terms of a $3(n-1)$-dimensional spherical oscillator.

4.2. Deformed oscillator

For the (an)harmonic oscillator, the number of oscillator quanta n_b is a good quantum number. However, when $v_0 \neq 0$ in Eq. (4), the oscillator shells with $\Delta n_b = \pm 2$ are mixed, and the eigenfunctions are spread over many different oscillator shells. A dynamical symmetry that involves the mixing between oscillator shells, is provided by the reduction

$$\left| \begin{array}{ccccc} U(3n-2) & \supset & SO(3n-2) & \supset & SO(3n-3) \\ N & , & \omega & , & \sigma \end{array} \right\rangle . \tag{19}$$

We consider now a dipole-dipole interaction which can be rewritten as the difference between the Casimir operators of $SO(3n-2)$ and $SO(3n-3)$

$$\begin{aligned} H_2 &= \kappa \left[\hat{N}(\hat{N}+3n-4) - \hat{C}_{2SO(3n-2)} + \hat{C}_{2SO(3n-3)} \right] \\ &= \kappa \left[(s^\dagger s^\dagger - \sum_k b_k^\dagger \cdot b_k^\dagger)\,(\text{h.c.}) + \hat{C}_{2SO(3n-3)} \right] , \end{aligned} \tag{20}$$

where $\hat{N}$ is the number operator of Eq. (3). The energy spectrum in this case, called the $SO(3n-2)$ limit, is given by the eigenvalues of the Casimir operators as

$$E_2 = \kappa \left[(N-\omega)(N+\omega+3n-4) + \sigma(\sigma+3n-5) \right] . \tag{21}$$

The label $\omega = N, N-2, \ldots, 1$ or 0 for N odd or even, respectively, characterizes the symmetric representations of $SO(3n-2)$, and $\sigma = 0, 1, \ldots, \omega$ those of $SO(3n-3)$.

The eigenstates are ordered in bands labeled by ω. Although the size of the model space, and hence the total number of states, is the same as for the harmonic oscillator, the ordering and classification of the states is different. For example, in the $U(3n-3)$ limit all states are vibrational, whereas the $SO(3n-2)$ limit gives rise to a rotational-vibrational spectrum, characterized by a series of rotational bands which are labeled by ω, or equivalently by the vibrational quantum number $v = (N-\omega)/2$

$$E_2 = 4\kappa N v \left(1 - \frac{2v-3n+4}{2N} \right) + \kappa\,\sigma(\sigma+3n-5) . \tag{22}$$

The classical limit of the $SO(3n-2)$ limit is given by

$$H_{2,\text{cl}} = \kappa(N-1) \left[q^2 p^2 + (1-q^2)^2 + \Lambda_{n-1}^2 \right] . \tag{23}$$

Here I have used the hyperspherical variables of Eq. (16). Just as for the harmonic oscillator, the potential only depends on the hyperradius q. In the limit of small oscillations around the minimum of the potential energy surface $q = 1 + \Delta q$ one finds to leading order in N a one-dimensional harmonic oscillator in the hyperradius q

$$H_{2,\text{cl}} \rightarrow \kappa N \left(p^2 + 4(\Delta q)^2 \right) , \tag{24}$$

with frequency $4\kappa N$. To leading order in N, the frequency coincides with that of Eq. (22). This analysis shows the explicit connection between the $SO(3n-2)$ dynamical symmetry and the deformed (or displaced) oscillator.

4.3. Three-body clusters: oblate top

The spectrum generating algebra of three identical clusters is given by $U(7)$. The potential energy surfaces of the $U(6)$ and $SO(7)$ limits (spherical and deformed oscillators) only depend on the hyperspherical radius q. The corresponding equilibrium shapes are characterized by $q_0 = 0$ and $q_0 = 1$, respectively. Another interesting case is provided by the Hamiltonian [8]

$$\begin{aligned} H_3 &= \xi_1 \, (R^2 \, s^\dagger s^\dagger - b_1^\dagger \cdot b_1^\dagger - b_2^\dagger \cdot b_2^\dagger) \, (R^2 \, \tilde{s}\tilde{s} - \tilde{b}_1 \cdot \tilde{b}_1 - \tilde{b}_2 \cdot \tilde{b}_2) \\ & +\xi_2 \left[(b_1^\dagger \cdot b_1^\dagger - b_2^\dagger \cdot b_2^\dagger) \, (\tilde{b}_1 \cdot \tilde{b}_1 - \tilde{b}_2 \cdot \tilde{b}_2) + 4 \, (b_1^\dagger \cdot b_2^\dagger) \, (\tilde{b}_2 \cdot \tilde{b}_1) \right] \\ & +\kappa \, \vec{L} \cdot \vec{L} \; . \end{aligned} \tag{25}$$

where $\vec{L}$ denotes the angular momentum in coordinate space. For $R^2 = 0$, this Hamiltonian has $U(7) \supset U(6)$ symmetry and corresponds to a spherical vibrator, whereas for $R^2 = 1$ and $\xi_2 = 0$ it has $U(7) \supset SO(7)$ symmetry and corresponds to a deformed oscillator. The general case with $R^2 \neq 0$ and ξ_1, $\xi_2 > 0$ does not correspond to a dynamical symmetry, and hence its energy spectrum cannot be obtained in closed analytic form. In this case, the energy eigenvalues and eigenvectors are calculated numerically by diagonalizing the Hamiltonian in an appropriate basis. However, an approximate energy formula can still be derived in a semiclassical mean-field analysis. The general expression of the classical limit of the Hamiltonian of Eq. (25) has a complicated structure. We first study the potential energy surface which is obtained by setting all momenta equal to zero. Its equilibrium configuration corresponds to coordinates that have equal length ($q_{1,0} = q_{2,0} = q_0/\sqrt{2}$)

$$q_0 = \sqrt{2R^2/(1+R^2)} \; , \qquad \chi_{1,0} = \pi/4 \; , \tag{26}$$

and are perpendicular

$$\zeta_{12,0} = \pi/4 \; , \tag{27}$$

where $2\zeta_{12}$ denotes the relative angle between $\vec{\alpha}_1$ and $\vec{\alpha}_2$. Geometrically, the three clusters are located at the vertices of an equilateral triangle.

In the limit of small oscillations around the equilibrium shape the intrinsic degrees of freedom decouple and become harmonic. The resulting energy spectrum is that of the vibrational excitations of an oblate symmetric top [8]

$$E_{3,\text{vib}} = \omega_1(v_1 + \frac{1}{2}) + \omega_2(v_2 + 1) \; , \tag{28}$$

with frequencies

$$\omega_1 = 4NR^2\xi_1 \; , \qquad \omega_2 = \frac{4NR^2}{1+R^2}\xi_2 \; , \tag{29}$$

in agreement with the results obtained in a normal mode analysis [5]. Here v_1 represents the vibrational quantum number for a symmetric stretching A vibration, and $v_2 = v_{2a} + v_{2b}$ for a degenerate E vibration. The rotational spectrum is given by

$$E_{3,\text{rot}} = \kappa_1 \, L(L+1) \; . \tag{30}$$

In Fig. 2 I show the structure of the spectrum of the oblate top according to the approximate energy formula of Eqs. (28) and (30). The energy spectrum consists of a series of rotational bands labeled by (v_1, v_2). It is assumed that the spin of the identical clusters is zero, as is relevant

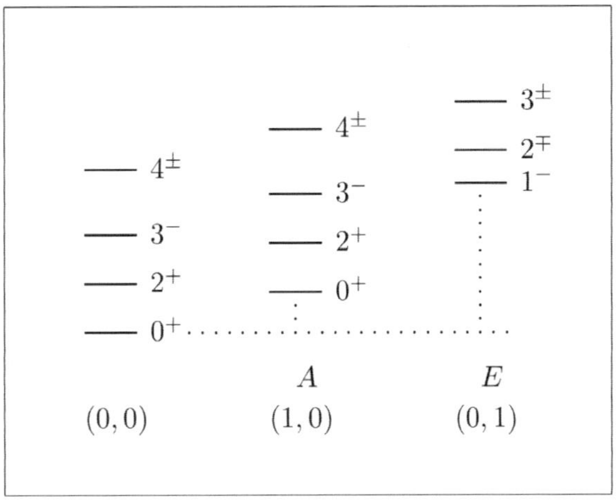

Figure 2. Schematic spectrum of an oblate symmetric top. The rotational bands are labeled by (v_1, v_2) (bottom). All states are symmetric under S_3.

for the description of the ^{12}C nucleus as a cluster of three α particles. As a consequence, all states in Fig. 2 are symmetric under S_3. For this case, the bands with $(v_1, 0)$ can have angular momenta and parity $L^P = 0^+, 2^+, 3^-, 4^\pm, \ldots$, whereas the angular momentum content of the doubly degenerate vibrations $(v_1, 1)$ is given by $L^P = 1^-, 2^\mp, 3^\mp, \ldots$, in agreement with Ref. [13]. For the schematic Hamiltonian of Eq. (25), the angular momentum states belonging to different K-bands are degenerate [8].

4.4. *Four-body clusters: spherical top*

As a last example, I discuss the spherical top with tetrahedral symmetry as a special case of the ACM for four identical clusters whose spectrum generating algebra is given by $U(10)$. Let us consider a S_4 invariant Hamiltonian of the form

$$\begin{aligned} H_4 &= \xi_1\,(R^2\, s^\dagger s^\dagger - b_1^\dagger \cdot b_1^\dagger - b_2^\dagger \cdot b_2^\dagger - b_3^\dagger \cdot b_3^\dagger)\,(\text{h.c.}) \\ &\quad +\xi_2 \Big[(-2\sqrt{2}\, b_1^\dagger \cdot b_3^\dagger + 2b_1^\dagger \cdot b_2^\dagger)\,(\text{h.c.}) \\ &\qquad\qquad + (-2\sqrt{2}\, b_2^\dagger \cdot b_3^\dagger + (b_1^\dagger \cdot b_1^\dagger - b_2^\dagger \cdot b_2^\dagger))\,(\text{h.c.})\Big] \\ &\quad +\xi_3 \Big[(2b_1^\dagger \cdot b_3^\dagger + 2\sqrt{2}\, b_1^\dagger \cdot b_2^\dagger)\,(\text{h.c.}) \\ &\qquad\qquad +(2b_2^\dagger \cdot b_3^\dagger + \sqrt{2}\,(b_1^\dagger \cdot b_1^\dagger - b_2^\dagger \cdot b_2^\dagger))\,(\text{h.c.}) \\ &\qquad\qquad +(b_1^\dagger \cdot b_1^\dagger + b_2^\dagger \cdot b_2^\dagger - 2b_3^\dagger \cdot b_3^\dagger)\,(\text{h.c.})\Big] \\ &\quad +\kappa_1\, \vec{L} \cdot \vec{L} + \kappa_2\, (\vec{L} \cdot \vec{L} - \vec{I} \cdot \vec{I})^2 \;, \end{aligned} \tag{31}$$

where $\vec{L}$ denotes the angular momentum in coordinate space and $\vec{I}$ the angular momentum in index space.

For $R^2 = 0$, this Hamiltonian has $U(10) \supset U(9)$ symmetry and corresponds to a spherical vibrator, whereas for $R^2 = 1$ and $\xi_2 = \xi_3 = 0$ it has $U(10) \supset SO(10)$ symmetry and corresponds to a deformed oscillator. The general case with $R^2 \neq 0$ and ξ_1, ξ_2, $\xi_3 > 0$ does not correspond to a dynamical symmetry, and hence its energy spectrum cannot be obtained in closed analytic form.

In this case, the energy eigenvalues and eigenvectors are calculated numerically by diagonalizing the Hamiltonian in an appropriate basis. However, an approximate energy formula can still be derived in a semiclassical mean-field analysis. The general expression of the classical limit of the Hamiltonian of Eq. (31) has a complicated structure. We first study the potential energy surface which is obtained by setting all momenta equal to zero. Its equilibrium configuration corresponds to coordinates that have equal length ($q_{1,0} = q_{2,0} = q_{3,0} = q_0/\sqrt{3}$)

$$q_0 = \sqrt{2R^2/(1+R^2)} , \qquad \chi_{1,0} = \arctan\sqrt{2} , \qquad \chi_{2,0} = \pi/4 , \tag{32}$$

and are mutually perpendicular

$$\zeta_{12,0} = \zeta_{23,0} = \zeta_{31,0} = \pi/4 , \tag{33}$$

where $2\zeta_{ij}$ denotes the relative angle between $\vec{\alpha}_i$ and $\vec{\alpha}_j$. Geometrically, the four clusters are located at the vertices of an regular tetrahedron.

In the limit of small oscillations around the equilibrium shape the intrinsic degrees of freedom decouple and become harmonic. The resulting energy spectrum is that of the vibrational excitations of a spherical top with tetrahedral symmetry

$$E_{4,\mathrm{vib}} = \omega_1(v_1+\frac{1}{2}) + \omega_2(v_2+1) + \omega_3(v_3+\frac{3}{2}) , \tag{34}$$

with frequencies

$$\omega_1 = 4NR^2\xi_1 , \qquad \omega_2 = \frac{8NR^2}{1+R^2}\xi_2 , \qquad \omega_3 = \frac{8NR^2}{1+R^2}\xi_3 . \tag{35}$$

in agreement with the results obtained in a normal mode analysis. Here v_1 represents the vibrational quantum number for a symmetric stretching A vibration, $v_2 = v_{2a} + v_{2b}$ denotes a doubly degenerate E vibration, and $v_3 = v_{3a} + v_{3b} + v_{3c}$ a three-fold degenerate F vibration.

Whereas the angular momentum L is an exact symmetry of H_4 of Eq. (31), the angular momentum in index space I in general does not commute with the Hamiltonian. Only if $\xi_2 = \xi_3$ in Eq. (31), does I become a good quantum number. The rotational excitations of the ground state vibrational band of H_4 with $(v_1, v_2, v_3) = (0, 0, 0)$ are characterized by $L = I$. This property is a consequence from the fact that the operator $\vec{L}\cdot\vec{L} - \vec{I}\cdot\vec{I}$ annihilates the coherent (or intrinisic) state corresponding to the rigid equilibrium shape of Eqs. (32,33). Therefore, the rotational energies of the ground state band are given by

$$E_{4,\mathrm{rot}} = \kappa_1 L(L+1) . \tag{36}$$

In Fig. 3 I show the structure of the spectrum of a spherical top with tetrahedral symmetry according to the approximate energy formula of Eqs. (34) and (36). The energy spectrum consists of a series of rotational bands labeled by (v_1, v_2, v_3). It is assumed that the spin of the identical clusters is zero, as is relevant for the description of the ^{16}O nucleus as a cluster of four α particles. As a consequence, all states in Fig. 3 are symmetric under S_4. For this case, the bands with $(v_1, 0, 0)$ can have angular momenta and parity $L^P = 0^+, 3^-, 4^+, 6^\pm, \ldots$, whereas the angular momentum content of the doubly degenerate E vibration $(v_1, 1, 0)$ is given by $L^P = 2^\pm, 4^\pm, 5^\pm, \ldots$, and for the triply degenerate F vibration $(v_1, 0, 1)$ by $L^P = 1^-, 2^+, 3^\pm, 4^\pm, \ldots$, in agreement with Ref. [13].

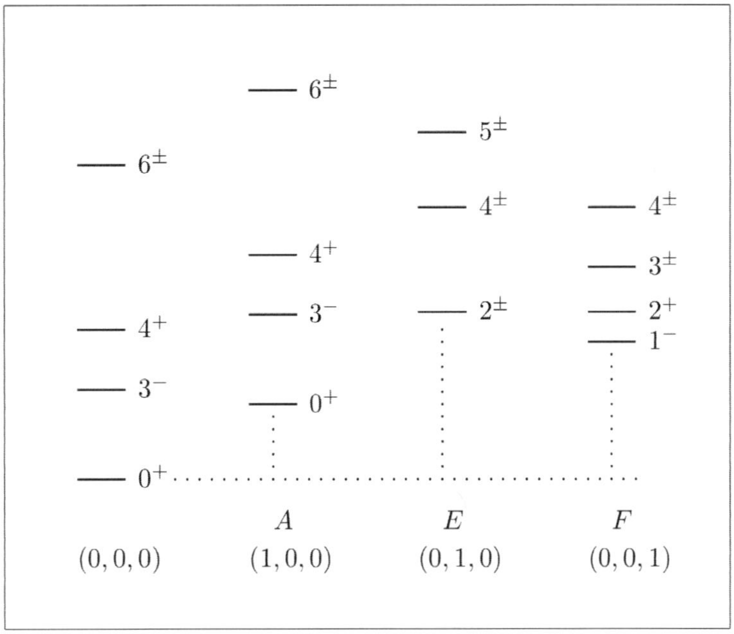

Figure 3. Schematic spectrum of a spherical top with tetrahedral symmetry. The rotational bands are labeled by (v_1, v_2, v_3) (bottom). All states are symmetric under S_4.

5. Summary and conclusions

In this contribution, I discussed an algebraic treatment of few-body systems in terms of the Algebraic Cluster Model which is based on the algebraic quantization of the relative Jacobi variables. The ensuing $U(3n-2)$ spectrum generating algebra incorporates all vibrational and rotational degrees of freedom from the beginning. It was shown that the permutation symmetry of identical clusters can be taken into account in an exact manner.

I reviewed four special solutions of S_n invariant ACM Hamiltonians for which the energy eigenvalue problem can be solved in closed analytic form. Two cases can be solved exactly. They correspond to dynamical symmetries, called the $U(3n-2) \supset U(3n-3)$ and $U(3n-2) \supset SO(3n-2)$ limits, which were interpreted as the spherical and deformed oscillator, respectively. In addition, there are some other cases of S_n invariant Hamiltonians for which approximate solutions can be obtained in a semiclassical mean-field analysis. For the case of three-body clusters I discussed the oblate symmetric top (equilateral triangle) and for four-body clusters the spherical top with tetrahedral symmetry (regular tetrahedron).

In this contribution, I focussed on the symmetry structure of the ACM. For two- and three-body clusters interesting applications were found both in molecular, nuclear and hadronic physics [2]-[8]. Future work includes the derivation of electromagnetic transition rates and form factors as well as the analysis of possible applications of the ACM for four-body systems in molecular physics (X_4 molecules), nuclear physics (^{16}O as a cluster of four α particles). and hadronic physics ($q^4 - \bar{q}$ multiquark configurations).

Finally, it is important to note that the ACM provides a general framework to study the full rotational and vibrational structure of many-body systems which is not restricted to the case of identical particles discussed in this contribution. It can be applied to other situations as well, such as nonidentical particles [7] and/or other geometric configurations.

Acknowledgments
This work was supported in part by CONACyT (grant 78833) and and PAPIIT-UNAM (grant IN113711).

References

[1] Iachello F and Arima A 1987 *The Interacting Boson Model* (Cambridge: Cambridge U. Press)
[2] Iachello F and Levine R D 1995 *Algebraic Theory of Molecules* (Oxford: Oxford U. Press)
[3] Iachello F and Jackson A D 1982 *Phys. Lett. B* **108** 151
Iachello F 1983 *Nucl. Phys. A* **396** 233c
Daley H J and Iachello F 1983 *Phys. Lett. B* **131** 281
[4] Iachello F, Mukhopadhyay N C and Zhang L 1991 *Phys. Rev. D* **44** 898
Iachello F and Kusnezov D 1992 *Phys. Rev. D* **45** 4156
[5] Bijker R, Iachello F and Leviatan A 1994 *Ann. Phys. (N.Y.)* **236** 69
Bijker R, Iachello F and Leviatan A 2000 *Ann. Phys. (N.Y.)* **284** 89
[6] Bijker R, Dieperink A E L and Leviatan A 1995 *Phys. Rev. A* **52** 2786
[7] Bijker R and Leviatan A 1998 *Few-Body Systems* **25** 89
[8] Bijker R and Iachello F 2000 *Phys. Rev. C* **61** 067305
Bijker R and Iachello F 2002 *Ann. Phys. (N.Y.)* **298** 334
[9] Kramer P and Moshinsky M 1966 *Nucl. Phys.* **82** 241
[10] Bijker R 2010 *AIP Conference Proceedings* **1323** 28
[11] Van Roosmalen O S and Dieperink A E L 1981 *Phys. Lett. B* **100** 299
Van Roosmalen O S and Dieperink A E L 1982 *Ann. Phys. (N.Y.)* **139** 198
[12] Levit S and Smilansky U 1982 *Nucl. Phys. A* **389** 56
[13] Herzberg G 1991 *Molecular Spectra and Molecular Structure. II. Infrared and Raman Spectra of Polyatomic Molecules* (Malabar Florida: Krieger)

Group Theory in Finite Hamiltonian Systems

Kurt Bernardo Wolf
Instituto de Ciencias Físicas
Universidad Nacional Autónoma de México
Av. Universidad s/n, Cuernavaca, Morelos 62251, México
E-mail: bwolf@fis.unam.mx

Abstract. We give a short review of the methods and aims of the *Óptica Matemática* projects in studying discrete Hamiltonian systems with group-theoretical methods.

There are several partially overlapping definitions of Hamiltonian systems in the applied mathematical literature —and of phase space. They have been proposed to be useful in classical mechanics, geometric optics, and others for wave and quantum systems. We study discrete Hamiltonian systems, where 'configuration' space is a denumerable collection of points, finite or infinite. They have corresponding phase spaces that are compact or noncompact manifolds stemming from the *mother* Lie algebra of the system.[1]

Three postulates

We propose three postulates to define Hamiltonian systems: the first two are the geometric and dynamic Hamilton equations, $\{\mathsf{K},\mathsf{X}\} =: -\mathrm{i}\mathsf{P}$ and $\{\mathsf{K},\mathsf{P}\} =: \mathrm{i}F(\mathsf{X})$ expressed with Lie brackets of operators associated with the classical observable of position X and its tangent operator P (momentum) in the direction determined by the generator of evolution K (Hamiltonian + any constant). The third postulate is algebraic; it determines the Lie bracket $\{\mathsf{X},\mathsf{P}\} = \mathrm{i}(\alpha 1+\beta\mathsf{K})$ between the canonical position and momentum operators. When they commute and the bracket is Poisson, the Hamiltonian system is classical; and when they are self-adjoint operators or matrices subject to commutators, it is a quantum system. When we postulate that $\alpha = \hbar$, $\beta = 0$, the Hamiltonian system will be mothered by the Heisenberg-Weyl algebra and group HW, yielding Schrödinger quantum-mechanical systems. We can postulate instead that the key commutator close into a different Lie algebra. In one space dimension we can have the spin $\mathsf{SU}(2)$ group ($\beta = 1$) [1, 2], the Lorentz group ($\beta = -1$) $\mathsf{SU}(1,1)$ [3], the group of Euclidean motions $\mathsf{ISO}(2)$ ($\beta = 0$) [4], or any three-parameter algebra, or q-algebra [5, 6, 7, 8], *which contracts to* HW. This postulate thus entails the correspondence principle: when the number and density of points increase without limit in a discrete Hamiltonian system, the limit should be some well-known continuous Hamiltonian system [9, 10].

We have found particularly fruitful the group $\mathsf{SU}(2)$, which is compact, and hence the spectrum of the position, momentum, and Hamiltonian (displaced energy) is finite: $\{x_m\}$ for $m \in \{-j, -j+1, \ldots, j\}$, comprising $N = 2j+1$ points. Since the rotation of a sphere projects its points to harmonic motion, the system will be a discrete harmonic oscillator. Its 'states' (or

[1] We apologize that most of the references below are to our own work, where further sources and differences can be found; we have tried to give a condensed overview of results published during the past few years.

N-point signals, or wavefunctions, or kets) are the overlap between the position and Hamiltonian bases of the complex Hilbert space $\mathcal{C}^N$. The special functions required are Wigner little-d functions, widely known in quantum angular momentum theory, with their expression in terms of binomial distributions and Kravchuk orthogonal discrete polynomials. With the position generator compact, as said above we have infinite discrete Hamiltonian systems: the free system mothered by $\mathsf{ISO}(2)$ and the repulsive oscillator mothered by $\mathsf{SU}(1,1)$. Finally, Hamiltonian systems mothered by q- and super- algebras yield discrete oscillators whose wavefunctions include q-Kravchuk [7], p-Kravchuk, Meixner, Pollaczek [11] and Hahn [12, 13] polynomials, all of which contract to the common quantum oscillator.

We should underline that the group action can also be bilateral: on ket-bra matrices that stand for *density distributions* of pure or entangled states. Hence, discrete Hamiltonian systems are actually quantum systems.

Phase space

The phase space of discrete Hamiltonian systems is a symplectic manifold embedded in the space of coadjoint orbits of the mother algebra [14, 15], which we may call *meta*-phase space. Clearly, if the spectra of the generators of the mother Lie algebra are discrete and their interval finite, phase space should be a compact manifold —a sphere in the case of $\mathsf{SU}(2)$. The traditional Fourier approach to finite systems considers discrete tori; yet, there one cannot rotate around a point (as a harmonic oscillator would) without tearing the torus somewhere [16, 17, 18]. In the ring of the generated Lie group we find a Wigner operator that is covariant under that group: recall that under HW and $\mathsf{Sp}(2,\Re)$ canonical transforms [19], the common HW-Wigner function transforms its phase space plane covariantly [20]. For compact groups such as $\mathsf{SU}(2)$, the Wigner operator is an $N \times N$ matrix, a covariant function on this group, and reducible to one of the coset spaces. Its expectation value in a state yields a quasiprobability distribution that is the $\mathsf{SU}(2)$-covariant Wigner *function*. We can thus see the music sheet of finite signals on a sphere (best projected on a rectangle), where we can classify the N^2 unitary $\mathsf{U}(N)$ transformations that linearly map and aberrate signals [21], in correspondence with geometric optical models. In the contraction limit to the HW-Wigner function, unitarity somehow becomes canonicity.

Two-dimensional systems

Two-dimensional discrete Hamiltonian systems are useful for the analysis of finite pixellated images, following cartesian or polar coordinates [22], and mantaining close relation with the continuous geometric-optical counterpart model. The N^2 complex pixel values of an $N \times N$-pixel screen form an irreducible representation space under the direct product group $\mathsf{SU}(2)_x \times \mathsf{SU}(2)_y = \mathsf{SO}(4)$ [23, 24]. Through symmetry *importation*, the images can be unitarily transformed under the four-parameter $\mathsf{U}(2)$-Fourier group of geometric optics that includes isotropic and anisotropic Fourier-Kravchuk transforms, as well as rotations and gyrations [25]. The latter are the discrete counterparts of the transformation between Hermite-Gauss (HG) and Laguerre-Gauss (LG) wave-optical beams [26]. Finally, we also examine the relative advantages of using sampled values of HG and LG functions *versus* our discrete Kravchuk functions to analyze images into their mode [27] and angular momentum [28] components, and synthesize the original image faithfully.

We are assembling these results in an orderly fashion for a volume in order to show the conceptual economy that group theory brings into the analysis of discrete Hamiltonian systems, revealing their symmetry and dynamics, classical as well as quantum.

We thank the *Óptica Matemática* projects DGAPA-UNAM IN101011 and CONACYT 79899 for their generous support.

[1] Atakishiyev N M and Wolf K B 1997 *J. Opt. Soc. Am.* A **14** 1467
[2] Atakishiyev N M, Pogosyan G S and Wolf K B 2005 *Phys. Part. Nuclei, Suppl. 3* **36** 521
[3] Muñoz C A, Rueda-Paz J and Wolf K B 2009 *J. Phys. A* **42** 485210
[4] Nieto L M, Atakishiyev N M, Chumakov S M and Wolf K B 1998 *J. Phys.* A **31** 3875
[5] Atakishiyev N M, Frank A and Wolf K B 1994 *J. Math. Phys.* **35** 3253
[6] Arık M, Atakishiyev N M and Wolf K B 1999 *J. Phys.* A **32** L371
[7] Atakishiyev N M, Klimyk A U and Wolf K B 2004 *J. Phys.* A **37** 5569
[8] Atakishiyev N M, Klimyk A U and Wolf K B 2008 *J. Phys.* A **41** 085201
[9] Atakishiyev N M, Pogosyan G S and Wolf K B 2003 *Int. J. Mod. Phys.* A **18** 317
[10] Atakishiyev N M, Pogosyan G S and Wolf K B 2003 *Int. J. Mod. Phys.* A **18** 329
[11] Atakishiyev N M, Jafarov E I, Nagiyev Sh M and Wolf K B 1998 *Rev. Mex. Fís.* **44** 135
[12] Jafarov E I, Stoilova N I and Van der Jeugt J 2011 *J. Phys.* A **44** 265203
[13] Jafarov E I, Stoilova N I and Van der Jeugt J 2011 (*Preprint* arXiv:1106.1083v1)
[14] Atakishiyev N M, Chumakov S M and Wolf K B 1998 *J. Math. Phys.* **39** 6247
[15] Ali S T, Atakishiyev N M, Chumakov S M, and Wolf K B 2000 *Ann. H. Poincaré* **1** 685
[16] Wolf K B and Krötzsch G 2007 *J. Opt. Soc. Am.* A **24** 651
[17] Wolf K B and Krötzsch G 2007 *J. Opt. Soc. Am.* A **24** 2568
[18] Wolf K B and Krötzsch G 2007 *J. Opt. Soc. Am.* A **24** 2871
[19] Moshinsky M and Quesne C 1971 *J. Math. Phys* **12** 1772
[20] García-Calderón G and MoshinskyM 1980 *J. Phys.* A **13** L185
[21] Wolf B K 2008 *J. Phys. A* **41** 304026
[22] Vicent L E and Wolf K B 2008 *J. Opt. Soc. Am.* A **25** 1875
[23] Atakishiyev N M, Pogosyan G S, Vicent L E and Wolf K B 2001 *J. Phys.* A **34** 9381
[24] Atakishiyev N M, Pogosyan G S, Vicent L E and Wolf K B 2011 *J. Phys.* A **34** 9399
[25] Vicent L E and Wolf K B 2011 *Sigma* **7** 053
[26] Wolf K B and Alieva T 2008 *J. Opt. Soc. Am.* A **25** 365
[27] Wolf K B 2009 *J. Opt. Soc. Am.* A **26** 509
[28] Vicent L E and Wolf K B 2011 *J. Opt. Soc. Am.* A **28** 808

The probability representation as a new formulation of quantum mechanics

Margarita A Man'ko and Vladimir I Man'ko

P. N. Lebedev Physical Institute, Leninskii Prospect 53, Moscow 119991, Russia

E-mail: `mmanko@sci.lebedev.ru`

Abstract. We present a new formulation of conventional quantum mechanics, in which the notion of a quantum state is identified via a fair probability distribution of the position measured in a reference frame of the phase space with rotated axes. In this formulation, the quantum evolution equation as well as the equation for finding energy levels are expressed as linear equations for the probability distributions that determine the quantum states. We also give the integral transforms relating the probability distribution (called the tomographic-probability distribution or the state tomogram) to the density matrix and the Wigner function and discuss their connection with the Radon transform. Qudit states are considered and the invertible map of the state density operators onto the probability vectors is discussed. The tomographic entropies and entropic uncertainty relations are reviewed. We demonstrate the uncertainty relations for the position and momentum and the entropic uncertainty relations in the tomographic-probability representation, which is suitable for an experimental check of the uncertainty relations.

1. Introduction

Recently a new formulation of conventional quantum mechanics was suggested [1–3]. In this formulation, the notion of a quantum state is identified with the probability distribution called the tomographic-probability distribution or the state tomogram. This new formulation of quantum mechanics is called probability representation of quantum mechanics, and it is closely related to quasiprobability distributions of quantum states like the Wigner function [4], the Husimi–Kano Q-quasidistribution [5, 6], or the quasidistributions determining the diagonal representation of the density operator ($P(\alpha)$ function [7] or $\varphi(z)$ function [8]). All these quasidistributions are different representations of the density operator, and the idea for their introduction was to find a quantum state representation that is similar to the classical probability density on the phase space determining the classical-system state. However, all quasiprobability distributions introduced have the same property: they are not fair probability densities on the phase space, which cannot exist in view of the uncertainty relations [9–12] that prohibit simultaneous measurements of two random variables like position and momentum.

It turned out to be possible to introduce a tomographic-probability distribution as a notion of quantum states, which does not contradict the uncertainty relations, since this distribution depends only on one random position; it does not contain a dependence on the conjugate momentum. All statistic characteristics of quantum observables like means, variances, covariances, ... can be calculated in terms of measurable tomographic-probability distributions.

The aim of this paper is to present a review of the probability representation of quantum mechanics and to discuss possible experiments to check the uncertainty relations and the other quantum inequalities like the entropic inequalities for tomographic entropies [13–18], which are based on the concepts of Shannon entropy [19] and Rényi entropy [20].

In section 2, we give the definition of optical and symplectic tomograms for both, classical- and quantum-system states. In section 3, we present the basic classical and quantum equations in the probability representation. In section 4, we consider the uncertainty relations in tomographic form and review the properties of entropy in the tomographic-probability representation, including quantum inequalities. In section 5, we demonstrate the probability-representation scheme for spin (qudit) systems. We conclude in section 6 with prospectives.

2. Optical and symplectic tomograms

Let us consider a single-mode field with density operators $\hat{\rho}$, which are Hermitian (i.e., $\hat{\rho}^\dagger = \hat{\rho}$), nonnegative (i.e., $\hat{\rho} \geq 0$), and normalized (i.e., $\mathrm{Tr}\,\hat{\rho} = 1$). The quantum states of this field can be identified with the probability density $w(X, \theta)$ of a random position $-\infty < X < \infty$ depending on an extra angle parameter $0 < \theta \leq 2\pi$ via the relation

$$\hat{\rho} = \frac{1}{2\pi} \int_0^\pi d\theta \int_{-\infty}^{+\infty} d\eta\, dX\, \omega(X, \theta)|\eta| \exp\left[i\eta(X - \hat{q}\cos\theta - \hat{p}\sin\theta)\right]. \tag{1}$$

$\hat{q}$ and $\hat{p}$ are the position and momentum operators, respectively, i.e., $[\hat{q}, \hat{p}] = i\,1$, and we assume the reduced Planck constant $\hbar = 1$.

The probability density $w(X, \theta)$ is called the optical tomogram; in quantum optics the random variable X is called homodyne photon quadrature and the angle θ is called the local oscillator phase. In experiments on homodyne detection of quantum photon states [21,22], the local oscillator phase is a parameter under control and the quadrature X is measured. The output in an experiment with homodyne photon-state detection is the function $w(X, \theta)$. It has the property [23]

$$w(X, \theta + \pi) - w(-X, \theta) = 0, \tag{2}$$

which can be obtained from the inverse relationship

$$w(X, \theta) = \mathrm{Tr}\,\hat{\rho}\delta\left(X - \hat{q}\cos\theta - \hat{p}\sin\theta\right). \tag{3}$$

The optical tomogram is a nonnegative and normalized function: $w(X,\theta) \geq 0$ and $\int w(X,\theta)\, dX = 1$. It was used in experiments with a homodyne photon detector for reconstructing the Wigner function

$$W(q,p) = \mathrm{Tr}\, \hat{\rho}\hat{U}(q,p), \tag{4}$$

where the operator $\hat{U}(q,p)$ is connected to the parity operator $\hat{P}$ by

$$\hat{U}(q,p) = 2\, \exp\left[2\, i\, (p\hat{q} - q\hat{p})\right] \hat{P}. \tag{5}$$

The density operator $\hat{\rho}$ is given in terms of the Wigner function as follows:

$$\hat{\rho} = \frac{1}{\pi} \int W(q,p) \exp\left[2\, i\, (p\hat{q} - q\hat{p})\right] \hat{P}\, dq\, dp. \tag{6}$$

The Wigner function is real and normalized, $(2\,\pi)^{-1} \int W(q,p)\, dq\, dp = 1$, and the relation of the Wigner function to the optical tomogram is

$$w(X,\theta) = \frac{1}{2\,\pi} \int W(q,p)\delta(X - q\cos\theta - p\sin\theta)\, dq\, dp, \tag{7}$$

with the inverse relation

$$W(q,p) = \frac{1}{2\,\pi} \int_0^{\pi} d\theta \int_{-\infty}^{+\infty} d\eta\, dX\, w(X,\theta)|\eta| \exp\left[i\eta(X - q\cos\theta - p\sin\theta)\right]. \tag{8}$$

Due to relations (7) and (8) found in [24,25], the optical tomogram can be used as a tool to reconstruct the Wigner function, which in turn was identified with the quantum state. A proposal to identify the tomogram itself with the quantum state was given in [1]. The physical meaning of the optical tomogram $w(X,\theta)$ can be elucidated if one considers this tomogram as the concept of a state in classical statistical mechanics. In fact, such a state is traditionally identified with the probability density $f(q,p)$ on the phase space. This function is nonnegative $f(q,p) \geq 0$ and normalized $\int f(q,p)\, dq\, dp = 1$. Applying the Radon transform to this function, we obtain the optical tomogram of the classical state as

$$w(X,\theta) = \int f(q,p)\delta\left(X - q\cos\theta - p\sin\theta\right)\, dq\, dp. \tag{9}$$

Since the position X is given as $X = q\cos\theta + p\sin\theta$, it has the physical meaning of the particle's position measured in a reference frame in the phase space with initial axes rotated by θ. Thus, the optical tomogram is the probability density of the particle position measured in a rotated reference frame in the phase space. The information on the classical state coded by the probability density $f(q,p)$ of two random variables q and p is equivalent to the information coded by the probability density of one random position X and the extra parameter θ labeling the reference frame where this position is measured. This physical meaning of the optical tomogram is preserved in the quantum domain as well. The statistical characteristics of the random variables in both, the classical and quantum domains, can be understood in view of this clarified interpretation of the optical tomogram.

In fact, for $\theta = 0$ the random position X has the physical meaning of the particle coordinate q, and

$$\int f(q,p)q^n dq\,dp = \int w(X,0)X^n dX. \tag{10}$$

For $\theta = \pi/2$, the random position X has the physical meaning of the particle momentum p, and

$$\int f(q,p)p^n dq\,dp = \int w(X,\pi/2)X^n dX. \tag{11}$$

Analogously, the same formulas exist in the quantum domain, namely,

$$\mathrm{Tr}\,\hat{\rho}\hat{q}^n = \int w(X,0)X^n dX, \qquad \mathrm{Tr}\,\hat{\rho}\hat{p}^n = \int w(X,\pi/2)X^n dX. \tag{12}$$

One can easily see that in the probability representation of quantum mechanics the formulas for highest moments of position and momentum are identical in both classical and quantum domains and have the standard form provided by conventional probability theory.

The mean value of energy in classical statistical mechanics for a system with the standard Hamiltonian $H = (p^2/2) + V(q)$ can be written in the tomographic form

$$\int f(q,p)\left[(p^2/2) + V(q)\right] dq\,dp = \int (X^2/2)\,w(X,0)\,dX + \int V(X)w(X,\pi/2)\,dX. \tag{13}$$

An analogous formula

$$\mathrm{Tr}\,\hat{\rho}\left[(\hat{p}^2/2) + V(\hat{q})\right] = \int (X^2/2)w(X,0)\,dX + \int V(X)w(X,\pi/2)\,dX \tag{14}$$

is valid for the quantum system.

Thus, the formulas for kinetic and potential energy means in classical and quantum mechanics are identical when expressed in terms of the tomographic-probability distributions.

One can introduce another tomographic-probability density called the symplectic tomogram $M(X,\mu,\nu)$, where a random variable X is measured in a specific reference frame and real parameters μ and ν can be interpreted as parameters that provide a scaling transform of the reference-frame axes in the particle phase space $q \to \mu q$ and $p \to \nu p$. Another interpretation of these parameters consists in the fact that they have the form $\mu = s\cos\theta$ and $\nu = s^{-1}\sin\theta$, where θ is the rotation angle used in the definition of optical tomogram and the scaling parameter s provides the scaling transform of the axes $q \to sq$ and $p \to s^{-1}p$. In such an interpretation, X is the particle position measured in a reference frame in the phase space with axes that were first rescaled (parameter s) and then rotated (parameter θ).

The symplectic tomogram of the quantum state is defined as

$$M(X,\mu,\nu) = \mathrm{Tr}\,\hat{\rho}\delta\left(X - \mu\hat{q} - \nu\hat{p}\right). \tag{15}$$

For the classical state, the symplectic tomogram is defined in view of the probability density $f(q,p)$ on the phase space

$$M(X,\mu,\nu) = \int f(q,p)\delta\left(X - \mu q - \nu p\right) dq\,dp. \tag{16}$$

Both formulas (15) and (16) have an inverse; in the quantum domain, one has

$$\hat{\rho} = \frac{1}{2\pi}\int M(X,\mu,\nu)\exp\left[i\left(X-\mu\hat{q}-\nu\hat{p}\right)\right]dX\,d\mu\,d\nu, \tag{17}$$

and in the classical domain, one has

$$f(q,p) = \frac{1}{4\pi^2}\int M(X,\mu,\nu)\exp\left[i\left(X-\mu q-\nu p\right)\right]dX\,d\mu\,d\nu. \tag{18}$$

Since the Dirac delta-function is a homogeneous function, i.e., $\delta(\lambda y) = (|\lambda|)^{-1}\delta(y)$, both quantum and classical tomograms have the homogeneity property $M(\lambda X,\lambda\mu,\lambda\nu) = (|\lambda|)^{-1}M(X,\mu,\nu)$. This means that the symplectic tomogram satisfies

$$\left(X\frac{\partial}{\partial X}+\mu\frac{\partial}{\partial\mu}+\nu\frac{\partial}{\partial\nu}+1\right)M(X,\mu,\nu) = 0. \tag{19}$$

The symplectic tomogram given for $\mu = \cos\theta$ and $\nu = \sin\theta$ yields the optical tomogram $w(X,\theta) = M(X,\cos\theta,\sin\theta)$ and, due to the homogeneity property, the optical tomogram provides the symplectic tomogram

$$M(X,\mu,\nu) = \frac{1}{\sqrt{\mu^2+\mu^2}}\,w\left(\frac{X}{\sqrt{\mu^2+\mu^2}},\,\tan^{-1}\frac{\nu}{\mu}\right), \tag{20}$$

where relation (2) must be taken into account. In view of relation (20), the measurement of the optical tomogram of the photon state also provides the symplectic tomogram, and both tomograms can be used for obtaining the statistics of photon quadratures.

The set of optical tomograms of classical states and the set of optical tomograms of quantum states are different sets. In fact, for the optical tomogram $w(X,\theta)$ of a given quantum state, integral (17) must be a nonnegative operator, and the nonnegativity condition for the operator provides the constrains for the set of quantum optical tomograms. Also for the optical tomogram $w(X,\theta)$ of a given classical state, the probability density $f(q,p)$ on the particle's phase space reads

$$f(q,p) = \frac{1}{4\pi^2}\int_0^{\pi}d\theta\int_{-\infty}^{+\infty}d\eta\,dX\,w(X,\theta)|\eta|\exp\left[i\eta(X-q\cos\theta-p\sin\theta)\right]. \tag{21}$$

The probability density must be a nonnegative function, the optical tomogram of the given classical state must have the above Radon integral as a nonnegative function, and this nonnegativity condition for the integral provides the constrains on the set of all optical tomograms of classical states. Both sets of optical tomograms (of classical and quantum states) have a common part. This intersection contains all normal probability densities which respect the quadrature uncertainty relations; this means that the normal distribution has the form

$$w(X,\theta) = \frac{1}{\sqrt{2\pi\sigma_{XX}}}\exp\left(-\frac{(X-\bar{X})^2}{2\sigma_{XX}}\right), \tag{22}$$

where

$$\sigma_{XX} = \cos^2\theta\,\sigma_{qq}+\sin^2\theta\,\sigma_{pp}+\sin 2\theta\,\sigma_{qp}. \tag{23}$$

The real variances σ_{qq} and σ_{pp} and covariance σ_{qp} have to satisfy the Schrödinger–Robertson uncertainty relation [9–12]

$$\sigma_{qq}\sigma_{pp} - \sigma_{qp}^2 \geq 1/4. \tag{24}$$

The symplectic tomogram of a classical Gaussian state with the probability density

$$f_G(q,p) = \left(4\pi^2 \det\sigma\right)^{-1/2} \exp\left[-\frac{1}{2}(\tilde{p},\tilde{q})\,\sigma^{-1}\begin{pmatrix}\tilde{p}\\ \tilde{q}\end{pmatrix}\right], \tag{25}$$

where $\tilde{p} = p - \langle p\rangle$, $\tilde{q} = q - \langle q\rangle$, and the symmetric dispersion matrix $\sigma = \begin{pmatrix}\sigma_{pp} & \sigma_{pq}\\ \sigma_{qp} & \sigma_{qq}\end{pmatrix}$, is given by a formula analogous to (22),

$$M_G(X,\mu,\nu) = \left[2\pi\sigma_{XX}(\mu,\nu)\right]^{-1/2} \exp\left[-\frac{\left(X - \bar{X}(\mu,\nu)\right)^2}{2\,\sigma_{XX}(\mu,\nu)}\right]. \tag{26}$$

Here, the mean value of position X is expressed in terms of the position and momentum means $\langle q\rangle$ and $\langle p\rangle$ as $\bar{X}(\mu,\nu) = \mu\langle q\rangle + \nu\langle p\rangle$ and the dispersion $\sigma_{XX}(\mu,\nu)$ is

$$\sigma_{XX}(\mu,\nu) = \mu^2\sigma_{qq} + \mu^2\sigma_{pp} + 2\,\mu\nu\sigma_{qp}. \tag{27}$$

For an arbitrary Gaussian quantum state with the density operator

$$\hat{\rho} = Z^{-1}\exp\left[A\hat{q}^2 + B\hat{p}^2 + C\left(\hat{q}\hat{p} + \hat{p}\hat{q}\right) + D\hat{q} + E\hat{p}\right], \tag{28}$$

in which the quadratic form under the exponent is a Hermitian operator and $Z = \mathrm{Tr}\left\{\exp\left[A\hat{q}^2 + B\hat{p}^2 + C\left(\hat{q}\hat{p} + \hat{p}\hat{q}\right) + D\hat{q} + E\hat{p}\right]\right\}$, the optical tomogram has the Gaussian form (22), and the symplectic tomogram has the Gaussian form (26) with the parameters (23) and (27) providing the normal distributions.

As an example of a Gaussian quantum state, we present the optical and symplectic tomograms of the coherent state $\mid\alpha\rangle$

$$w_\alpha(X,\theta) = \pi^{-1/2}\exp\left[-\left(X - \sqrt{2}\,\mathrm{Re}\,\alpha\cos\theta - \sqrt{2}\,\mathrm{Im}\,\alpha\sin\theta\right)^2\right] \tag{29}$$

and

$$M_\alpha(X,\mu,\nu) = \frac{1}{\sqrt{\pi\left(\mu^2+\nu^2\right)}}\exp\left[-\frac{\left(X - \sqrt{2}\,\mathrm{Re}\,\alpha\,\mu - \sqrt{2}\,\mathrm{Im}\,\alpha\,\nu\right)^2}{\mu^2+\nu^2}\right]. \tag{30}$$

One can see that, in the tomographic-probability representation, both classical and quantum Gaussian states are given by the same normal distributions. The difference in these Gaussian states is described by the difference in the set of parameters σ_{qq}, σ_{pp}, and σ_{qp}, which in the quantum domain must satisfy the Schrödinger–Robertson inequality (24).

The symplectic tomogram of a quantum state can be presented in terms of the wave function $\psi(y)$ [3, 26]

$$M_\psi(X,\mu,\nu) = \frac{1}{2\pi|\nu|}\left|\int \psi(y)\exp\left[i\left(\frac{\mu}{2\nu}y^2 - \frac{Xy}{\nu}\right)\right]dy\right|^2, \tag{31}$$

and, analogously, the optical tomogram is

$$W_\psi(X,\theta) = \frac{1}{2\pi|\sin\theta|}\left|\int \psi(y)\exp\left[i\left(\frac{\cos\theta}{2\sin\theta}y^2 - \frac{Xy}{\sin\theta}\right)\right]dy\right|^2. \tag{32}$$

The tomogram of any quantum pure state (31) satisfies the condition

$$\frac{1}{2\pi}\int M_\psi(X,\mu,\nu)M_\psi(-Y,\mu,\nu)e^{i(X-Y)}dX\,dY\,d\mu\,d\nu = 1, \tag{33}$$

where the integral on the left-hand side coincides with the state purity parameter $\tilde{\pi} = \mathrm{Tr}\,\hat{\rho}^2$. For other quantum states, this integral must be nonnegative and smaller than unity, since the purity parameter belongs to the domain $0 < \tilde{\pi} \leq 1$.

Since the density operator $\hat{\rho}$ is expressed in terms of symplectic tomogram by equation (17), for tomograms $M_\psi(X,\mu,\nu)$ satisfying equality (33) one can introduce the wave function, both in classical and quantum mechanics, viy the relation

$$\psi(x)\psi^*(x') = \frac{1}{2\pi}\int M_\psi(X,\mu,\nu)e^{iX}\langle x \mid e^{-i\mu\hat{q}-i\nu\hat{p}} \mid x'\rangle dX\,d\mu\,d\nu. \tag{34}$$

Employing the equality

$$\langle x \mid e^{-i\mu\hat{q}-i\nu\hat{p}} \mid x'\rangle = e^{-i(\mu\nu/2)-i\mu x'}\delta(x-x'-\nu), \tag{35}$$

one obtains the wave function from the expression

$$\psi(x)\psi^*(x') = \frac{1}{2\pi}\int M_\psi(X,\mu,x-x')\exp\left(iX - i\mu\frac{x+x'}{2}\right]dX\,d\mu. \tag{36}$$

This equality (36) can take place only for the cases where the matrix on its right-hand side is nonnegative since the density matrix $\psi(x)\psi^*(x')$ is the matrix of a nonegative operator $\mid \psi\rangle\langle\psi \mid$, which is a rank-one projector.

3. Quantum and classical evolution equations in the tomographic probability representation

In the previous section, we considered classical and quantum single-mode field states in the probability representation. Since this representation is a simple change of variables given by an integral transform of the density matrix or the probability distribution function on the phase space, one can easily obtain all the classical and quantum equations, employing this change of variables.

First, we generalize the above considerations to multimode field states. In the conventional form, the multimode classical state is described by normalized probability density $f(\vec{q},\vec{p})$, where $\vec{q} = \{q_k\}$, $\vec{p} = \{p_k\}$, $k = 1,2,\ldots,N$. The optical tomogram of this classical state is

$$w(\vec{X},\vec{\theta}) = \int f(\vec{q},\vec{p})\left[\prod_{k=1}^{N}\delta\left(X_k - q_k\cos\theta_k - p_k\sin\theta_k\right)\right]d\vec{q}\,d\vec{p}, \tag{37}$$

where $\vec{X} = \{X_k\}$ and $\vec{\theta} = \{\theta_k\}$.

The symplectic tomogram of a classical state is defined as

$$M(\vec{X},\vec{\mu},\vec{\nu}) = \int f(\vec{q},\vec{p})\left[\prod_{k=1}^{N}\delta\left(X_k-\mu_k q_k-\nu_k p_k\right)\right]d\vec{q}\,d\vec{p}, \tag{38}$$

where $\vec{\mu}=\{\mu_k\}$ and $\vec{\nu}=\{\nu_k\}$.

For quantum multimode field states with the density operator $\hat{\rho}$, the optical tomogram is defined as

$$w(\vec{X},\vec{\theta}) = \mathrm{Tr}\left[\hat{\rho}\prod_{k=1}^{N}\delta\left(X_k-\hat{q}_k\cos\theta_k-\hat{p}_k\sin\theta_k\right)\right], \tag{39}$$

and the symplectic tomogram reads

$$M(\vec{X},\vec{\mu},\vec{\nu}) = \mathrm{Tr}\left[\hat{\rho}\prod_{k=1}^{N}\delta\left(X_k-\mu_k\hat{q}_k-\nu_k\hat{p}_k\right)\right]. \tag{40}$$

The conventional classical probability density obeys the Liouville evolution equation

$$\left[\frac{\partial}{\partial t}+\sum_{k=1}^{N}\left(p_k\frac{\partial}{\partial q_k}-\frac{\partial V}{\partial q_k}(\vec{q})\frac{\partial}{\partial p_k}\right)\right]f(\vec{q},\vec{p},t)=0. \tag{41}$$

In the position representation, the conventional von Neumann equation for the density matrix $\rho(\vec{x},\vec{x}',t)$ reads

$$i\frac{\partial}{\partial t}\rho(\vec{x},\vec{x}',t) = \left[-\frac{1}{2}\sum_{j=1}^{N}\left(\frac{\partial^2}{\partial x_j^2}-\frac{\partial^2}{\partial x_j'^2}\right)+\Big[V(\vec{x})-V(\vec{x}')\Big]\right]\rho(\vec{x},\vec{x}',t). \tag{42}$$

For both classical and quantum evolution equations, we assume that in the corresponding Hamiltonian the masses are one in the kinetic energy term and V is the potential energy term.

Applying transforms (37) and (38) to the probability density $f(\vec{q},\vec{p},t)$, we arrive at the optical and symplectic tomographic forms of the Liouville evolution equation

$$\begin{aligned}&\left\{\frac{\partial}{\partial t}-\sum_{j=1}^{N}\left[\cos^2\theta_j\frac{\partial}{\partial\theta_j}-\frac{\sin 2\theta_j}{2}\left(1+X_j\frac{\partial}{\partial X_j}\right)\right]-\sum_{k=1}^{N}\left[\frac{\partial V}{\partial q_k}\left(q_1\rightarrow\sin\theta_1\frac{\partial}{\partial\theta_1}\left(\frac{\partial}{\partial X_1}\right)^{-1}\right.\right.\right.\\&+X_1\cos\theta_1+i\,\frac{\sin\theta_1}{2}\frac{\partial}{\partial X_1},\,q_2\rightarrow\sin\theta_2\frac{\partial}{\partial\theta_2}\left(\frac{\partial}{\partial X_2}\right)^{-1}+X_2\cos\theta_2+i\,\frac{\sin\theta_2}{2}\frac{\partial}{\partial X_2},\ldots,\\&q_j\rightarrow\sin\theta_j\frac{\partial}{\partial\theta_j}\left(\frac{\partial}{\partial X_j}\right)^{-1}+X_j\cos\theta_j+i\,\frac{\sin\theta_j}{2}\frac{\partial}{\partial X_j},\ldots,q_N\rightarrow\sin\theta_N\frac{\partial}{\partial\theta_N}\left(\frac{\partial}{\partial X_N}\right)^{-1}\\&\left.\left.\left.+X_N\cos\theta_N+i\,\frac{\sin\theta_N}{2}\frac{\partial}{\partial X_N}\right)\sin\theta_k\frac{\partial}{\partial X_k}\right]\right\}w(\vec{X},\vec{\theta},t)=0\end{aligned} \tag{43}$$

and

$$\begin{aligned}&\left\{\frac{\partial}{\partial t}-\sum_{j=1}^{N}\nu_j\frac{\partial}{\partial\mu_j}-\sum_{k=1}^{N}\frac{\partial V}{\partial q_k}\left[q_1\rightarrow-\frac{\partial}{\partial\mu_1}\left(\frac{\partial}{\partial X_1}\right)^{-1},q_2\rightarrow-\frac{\partial}{\partial\mu_2}\left(\frac{\partial}{\partial X_2}\right)^{-1},\ldots,\right.\right.\\&\left.\left.q_j\rightarrow-\frac{\partial}{\partial\mu_j}\left(\frac{\partial}{\partial X_j}\right)^{-1},\ldots,q_N\rightarrow-\frac{\partial}{\partial\mu_N}\left(\frac{\partial}{\partial X_N}\right)^{-1}\right]\nu_k\frac{\partial}{\partial X_k}\right\}w(\vec{X},\vec{\mu},\vec{\nu},t)=0,\end{aligned} \tag{44}$$

respectively.

Applying transforms (39) and (40) to the density operator in the position representation, we arrive at the tomographic form of equation (42); thus the evolution equation for the optical tomogram reads [27]

$$\left\{\frac{\partial}{\partial t}-\sum_{j=1}^{N}\left[\cos^2\theta_j\frac{\partial}{\partial\theta_j}-\frac{1}{2}\sin 2\theta_j\left(1+X_j\frac{\partial}{\partial X_j}\right)\right]-\frac{1}{i}\left[V\left(q_1\to\sin\theta_1\frac{\partial}{\partial\theta_1}\left(\frac{\partial}{\partial X_1}\right)^{-1}\right.\right.\right.$$
$$+X_1\cos\theta_1+i\frac{\sin\theta_1}{2}\frac{\partial}{\partial X_1},\quad q_2\to\sin\theta_2\frac{\partial}{\partial\theta_2}\left(\frac{\partial}{\partial X_2}\right)^{-1}+X_2\cos\theta_2+i\frac{\sin\theta_2}{2}\frac{\partial}{\partial X_2},\ldots,$$
$$q_j\to\sin\theta_j\frac{\partial}{\partial\theta_j}\left(\frac{\partial}{\partial X_j}\right)^{-1}+X_j\cos\theta_j+i\frac{\sin\theta_j}{2}\frac{\partial}{\partial X_j},\ldots,\quad q_N\to\sin\theta_N\frac{\partial}{\partial\theta_N}\left(\frac{\partial}{\partial X_N}\right)^{-1}$$
$$\left.\left.\left.+X_N\cos\theta_N+i\frac{\sin\theta_N}{2}\frac{\partial}{\partial X_N}\right)-\text{c.c.}\right]\right\}w(\vec{X},\vec{\theta},t)=0, \tag{45}$$

and the evolution equation for symplectic tomogram is [1–3]:

$$\left\{\frac{\partial}{\partial t}-\sum_{j=1}^{N}\nu_j\frac{\partial}{\partial\mu_j}-\frac{1}{i}\left[V\left(q_1\to\frac{\partial}{\partial\mu_1}\left(\frac{\partial}{\partial X_1}\right)^{-1}+\frac{i}{2}\nu_1\frac{\partial}{\partial X_1},\right.\right.\right.$$
$$q_2\to\frac{\partial}{\partial\mu_2}\left(\frac{\partial}{\partial X_2}\right)^{-1}+\frac{i}{2}\nu_2\frac{\partial}{\partial X_2},\ldots,q_j\to\frac{\partial}{\partial\mu_j}\left(\frac{\partial}{\partial X_j}\right)^{-1}+\frac{i}{2}\nu_j\frac{\partial}{\partial X_j},\ldots,$$
$$\left.\left.\left.q_N\to\frac{\partial}{\partial\mu_N}\left(\frac{\partial}{\partial X_N}\right)^{-1}+\frac{i}{2}\nu_N\frac{\partial}{\partial X_N}\right)-\text{c.c.}\right]\right\}w(\vec{X},\vec{\mu},\vec{\nu},t)=0. \tag{46}$$

By an analogous change of variables, one can obtain the tomographic form of the equation providing the energy spectrum of the quantum states [27]. In fact, one has the Schrödinger equation for the eigenvalues of the Hamiltonian

$$\left[\frac{\hat{p}^2}{2}+V(\hat{q})\right]\mid\psi_E\rangle=E\mid\psi_E\rangle, \tag{47}$$

and this equation can be rewritten in the form of an equation for the density operator $\hat{\rho}_E=\mid\psi_E\rangle\langle\psi_E\mid$ as follows:

$$\frac{1}{2}\left\{\frac{\hat{p}^2}{2}+V(\hat{q}),\hat{\rho}_T\right\}=E\hat{\rho}_E. \tag{48}$$

Applying transforms (39) and (40) to the density operator in the position representation, we arrive at the integro-differential equations for the energy spectrum of the Hamiltonian (47). These equations are equivalent to the standard Schrödinger equation for a fair probability density; they read

$$Ew_E(\vec{X},\vec{\theta})=\left[\sum_{\sigma=1}^{N}\left\{\frac{\cos^2\theta_\sigma}{2}\left[\frac{\partial}{\partial X_\sigma}\right]^{-2}\left(\frac{\partial^2}{\partial\theta_\sigma^2}+1\right)-\frac{X_\sigma}{2}\left[\frac{\partial}{\partial X_\sigma}\right]^{-1}\right.\right.$$
$$\left.\left.\left(\cos^2\theta_\sigma+\sin 2\theta_\sigma\frac{\partial}{\partial\theta_\sigma}\right)+\frac{X_\sigma^2}{2}\sin^2\theta_\sigma-\frac{\cos^2\theta_\sigma}{8}\frac{\partial^2}{\partial X_\sigma^2}\right\}\right]w_E(\vec{X},\vec{\theta})$$
$$+\left[\operatorname{Re}V\left\{\sin\theta_\sigma\frac{\partial}{\partial\theta_\sigma}\left[\frac{\partial}{\partial X_\sigma}\right]^{-1}+X_\sigma\cos\theta_\sigma+i\frac{\sin\theta_\sigma}{2}\frac{\partial}{\partial X_\sigma}\right\}\right]w_E(\vec{X},\vec{\theta}) \tag{49}$$

for the optical tomogram and

$$EM_E(\vec{X},\vec{\mu},\vec{\nu}) = \left[\sum_{\sigma=1}^{N}\left\{\frac{1}{2}\left[\frac{\partial}{\partial X_\sigma}\right]^{-2}\left[\frac{\partial}{\partial \nu_\sigma}\right]^{2} - \frac{\mu_\sigma^2}{8}\left[\frac{\partial}{\partial X_\sigma}\right]^{2}\right\}\right] M_E(\vec{X},\vec{\mu},\vec{\nu})$$
$$+\left[\operatorname{Re} V\left\{-\left[\frac{\partial}{\partial X_\sigma}\right]^{-1}\frac{\partial}{\partial \mu_\sigma} + \frac{i\nu_\sigma}{2}\frac{\partial}{\partial X_\sigma}\right\}\right] M_E(\vec{X},\vec{\mu},\vec{\nu}) \qquad (50)$$

for symplectic tomogram.

4. Uncertainty relations in the tomographic form

The Schrödinger–Robertson uncertainty relations (24) can be written for an arbitrary local oscillator phase θ; in this case, they take a form of the integral inequality for the measurable optical tomogram of the quantum state [3, 28, 29]:

$$F(\theta) = \Big(\int X^2 w(X,\theta)dX - \Big[\int Xw(X,\theta)dX\Big]^2\Big)\Big(\int X^2 w\Big(X,\theta+\frac{\pi}{2}\Big)dX$$
$$-\Big[\int Xw\Big(X,\theta+\frac{\pi}{2}\Big)dX\Big]^2\Big) - \Big\{\int X^2 w\Big(X,\theta+\frac{\pi}{4}\Big)dX - \Big[\int Xw\Big(X,\theta+\frac{\pi}{4}\Big)dX\Big]^2$$
$$-\frac{1}{2}\Big[\int X^2 w(X,\theta)dX - \Big[\int Xw(X,\theta)dX\Big]^2 + \int X^2 w\Big(X,\theta+\frac{\pi}{2}\Big)dX$$
$$-\Big[\int Xw\Big(X,\theta+\frac{\pi}{2}\Big)dX\Big]^2\Big]\Big\}^2 - \frac{1}{4} \geq 0. \qquad (51)$$

The positivity of the function $F(\theta)$ for an arbitrary local oscillator phase cannot be violated in the quantum domain, but the optical tomograms of classical states can violate the positivity of $F(\theta)$.

The uncertainty relations for position and momentum follow from the nonnegativity of the Hermitian density operator, which must have only nonnegative eigenvalues, but for systems like the oscillator the number of eigenvalues is infinite. In view of this, there exists an infinite hierarchy of the uncertainty relations, which include the inequalities not only for dispersions and means of the position and momentum, but also those for all highest moments of these quantum observables. The experimental check of the inequalities became possible only recently, when the homodyne photon detection was employed to obtain the optical tomogram of the photon quantum state [21].

The following inequality exists for quadratures that depend on two quantum states [30]

$$\frac{1}{2}\Big[\int w_1(X,\theta)X^2dX - \Big(\int w_1(X,\theta)X\,dX\Big)^2\Big]\Big[\int w_2(X,\theta+\pi/2)X^2dX$$
$$-\Big(\int w_2(X,\theta+\pi/2)X\,dX\Big)^2\Big] + \frac{1}{2}\Big[\int w_2(X,\theta)X^2dX - \Big(\int w_2(X,\theta)X\,dX\Big)^2\Big]$$
$$\times\Big[\int w_1(X,\theta+\pi/2)X^2dX - \Big(\int w_1(X,\theta+\pi/2)X\,dX\Big)^2\Big] - \Big\{\int w_1(X,\theta+\pi/4)X^2dX$$
$$-\Big(\int w_1(X,\theta+\pi/4)X\,dX\Big)^2 - \frac{1}{2}\Big[\int w_1(X,\theta)X^2dX - \Big(\int w_1(X,\theta)X\,dX\Big)^2\Big]$$

$$-\frac{1}{2}\Big[\int w_1(X,\theta+\pi/2)X^2dX-\Big(\int w_1(X,\theta+\pi/2)X\,dX\Big)^2\Big]\Big\}\Big\{\int w_2(X,\theta+\pi/4)X^2dX$$
$$-\Big(\int w_2(X,\theta+\pi/4)X\,dX\Big)^2-\frac{1}{2}\Big[\int w_2(X,\theta)X^2dX-\Big(\int w_2(X,\theta)X\,dX\Big)^2\Big]$$
$$-\frac{1}{2}\Big[\int w_2(X,\theta+\pi/2)X^2dX-\Big(\int w_2(X,\theta+\pi/2)X\,dX\Big)^2\Big]\Big\}\geq\frac{1}{4}, \qquad (52)$$

which is a tomographic form of the state-extended uncertainty relations introduced by D. A. Trifonov [31, 32].

Other quantum inequalities exist, which must be fulfilled by measurable optical and symplectic tomograms.

For example, the Shannon entropy $S_\theta = -\int w(X,\theta)\ln w(X,\theta)\,dX$ is associated with the optical tomogram $w(X,\theta)=|\psi(X,\theta)|^2$, where the complex function $\psi(X,\theta)$ is given in the form of propagating initial wave function $\psi(y)$ under the influence of the harmonic-oscillator potential [3]

$$\psi(X,\theta)=\frac{1}{\sqrt{2\pi i\sin\theta}}\int\exp\left[\frac{i}{2}\left(\cot\theta\;(y^2+X^2)-\frac{2X}{\sin\theta}\,y\right)\right]\psi(y)\,dy, \qquad (53)$$

must satisfy the inequality

$$-\int w(X,\theta)\ln w(X,\theta)\,dX-\int w(X,\theta+\pi/2)\ln w(X,\theta+\pi/2)\,dX\geq\ln\pi e. \qquad (54)$$

The other inequality must be fulfilled as a property of Rényi entropy [3]; it reads

$$(q-1)\ln\left\{\int_{-\infty}^{\infty}dX\left[w\left(X,\theta+\frac{\pi}{2}\right)\right]^{1/(1-q)}\right\}+(q+1)\ln\left\{\int_{-\infty}^{\infty}dX[w(X,\theta)]^{1/(1+q)}\right\}$$
$$\geq\frac{1}{2}\Big\{(q-1)\ln[\pi(1-q)]+(q+1)\ln[\pi(1+q)]\Big\}. \qquad (55)$$

The quantum entropic inequality for the symplectic tomogram can be written with respect to the photon distribution function as follows:

$$H=-\sum_{n=0}^{\infty}\frac{1}{2\pi}\int\frac{e^{-X^2/(\mu^2+\nu^2)}}{\sqrt{\pi(\mu^2+\nu^2)}}\frac{1}{2^n n!}H_n^2\left(\frac{X}{\sqrt{\mu^2+\nu^2}}\right)M(-Y,\mu,\nu)e^{i(X-Y)}dX\,dY\,d\mu\,d\nu$$
$$\times\ln\left\{\frac{1}{2\pi}\int\frac{e^{-X^2/(\mu^2+\nu^2)}}{\sqrt{\pi(\mu^2+\nu^2)}}\frac{1}{2^n n!}H_n^2\left(\frac{X}{\sqrt{\mu^2+\nu^2}}\right)M(-Y,\mu,\nu)e^{i(X-Y)}dX\,dY\,d\mu\,d\nu\right\}\geq 0, \qquad (56)$$

which means that the quantum-state tomogram must satisfy this inequality since the photon distribution in any quantum state must be nonnegative normalized function of discrete variable which is the photon number. In the classical domain, the function expressed in terms of the classical-state tomogram can be negative or complex. In view of this fact, the entropy value calculated in terms of the measurable tomogram can provide an information on accuracy of the measurement, as well as an information on the classicality or quantumness of the state.

5. Qubit states in the probability representation

The spin states can also be described by probability distributions instead of spinors and density matrices [33, 34]. The spin tomogram of a qudit state is determined by the formula

$$w(m,u) = \langle m \mid u^{\dagger} \rho u \mid m \rangle, \tag{57}$$

where ρ is the density matrix and $\mid m\rangle$ is the state corresponding to the mth spin projection, $m = -j, -j+1, \ldots, j$. The matrix u in (57) is the unitary matrix rotating the vectors in the Hilbert space of qudit states, and the normalization condition

$$\sum_{m=-j}^{j} w(m,u) = 1 \tag{58}$$

must hold.

The spin tomogram is a fair probability distribution of a random spin projection depending on extra parameters coded by unitary matrix u. As shown in [2, 33, 34], a reconstruction formula for the density matrix $\langle m \mid \hat{\rho} \mid m'\rangle$ exists, which can be expressed in terms of the spin tomogram $w(m,u)$. Thus all statistical properties of spin observables can be obtained from this tomogram. For example, the mean value of the spin projection on the z axis is

$$\mathrm{Tr}\left(\rho \hat{J}_z\right) = \sum_{m=-j}^{j} m w(m, u=1). \tag{59}$$

Formula (57) can be presented in another form: if the density matrix ρ has nonnegative eigenvalues $\rho_1, \rho_2, \ldots, \rho_N$, corresponding to the vector $\hat{\rho}$, and N eigenvectors $\vec{u}_1, \vec{u}_2, \ldots, \vec{u}_N$, which can be considered as columns of the unitary matrix u_0, the probabilities $w(-j,u), w(-j+1,u), \ldots, w(j,u)$ can be considered as components of a probability vectors $\vec{w}(u)$ which, in this case, reads

$$\vec{w}(u) = |uu_0|^2 \vec{\rho}. \tag{60}$$

A matrix $|a|^2$, by definition, has the matrix elements expressed as $|a|^2_{kj} = |a_{kj}|^2$. Thus, the state density matrix ρ, determined by its eigenvalues ρ_k and eigenvectors $\vec{w}(u)$, is mapped onto the probability vector $\vec{w}(u)$. This map is invertible and therefore the spin state can be identified with the probability distribution (or the probability vector) depending on extra parameters coded by the unitary matrix u, which rotates the reference frame in the Hilbert space of spin states.

One can introduce the tomographic entropy using Shannon formula

$$H_u = -\sum_{m=-j}^{j} w(m,u) \ln w(m,u). \tag{61}$$

The von Neumann entropy of the spin state is the minimum of the tomographic entropy with respect to the parameters of the unitary group u

$$\min H_u = -\mathrm{Tr}\, \rho \ln \rho. \tag{62}$$

One can also introduce the tomographic Rényi entropy of the qudit state

$$R_u = \frac{1}{1-q} \ln \left(\sum_{m=-j}^{j} (w((m,u))^q \right) \tag{63}$$

and show [35] that its minimum yields the conventional Rényi entropy of the quantum state

$$\min R_u = \frac{1}{1-q} \ln \operatorname{Tr} \rho^q. \tag{64}$$

The relation of the unitary matrices with spin tomograms provides the following new inequality for the $N \times N$ matrices:

$$-\int \left(\sum_{j=1}^{N} |u_{jk}|^2 \ln |u_{jk}|^2 \right) du \geq \frac{1}{2} \ln N, \tag{65}$$

where du is the Haar measure on the unitary group normalized as $\int du = 1$.

6. Conclusions

We showed the tomographic approach to describe the classical and quantum states by means of tomographic probability distributions, which are positive and normalized. These probability distributions were suggested for continuous and discrete variables like photon quadratures (position and momentum) and spin projections (qudits).

The basic classical and quantum evolution equations like the state evolution equation and the energy level equation were obtained in the form of equations for fair probability distributions. The unified classical–quantum state description appeared to be appropriate for introducing the joint quantum–classical dynamics, using the tomographic picture of states [36, 37].

The quantum inequalities like the basic uncertainty relations and entropic inequalities were expressed in terms of measurable optical tomograms; thus, all these inequalities can be checked in the experiments with homodyne photon detection [21, 29]. This check provides a new possibility to control the accuracy of homodyne detecting of the photon states.

The spin-tomographic probability distributions describe the qudit states and can be useful for studying the Bell inequalities and quantum correlations in multipartite spin systems.

Acknowledgements

This study was partially supported by the Russian Foundation for Basic Research under Projects Nos. 09-00-00142, 10-02-00312 and 11-02-00456. The authors are grateful to the organizers of the Symposium "Symmetries in Science XV" (Kloster Mehrerau, Bregenz, Austria, 31 July – 5 August, 2011) and especially Profs. Dieter Schuch and Michael Ramek for the invitation and their kind hospitality.

References

[1] Mancini S, Man'ko V I and Tombesi P 1996 *Phys. Lett.* A **213** 1
[2] Ibort A, Man'ko V I, Marmo G, Simoni A and Ventriglia F 2009 *Phys. Scr.* **79** 065013
[3] Man'ko M A and Man'ko V I 2009 *Found. Phys.* **DOI** 10.1007/s10701-009-9403-9 **41** 330 (2011)
[4] Wigner E 1932 *Phys. Rev.* **40** 749
[5] Husimi K 1940 *Proc. Phys. Math. Soc. Jpn* **22** 264
[6] Kano Y 1965 *J. Math. Phys.* **6** 1913
[7] Glauber R J 1963 *Phys. Rev. Lett.* **10** 84
[8] Sudarshan E C G 1963 *Phys. Rev. Lett.* **10** 277
[9] Heisenberg W 1927 *Z. Phys.* **43** 172
[10] Schrödinger E 1930 *Ber. Kgl. Akad. Wiss. Berlin* Z. 296
[11] Robertson H P 1930 *Phys. Rev.* **35** 667
[12] Robertson H P 1934 **46** 794
[13] Man'ko M A, Man'ko V I, De Nicola S and Fedele R 2006 *Acta Phys. Hung.* B **26/1-2** 71
[14] De Nicola S, Fedele R, Man'ko M A and Man'ko V I 2006 *Eur. J. Phys.* B **52** 191
[15] Man'ko M A 2006 *J. Russ. Laser Res.* **27** 405
[16] De Nicola S, Fedele R, Man'ko M A and Man'ko V I 2007 *Theor. Math. Phys.* **152** 1081 [tr. 2007 *Teor. Mat. Fiz.* **152** 241]
[17] De Nicola S, Fedele R, Man'ko M A and Man'ko V I 2007 *J. Phys. Conf. Ser.* **70** 012007
[18] De Nicola S, Fedele R, Man'ko M A and Man'ko V I 2009 *Phys. Scr.* **T135** 014053
[19] Shannon C E 1948 *Bell Syst. Tech. J.* **27** 379
[20] Rényi A 1970 *Probability Theory* (Amsterdam: North-Holland)
[21] Smithey D T, Beck M, Raymer M G and Faridani A 1993 *Phys. Rev. Lett.* **70** 1244
[22] Parigi V, Zavatta A, Kim M, Bellini M et al. 2007 *Science* **317** 1890
[23] Filippov S N and Man'ko V I 2011 *Phys. Rev.* A **84** 033827
[24] Bertrand J and Bertrand P 1987 *Found. Phys.* **17** 397
[25] Vogel K and Risken H 1989 *Phys. Rev.* A **40** 2847
[26] Man'ko V I and Mendes R V 1999 *Phys. Lett.* A **263** 53
[27] Korennoy Ya A and Man'ko V I 2011 *J. Russ. Laser Res.* **32** 74; 338
[28] Man'ko V I, Marmo G, Simoni A and Ventriglia F 2009 *Adv. Sci. Lett.* **2** 517
[29] Man'ko V I, Marmo G, Porzio A, Solimeno S and Ventriglia F 2011 *Phys. Scr.* **83** 045001
[30] Chernega V N 2012 *Phys. Scr.* **T147** 014006
[31] Trifonov D A 2000 *J. Phys. A: Math. Gen.* **33** L299
[32] Trifonov D A 2002 *Eur. Phys. J. B – Cond. Matter Complex Syst.* **29** 349
[33] Man'ko V I and Man'ko O V 1997 *Zh. Éksp. Teor. Fiz.* **112** 796 [tr. *J. Exp. Theor. Phys.* **85** 430]
[34] Dodonov V V and Man'ko V I 1997 *Phys. Lett.* A **239** 335
[35] Man'ko M A, Man'ko V I and Vilela Mendes R 2006 *J. Russ. Laser Res.* **27** 507
[36] Man'ko O V and Man'ko V I 2004 *J. Russ. Laser Res.* **25** 115
[37] Chernega V N and Man'ko V I 2011 *AIP Conf. Proc.* **1424** 39

Living in a world without imaginaries

Martin Erik Horn

Johann Wolfgang Goethe University Frankfurt, Department of Physics, Institute of Physics Education, Max-von-Laue-Str. 1, D – 60438 Frankfurt on the Main, Germany

E-mail: m.horn@physik.uni-frankfurt.de

Abstract. We need complex numbers to compress and condense long mathematical formulae into shorter and more abstract ones. But we do not need them to describe the world we live in. Such a mathematical world is presented is this paper. It is a world which can be explained to and understood by learners much easier than the usual obscured world full of imaginaries. And it is a geometrical world, not an algebraic one.

1. Introduction

In his great book *The Road to Reality* Roger Penrose writes: "Yet, there are other kinds of number which, according to accepted theory, do appear to play a fundamental role in the workings of the universe. The most important and striking of these are the *complex numbers*, in which the seemingly mystical quantity $\sqrt{-1}$, usually denoted by 'i', is introduced and adjoined to the real-number system. First encountered in the 16th century, but treated for hundreds of years with distrust, the mathematical utility of complex numbers gradually impressed the mathematical community to a greater and greater degree, until complex numbers became an indispensable, even magical, ingredient of our mathematical thinking. Yet we now find that they are fundamental not just to mathematics: these strange numbers also play an extraordinary and very basic role in the operation of the physical universe at its tiniest scales. This is a cause for wonder, and it is an even more striking instance of the convergence between mathematical ideas and the deeper workings of the physical universe than is the system of real numbers" [1].

Although everything written in this short description about the magical effects of the imaginary unit i is true, the imaginary unit i is not necessary to describe the world we live in. This magical tiny i might even be confusing and obscuring our picture of the world. The imaginary unit is surely helpful in compressing and condensing long mathematical formulae into shorter ones, which is a clear indication of high mathematical expertise.

But is it also helpful to present mathematical formulae full of this abstract symbol i to students at school, high school or universities who for the first time deal with these not always simple descriptions of space, time and the relations of physics? I am sure, it is not really helpful for learners.

Reading the lines of Penrose students learn that physics and mathematics contain mystical, even magical ingredients. But I want to live in a world which is understandable and not mystical. I like magicians and their shows. But I like these shows not because of the unphysical phenomena which are shown. I like them because I love to find out how the trick behind all this works.

Penrose and many others present the quantity $\sqrt{-1}$ like magicians on the stage of physics and mathematics. But there is a trick behind complex numbers! Let's find out this trick.

One way to understand imaginary numbers is to construct a world without them and look, which mathematical objects occupy the position of them. Such a world without an imaginary unit i will be presented and discussed in this paper.

It is a world which consists of $+1$, -1, and matrices.

2. Historical background

In mathematics there are two contrasting, opposite tendencies: the tendency to explain and understand mathematical structures as algebraic relations and opposed to that as geometric relations. We can take a more geometric standpoint or we can take a more algebraic standpoint.

Historically de-geometrisation[1], i.e. leaving the geometric standpoint and approaching a more algebraic standpoint, is closely connected with the algebraisation[2] of geometric constructions. And vice versa promoting geometrisation means de-algebraisation [2].

But these two tendencies not only influence the way we work in mathematics, it very deeply influences the way we see and interpret our world, as the following conversation between Dirac and Abdus Salam indicates. "Once Dirac asked me if I thought algebraically or geometrically," Abdus Salam told and went on: "I did not know what he meant, but after further questioning, Dirac said 'Precisely as I thought. You think algebraically as most people in the Indian subcontinent do.' Dirac, it appeared, thought geometrically" [3].

Dirac even claimed, that he can "picture, without effort, the de Sitter space as a four-dimensional surface in a five-dimensional space" [4], which is something a convinced algebraist never would do, because "mathematics may be defined as the subject in which we never know what we are talking about" [5]. This explanation of Russell shows his clear focus on an algebraic world view.

But modern physics is deeply connected with a geometric world view. The geometrisation of the world which surrounds us is an elementary conceptual principle starting with the unification of space and time into a geometrically united spacetime.

Einstein explained this inherent connection between physics and geometry in his academy lesson in 1921: "Yet on the other hand it is certain that mathematics generally, and particularly geometry, owes its existence to the need which was felt of learning something about the relations of real things to one another. (...) It is clear that the system of concepts of axiomatic geometry alone cannot make any assertions as to the relations of real objects of this kind, which we will call practically-rigid bodies. To be able to make such assertions, geometry must be stripped of its merely logical-formal character by the co-ordination of real objects of experience with the empty conceptual frame-work of axiomatic geometry. (...) Geometry thus completed is evidently a natural science; we may in fact regard it as the most ancient branch of physics" [6][2]. Thus physics exerts a considerable stimulus on mathematics to geometrisation.

Hestenes even comes to the conclusion: "Mathematics is too important to be left to the mathematicians!" [8] He therefore supports the use of geometric algebra as a mathematical tool which unites algebraic and geometric ideas in a unified mathematical language.

Euler introduced the symbol i for the imaginary unit $\sqrt{-1}$ in 1777 as an algebraic symbol [9]. Algebraic equations can be solved much easier using this symbol, and the story of i, this imaginary tale [10], surely is a story of algebra. The path the history of European mathematics had chosen with respect to imaginary and complex numbers, was an algebraic path.

---

[1] In this context Scriba and Schreiber use the German words "Entgeometrisierung" and "Algebraisierung" [2].

[2] The complete German quotation is: "Andererseits ist es aber doch sicher, daß die Mathematik überhaupt und im speziellen auch die Geometrie ihre Entstehung dem Bedürfnis verdankt, etwas zu erfahren über das Verhalten wirklicher Dinge. (...) Es ist klar, daß das Begriffssystem der axiomatischen Geometrie allein über das Verhalten derartiger Gegenstände der Wirklichkeit, die wir als praktisch-starre Körper bezeichnen wollen, keine Aussagen liefern kann. Um derartige Aussagen liefern zu können, muß die Geometrie dadurch ihres nur logisch-formalen Charakters entkleidet werden, daß den leeren Begriffschemata der axiomatischen Geometrie erlebbare Gegenstände der Wirklichkeit (Erlebnisse) zugeordnet werden. (...) Die so ergänzte Geometrie ist offenbar eine Naturwissenschaft; wir können sie geradezu als den ältesten Zweig der Physik betrachten" [7].

But is there a second possible path, not followed by our mathematical ancestors, which opens a geometric entrance to imaginary and complex numbers? I am convinced**:** a development of mathematics with a more geometrical focus could have led to a mathematical world without imaginaries. In this (till now not really appreciated) geometric mathematical world, geometric objects nearly automatically appear which take the function of the imaginary unit. But these objects are nothing artificial which has to be invented. They are objects which are there because they are inherent to geometry. They are as "real" as other real geometric objects.

Hamilton opened the door to this mathematical structures a little bit, when he discussed Pauli-like "quadruples of numbers that form 2 x 2 matrices" [10]. But it seems that the Quaternionists and Hamiltonians nearly immediately closed this door again by neglecting the geometric meaning of quarternion calculus and by "presenting their views in a rather one-sided and metaphysical bearing" only acknowledging algebraic definitions [11][3].

One can even conclude that history of mathematics suffered from a subvirus of the MV/K-type virus Hestenes identified [12]. This virus presses mathematicians to hallucinate that "complex Clifford algebras are more general than real Clifford algebras." And this "suggests that the standard mathematical practice of regarding complex numbers as scalars is an egregious case of mistaken identity, a mistake which can be corrected by recognizing the geometric primacy of the real geometric algebra" [12].

3. Two-dimensional worlds

Two of the thee Pauli matrices do not include imaginary units and thus are quadrupoles of the real numbers + 1, – 1, and 0 only. These two Pauli matrices are the base vectors

$$\sigma_x = \begin{pmatrix} 0 & 1 \\ 1 & 0 \end{pmatrix} \tag{1}$$

$$\sigma_z = \begin{pmatrix} 1 & 0 \\ 0 & -1 \end{pmatrix} \tag{2}$$

of a two-dimensional space with signature (+, +). The other two basic elements of this space are a unit scalar

$$\mathbf{1}^2 = \sigma_x{}^2 = \sigma_z{}^2 = \mathbf{1} = \begin{pmatrix} 1 & 0 \\ 0 & 1 \end{pmatrix} \tag{3}$$

and a base bivector

$$\sigma_{zx} = \sigma_z\sigma_x = -\,\sigma_x\sigma_z = \begin{pmatrix} 0 & 1 \\ -1 & 0 \end{pmatrix} \tag{4}$$

which clearly can take the place of an imaginary unit because it squares to minus one.

$$\sigma_{zx}{}^2 = -\,\mathbf{1} = \begin{pmatrix} -1 & 0 \\ 0 & -1 \end{pmatrix} \tag{5}$$

---

[3] The complete German quotation is: "Trotzdem diese Theorie nicht mehr jung ist, gehen die Ansichten über ihren Wert bis heute stark auseinander. Der Grund hierfür dürfte darin zu suchen sein, daß die Vertreter der Quaternionentheorie ihre Lehren meistens einseitig und mit einem metaphysischen Anfluge behaftet darstellen, wobei die einfache geometrische Deutung, welche man den Operationen ihres Kalkuls geben kann, nicht immer genügend hervortritt" [11].

This base bivector $\sigma_x\sigma_z$ is in the first place a geometrical object: an oriented plane element in the direction of the vectors σ_x and σ_z with a unit area. But as we are able to calculate directly with geometrical objects in geometric algebra, it is an algebraic object too. This area is as real as a vector: it exists in our conceptual scheme of space. But it automatically takes the position of the now superfluous imaginary unit i.

The two vectors and the bivector anticommute in this two-dimensional space. This feature is shown in the commutation table of these real (2x2) matrices (see table 1), where "com" means that the two elements commute while "A" means that the two elements anticommute.

Table 1. Commutation table for real (2x2) matrices.

	1	σ_x	σ_z	$\sigma_z\sigma_x$
1: **1**	**+1**	com	com	com
2: σ_x	com	**+1**	**A**	**A**
3: σ_z	com	**A**	**+1**	**A**
4: $\sigma_z\sigma_x$	com	**A**	**A**	**−1**

In the diagonal line the squares (3) and (5) of these elements are shown, which give the signature (+, +). If σ_x and $(\sigma_z\,\sigma_x)$ were employed as base vectors, the signature would be (+, –). Of course, the basic elements σ_x and σ_z are in the first place geometric objects too in this two-dimensional world. Every vector **r** can be written as a linear combination of these two basic elements:

$$\mathbf{r} = x\sigma_x + z\sigma_z = \begin{pmatrix} z & x \\ x & -z \end{pmatrix} \tag{6}$$

Therefore σ_x and σ_z are correctly identified as base vectors. But they have a second, operational geometric meaning. They are basic reflections (see [1]), for a left- and right-sided multiplication of a vector **r** with σ_x or σ_z gives a vector **r'** which is reflected at the axes of the corresponding base vector:

$$\mathbf{r}' = \sigma_x\,\mathbf{r}\,\sigma_x = x\sigma_x - z\sigma_z \tag{7}$$

$$\mathbf{r}'' = \sigma_z\,\mathbf{r}\,\sigma_z = -x\sigma_x + z\sigma_z \tag{8}$$

Only the coordinates perpendicular to the reflection axes change the sign. A reflection at the $(\sigma_z\sigma_x)$-plane is an identity operation given by

$$\mathbf{r} = \sigma_z\sigma_x\,\mathbf{r}\,\sigma_z\sigma_x = x\sigma_x + z\sigma_z \tag{9}$$

while a reflection at the origin

$$\mathbf{r}''' = -\mathbf{1}\,\mathbf{r}\,\mathbf{1} = -x\sigma_x - z\sigma_z \tag{10}$$

reverses all signs.

Of course the commutation relations are a consequence of the algebra of these matrices. The matrix representation chosen here is only one of an infinite number of possible matrix representations. For example one could chose the two base vectors

$$\mathbf{e}_1 = \sigma_x + \sigma_z + \sigma_z\sigma_x = \begin{pmatrix} 1 & 2 \\ 0 & -1 \end{pmatrix} \tag{11}$$

$$\mathbf{e}_2 = -\frac{1}{2}\sigma_x + \sigma_z + \frac{1}{2}\sigma_z\sigma_x = \begin{pmatrix} 1 & 0 \\ -1 & -1 \end{pmatrix} \tag{12}$$

with

$$\mathbf{e}_1^{\ 2} = \mathbf{e}_2^{\ 2} = \mathbf{1} = \begin{pmatrix} 1 & 0 \\ 0 & 1 \end{pmatrix} \tag{13}$$

and

$$\mathbf{e}_1 \mathbf{e}_2 = -\frac{1}{2}\sigma_x - \sigma_z - \frac{3}{2}\sigma_z\sigma_x = \begin{pmatrix} -1 & -2 \\ 1 & 1 \end{pmatrix} \tag{14}$$

with

$$(\mathbf{e}_1 \mathbf{e}_2)^2 = -\mathbf{1} = \begin{pmatrix} -1 & 0 \\ 0 & -1 \end{pmatrix} \tag{15}$$

Again all vectors $\mathbf{e}_1$, $\mathbf{e}_2$ and the bivector $\mathbf{e}_1\mathbf{e}_2$ anticommute, giving the same commutation table (see table 1). Therefore it is not necessary to use a special matrix representation. In some situations it might be helpful using a matrix representation, but in other situations it might be confusing. All information about the geometrical content is encoded in the algebra.

4. Four-dimensional worlds

Snygg already presented the base vectors of three-dimensional Euclidean space as real (4x4) matrices [13]. But real (4x4) matrices can be used to construct four linear independent base vectors, thus spanning a four-dimensional space.

The mathematics to construct (4x4) matrices using (2x2) matrices was already developed in 1858 by Zehfuss [14], although his direct product was named in a strange twist after Kronecker [15] who only much later mentioned this direct matrix product. The Zehfuss-Kronecker product now plays an important role in describing quantum computing (see [16]), then sometimes even more unsatisfactorily called tensor product (see e.g. [17]).

The Zehfuss-Kronecker product is defined as

$$\mathbf{A} \otimes \mathbf{B} = \begin{pmatrix} a_{11} & a_{12} \\ a_{21} & a_{22} \end{pmatrix} \otimes \mathbf{B} = \begin{pmatrix} a_{11}\mathbf{B} & a_{12}\mathbf{B} \\ a_{21}\mathbf{B} & a_{22}\mathbf{B} \end{pmatrix} \tag{16}$$

in the special case of (2x2) matrices and generally

$$\mathbf{A} \otimes \mathbf{B} = \begin{pmatrix} a_{11}\mathbf{B} & a_{12}\mathbf{B} & \cdots & a_{1n}\mathbf{B} \\ a_{21}\mathbf{B} & a_{22}\mathbf{B} & \cdots & a_{2n}\mathbf{B} \\ \vdots & \vdots & \ddots & \vdots \\ a_{m1}\mathbf{B} & a_{m2}\mathbf{B} & \cdots & a_{mn}\mathbf{B} \end{pmatrix} \tag{17}$$

The 16 new basic elements of a four-dimensional space can then be constructed using the four basic elements of the two-dimensional space (1) – (4) and multiplying them according to the direct Zehfuss-Kronecker product (17). The results are given in table 2, which of course show only one possible matrix representation. Taking (13), (11), (12), and (14) as starting point would give a set of 16 different matrices which nevertheless are subject to the same algebra.

These new basic elements must now be assigned to one scalar, four vectors, six bivectors, four trivectors and one tetravector.

Table 2. Zehfuss-Kronecker multiplication table for real (2x2) matrices.

$\otimes$	$\mathbf{1} = \begin{pmatrix} 1 & 0 \\ 0 & 1 \end{pmatrix}$	$\sigma_x = \begin{pmatrix} 0 & 1 \\ 1 & 0 \end{pmatrix}$	$\sigma_z = \begin{pmatrix} 1 & 0 \\ 0 & -1 \end{pmatrix}$	$\sigma_z\sigma_x = \begin{pmatrix} 0 & 1 \\ -1 & 0 \end{pmatrix}$
$\mathbf{1} = \begin{pmatrix} 1 & 0 \\ 0 & 1 \end{pmatrix}$	$\mathbf{1} \otimes \mathbf{1}$ $= \begin{pmatrix} \mathbf{1} & 0 \\ 0 & \mathbf{1} \end{pmatrix}$ $= \begin{pmatrix} 1 & 0 & 0 & 0 \\ 0 & 1 & 0 & 0 \\ 0 & 0 & 1 & 0 \\ 0 & 0 & 0 & 1 \end{pmatrix}$	$\mathbf{1} \otimes \sigma_x$ $= \begin{pmatrix} \sigma_x & 0 \\ 0 & \sigma_x \end{pmatrix}$ $= \begin{pmatrix} 0 & 1 & 0 & 0 \\ 1 & 0 & 0 & 0 \\ 0 & 0 & 0 & 1 \\ 0 & 0 & 1 & 0 \end{pmatrix}$	$\mathbf{1} \otimes \sigma_z$ $= \begin{pmatrix} \sigma_z & 0 \\ 0 & \sigma_z \end{pmatrix}$ $= \begin{pmatrix} 1 & 0 & 0 & 0 \\ 0 & -1 & 0 & 0 \\ 0 & 0 & 1 & 0 \\ 0 & 0 & 0 & -1 \end{pmatrix}$	$\mathbf{1} \otimes (\sigma_z\sigma_x)$ $= \begin{pmatrix} \sigma_z\sigma_x & 0 \\ 0 & \sigma_z\sigma_x \end{pmatrix}$ $= \begin{pmatrix} 0 & 1 & 0 & 0 \\ -1 & 0 & 0 & 0 \\ 0 & 0 & 0 & 1 \\ 0 & 0 & -1 & 0 \end{pmatrix}$
$\sigma_x = \begin{pmatrix} 0 & 1 \\ 1 & 0 \end{pmatrix}$	$\sigma_x \otimes \mathbf{1}$ $= \begin{pmatrix} 0 & \mathbf{1} \\ \mathbf{1} & 0 \end{pmatrix}$ $= \begin{pmatrix} 0 & 0 & 1 & 0 \\ 0 & 0 & 0 & 1 \\ 1 & 0 & 0 & 0 \\ 0 & 1 & 0 & 0 \end{pmatrix}$	$\sigma_x \otimes \sigma_x$ $= \begin{pmatrix} 0 & \sigma_x \\ \sigma_x & 0 \end{pmatrix}$ $= \begin{pmatrix} 0 & 0 & 0 & 1 \\ 0 & 0 & 1 & 0 \\ 0 & 1 & 0 & 0 \\ 1 & 0 & 0 & 0 \end{pmatrix}$	$\sigma_x \otimes \sigma_z$ $= \begin{pmatrix} 0 & \sigma_z \\ \sigma_z & 0 \end{pmatrix}$ $= \begin{pmatrix} 0 & 0 & 1 & 0 \\ 0 & 0 & 0 & -1 \\ 1 & 0 & 0 & 0 \\ 0 & -1 & 0 & 0 \end{pmatrix}$	$\sigma_x \otimes (\sigma_z\sigma_x)$ $= \begin{pmatrix} 0 & \sigma_z\sigma_x \\ \sigma_z\sigma_x & 0 \end{pmatrix}$ $= \begin{pmatrix} 0 & 0 & 0 & 1 \\ 0 & 0 & -1 & 0 \\ 0 & 1 & 0 & 0 \\ -1 & 0 & 0 & 0 \end{pmatrix}$
$\sigma_z = \begin{pmatrix} 1 & 0 \\ 0 & -1 \end{pmatrix}$	$\sigma_z \otimes \mathbf{1}$ $= \begin{pmatrix} \mathbf{1} & 0 \\ 0 & -\mathbf{1} \end{pmatrix}$ $= \begin{pmatrix} 1 & 0 & 0 & 0 \\ 0 & 1 & 0 & 0 \\ 0 & 0 & -1 & 0 \\ 0 & 0 & 0 & -1 \end{pmatrix}$	$\sigma_z \otimes \sigma_x$ $= \begin{pmatrix} \sigma_x & 0 \\ 0 & -\sigma_x \end{pmatrix}$ $= \begin{pmatrix} 0 & 1 & 0 & 0 \\ 1 & 0 & 0 & 0 \\ 0 & 0 & 0 & -1 \\ 0 & 0 & -1 & 0 \end{pmatrix}$	$\sigma_z \otimes \sigma_z$ $= \begin{pmatrix} \sigma_z & 0 \\ 0 & -\sigma_z \end{pmatrix}$ $= \begin{pmatrix} 1 & 0 & 0 & 0 \\ 0 & -1 & 0 & 0 \\ 0 & 0 & -1 & 0 \\ 0 & 0 & 0 & 1 \end{pmatrix}$	$\sigma_z \otimes (\sigma_z\sigma_x)$ $= \begin{pmatrix} \sigma_z\sigma_x & 0 \\ 0 & -\sigma_z\sigma_x \end{pmatrix}$ $= \begin{pmatrix} 0 & 1 & 0 & 0 \\ -1 & 0 & 0 & 0 \\ 0 & 0 & 0 & -1 \\ 0 & 0 & 1 & 0 \end{pmatrix}$
$\sigma_z\sigma_x = \begin{pmatrix} 0 & 1 \\ -1 & 0 \end{pmatrix}$	$(\sigma_z\sigma_x) \otimes \mathbf{1}$ $= \begin{pmatrix} 0 & \mathbf{1} \\ -\mathbf{1} & 0 \end{pmatrix}$ $= \begin{pmatrix} 0 & 0 & 1 & 0 \\ 0 & 0 & 0 & 1 \\ -1 & 0 & 0 & 0 \\ 0 & -1 & 0 & 0 \end{pmatrix}$	$(\sigma_z\sigma_x) \otimes \sigma_x$ $= \begin{pmatrix} 0 & \sigma_x \\ -\sigma_x & 0 \end{pmatrix}$ $= \begin{pmatrix} 0 & 0 & 0 & 1 \\ 0 & 0 & 1 & 0 \\ 0 & -1 & 0 & 0 \\ -1 & 0 & 0 & 0 \end{pmatrix}$	$(\sigma_z\sigma_x) \otimes \sigma_z$ $= \begin{pmatrix} 0 & \sigma_z \\ -\sigma_z & 0 \end{pmatrix}$ $= \begin{pmatrix} 0 & 0 & 1 & 0 \\ 0 & 0 & 0 & -1 \\ -1 & 0 & 0 & 0 \\ 0 & 1 & 0 & 0 \end{pmatrix}$	$(\sigma_z\sigma_x) \otimes (\sigma_z\sigma_x)$ $= \begin{pmatrix} 0 & \sigma_z\sigma_x \\ -\sigma_z\sigma_x & 0 \end{pmatrix}$ $= \begin{pmatrix} 0 & 0 & 0 & 1 \\ 0 & 0 & -1 & 0 \\ 0 & -1 & 0 & 0 \\ 1 & 0 & 0 & 0 \end{pmatrix}$

The new scalar surely is the (4x4) identity matrix:

$$\mathbf{1}_{(4\text{-dim})} := \mathbf{1} \otimes \mathbf{1} = \begin{pmatrix} 1 & 0 & 0 & 0 \\ 0 & 1 & 0 & 0 \\ 0 & 0 & 1 & 0 \\ 0 & 0 & 0 & 1 \end{pmatrix} \tag{18}$$

If we want a space with only positive signature elements, it is necessary to know the squares of the basic elements. They are:

$$(\mathbf{1} \otimes \sigma_x)^2 = (\mathbf{1} \otimes \sigma_z)^2 = \mathbf{1}_{(4\text{-dim})} \tag{19}$$

$$(\sigma_x \otimes \mathbf{1})^2 = (\sigma_z \otimes \mathbf{1})^2 = \mathbf{1}_{(4\text{-dim})} \tag{20}$$

But these four Zehfuss-Kronecker products ($\mathbf{1} \otimes \sigma_x$), ($\mathbf{1} \otimes \sigma_z$), ($\sigma_x \otimes \mathbf{1}$), and ($\sigma_z \otimes \mathbf{1}$) should not be chosen as base vectors because they do not form a set of anticommuting matrices.

$$(\mathbf{1} \otimes \sigma_z\sigma_x)^2 = (\sigma_z\sigma_x \otimes \mathbf{1})^2 = -\,\mathbf{1}_{(4\text{-dim})} \tag{21}$$

$$(\sigma_x \otimes \sigma_x)^2 = (\sigma_z \otimes \sigma_z)^2 = \mathbf{1}_{(4\text{-dim})} \tag{22}$$

$$(\sigma_x \otimes \sigma_z)^2 = (\sigma_z \otimes \sigma_x)^2 = \mathbf{1}_{(4\text{-dim})} \tag{23}$$

$$(\sigma_x \otimes \sigma_z\sigma_x)^2 = (\sigma_z \otimes \sigma_z\sigma_x)^2 = -\,\mathbf{1}_{(4\text{-dim})} \tag{24}$$

$$(\sigma_z\sigma_x \otimes \sigma_x)^2 = (\sigma_z\sigma_x \otimes \sigma_z)^2 = -\,\mathbf{1}_{(4\text{-dim})} \tag{25}$$

$$(\sigma_z\sigma_x \otimes \sigma_z\sigma_x)^2 = \mathbf{1}_{(4\text{-dim})} \tag{26}$$

All these relations can be found directly by evaluating the matrix representations. But didactically it makes more sense to derive (19) – (26) by using the multiplication rule of the Kronecker-Zehfuss product [18]

$$(\mathbf{A}_1 \otimes \mathbf{B}_1)\,(\mathbf{A}_2 \otimes \mathbf{B}_2) = (\mathbf{A}_1\,\mathbf{A}_2) \otimes (\mathbf{B}_1\,\mathbf{B}_2) \tag{27}$$

because then it is clear that the results are a direct consequence of the underlying algebra and not of a special matrix representation.

The commutation rules can be found in an equally simple way using equation (27). For example the basic elements $(\sigma_z\sigma_x) \otimes \sigma_x$ (#14 in table A1, see appendix) and $\sigma_x \otimes \sigma_x$ (#6 in table A1) anticommute because

$$((\sigma_z\sigma_x) \otimes \sigma_x)\,(\sigma_x \otimes \sigma_x) = ((\sigma_z\sigma_x\sigma_x) \otimes (\sigma_x\sigma_x) = \sigma_z \otimes \mathbf{1} \tag{28}$$

and

$$(\sigma_x \otimes \sigma_x)\,((\sigma_z\sigma_x) \otimes \sigma_x) = ((\sigma_x\sigma_z\sigma_x) \otimes (\sigma_x\sigma_x) = -\,\sigma_z \otimes \mathbf{1} \tag{29}$$

thus giving

$$(\sigma_x \otimes \sigma_x)\,((\sigma_z\sigma_x) \otimes \sigma_x) = -\,((\sigma_z\sigma_x) \otimes \sigma_x)\,(\sigma_x \otimes \sigma_x) \tag{30}$$

Again the commutation rules are a direct consequence of the underlying algebra and do not depend on the matrix representation. The commutation rules are shown in table A1 (see appendix) where "com" means that the two elements commute while "A" means that the two elements anticommute.

5. Some philosophy

Analyzing table A1 we now try to find sets of four basic elements which can be used as base vectors. Because each base vector has to anticommute with every other base vector, we need a set of four anticommuting base elements.

Astonishingly it is not possible to find a set with four anticommuting base vectors which all have a positive square of + 1. And it is not possible to find a set with four anticommuting base vectors which all have a negative square of – 1.

We therefore fail to construct a four-dimensional space which has four spacelike base vectors. And we fail to construct a four-dimensional space which has four timelike base vectors. What is the philosophical meaning of this?

We all agree that we are living in a four-dimensional spacetime. But why isn't it a four-dimensional space only? Why isn't our world a four-dimensional time? Is there a reason for this?

We can not be sure that nature does not make use of an artificial imaginary unit. We can not be sure that nature only makes use of the geometrically based imaginary elements #4, #8, #12, #13, #14, and #15 of table A1.

But we can conclude: If nature does not make use of an artificial imaginary unit, we will live in a world which automatically consists of space and time dimensions. Table A1 clearly shows that we are only able to construct spaces with a signature of (+, +, +, –) or (+, +, –, –). There are no pure spacelike four-dimensional worlds, and there are no pure timelike four-dimensional worlds in a totally geometrically based world.

We are not able to measure imaginary or complex values in physics. We can walk 7 km, but we never walk (7 + 3i) km. We use this imaginary unit called i because it simplifies our calculations. But it might be something artificial. Therefore it is not that stupid to construct a world without an artificial imaginary unit i. The world we live in might be structured like that, and the conclusion again is:

If nature does not make use of an artificial imaginary unit, we will live in a world which automatically consists of space and time.

6. Dirac matrices

Dirac matrices are matrix representations of base vectors of four-dimensional spacetime [8], [19], [20]. As we are living in such a spacetime of signature (+, –, –, –) it makes sense to try to identify some of the basic elements of table 2 with Dirac matrices.

Comparing table 2 with the Dirac matrices given in [21] shows that the timelike Dirac matrix and base vector γ_t can be identified with element #9:

$$\gamma_t = \sigma_z \otimes \mathbf{1} = \begin{pmatrix} 1 & 0 & 0 & 0 \\ 0 & 1 & 0 & 0 \\ 0 & 0 & -1 & 0 \\ 0 & 0 & 0 & -1 \end{pmatrix} \tag{31}$$

The spacelike Dirac matrix and base vector in x direction γ_x can be identified with element #14:

$$\gamma_x = (\sigma_z\sigma_x) \otimes \sigma_x = \begin{pmatrix} 0 & 0 & 0 & 1 \\ 0 & 0 & 1 & 0 \\ 0 & -1 & 0 & 0 \\ -1 & 0 & 0 & 0 \end{pmatrix} \tag{32}$$

Unfortunately the Dirac matrix and base vector in y direction γ_y contains imaginary units i. But the spacelike Dirac matrix and base vector in z direction γ_z can again be identified with element #15:

$$\gamma_z = (\sigma_z\sigma_x) \otimes \sigma_z = \begin{pmatrix} 0 & 0 & 1 & 0 \\ 0 & 0 & 0 & -1 \\ -1 & 0 & 0 & 0 \\ 0 & 1 & 0 & 0 \end{pmatrix} \tag{33}$$

The timelike Dirac matrix and base vector in v direction of cosmological relativity [22] can be identified with element #5:

$$\gamma_v = \sigma_x \otimes \mathbf{1} = \begin{pmatrix} 0 & 0 & 1 & 0 \\ 0 & 0 & 0 & 1 \\ 1 & 0 & 0 & 0 \\ 0 & 1 & 0 & 0 \end{pmatrix} \tag{34}$$

Hence it is clear that real Dirac matrices without imaginary unit i can be interpreted as the base vectors of a spacetime with signature (+, +, –, –) only.

It is even not possible to construct a space with signature (+, –, –, –). But there are several possibilities to find adequate base vectors for a world with signature (–, +, +, +). When time is assigned to a negatively squaring base vector and the three space directions are assigned to positively squaring base vectors this mathematical model can reproduce the geometric structure of our special relativistic world we live in.

For examples the basic elements #8, #2, #3, and #16 of table A1 show the necessary commutation relations and can be chosen as a set of base vectors.

Base vector $\mathbf{e}_1$ with $\mathbf{e}_1^2 = -1$:

$$\mathbf{e}_1 = \sigma_x \otimes (\sigma_z\sigma_x) = \begin{pmatrix} 0 & 0 & 0 & 1 \\ 0 & 0 & -1 & 0 \\ 0 & 1 & 0 & 0 \\ -1 & 0 & 0 & 0 \end{pmatrix} \tag{35}$$

Base vector $\mathbf{e}_2$ with $\mathbf{e}_2^2 = +1$:

$$\mathbf{e}_2 = \mathbf{1} \otimes \sigma_x = \begin{pmatrix} 0 & 1 & 0 & 0 \\ 1 & 0 & 0 & 0 \\ 0 & 0 & 0 & 1 \\ 0 & 0 & 1 & 0 \end{pmatrix} \tag{36}$$

Base vector $\mathbf{e}_3$ with $\mathbf{e}_3^2 = +1$:

$$\mathbf{e}_3 = \mathbf{1} \otimes \sigma_z = \begin{pmatrix} 1 & 0 & 0 & 0 \\ 0 & -1 & 0 & 0 \\ 0 & 0 & 1 & 0 \\ 0 & 0 & 0 & -1 \end{pmatrix} \tag{37}$$

Base vector $\mathbf{e}_4$ with $\mathbf{e}_4^2 = +1$:

$$\mathbf{e}_4 = (\sigma_z\sigma_x) \otimes (\sigma_z\sigma_x) = \begin{pmatrix} 0 & 0 & 0 & 1 \\ 0 & 0 & -1 & 0 \\ 0 & -1 & 0 & 0 \\ 1 & 0 & 0 & 0 \end{pmatrix} \tag{38}$$

7. Outlook

"Are imaginary numbers real?" is asked by Gull, Lasenby, and Doran [19]. This question can be treated from epistemological totally different positions. And in many situations it is far from clear how we could or should or have to interpret imaginary and complex numbers.

There are still some secrets hidden behind the geometrical features of our world. We can try to find them. And the only way to do that is to follow geometry and algebra. Both fields should be set into proper and balanced relations. This means that we should not follow algebraic paths only. Perhaps algebra is the best instrument we possess to understand geometry, and perhaps geometry is the best instrument we possess to understand algebra.

The Zehfuss-Kronecker product can be applied again and again, as it is done in the theory of quantum computing. There must be a lot of geometry behind this important emerging field. It is surely promising to analyze higher-dimensional spaces which were constructed in a geometrical way. Even quantum computing is a geometrical subject. Quantum algebra is geometric algebra. And it can be reformulated without this strange and sometimes superfluous symbol i.

Appendix: **Table A1.** Commutation table for constructed real (4x4) matrices.

	1	2	3	4	5	6	7	8	9	10	11	12	13	14	15	16
1: $\mathbf{1}\otimes\mathbf{1}$	+1	com	com	com	com	com	com	com	com	com	com	com	com	com	com	com
2: $\mathbf{1}\otimes\sigma_x$	com	+1	A	A	com	com	A	A	com	com	A	A	com	com	A	A
3: $\mathbf{1}\otimes\sigma_z$	com	A	+1	A	com	A	com	A	com	A	com	A	com	A	com	A
4: $\mathbf{1}\otimes(\sigma_z\sigma_x)$	com	A	A	−1	com	A	A	com	com	A	A	com	com	A	A	com
5: $\sigma_x\otimes\mathbf{1}$	com	com	com	com	+1	com	com	com	A	A	A	A	A	A	A	A
6: $\sigma_x\otimes\sigma_x$	com	com	A	A	com	+1	A	A	A	A	com	com	A	A	com	com
7: $\sigma_x\otimes\sigma_z$	com	A	com	A	com	A	+1	A	A	com	A	com	A	com	A	com
8: $\sigma_x\otimes(\sigma_z\sigma_x)$	com	A	A	com	com	A	A	−1	A	com	com	A	A	com	cm	A
9: $\sigma_z\otimes\mathbf{1}$	com	com	com	com	A	A	A	A	+1	com	com	com	A	A	A	A
10: $\sigma_z\otimes\sigma_x$	com	com	A	A	A	A	com	com	com	+1	A	A	A	A	com	com
11: $\sigma_z\otimes\sigma_z$	com	A	com	A	A	com	A	com	com	A	+1	A	A	com	A	com
12: $\sigma_z\otimes(\sigma_z\sigma_x)$	com	A	A	com	A	com	com	A	com	A	A	−1	A	com	com	A
13: $(\sigma_z\sigma_x)\otimes\mathbf{1}$	com	com	com	com	A	A	A	A	A	A	A	A	−1	com	com	com
14: $(\sigma_z\sigma_x)\otimes\sigma_x$	com	com	A	A	A	A	com	com	A	A	com	com	com	−1	A	A
15: $(\sigma_z\sigma_x)\otimes\sigma_z$	com	A	com	A	A	com	A	com	A	com	A	com	com	A	−1	A
16: $(\sigma_z\sigma_x)\otimes(\sigma_z\sigma_x)$	com	A	A	com	A	com	com	A	A	com	com	A	com	A	A	+1

References

[1] Penrose R 2005 *The Road to Reality. A Complete Guide to the Laws of the Universe* (London: Vintage Books) p 67 & p 209

[2] Scriba C J and Schreiber P 2010 *5000 Jahre Geometrie* 3rd edition *(Vom Zählstein zum Computer)* eds H-W Alten, A Djafari Naini and H Wesemüller-Kock (Berlin, Heidelberg: Springer) pp 404–406

[3] Fraser G 2008 *Cosmic Anger. Abdus Salam – The First Muslim Nobel Scientist* (Oxford: Oxford University Press) p 116

[4] Salam A 1990 *Dirac and Finite Field Theories (Reminiscences about a great physicist: Paul Adrien Maurice Dirac)* first paperback edition, eds B N Kursunoglu and E P Wigner (Cambridge: Cambridge University Press) chapter 22 p 269

[5] Hellman H 2006 *Great Feuds in Mathematics. Ten of the Liveliest Disputes Ever* (Hoboken, New Jersey: John Wiley & Sons) p 158

[6] Einstein A 1922 *Geometry and Experience (Sidelights on Relativity, part II)* An expanded form of an Address to the Prussian Academy of Sciences in Berlin on January 27th 1921 (London: Methuen) pp 31–32

Reprint in: Hawking S 2009 *A Stubbornly Persistent Illusion: The Essential Scientific Writings of Albert Einstein* first paperback edition (Philadelphia, London: Running Press) p 251

[7] Einstein A 1921 *Geometrie und Erfahrung*, erweiterte Fassung des Festvortrages gehalten an der Preußischen Akademie der Wissenschaften zu Berlin am 27. Januar 1921 (Berlin: Verlag von Julius Springer) pp 5–6

[8] Hestenes D 2003 Oersted Medal Lecture 2002: Reforming the Mathematical Language of Physics *Am. J. Phys.* **71** (2) 104–121

[9] Gericke H 1992 *Mathematik in Antike und Orient / Mathematik im Abendland. Von den römischen Feldmessern bis zu Descartes* Special edition in one volume (Wiesbaden: Fourier Verlag) part II chaper 4.1.3 p 236

[10] Nahin P J 2007 *An Imaginary Tale. The Story of* $\sqrt{-1}$ 15th printing and 1st paperback printing with a new preface of the author (Princeton, Oxford: Princeton University Press) p 252

[11] Klein F and Sommerfeld A 1897 *Über die Theorie des Kreisels* (Leipzig: B. G. Teubner Verlag) Heft I p 55

[12] Hestenes D 1992 *Mathematical Viruses (Clifford Algebras and Their Applications in Mathematical Physics)* eds A Micali, R Boudet and J Helmstetter (Dordrecht: Kluwer Academic Publishers) pp 3–16

[13] Snygg J 1997 *Clifford Algebra. A Computational Tool for Physicists* (New York, Oxford: Oxford University Press) chapter 1 pp 3–24

[14] Zehfuss J G 1858 Über eine gewisse Determinante *Zeitschrift für Mathematik und Physik* **3** pp 298–301

[15] Henderson H V, Pukelsheim F and Searle S R 1983 On the History of the Kronecker Product *Linear and Multilinear Algebra* **14** (2) pp 113–120

[16] Steeb W-H and Hardy Y 2004 *Problems and Solutions in Quantum Computing and Quantum Information* (New Jersey, London, Singapore: World Scientific) p 14–23

[17] Homeister M 2008 *Quantum Computing verstehen. Grundlagen – Anwendungen – Perspektiven* second edition (Wiesbaden: Friedrich Vieweg & Sohn Verlag/GWV Fachverlage) p 37–43

[18] Steeb W-H 1991 *Kronecker Product of Matrices and Applications* (Mannheim, Wien, Zürich: Bibliographisches Institut/Wissenschaftsverlag) p 17

[19] Gull S, Lasenby A and Doran C 1993 Imaginary Numbers are not Real – The Geometric Algebra of Spacetime *Foundations of Physics* **23** (9) pp 1175–1201

[20] Horn M E 2010 Reconsidering and Rethinking Quaternionic Special Relativity *Gravisma 2010 Workshop Proceedings. 2nd International Workshop on Computer Graphics, Computer Vision and Mathematics in Brno* eds V Skala and E M Hitzer (Plzen: Vaclav Skala – Union Agency Publishing) pp 123–129

[21] Horn M E 2011 Geometrische Algebra in höheren Dimensionen *PhyDid B – Didaktik der Physik, Beiträge zur DPG-Frühjahrstagung des Fachverbands Didaktik der Physik in Münster* paper 17.1 p 3 URL: http://www.phydid.de [30 Nov 2011]
[22] Carmeli M 2002 *Cosmological Special Relativity. The Large-Scale Structure of Space, Time and Velocity* 2nd edition (Signapore: World Scientific)

Journal of Physics: Conference Series **380** (2012) 012007 doi:10.1088/1742-6596/380/1/012007

Ultimate generalization of Noether's theorem in the realm of Hamiltonian point dynamics

Jürgen Struckmeier

GSI Helmholtzzentrum für Schwerionenforschung GmbH, Planckstr. 1, D-64291 Darmstadt, Germany

E-mail: `j.struckmeier@gsi.de`, `www.gsi.de/~struck`

Abstract. Noether's theorem in the realm of point dynamics establishes the correlation of a constant of motion of a Hamilton-Lagrange system with a particular symmetry transformation that preserves the form of the action functional. Although usually derived in the Lagrangian formalism [1, 2], the natural context for deriving Noether's theorem for first-order Lagrangian systems is the Hamiltonian formalism. The reason is that the class of transformations that leave the action functional invariant coincides with the class of canonical transformations. As a result, any invariant of a Hamiltonian system can be correlated with a symmetry transformation simply by means of the canonical transformation rules. As this holds for any invariant, we thereby obtain the most general representation of Noether's theorem. In order to allow for symmetry mappings that include a transformation of time, we must refer to the extended Hamiltonian formalism. This formalism enables us to define generating functions of canonical transformations that also map time and energy in addition to the conventional mappings of canonical space and momentum variables.

As an example for the generalized Noether theorem, a manifest representation of the symmetry transformation is derived that corresponds to the Runge-Lenz invariant of the Kepler system.

1. Introduction

Even more than hundred years after the emerging of Einstein's special theory of relativity, the presentation of classical dynamics in terms of the Lagrangian and the Hamiltonian formalism is still usually based in literature on the Newtonian absolute time as the system evolution parameter. The idea how the Hamilton-Lagrange formalism is to be generalized in order to be compatible with special relativity is obvious and well-established. It consists of introducing a system evolution parameter, s, as the new independent variable, and of subsequently treating the time $t = t(s)$ as a *dependent* variable of s, in parallel to all configuration space variables $q^i(s)$.

In order to preserve the canonical form of the the action functional, we must introduce an *extended* Hamiltonian, H_e. Setting up the correlation of extended and conventional Hamiltonians, the *crucial point* is that we must not confuse the conventional Hamilton *function*, H, with its *value*, e. Its negative, $-e(s)$, then plays the role of the *additional canonical variable* that is conjugate to the time variable, $t(s)$.

With our relation of H_e and H in place, we find the subsequent extended set of canonical equations to perfectly coincide in its *form* with the conventional one, which means that no additional functions are involved. This is also true for the theory of extended canonical transformations.

It will be shown that the most general form of Noether's theorem for Hamiltonian point dynamics can be represented by a one-parameter *infinitesimal canonical transformation*. Namely, the characteristic function of the generator of an infinitesimal transformation must be a constant of motion in order for the subsequent transformation to be *canonical*, hence to preserve the action functional. Then the canonical transformation rules embody the symmetry relations of the dynamical system that correspond to this constant of motion.

As a non-trivial example of the correlation of a system's invariant with a symmetry transformation that leaves the action functional invariant, we present a particular symmetry for the Runge-Lenz invariant of the classical Kepler system that is associated with a non-zero time shift.

2. Generalized action functional

The state of a classical dynamical system of n degrees of freedom at time t is completely described by $\boldsymbol{q} = (q^1, \ldots, q^n)$ the vector of generalized space coordinates and $\boldsymbol{p} = (p_1, \ldots, p_n)$ the covector of generalized momenta. We assume the system to be described by a Hamiltonian

$$H : \mathbb{R}^{2n} \times \mathbb{R} \to \mathbb{R}, \qquad e = H(\boldsymbol{q}, \boldsymbol{p}, t). \tag{1}$$

A Hamiltonian H contains the *complete information* on the given dynamical system through the dependence of its *value* e on each q^i and each p_i along the time axis t.

In order to formulate the principle of least action — originating from Leibniz, Maupertuis, Euler, and Lagrange — we define the action functional $\Phi(\gamma)$ as the line integral

$$\Phi(\gamma) = \int_\gamma \sum_{i=1}^{n} p_i \mathrm{d}q^i - H(\boldsymbol{q}, \boldsymbol{p}, t)\, \mathrm{d}t, \tag{2}$$

hence as a mapping of the set of phase-space paths $\gamma \subset \mathbb{R}^{2n} \times \mathbb{R}$ into $\mathbb{R}$. A phase-space path is defined as the smooth mapping that connects a system's *initial state* $(\boldsymbol{q}_0, \boldsymbol{p}_0, t_0)$ with a fixed *final state* $(\boldsymbol{q}_1, \boldsymbol{p}_1, t_1)$,

$$\gamma : \left\{ \left(\boldsymbol{q}, \boldsymbol{p}, t\right) \in \mathbb{R}^{2n+1} \,\middle|\, \left(\boldsymbol{q}_0, \boldsymbol{p}_0, t_0\right) \mapsto \left(\boldsymbol{q}_1, \boldsymbol{p}_1, t_1\right) \right\}.$$

The phase-space path γ_{ext} the dynamical system actually realizes follows from the *principle of least action*. It states that the variation of the action functional (2) vanishes for γ_{ext}, hence $\delta\Phi(\gamma_{\mathrm{ext}}) = 0$. Commonly, a restricted path $\gamma_{\mathrm{r}} \subset \mathbb{R}^{2n}$ is defined by parameterizing (2) in terms of the system's *time*, t

$$\gamma_{\mathrm{r}} : \left\{ \left(\boldsymbol{q}, \boldsymbol{p}\right) \in \mathbb{R}^{2n} \,\middle|\, q^i = q^i(t), p_i = p_i(t); t_0 \le t \le t_1 \right\}.$$

The line integral (2) is thus converted into a conventional integral. The action principle then writes

$$\delta\Phi(\gamma) = \delta \int_{t_0}^{t_1} \left[\sum_{i=1}^{n} p_i(t) \frac{\mathrm{d}q^i(t)}{\mathrm{d}t} - H\big(\boldsymbol{q}(t), \boldsymbol{p}(t), t\big) \right] \mathrm{d}t \stackrel{!}{=} 0. \tag{3}$$

From the calculus of variations, one finds that the functional $\Phi(\gamma_{\mathrm{ext}})$ takes on an *extreme* $(\delta\Phi(\gamma_{\mathrm{ext}}) = 0)$, exactly if the phase-space path $\big(\boldsymbol{q}(t), \boldsymbol{p}(t)\big)$ satisfies the "canonical equations" $(i = 1, \ldots, n)$,

$$\frac{\mathrm{d}q^i}{\mathrm{d}t} = \frac{\partial H}{\partial p_i}, \qquad \frac{\mathrm{d}p_i}{\mathrm{d}t} = -\frac{\partial H}{\partial q^i}, \qquad \frac{\mathrm{d}e}{\mathrm{d}t} = \frac{\partial H}{\partial t}, \tag{4}$$

where $e(t)$ denotes according to Eq. (1) the *instantaneous value* of the Hamiltonian H.

If the Hamiltonian H depends explicitly on time t, then the parametrization of the line integral (2) in terms of t as in Eq. (3) is of restricted utility. The most general parametrization of the variational problem $\delta\Phi(\gamma) \stackrel{!}{=} 0$ is encountered if we treat the time $t = t(s)$ as a canonical variable and parameterize the line integral in terms of a system evolution parameter, s [3]

$$\delta \int_{s_0}^{s_1} \left[\sum_{i=1}^{n} p_i(s) \frac{\mathrm{d}q^i(s)}{\mathrm{d}s} - H\big(\boldsymbol{q}(s), \boldsymbol{p}(s), t(s)\big) \frac{\mathrm{d}t(s)}{\mathrm{d}s} \right] \mathrm{d}s \stackrel{!}{=} 0.$$

The q^i and time t are now treated on *equal footing*. The symmetric form of the integrand suggests to define the $2n+2$ dimensional *extended phase space* by introducing

$$q^0(s) \equiv ct(s), \qquad p_0(s) \equiv -e(s)/c$$

as an *additional pair* of canonically conjugate coordinates. Herein, $e = e(s) \in \mathbb{R}$ is the *instantaneous value* of the Hamiltonian $H(\boldsymbol{q}, \boldsymbol{p}, t)$ at s, but *not* the *function* H. In contrast to H, the canonical coordinate $p_0 = -e/c$ constitutes a function of the independent variable s only, hence exhibits no derivative other than that with respect to s,

$$e(s) \stackrel{\not\equiv}{=} H\big(\boldsymbol{q}(s), \boldsymbol{p}(s), t(s)\big).$$

The s-parametrized action functional can be converted into the standard form of Eq. (3)

$$\delta \int_{s_0}^{s_1} \left[\sum_{\alpha=0}^{n} p_\alpha(s) \frac{\mathrm{d}q^\alpha(s)}{\mathrm{d}s} - H_\mathrm{e}\big(\boldsymbol{q}(s), \boldsymbol{p}(s), t(s), e(s)\big) \right] \mathrm{d}s \stackrel{!}{=} 0 \tag{5}$$

if we define the *extended* Hamiltonian H_e as [4, 5]

$$H_\mathrm{e}\big(\boldsymbol{q}, \boldsymbol{p}, t, e\big) \equiv \Big[H(\boldsymbol{q}, \boldsymbol{p}, t) - e\Big] \frac{\mathrm{d}t}{\mathrm{d}s}. \tag{6}$$

With H_e, we encounter the *extended* functional (5) exactly in the form of the *conventional* functional (3). Note that the sum in (5) now includes terms related to $q^0 = ct$ and $p_0 = -e/c$. As $H(\boldsymbol{q}, \boldsymbol{p}, t) \stackrel{\not\equiv}{=} e$, the extended Hamiltonian H_e actually represents an *implicit function*,

$$H_\mathrm{e}\big(\boldsymbol{q}, \boldsymbol{p}, t, e\big) \stackrel{\not\equiv}{=} 0. \tag{7}$$

Of course, this does *not* mean that H_e may be eliminated from the action functional (5) since the partial derivatives of H_e do not vanish and hence enter into the calculation of the variation.

The action functional (2) can now be written equivalently in terms of the extended Hamiltonian H_e as

$$\Phi(\gamma) = \int_\gamma \sum_{\alpha=0}^{n} p_\alpha \mathrm{d}q^\alpha - H_\mathrm{e}(\boldsymbol{q}, \boldsymbol{p}, t, e)\, \mathrm{d}s,$$

with the paths γ being defined as the set of smooth mappings

$$\gamma : \left\{ \big(\boldsymbol{q}, \boldsymbol{p}, t, e\big) \in \mathbb{R}^{2n+1} \,\big|\, \big(\boldsymbol{q}_0, \boldsymbol{p}_0, t_0, e_0\big) \mapsto \big(\boldsymbol{q}_1, \boldsymbol{p}_1, t_1, e_1\big);\ H_\mathrm{e} = 0 \right\}.$$

Similar to the case of Eq. (3) but now taking s as the system's parameter, the variation of the generalized functional (5) vanishes if the *non-restricted* phase-space path $\gamma \subset \mathbb{R}^{2n+1}$

$$\gamma : \left\{ (\boldsymbol{q}, \boldsymbol{p}, t, e) \in \mathbb{R}^{2n+1} \,\middle|\, q^i = q^i(s), p_i = p_i(s), t = t(s);\ e = e(s), H_\mathrm{e} = 0;\ s_0 \leq s \leq s_1 \right\}$$

satisfies the *extended* set of canonical equations

$$\frac{\mathrm{d}q^i}{\mathrm{d}s} = \frac{\partial H_\mathrm{e}}{\partial p_i}, \qquad \frac{\mathrm{d}p_i}{\mathrm{d}s} = -\frac{\partial H_\mathrm{e}}{\partial q^i}, \qquad \frac{\mathrm{d}t}{\mathrm{d}s} = -\frac{\partial H_\mathrm{e}}{\partial e}, \qquad \frac{\mathrm{d}e}{\mathrm{d}s} = \frac{\partial H_\mathrm{e}}{\partial t}. \tag{8}$$

The number of canonical equations is now *even*. We have thus converted the *pre-symplectic* conventional Hamiltonian formalism into an extended *symplectic* description. Geometrically, the system's motion now takes place on a *hyper-surface*, defined by $H_\mathrm{e} = 0$, within the cotangent bundle $T^*(\mathbb{M} \times \mathbb{R})$ over the space-time configuration manifold $\mathbb{M} \times \mathbb{R}$. This contrasts with the conventional, unconstrained Hamiltonian description on the pre-symplectic cotangent bundle $(T^*\mathbb{M}) \times \mathbb{R}$. On the other hand, the extended description parallels that of a conventional Hamiltonian with no *explicit* time dependence, $H(\boldsymbol{q}, \boldsymbol{p}) = e_0$, where the system's initial energy e_0 embodies a *constant of motion*. In that case, the system's motion again takes place on a *hyper-surface* that is now defined by $H(\boldsymbol{q}, \boldsymbol{p}) = e_0$ within the cotangent bundle $T^*\mathbb{M}$ over the configuration manifold $\mathbb{M}$.

3. Example: Relativistic particle in an external potential V

An example for a *non-trivial* extended Hamiltonian is furnished by H_e of a relativistic particle in an external potential,

$$H_\mathrm{e}(\boldsymbol{q}, \boldsymbol{p}, t, e) = \frac{1}{2m}\left[\boldsymbol{p}^2 - \left(\frac{e - V(\boldsymbol{q}, t)}{c}\right)^2\right] + \tfrac{1}{2}mc^2. \tag{9}$$

Due to the constraint $H_\mathrm{e} = 0$ from Eq. (7), we can solve Eq. (9) for e to find the equivalent *conventional* Hamiltonian H as the right-hand side of the equation $e = H$,

$$e = \sqrt{\boldsymbol{p}^2 c^2 + m^2 c^4} + V(\boldsymbol{q}, t) = H(\boldsymbol{q}, \boldsymbol{p}, t). \tag{10}$$

The conventional Hamiltonian H corresponding to H_e from Eq. (9) is no longer a *quadratic form* in the canonical momenta. To derive the corresponding quantum equation, the *canonical quantization rules*

$$p_\mu \mapsto \hat{p}_\mu = -i\hbar\frac{\partial}{\partial q^\mu}, \qquad e \mapsto \hat{e} = i\hbar\frac{\partial}{\partial t}, \qquad q^\mu \mapsto \hat{q}^\mu = q^\mu \mathbb{1}, \qquad t \mapsto \hat{t} = t\mathbb{1}$$

may thus be applied to the extended Hamiltonian $H_\mathrm{e} = 0$ only, which here yield the Klein-Gordon equation.

The canonical equation for $\mathrm{d}t/\mathrm{d}s$ is obtained as

$$\frac{\mathrm{d}t}{\mathrm{d}s} = -\frac{\partial H_\mathrm{e}}{\partial e} = \frac{e - V}{mc^2} = \frac{\sqrt{\boldsymbol{p}^2 c^2 + m^2 c^4}}{mc^2} = \sqrt{1 + \left(\frac{\boldsymbol{p}}{mc}\right)^2} = \gamma.$$

Thus, if t quantifies the laboratory time, then s measures the particle's *proper time*. We easily convince ourselves that the other three canonical equations emerging from H_e according to Eqs. (8) coincide with the conventional canonical equations emerging from H according to Eqs. (4). Thus, H_e from Eq. (9) and H from Eq. (10) indeed describe *the same physical system*.

4. Extended canonical transformations

As usual, the general condition for a transformation to be *canonical* is to preserve the *form* of the action functional. In the extended description, where time $q^0 \equiv ct$ and $p_0 \equiv -e/c$ are canonical conjugate dynamical variables, this means that now the form of the *extended* action principle from Eq. (5) must be preserved, hence

$$\delta \int_{s_1}^{s_2} \left[\sum_{\alpha=0}^{n} p_\alpha \frac{\mathrm{d}q^\alpha}{\mathrm{d}s} - H_\mathrm{e} \right] \mathrm{d}s = \delta \int_{s_1}^{s_2} \left[\sum_{\alpha=0}^{n} P_\alpha \frac{\mathrm{d}Q^\alpha}{\mathrm{d}s} - H_\mathrm{e}' \right] \mathrm{d}s.$$

For this requirement to hold, the *integrands* may differ at most by the total derivative $\mathrm{d}F_1/\mathrm{d}s$ of a function $F_1(\boldsymbol{q}, \boldsymbol{Q}, t, T)$, with $q^0 \equiv ct, Q^0 \equiv cT$. Comparing the coefficients of the derivatives in the action functionals with

$$\frac{\mathrm{d}F_1}{\mathrm{d}s} = \sum_{\alpha=0}^{n} \left(\frac{\partial F_1}{\partial q^\alpha} \frac{\mathrm{d}q^\alpha}{\mathrm{d}s} + \frac{\partial F_1}{\partial Q^\alpha} \frac{\mathrm{d}Q^\alpha}{\mathrm{d}s} \right),$$

we find the canonical transformation rules for an extended generating function of type $F_1(\boldsymbol{q}, \boldsymbol{Q}, t, T)$,

$$p_i = \frac{\partial F_1}{\partial q^i}, \qquad P_i = -\frac{\partial F_1}{\partial Q^i}, \qquad e = -\frac{\partial F_1}{\partial t}, \qquad E = \frac{\partial F_1}{\partial T}, \qquad H_\mathrm{e}' = H_\mathrm{e}.$$

The value of the extended Hamiltonian H_e is thus conserved under extended canonical transformations, which means that the physical motion is kept being confined to the surface $H_\mathrm{e}' = 0$. Of course, the functional dependence on the respective set of canonical variables will be different for H_e and H_e', in general.

The Legendre transformation

$$F_2(\boldsymbol{q}, \boldsymbol{P}, t, E) = F_1(\boldsymbol{q}, \boldsymbol{Q}, t, T) + \sum_{i=1}^{n} Q^i P_i - TE$$

yields an *equivalent*, more useful set of canonical transformation rules

$$p_i = \frac{\partial F_2}{\partial q^i}, \qquad Q^i = \frac{\partial F_2}{\partial P_i}, \qquad e = -\frac{\partial F_2}{\partial t}, \qquad T = -\frac{\partial F_2}{\partial E}, \qquad H_\mathrm{e}' = H_\mathrm{e}. \tag{11}$$

According to Eq. (6), the transformation rule $H_\mathrm{e}' = H_\mathrm{e}$ for the extended Hamiltonians can be expressed in terms of conventional Hamiltonians as

$$\Big[H'(\boldsymbol{Q}, \boldsymbol{P}, T) - E \Big] \frac{\mathrm{d}T}{\mathrm{d}s} = \Big[H(\boldsymbol{q}, \boldsymbol{p}, t) - e \Big] \frac{\mathrm{d}t}{\mathrm{d}s}.$$

Eliminating the evolution parameter s, we arrive at the following two equivalent transformation rules for the conventional Hamiltonians under extended canonical transformations

$$\begin{aligned} \Big[H'(\boldsymbol{Q}, \boldsymbol{P}, T) - E \Big] \frac{\partial T}{\partial t} &= H(\boldsymbol{q}, \boldsymbol{p}, t) - e \\ \Big[H(\boldsymbol{q}, \boldsymbol{p}, t) - e \Big] \frac{\partial t}{\partial T} &= H'(\boldsymbol{Q}, \boldsymbol{P}, T) - E. \end{aligned} \tag{12}$$

The transformation rules are generalizations of the rule for conventional canonical transformations as cases with $T \neq t$ are now included.

4.1. Extended generating function of a conventional canonical transformation

An important example of an extended generating function is the particular F_2 that defines a *conventional* canonical transformation. Consider the particular extended generating function

$$F_2(\boldsymbol{q}, \boldsymbol{P}, t, E) = f_2(\boldsymbol{q}, \boldsymbol{P}, t) - tE, \tag{13}$$

with $f_2(\boldsymbol{q}, \boldsymbol{P}, t)$ denoting a conventional generating function. The coordinate transformation rules (11) for this F_2 follow as

$$p_i = \frac{\partial f_2}{\partial q^i}, \qquad Q^i = \frac{\partial f_2}{\partial P_i}, \qquad e = -\frac{\partial f_2}{\partial t} + E, \qquad T = t.$$

Since $\partial T/\partial t = 1$, the extended transformation rule for conventional Hamiltonians from Eq. (12) simplifies to

$$H' - E = H - e \qquad \Longrightarrow \qquad H'(\boldsymbol{Q}, \boldsymbol{P}, t) = H(\boldsymbol{q}, \boldsymbol{p}, t) + \frac{\partial f_2}{\partial t}.$$

The partial derivatives of f_2 obviously yield the usual conventional canonical transformation rules. The particular extended generating function F_2 from Eq. (13) thus defines the conventional canonical transformation generated by f_2. We conclude that the group of conventional canonical transformations establishes a *subgroup* of the group of extended canonical transformations.

5. Generalized Noether theorem

We are now prepared to derive the generalized Noether theorem in the Hamiltonian formalism on the basis of an extended infinitesimal canonical transformation. The extended generating function of an *infinitesimal* canonical transformation is

$$F_2(\boldsymbol{q}, \boldsymbol{P}, t, E) = \sum_{i=1}^{n} q^i P_i - tE + \delta\epsilon\, I(\boldsymbol{q}, \boldsymbol{p}, t, e), \tag{14}$$

with $\delta\epsilon \neq 0$ a small parameter and $I(\boldsymbol{q}, \boldsymbol{p}, t, e)$ a function of the set of extended phase-space variables. The subsequent transformation rules (11) are

$$\begin{aligned} p_i &= \frac{\partial F_2}{\partial q^i} = P_i + \delta\epsilon \frac{\partial I}{\partial q^i}, & e &= -\frac{\partial F_2}{\partial t} = E - \delta\epsilon \frac{\partial I}{\partial t} \\ Q^i &= \frac{\partial F_2}{\partial P_i} = q^i + \delta\epsilon \frac{\partial I}{\partial P_i}, & T &= -\frac{\partial F_2}{\partial E} = t - \delta\epsilon \frac{\partial I}{\partial E} \\ H_\mathrm{e}' &= H_\mathrm{e}. \end{aligned}$$

To *first order* in $\delta\epsilon$, the variations δp_i, δq^i, δe, δt, and δH_e follow as

$$\begin{aligned} \delta p_i \equiv P_i - p_i &= -\delta\epsilon \frac{\partial I}{\partial q^i}, & \delta e \equiv E - e &= \delta\epsilon \frac{\partial I}{\partial t} \\ \delta q^i \equiv Q^i - q^i &= \delta\epsilon \frac{\partial I}{\partial p_i}, & \delta t \equiv T - t &= -\delta\epsilon \frac{\partial I}{\partial e} \\ \delta H_\mathrm{e} \equiv H_\mathrm{e}' - H_\mathrm{e} &= 0. \end{aligned} \tag{15}$$

The transformation rule $H_\mathrm{e}' = H_\mathrm{e}$ for the extended Hamiltonian ensures that the constraint $H_\mathrm{e} = 0$ is maintained in the transformed system, hence that $H_\mathrm{e}' = 0$. On the other hand, the variation of H_e due to variations of the canonical variables is

$$\delta H_\mathrm{e} = \sum_{i=1}^{n} \left(\frac{\partial H_\mathrm{e}}{\partial q^i} \delta q^i + \frac{\partial H_\mathrm{e}}{\partial p_i} \delta p_i \right) + \frac{\partial H_\mathrm{e}}{\partial t} \delta t + \frac{\partial H_\mathrm{e}}{\partial e} \delta e.$$

Inserting the variations from the transformation rules (15), we must make sure that the requirement $\delta H_\mathrm{e} = 0$ actually holds in order for the transformation to be *canonical*,

$$\begin{aligned}\delta H_\mathrm{e} &= \delta\epsilon\left[\sum_{i=1}^{n}\left(\frac{\partial H_\mathrm{e}}{\partial q^i}\frac{\partial I}{\partial p_i} - \frac{\partial H_\mathrm{e}}{\partial p_i}\frac{\partial I}{\partial q^i}\right) - \frac{\partial H_\mathrm{e}}{\partial t}\frac{\partial I}{\partial e} + \frac{\partial H_\mathrm{e}}{\partial e}\frac{\partial I}{\partial t}\right] \\ &= \delta\epsilon\left[H_\mathrm{e}, I\right]_\mathrm{ext} \stackrel{!}{=} 0. \end{aligned} \tag{16}$$

Herein $[H_\mathrm{e}, I]_\mathrm{ext}$ defines the extended Poisson bracket. Thus, the requirement $\delta H_\mathrm{e} = 0$ from Eqs. (15) for a transformation to be *canonical* is satisfied if and only if the function $I(\boldsymbol{q}, \boldsymbol{p}, t, e)$ in the generating function "commutes" with the system's extended Hamiltonian H_e. Along the system's phase-space trajectory, the canonical equations (8) apply, hence

$$\begin{aligned}\delta H_\mathrm{e} &= \delta\epsilon\left[\sum_{i=1}^{n}\left(-\frac{\mathrm{d}p_i}{\mathrm{d}s}\frac{\partial I}{\partial p_i} - \frac{\mathrm{d}q^i}{\mathrm{d}s}\frac{\partial I}{\partial q^i}\right) - \frac{\mathrm{d}e}{\mathrm{d}s}\frac{\partial I}{\partial e} - \frac{\mathrm{d}t}{\mathrm{d}s}\frac{\partial I}{\partial t}\right] \\ &= -\delta\epsilon\frac{\mathrm{d}I}{\mathrm{d}s} \stackrel{!}{=} 0. \end{aligned}$$

We can now express the generalized Noether theorem and its inverse in the extended Hamiltonian formalism as:

Theorem 1 (generalized Noether) *The characteristic function $I(\boldsymbol{q}, \boldsymbol{p}, t, e)$ in the extended generating function F_2 from Eq. (14) must be a* constant of motion *in order to define a canonical transformation. The subsequent transformation rules (15) then comprise an infinitesimal one-parameter symmetry transformation that preserves the form of the action functional (5).*

Conversely, if a one-parameter symmetry transformation is known to preserve the form of the extended action functional (5), then the transformation is canonical, *and hence can be derived from a generating function. The characteristic function $I(\boldsymbol{q}, \boldsymbol{p}, t, e)$ in the corresponding* infinitesimal *generating function (14) then represents a constant of motion.*

We may reformulate the generalized Noether theorem in terms of a conventional Hamiltonian H with the time t the independent variable.

From the correlation (6) of extended and conventional Hamiltonians, one finds

$$\frac{\partial H_\mathrm{e}}{\partial t} = \frac{\partial H}{\partial t}\frac{\mathrm{d}t}{\mathrm{d}s}, \qquad \frac{\partial H_\mathrm{e}}{\partial e} = -\frac{\mathrm{d}t}{\mathrm{d}s}, \qquad \frac{\partial H_\mathrm{e}}{\partial q^i} = \frac{\partial H}{\partial q^i}\frac{\mathrm{d}t}{\mathrm{d}s}, \qquad \frac{\partial H_\mathrm{e}}{\partial p_i} = \frac{\partial H}{\partial p_i}\frac{\mathrm{d}t}{\mathrm{d}s}.$$

In terms of a conventional Hamiltonian H, the commutation condition from Eq. (16) for $\delta H_\mathrm{e} = 0$ is converted into

$$\frac{\partial I}{\partial t} + \frac{\partial I}{\partial e}\frac{\partial H}{\partial t} + \sum_{i=1}^{n}\left(\frac{\partial I}{\partial q^i}\frac{\partial H}{\partial p_i} - \frac{\partial I}{\partial p_i}\frac{\partial H}{\partial q^i}\right) = 0.$$

Due to the conventional canonical equations (4) this is equivalent to

$$\frac{\mathrm{d}}{\mathrm{d}t}I(\boldsymbol{q}, \boldsymbol{p}, t, e) = 0. \tag{17}$$

The infinitesimal symmetry transformation rules (15) that are associated with an invariant I are

$$\delta p_i = -\delta\epsilon\frac{\partial I}{\partial q^i}, \qquad \delta q^i = \delta\epsilon\frac{\partial I}{\partial p_i}, \qquad \delta e = \delta\epsilon\frac{\partial I}{\partial t}, \qquad \delta t = -\delta\epsilon\frac{\partial I}{\partial e}. \tag{18}$$

The condition (17) in conjunction with the one-parameter infinitesimal symmetry transformation (18) comprises the mathematical kernel of the generalized Noether theorem in the realm of point dynamics.

Journal of Physics: Conference Series **380** (2012) 012007 doi:10.1088/1742-6596/380/1/012007

6. Examples: Kepler system invariants and their symmetries

6.1. Rotational symmetry and angular momentum conservation

The classical Kepler system is a two-body problem with the mutual interaction following an inverse square force law. In the frame of the reference body, the Cartesian coordinates q_1, q_2 of its counterpart may be described in the plane of motion by

$$\ddot{q}_i + \mu(t)\frac{q_i}{\sqrt{\left(q_1^2+q_2^2\right)^3}} = 0, \qquad i = 1, 2, \tag{19}$$

with $\mu(t) = G\big[m_1(t) + m_2(t)\big]$ the possibly time-dependent gravitational coupling strength that is induced by possibly time-dependent masses $m_1(t)$ and $m_2(t)$ of the interacting bodies. We may regard the equation of motion (19) to originate from the Hamiltonian

$$H(\boldsymbol{q}, \boldsymbol{p}, t) = \tfrac{1}{2}p_1^2 + \tfrac{1}{2}p_2^2 + V(\boldsymbol{q}, t) \tag{20}$$

containing the interaction potential

$$V(\boldsymbol{q}, t) = -\frac{\mu(t)}{\sqrt{q_1^2+q_2^2}} = -\frac{\mu(t)}{r}.$$

As the potential spatially depends on the distance $r = \sqrt{q_1^2+q_2^2}$ only, it is obviously invariant with respect to rotations in configuration space, (q_1, q_2)

$$\begin{pmatrix} Q_1 \\ Q_2 \end{pmatrix} = \begin{pmatrix} \cos\epsilon & \sin\epsilon \\ -\sin\epsilon & \cos\epsilon \end{pmatrix} \begin{pmatrix} q_1 \\ q_2 \end{pmatrix}, \tag{21}$$

with the parameter ϵ denoting the counterclockwise rotation angle. As the transformation depends on the parameter ϵ only and not on the canonical coordinates, we refer to it as a *global* symmetry transformation.

This symmetry is maintained if we choose $\epsilon \equiv \delta\epsilon$ to be very small. Then $\cos\delta\epsilon \approx 1$, $\sin\delta\epsilon \approx \delta\epsilon$, and the infinitesimal rules are

$$\delta q_1 \equiv Q_1 - q_1 = \delta\epsilon\, q_2, \qquad \delta q_2 \equiv Q_2 - q_2 = -\delta\epsilon\, q_1.$$

These rules can be regarded as being derived from the generating function of the *infinitesimal* canonical transformation

$$F_2(q_1, q_2, P_1, P_2, t, E) = -tE + q_1 P_1 + q_2 P_2 + \delta\epsilon\,(p_1 q_2 - p_2 q_1). \tag{22}$$

According to Noether's theorem, the expression proportional to the parameter $\delta\epsilon$ must be a constant of motion in order for F_2 to define a canonical transformation, and hence to preserve the physical system. Thus

$$I = p_1 q_2 - p_2 q_1, \qquad \frac{\mathrm{d}I}{\mathrm{d}t} = 0,$$

which establishes the well-known *conservation law of angular momentum* in — possibly time-dependent — central-force fields. With a given symmetry transformation, we have thus determined the corresponding constant of motion.

As with any generating function of a canonical transformation, we can derive from the generating function (22) the rules of both the configuration space coordinates and the respective canonical momenta. In matrix form, the infinitesimal rules for the momenta are

$$\begin{pmatrix} P_1 \\ P_2 \end{pmatrix} = \left[\mathbb{1} + \mathbb{A}_{\delta\epsilon}\right] \begin{pmatrix} p_1 \\ p_2 \end{pmatrix}, \qquad \mathbb{A}_{\delta\epsilon} = \delta\epsilon \begin{pmatrix} 0 & 1 \\ -1 & 0 \end{pmatrix},$$

with $\mathbb{1}$ denoting the 2×2 unit matrix. The corresponding *finite* transformation is then

$$\begin{pmatrix} P_1 \\ P_2 \end{pmatrix} = \exp\left(\mathbb{A}_\epsilon\right) \begin{pmatrix} p_1 \\ p_2 \end{pmatrix}, \qquad \exp\left(\mathbb{A}_\epsilon\right) = \begin{pmatrix} \cos\epsilon & \sin\epsilon \\ -\sin\epsilon & \cos\epsilon \end{pmatrix},$$

which coincides with the transformation of the configuration space variables from Eq. (21). This reflects the fact that the Hamiltonian (20) is equally invariant under rotations in momentum space.

6.2. Symmetry transformations associated with the Runge-Lenz invariant

In the Hamiltonian formulation, the converse is also true: if I denotes a constant of motion of a dynamical system, then the associated infinitesimal symmetry transformation is given by the canonical transformation rules emerging from the one-parameter generating function (14). As an example, a particularly transparent representation of the symmetry transformation that corresponds to the Runge-Lenz invariant of the time-independent case of the Kepler system (20) is derived in the following by admitting a symmetry transformation $(q_1, q_2)|_t \mapsto (Q_1, Q_2)|_{t+\delta t}$ that is associated with a *time shift* δt.

For constant μ, hence constant masses m_1 and m_2 of the interacting bodies, one component of the constant Runge-Lenz vector is expressed in terms of canonicals variables as

$$I(q_1, q_2, p_1, p_2) = -q_1 p_2^2 + q_2 p_1 p_2 + \mu \frac{q_1}{\sqrt{q_1^2 + q_2^2}}. \tag{23}$$

Inserting directly the invariant (23) as the characteristic function I into the infinitesimal generating function (14), the subsequent canonical transformation rules (18) then define the — rather intricate — corresponding infinitesimal symmetry transformation that preserves the action functional (5). As I in the form of Eq. (23) does not depend on e, we have $\delta t = 0$, hence *no time shift* is associated with the symmetry transformation.

A more conspicuous representation of the symmetry transformation emerges if we express the invariant I in extended phase-space variables. With the canonical variable e being defined as the *value* of H, hence

$$e = \tfrac{1}{2}p_1^2 + \tfrac{1}{2}p_2^2 - \frac{\mu}{\sqrt{q_1^2 + q_2^2}},$$

we can replace the square root term in I from Eq. (23). The invariant then acquires the *equivalent* form

$$I(q_1, q_2, p_1, p_2, e) = \tfrac{1}{2}q_1 p_1^2 + q_2 p_1 p_2 - \tfrac{1}{2}q_1 p_2^2 - q_1 e. \tag{24}$$

In contrast to the conventional symmetry analysis (cf, for instance, Ref. [6], p. 121), the invariant I now depends on the canonical energy variable, e, which entails a representation of the symmetry transformation with $\delta t \neq 0$. Thus, the one-parameter symmetry transformation is now associated with both, a shift of the p_i, q_i *and* a shift of time t. Explicitly, the infinitesimal transformation rules (18) associated with I from Eq. (24) are

$$\begin{aligned} \delta p_1 &= \delta\epsilon\left(\tfrac{1}{2}p_2^2 - \tfrac{1}{2}p_1^2 + e\right), & \delta p_2 &= \delta\epsilon\, p_1 p_2 \\ \delta q_1 &= \delta\epsilon\left(q_1 p_1 + q_2 p_2\right), & \delta q_2 &= \delta\epsilon\left(p_1 q_2 - p_2 q_1\right) \\ \delta e &= 0, & \delta t &= \delta\epsilon\, q_1. \end{aligned}$$

The transformation rules for the new configuration space Q_1, Q_2 variables depend *linearly* on the original ones, q_1, q_2. We may thus rewrite the infinitesimal configuration space transformation $Q_i = q_i + \delta q_i$, $i = 1, 2$ in matrix form as

$$\left.\begin{pmatrix} Q_1 \\ Q_2 \end{pmatrix}\right|_T = \left[\mathbb{1} + \mathbb{A}_{\delta\epsilon}\right] \left.\begin{pmatrix} q_1 \\ q_2 \end{pmatrix}\right|_t, \qquad \mathbb{A}_{\delta\epsilon} = \delta\epsilon \left.\begin{pmatrix} p_1 & p_2 \\ -p_2 & p_1 \end{pmatrix}\right|_t,$$

wherein $\mathbb{1}$ denotes the 2×2 unit matrix. With $\delta\epsilon$ still an *infinitesimal* quantity, this transformation can be written equivalently in terms of the matrix exponential $\exp(\mathbb{A}_{\delta\epsilon})$,

$$\begin{pmatrix} Q_1 \\ Q_2 \end{pmatrix}\Bigg|_{t+\delta t} = e^{\delta\phi} \begin{pmatrix} \cos\delta\psi & \sin\delta\psi \\ -\sin\delta\psi & \cos\delta\psi \end{pmatrix} \begin{pmatrix} q_1 \\ q_2 \end{pmatrix}\Bigg|_{t},$$

with $\delta t = q_1\delta\epsilon$ and $\delta\phi = p_1\delta\epsilon$, $\delta\psi = p_2\delta\epsilon$. The system symmetry that corresponds to the Runge-Lenz invariant is thus that the configuration space variables Q_i at time $T = t + \delta t$ are correlated with the q_i at time t by a *local scaled rotation.* The infinitesimal transformation depends on the actual system coordinates. It is, therefore, referred to as a *local* symmetry transformation.

7. Conclusions

Parameterizing the action functional $\Phi(\gamma)$ in terms of a "system evolution parameter", s, enables us to put the space-time variables $t = t(s)$ and $q^i = q^i(s)$ on equal footing. The generalization of the Hamiltonian description of dynamics completely retains its canonical form: the conventional set of canonical equations follows from the extended set for the particular extended Hamiltonian $H_\mathrm{e} \equiv H - e = 0$, whereas conventional canonical transformations simply constitute a subgroup of extended canonical transformations. In the context of the extended Hamiltonian formalism, we can define canonical transformations that additionally map the *time scales* of source and target systems. This sets the stage for deriving the generalized Noether theorem by establishing the connection of a system's constant of motion with a corresponding canonical transformation that may include a transformation of time. The constant of motion enters as the characteristic function I into the generating function F_2 of an extended infinitesimal canonical transformation. The subsequent set of canonical transformation rules emerging from this generating function F_2 then establishes the pertaining infinitesimal one-parameter symmetry transformation that preserves the action functional. As this correlation holds for *all* system invariants I, the distinction between Noether- and non-Noether invariants is obsolete. We thus encounter the *most general form* of Noether's theorem in the realm of Hamiltonian point dynamics.

References

[1] Noether E 1918 *Nachr. Ges. Wiss. Göttingen, Math.-Phys. Kl.* **57** 235
[2] José J V and Saletan E J 1998 *Classical Dynamics* (Cambridge: Cambridge University Press)
[3] Greiner W 2010 *Classical Mechanics* (Berlin, Heidelberg: Springer Verlag)
[4] Struckmeier J 2005 *J. Phys. A: Math. Gen.* **38** 1257
[5] Struckmeier J 2009 *Int. J. Mod. Phys.* E **18** 79
[6] Stephani H 1989 *Differential equations — Their solutions using symmetries* (Cambridge: Cambridge University Press)

Symmetries in Science XV IOP Publishing
Journal of Physics: Conference Series **380** (2012) 012008 doi:10.1088/1742-6596/380/1/012008

From Lagrangian to Quantum Mechanics with Symmetries

M C Nucci

Dipartimento di Matematica e Informatica, Università degli Studi di Perugia & INFN Sezione Perugia, 06123 Perugia, Italy

E-mail: nucci@unipg.it

Abstract. We present an old and regretfully forgotten method by Jacobi which allows one to find many Lagrangians of simple classical models and also of nonconservative systems. We underline that the knowledge of Lie symmetries generates Jacobi last multipliers and each of the latter yields a Lagrangian. Then it is shown that Noether's theorem can identify among those Lagrangians the physical Lagrangian(s) that will successfully lead to quantization. The preservation of the Noether symmetries as Lie symmetries of the corresponding Schrödinger equation is the key that takes classical mechanics into quantum mechanics. Some examples are presented.

1. Introduction

The inverse problem of calculus of variation has attracted a lot of interest since in the second half of the XVIII century Euler [8] and then Lagrange [17] introduced the direct problem, namely the idea of linking the solution of a differential equation to the maximum/minimum of a functional, the celebrated problem of the brachistochrone being indeed the most famous classical example. It will take hundred of pages to cite all the papers and books that have been published since up to date. Most authors mark the birthdate of the inverse problem with the 1887-papers by either Helmholtz [11] or Volterra [33]. Some other especially among the Russian speaking researchers pushes the date slightly back to the 1886-paper by Sonin [32]. Very few recognize the seminal work by Jacobi, namely his 1845-paper [14] and his 1842-1843 Dynamics Lectures published posthumously in 1884 [15], available in English since 2009 [16], where he links his last multiplier to the Lagrangian for any even-order ordinary differential equation (ODE). Actually both Volterra and Sonin recognize the contribution of Jacobi last multiplier in their papers, Sonin more explicitly than Volterra since he showed that his own method involves the Jacobi last multiplier (p.10 in [32]).

The method of Jacobi last multiplier was enhanced when Lie determined the link with his symmetries [18], a link very easy to implement that allows to derive many multipliers and therefore Lagrangians.

It is known that a single second-order ODE admits different many Lagrangians [34], but so far there is not a method that can discern the physical Lagrangian among them, although some were proposed, e.g. [6]. The same is true for systems of second-order ODEs that admit more that one Lagrangian. It has been shown that some systems of second-order ODEs do not admit a Lagrangian [7] although in [27] the Lagrangian of some of those systems were determined by following Bateman's statement [2], namely *finding a set of equations equal in number to a given set, compatible with it and derivable from a variational principle* without recourse to any additional set of equations.

We propose that the physical Lagrangian should be the one that admits the highest possible number of Noether symmetries [26], [23], [24]. It was proven in [10] that the maximal dimension of the Lie symmetry algebra of a system of n ODEs of second order is $n^2 + 4n + 3$, and the highest number of corresponding Noether symmetries is $(n^2 + 3n + 6)/2$. In particular for a single second-order ODE the highest number of Noether symmetries is five [20].

Consequently we conjecture that the passage from a classical system to its quantum analogue should preserve exactly those Noether symmetries, namely the Noether symmetries of the physical Lagrangian shall become the Lie symmetries of the corresponding Schrödinger equation [26], [23], [24].

In this paper after recalling the method of Jacobi last multiplier and its link to Lie symmetries, we present the simple example of the one-dimensional free particle: ten inequivalent[1] Lagrangians are presented, their Noether symmetries identify two Lagrangians that admit the highest number of Noether symmetry: one independent and one dependent on time. Then the corresponding Schrödinger equations are obtained. Also the case of the nonlinear second-order Riccati equation is illustrated and its corresponding Schrödinger equation obtained.

2. Jacobi last multiplier

The method of the Jacobi Last Multiplier [12], [13], [14], [15] provides a means to determine all the solutions of the partial differential equation

$$\mathcal{A}f = \sum_{i=1}^{n} a_i(x_1, \ldots, x_n)\frac{\partial f}{\partial x_i} = 0 \tag{1}$$

or its equivalent associated Lagrange system

$$\frac{\mathrm{d}x_1}{a_1} = \frac{\mathrm{d}x_2}{a_2} = \ldots = \frac{\mathrm{d}x_n}{a_n}. \tag{2}$$

In fact, if one knows the Jacobi Last Multiplier and all but one of the solutions, then the last solution can be obtained by a quadrature. The Jacobi Last Multiplier M is given by

$$\frac{\partial(f, \omega_1, \omega_2, \ldots, \omega_{n-1})}{\partial(x_1, x_2, \ldots, x_n)} = M\mathcal{A}f, \tag{3}$$

where

$$\frac{\partial(f, \omega_1, \omega_2, \ldots, \omega_{n-1})}{\partial(x_1, x_2, \ldots, x_n)} = \det\begin{bmatrix} \dfrac{\partial f}{\partial x_1} & \cdots & \dfrac{\partial f}{\partial x_n} \\ \dfrac{\partial \omega_1}{\partial x_1} & & \dfrac{\partial \omega_1}{\partial x_n} \\ \vdots & & \vdots \\ \dfrac{\partial \omega_{n-1}}{\partial x_1} & \cdots & \dfrac{\partial \omega_{n-1}}{\partial x_n} \end{bmatrix} = 0 \tag{4}$$

and $\omega_1, \ldots, \omega_{n-1}$ are $n-1$ solutions of (1) or, equivalently, first integrals of (2) independent of each other. This means that M is a function of the variables $(x_1, \ldots, x_n)$ and depends on the chosen $n-1$ solutions, in the sense that it varies as they vary. The essential properties of the Jacobi Last Multiplier are:

[1] Namely they do not differ by a total derivative.

(a) If one selects a different set of $n-1$ independent solutions $\eta_1, \ldots, \eta_{n-1}$ of equation (1), then the corresponding Last Multiplier N is linked to M by the relationship:

$$N = M\frac{\partial(\eta_1, \ldots, \eta_{n-1})}{\partial(\omega_1, \ldots, \omega_{n-1})}.$$

(b) Given a non-singular transformation of variables

$$\tau: \quad (x_1, x_2, \ldots, x_n) \longrightarrow (x'_1, x'_2, \ldots, x'_n),$$

then the Last Multiplier M' of $\mathcal{A}'F = 0$ is given by:

$$M' = M\frac{\partial(x_1, x_2, \ldots, x_n)}{\partial(x'_1, x'_2, \ldots, x'_n)},$$

where M obviously comes from the $n-1$ solutions of $\mathcal{A}F = 0$ which correspond to those chosen for $\mathcal{A}'F = 0$ through the inverse transformation τ^{-1}.

(c) One can prove that each multiplier M is a solution of the following linear partial differential equation:

$$\sum_{i=1}^{n} \frac{\partial(Ma_i)}{\partial x_i} = 0, \tag{5}$$

or equivalently:

$$\frac{\mathrm{d}}{\mathrm{d}t}(\log M) + \sum_{i=1}^{n} \frac{\partial a_i}{\partial x_i} = 0; \tag{6}$$

viceversa every solution M of this equation is a Jacobi Last Multiplier.

(d) If one knows two Jacobi Last Multipliers M_1 and M_2 of equation (1), then their ratio is a solution ω of (1), or, equivalently, a first integral of (2). Naturally the ratio may be quite trivial, namely a constant. Viceversa the product of a multiplier M_1 times any solution ω yields another last multiplier $M_2 = M_1\omega$.

Since the existence of a solution/first integral is consequent upon the existence of symmetry, an alternate formulation in terms of symmetries was provided by Lie [18], [19]. A clear treatment of the formulation in terms of solutions/first integrals and symmetries is given by Bianchi [3]. If we know $n-1$ symmetries of (1)/(2), say

$$\Gamma_i = \sum_{j=1}^{n} \xi_{ij}(x_1, \ldots, x_n)\partial_{x_j}, \quad i = 1, n-1, \tag{7}$$

Jacobi's last multiplier is given by $M = \Delta^{-1}$, provided that $\Delta \neq 0$, where

$$\Delta = \det\begin{bmatrix} a_1 & \cdots & a_n \\ \xi_{1,1} & & \xi_{1,n} \\ \vdots & & \vdots \\ \xi_{n-1,1} & \cdots & \xi_{n-1,n} \end{bmatrix}. \tag{8}$$

There is an obvious corollary to the results of Jacobi mentioned above. In the case that there exists a constant multiplier, the determinant is a first integral. This result is potentially very useful in the search for first integrals of systems of ordinary differential equations. In particular, if each component of the vector field of the equation of motion is missing the variable associated with that component, i.e., $\partial a_i/\partial x_i = 0$, the last multiplier is a constant, and any other Jacobi Last Multiplier is a first integral.

Another property of the Jacobi Last Multiplier is its (almost forgotten) relationship with the Lagrangian, $L = L(t, q, \dot{q})$, for any second-order equation

$$\ddot{q} = F(t, q, \dot{q}) \tag{9}$$

i.e. [15] (Lecture 10)[2], [34]

$$M = \frac{\partial^2 L}{\partial \dot{q}^2} \tag{10}$$

where $M = M(t, q, \dot{q})$ satisfies the following equation

$$\frac{\mathrm{d}}{\mathrm{d}t}(\log M) + \frac{\partial F}{\partial \dot{q}} = 0. \tag{11}$$

Then equation (9) becomes the Euler-Lagrange equation:

$$-\frac{\mathrm{d}}{\mathrm{d}t}\left(\frac{\partial L}{\partial \dot{q}}\right) + \frac{\partial L}{\partial q} = 0. \tag{12}$$

The proof is based on taking the derivative of (12) with respect to $\dot{q}$ and showing that this yields (11). If one knows a Jacobi last multiplier, then L can be easily obtained by a double integration, i.e.:

$$L = \int \left(\int M\, \mathrm{d}\dot{q} \right) \mathrm{d}\dot{q} + f_1(t, q)\dot{q} + f_2(t, q), \tag{13}$$

where f_1 and f_2 are functions of t and q which have to satisfy a single partial differential equation related to (9) [25]. As it was shown in [25], f_1, f_2 are related to the gauge function $g = g(t, q)$. In fact, we may assume

$$\begin{aligned} f_1 &= \frac{\partial g}{\partial q} \\ f_2 &= \frac{\partial g}{\partial t} + f_3(t, q) \end{aligned} \tag{14}$$

where f_3 has to satisfy the mentioned partial differential equation and g is obviously arbitrary. We remark the importance of the gauge function in order to apply Noether's theorem [21] correctly. Therefore we do not annihilate the gauge function.

In [25] it was shown that if one knows several (at least two) Lie symmetries of the second-order differential equation (9), i.e.

$$\Gamma_j = V_j(t, q)\partial_t + G_j(t, q)\partial_q, \quad j = 1, r, \tag{15}$$

then many Jacobi Last Multipliers could be derived by means of (8), i.e.

$$\frac{1}{M_{nm}} = \Delta_{nm} = \det \begin{bmatrix} 1 & \dot{q} & F(t, q, \dot{q}) \\ V_n & G_n & \dfrac{\mathrm{d}G_n}{\mathrm{d}t} - \dot{q}\dfrac{\mathrm{d}V_n}{\mathrm{d}t} \\ V_m & G_m & \dfrac{\mathrm{d}G_m}{\mathrm{d}t} - \dot{q}\dfrac{\mathrm{d}V_m}{\mathrm{d}t} \end{bmatrix}, \tag{16}$$

with $(n, m = 1, r)$, and therefore many Lagrangians can be obtained by means of (13).

[2] *Jacobi's Lectures on Dynamics* are finally available in English [16].

3. Lagrangians for the free particle

It is well-known since Lie's seminal work [19] that the equation of a free particle

$$\ddot{q} = 0 \tag{17}$$

admits an eight-dimensional Lie symmetry algebra[3], $sl(3, I\!R)$, generated by the following operators:

$$X_1 = qt\partial_t + q^2\partial_q, \qquad X_2 = q\partial_t, \qquad X_3 = t^2\partial_t + qt\partial_q, \qquad X_4 = q\partial_q,$$
$$X_5 = t\partial_t, \qquad X_6 = \partial_t, \qquad X_7 = t\partial_q, \qquad X_8 = \partial_q. \tag{18}$$

An obvious Jacobi Last multiplier (JLM) of (17) is a constant since $\dot{q}$ does not appear in its right-hand side. This also implies that any JLM is a first integral of (17). We can found this trivial JLM and many others by calculating the inverse of the determinant of the matrix (16) for all the possible combinations of two different operators in (18). It results that ten different JLM and consequently as many Lagrangians, by means of (13), can be obtained[4], i.e.:

$$M_{13} = -\frac{1}{(t\dot{q}-q)^3} \quad\Rightarrow\quad L_{13} = -\frac{1}{2t^2(t\dot{q}-q)} + \frac{\mathrm{d}g}{\mathrm{d}t}(t,q)$$
$$M_{15} = -\frac{1}{\dot{q}(t\dot{q}-q)^2} \quad\Rightarrow\quad L_{15} = \frac{\dot{q}}{q^2}\left(\log(t\dot{q}-q) - \log(\dot{q})\right) + \frac{\mathrm{d}g}{\mathrm{d}t}(t,q)$$
$$M_{16} = \frac{1}{\dot{q}^2(t\dot{q}-q)} \quad\Rightarrow\quad L_{16} = \left(\frac{t\dot{q}}{q^2} - \frac{1}{q}\right)\left(\log(\dot{q}) - \log(t\dot{q}-q)\right) + \frac{\mathrm{d}g}{\mathrm{d}t}(t,q)$$
$$M_{17} = -\frac{1}{(t\dot{q}-q)^2} \quad\Rightarrow\quad L_{17} = -\frac{1}{t^2}\log(t\dot{q}-q) + \frac{\mathrm{d}g}{\mathrm{d}t}(t,q)$$
$$M_{18} = \frac{1}{\dot{q}(t\dot{q}-q)} \quad\Rightarrow\quad L_{18} = -\frac{\dot{q}}{q}\log(\dot{q}) - \left(\frac{1}{t} - \frac{\dot{q}}{q}\right)\log(t\dot{q}-q) + \frac{1}{t}(1+\log(q)) + \frac{\mathrm{d}g}{\mathrm{d}t}(t,q)$$
$$M_{26} = -\frac{1}{\dot{q}^3} \quad\Rightarrow\quad L_{26} = -\frac{1}{2\dot{q}} + \frac{\mathrm{d}g}{\mathrm{d}t}(t,q)$$
$$M_{28} = \frac{1}{\dot{q}^2} \quad\Rightarrow\quad L_{28} = -\log(\dot{q}) + \frac{\mathrm{d}g}{\mathrm{d}t}(t,q)$$
$$M_{38} = \frac{1}{t\dot{q}-q} \quad\Rightarrow\quad L_{38} = \left(\frac{\dot{q}}{t} - \frac{q}{t^2}\right)\left(\log(t\dot{q}-q) - 1\right) + \frac{\mathrm{d}g}{\mathrm{d}t}(t,q)$$
$$M_{48} = -\frac{1}{\dot{q}} \quad\Rightarrow\quad L_{48} = \dot{q}(1-\log(\dot{q})) + \frac{\mathrm{d}g}{\mathrm{d}t}(t,q)$$
$$M_{87} = 1 \quad\Rightarrow\quad L_{87} = \tfrac{1}{2}\dot{q}^2 + \frac{\mathrm{d}g}{\mathrm{d}t}(t,q) \tag{19}$$

The ten Lagrangians are NOT linked by the gauge function $g = g(t,q)$. We note that seven of the matrices (16) have determinant equal to zero, i.e.:

$$\Delta_{12}, \qquad \Delta_{14}, \qquad \Delta_{24}, \qquad \Delta_{35}, \qquad \Delta_{37}, \qquad \Delta_{57}, \qquad \Delta_{68}. \tag{20}$$

[3] Also reported in [9].

[4] Since $M_{nm} = -M_{mn}$ we have arbitrarily chosen the sign as we wished.

This is in agreement with the application of Jacobi's method to the linear harmonic oscillator [25] that yielded fourteen different Lagrangians and three matrices having determinant equal to zero.
Applying Noether's theorem yields that both L_{13} and L_{87} admit five Noether point symmetries, the maximum possible [20] in the case of equation (17). The main difference between the two Lagrangians is that one Lagrangian only is independent on time and is the traditional Lagrangian, namely the kinetic energy L_{87}. Moreover the two Lagrangians admit different Noether point symmetries. In fact Lagrangian L_{13} admits the following five Noether symmetries and corresponding first integrals of equation (17)

$$\begin{aligned} X_1 &\implies Int_1 = -\frac{\dot{q}}{q-t\dot{q}}, \\ X_2 &\implies Int_2 = \frac{\dot{q}^2}{2(q-t\dot{q})^2}, \\ X_3 &\implies Int_3 = -\frac{1}{q-t\dot{q}}, \\ X_4 - X_5 &\implies Int_4 = -\frac{\dot{q}}{(q-t\dot{q})^2}, \\ X_7 &\implies Int_7 = -\frac{1}{2(q-t\dot{q})^2}, \end{aligned} \tag{21}$$

while Lagrangian L_{87} admits the following five Noether symmetries and corresponding first integrals of equation (17)

$$\begin{aligned} X_3 &\implies In_3 = -\tfrac{1}{2}(q-t\dot{q})^2, \\ X_4 + 2X_5 &\implies In_4 = -\dot{q}(q-t\dot{q}), \\ X_6 &\implies In_6 = \tfrac{1}{2}\dot{q}^2, \\ X_7 &\implies In_7 = q - t\dot{q}, \\ X_8 &\implies In_8 = -\dot{q}. \end{aligned} \tag{22}$$

4. Schrödinger equations for the free particle

The Schrödinger equation for the free particle is

$$2i\psi_t + \psi_{xx} = 0. \tag{23}$$

We show that this equation can be obtained by considering a generic linear parabolic equation

$$2i\psi_t + f_1(x)\psi_{xx} + f_2(x)\psi_x + f_3(x)\psi = 0 \tag{24}$$

with f_k, $(k = 1, 3)$ functions of x to be determined in such a way that equation (24) admits the following five Lie symmetries[5]

$$\begin{aligned} X_3 &\Rightarrow \Omega_1 = t^2\partial_t + xt\partial_x + \omega_1\partial_\psi, \\ X_4 + 2X_5 &\Rightarrow \Omega_2 = 2t\partial_t + x\partial_x + \omega_2\partial_\psi, \\ X_6 &\Rightarrow \Omega_3 = \partial_t + \omega_3\partial_\psi, \\ X_7 &\Rightarrow \Omega_4 = t\partial_x + \omega_4\partial_\psi, \\ X_8 &\Rightarrow \Omega_5 = \partial_x + \omega_5\partial_\psi. \end{aligned} \tag{25}$$

[5] We have identified q with x.

where $\omega_i = \omega_i(t, x, \psi), (i = 1, 5)$ are functions of t, x, ψ that have to be determined. Equation (24) also admits the following two symmetries

$$\Omega_6 = \psi\partial_\psi, \qquad \Omega_\alpha = \alpha(t,x)\partial_\psi \tag{26}$$

with α any solution of equation (24) itself, since any linear partial differential equation possesses these two symmetries.

Using the interactive REDUCE programs [22], we obtain that

$$f_1 = 1, \quad f_2 = f_3 = 0, \tag{27}$$

and

$$\omega_1 = \tfrac{1}{2}(ix^2 - t)\psi, \quad \omega_2 = \omega_3 = \omega_5 = 0, \quad \omega_4 = ix\psi. \tag{28}$$

Therefore the Noether symmetries admitted by the Lagrangian L_{87}, namely the physical Lagrangian for the free particle, yield the right Schrödinger's equation (23) and thus the correct quantization procedure is achieved.

We now show that a quantum-correct Schrödinger equation can be derived even from the five Noether symmetries admitted by the time-dependent Lagrangian L_{13}. We consider a generic linear partial differential equation

$$f_{11}(t,x)\psi_{tt} + f_{12}(t,x)\psi_{tx} + f_{22}(t,x)\psi_{xx} + f_1(t,x)\psi_t + f_2(t,x)\psi_x + f_0(t,x)\psi = 0 \tag{29}$$

with $f_{rs}, f_r, (r, s = 1, 2), f_0$ functions of t, x to be determined in such a way that equation (29) admits the following five Lie symmetries[6]

$$\begin{aligned}
X_1 &\Rightarrow W_1 = xt\partial_t + x^2\partial_x + w_1\partial_\psi,\\
X_2 &\Rightarrow W_2 = x\partial_t + w_2\partial_\psi,\\
X_3 &\Rightarrow W_3 = t^2\partial_t + xt\partial_x + w_3\partial_\psi,\\
X_4 - X_5 &\Rightarrow W_4 = -t\partial_t + x\partial_x + w_4\partial_\psi,\\
X_7 &\Rightarrow W_5 = t\partial_x + w_5\partial_\psi.
\end{aligned} \tag{30}$$

where $w_i = w_i(t, x, \psi), (i = 1, 5)$ are functions of t, x, ψ that have to be determined. Equation (29) also admits the two symmetries

$$W_6 = \psi\partial_\psi, \qquad W_\beta = \beta(t,x)\partial_\psi \tag{31}$$

with β any solution of equation (29) itself.

Using the interactive REDUCE programs [22], we obtain that equation (29) becomes

$$4t^2\psi_{tt} + 8tx\psi_{tx} + 4x^2\psi_{xx} + 12t\psi_t + 12x\psi_x + 3\psi = 0 \tag{32}$$

with

$$w_1 = -\tfrac{1}{2}x\psi, \quad w_2 = w_4 = w_5 = 0, \quad w_3 = -\tfrac{1}{2}t\psi. \tag{33}$$

Equation (32) is parabolic and therefore can be put into its normal form since its characteristic coordinates are

$$\xi = \frac{x}{t}, \quad x = x, \quad \psi = \phi(\xi, x). \tag{34}$$

[6] We have identified q with x.

Journal of Physics: Conference Series **380** (2012) 012008 doi:10.1088/1742-6596/380/1/012008

Thus equation (32) transforms into

$$4x^2\phi_{xx} + 12x\phi_x + 3\phi = 0 \tag{35}$$

with solution:

$$\phi(\xi, x) = \alpha_1(\xi)x^{-1/2} + \alpha_2(\xi)x^{-3/2}, \tag{36}$$

where α_1, α_2 are arbitrary functions of ξ. This is obviously in agreement with the quantization of the free particle.

We remark that there is some freedom when imposing equation (29) to admit the Lie symmetries (30) but in any possible cases a parabolic equation with characteristic coordinates (34) is always obtained. For example another possibility is

$$4t^4\psi_{tt} + 8t^2x\psi_{tx} + 4t^2x^2\psi_{xx} + 4t^2(3t+x)\psi_t + 4tx(3t+x)\psi_x + (3t^2+4tx+x^2)\psi = 0 \tag{37}$$

with

$$w_1 = -\tfrac{1}{2}\left(1+\frac{x}{t}\right)x\psi, \quad w_2 = \frac{x^2}{2t^2}\log(t)\psi, \quad w_3 = -\tfrac{1}{2}(t+x)\psi,$$
$$w_4 = \frac{x}{t}\log(t)\psi, \quad w_5 = -\tfrac{1}{2}\log(t)\psi. \tag{38}$$

The normal form of equation (37) is

$$4x^2\phi_{xx} + (12+4\xi)x\phi_x + (3+4\xi+\xi^2)\phi = 0 \tag{39}$$

with solution:

$$\phi(\xi, x) = \alpha_1(\xi)x^{-3/2-\xi/2} + \alpha_2(\xi)x^{-1/2-\xi/2}, \tag{40}$$

also in agreement with the quantization of the free particle.

5. Schrödinger equation for the second-order Riccati equation

In [28] it was shown that the linearizable second-order Riccati equation, a member of the Riccati-chain [1], i.e.

$$\ddot{x} + 3x\dot{x} + x^3 = 0 \tag{41}$$

possesses many JLM and therefore Lagrangians. In particular the following time-independent Lagrangian[7]

$$Lagr = -\frac{1}{2(\dot{x}+x^2)} \tag{42}$$

was shown to admit five Noether point symmetries, i.e.

$$\Gamma_2 - \Gamma_8, \quad \Gamma_3 - \frac{2}{3}\Gamma_7, \quad \Gamma_4, \quad \Gamma_5, \quad \Gamma_6 \tag{43}$$

among the following eight Lie symmetries admitted by equation (41):

$$\begin{aligned}
\Gamma_1 &= t^3(tx-2)\partial_t - t(xt-2)(x^2t^2+2-2xt)\partial_x \\
\Gamma_2 &= xt^3\partial_t - (xt-1)(x^2t^2+4-2xt)\partial_x \\
\Gamma_3 &= xt^2\partial_t - x(x^2t^2+2-2xt)\partial_x \\
\Gamma_4 &= xt\partial_t - x^2(xt-1)\partial_x \\
\Gamma_5 &= x\partial_t - x^3\partial_x \\
\Gamma_6 &= \partial_t \\
\Gamma_7 &= t\partial_t - x\partial_x \\
\Gamma_8 &= t^2\partial_t - 2(xt-1)\partial_x.
\end{aligned} \tag{44}$$

[7] This Lagrangian has also been studied in [5].

Lagrangian (42) was obtained from the inverse of the determinant of the matrix (16) with the two symmetries Γ_5 and Γ_6 and the application of formula (13).

It interesting to show that equation (41) is linked to the following cubic-power dissipative equation

$$\frac{\mathrm{d}^2\tilde{x}}{\mathrm{d}\tilde{t}^2}+\left(\frac{\mathrm{d}\tilde{x}}{\mathrm{d}\tilde{t}}\right)^3=0 \tag{45}$$

by means of the canonical form of the two-dimensional Lie algebra generated by Γ_5 and Γ_6. In fact this Lie algebra corresponds to Type I in Lie's classification of the real two-dimensional Lie algebras in the plane [19], [3], namely it is abelian and transitive[8] and therefore its canonical coordinates are

$$\partial_{\tilde{t}}, \qquad \partial_{\tilde{x}}. \tag{46}$$

Therefore the following identification

$$\Gamma_5=\partial_{\tilde{t}}, \qquad \Gamma_6=\partial_{\tilde{x}} \tag{47}$$

yields the transformation

$$\tilde{t}=\frac{1}{2x^2}, \qquad \tilde{x}=\frac{tx-1}{x} \tag{48}$$

that takes (41) into (45).

Equation (45) is also linearizable and admits an eight-dimensional Lie symmetry algebra. Work is in progress to quantize n-power dissipative equations with the method illustrated here and to compare it to other proposed method, e.g. [29], [30], [31].

We now quantize the second-order Riccati equation (41) by imposing equation (29) to admit the following five Lie symmetries

$$\begin{aligned}
\Gamma_2-\Gamma_8 \quad &\Rightarrow \quad \Lambda_1=(xt-1)t^2\partial_t-(xt-1)(x^2t^2-2-2xt)\partial_x+\lambda_1\partial_\psi,\\
\Gamma_3-\frac{2}{3}\Gamma_7 \quad &\Rightarrow \quad \Lambda_2=\left(xt-\frac{2}{3}\right)t\partial_t-x\left(x^2t^2+\frac{4}{3}-2xt\right)\partial_x+\lambda_2\partial_\psi,\\
\Gamma_4 \quad &\Rightarrow \quad \Lambda_3=xt\partial_t-x^2(xt-1)\partial_x+\lambda_3\partial_\psi,\\
\Gamma_5 \quad &\Rightarrow \quad \Lambda_4=x\partial_t-x^3\partial_x+\lambda_4\partial_\psi,\\
\Gamma_6 \quad &\Rightarrow \quad \Lambda_5=\partial_t+\lambda_5\partial_\psi.
\end{aligned} \tag{49}$$

where $\lambda_i=\lambda_i(t,x,\psi),(i=1,5)$ are functions of t,x,ψ that have to be determined. We remind that equation (29) also admits the two symmetries (31).

Using the interactive REDUCE programs [22], we obtain that equation (29) becomes

$$4\psi_{tt}-8x^2\psi_{tx}+4x^4\psi_{xx}+8x^3\psi_x-3x^2\psi=0 \tag{50}$$

with

$$\begin{aligned}
\lambda_1=-\frac{(tx-1)^3}{2x}\psi, \quad \lambda_2=-\tfrac{1}{2}(tx-1)^2\psi, \quad \lambda_3=-\tfrac{1}{2}(tx-1)x\psi\\
\lambda_4=-\tfrac{1}{2}x^2\psi, \quad \lambda_5=-\frac{tx-1}{x}\psi.
\end{aligned} \tag{51}$$

[8] In [28] a missprint led to the erroneously statement that Γ_5 and Γ_6 generate an intransitive Lie algebra.

Equation (50) is parabolic and therefore can be put into its normal form since its characteristic coordinates are

$$\varrho = t - \frac{1}{x}, \quad x = x, \quad \psi = \phi(\varrho, x). \tag{52}$$

Thus equation (50) transforms into

$$4x^2\phi_{xx} + 8x\phi_x - 3\phi = 0 \tag{53}$$

with solution:

$$\phi(\varrho, x) = \beta_1(\varrho)x^{1/2} + \beta_2(\varrho)x^{-3/2}, \tag{54}$$

where β_1, β_2 are arbitrary functions of ϱ. Is this the correct quantization of equation (41)? It seems so but further insight is needed especially from the experimentalists.

6. Final remarks

In [23] a method was proposed to overcome the deadlock of nonlinear canonical transformations when quantizing with the known procedures [26]. It consists of the following steps to be applied to a classical Lagrangian equation[9]:

- Find the Lie symmetries of the Lagrangian equation

$$\Upsilon = W_0(t,x)\partial_t + W_1(t,x)\partial_x$$

- Among the Lie symmetries find the Noether symmetries admitted by the given Lagrangian

$$\Gamma = V_0(t,x)\partial_t + V_1(t,x)\partial_x, \qquad \Gamma \subset \Upsilon$$

- Construct the Schrödinger equation admitting these Noether symmetries as Lie symmetries

$$2i\psi_t + f_1(x)\psi_{xx} + f_2(x)\psi_x + f_3(x)\psi = 0$$

$$\Omega = V_0(t,x)\partial_t + V_1(t,x)\partial_x + G(t,x,\psi)\partial_\psi$$

- Summarizing: quantize preserving the Noether symmetries

In [24] this method has been applied to classical Lagrangian systems. In particular it led to the Schrödinger equation of a known completely integrable and solvable many-body problem, the so-called 'goldfish' [4]. In the case of Lagrangian systems the method consists of the following steps[10]:

- Find the Lie symmetries of the Lagrangian system

$$\Upsilon = W(t,\underline{x})\partial_t + \sum_{k=1}^{N} W_k(t,\underline{x})\partial_{x_k}$$

- Among the Lie symmetries find the Noether symmetries admitted by the given Lagrangian

$$\Gamma = V(t,\underline{x})\partial_t + \sum_{k=1}^{N} V_k(t,\underline{x})\partial_{x_k}, \quad \Gamma \subset \Upsilon$$

[9] The physical Lagrangian turns out to admit the highest number of Noether point symmetries.
[10] The physical Lagrangian turns out to admit the highest number of Noether point symmetries.

- Construct the Schrödinger equation admitting these Noether symmetries as Lie symmetries

$$2i\psi_t + \sum_{k,j=1}^{N} f_{kj}(\underline{x})\psi_{x_j x_k} + \sum_{k=1}^{N} h_k(\underline{x})\psi_{x_k} + f_3(\underline{x})\psi = 0$$

$$\Omega = V(t,\underline{x})\partial_t + \sum_{k=1}^{N} V_k(t,\underline{x})\partial_{x_k} + G(t,\underline{x},\psi)\partial_\psi$$

- Summarizing: quantize preserving the Noether symmetries

In this paper we have proposed a method that takes classical equations into the quantum realm by means of the Schrödinger equation regardless of the number of Lagrangians that may classically exist. The method consists of the following steps:

- Find the Lie symmetries of the given equation

$$\Upsilon = W_0(t,x)\partial_t + W_1(t,x)\partial_x$$

- Among many Lagrangians identify the one[11] that admits the highest number of Noether symmetries among the Lie symmetries

$$\Gamma = V_0(t,x)\partial_t + V_1(t,x)\partial_x, \qquad \Gamma \subset \Upsilon$$

- Construct a linear partial differential equation admitting these Noether symmetries as Lie symmetries
$f_{11}(t,x)\psi_{tt} + f_{12}(t,x)\psi_{tx} + f_{22}(t,x)\psi_{xx} + f_1(t,x)\psi_t + f_2(t,x)\psi_x + f_0(t,x)\psi = 0$

$$\Omega = V_0(t,x)\partial_t + V_1(t,x)\partial_x + G(t,x,\psi)\partial_\psi$$

- Since this equation is parabolic, determine its canonical coordinates in order to transform it into its normal form, namely the Schrödinger equation
- Summarizing: quantize preserving the Noether symmetries.

References

[1] Ames W F 1968 *Nonlinear Ordinary Differential Equations in Transport Processes* (New York: Academic Press)
[2] Bateman H 1931 On dissipative systems and related variational principles *Phys. Rev.* **38** 815
[3] Bianchi L 1918 *Lezioni sulla teoria dei gruppi continui finiti di trasformazioni* (Pisa: Enrico Spoerri)
[4] Calogero F 2001 *Physica* D **152-153** 78
[5] Cariñena J F, Rañada M F and Santander M 2005 *J. Math. Phys.* **46** 062703
[6] Dodonov V V, Man'ko V I and Skarzhinsky V D 1981 *Hadronic J* **4** 1734
[7] Douglas J 1941 *Trans. Am. Math. Soc.* **50** 71
[8] Euler L 1744 *Methodus inveniendi lineas curvas maximi minimive proprietate gaudentes, sive solutio problematis isoperimetrici latissimo sensu accepti* (Lausanne and Geneva: Bousquet)
[9] González-Gascón F and González-López A 1983 *J. Math. Phys.* **24** 2006
[10] González-López A 1988 *J. Math. Phys.* **29** 1097
[11] Helmholtz H 1887 *J. Reine Angew. Math.* **100** 137
[12] Jacobi C G J 1844 *Giornale Arcadico di Scienze, Lettere ed Arti* **99** 129
[13] Jacobi C G J 1844 *J. Reine Angew. Math.* **27** 199
[14] Jacobi C G J 1845 *J. Reine Angew. Math.* **29** 213 and 333
[15] Jacobi C G J 1884 *Vorlesungen über Dynamik. Nebst fünf hinterlassenen Abhandlungen desselben* ed A Clebsch (Berlin: Druck und Verlag von Georg Reimer)

[11] They could be more than one as we have shown in the present paper.

[16] Jacobi C G J 2009 *Jacobi's Lectures on Dynamics* tr K Balagangadharan, ed Biswarup Banerjee, Industan Book Agency, (Providence: American Mathematical Society)
[17] Lagrange J L 1760-1761 *Miscellanea physico-mathematica societatis Taurinensia* **2** 407
[18] Lie S 1874 *Forhandlinger I Videnskabs-Selskabet I Christiania* 255
[19] Lie S 1912 *Vorlesungen über Differentialgleichungen mit bekannten infinitesimalen Transformationen* (Leipzig: Teubner)
[20] Mahomed F M and Leach P G L 1990 *J. Math. An. Appl.* **151** 80
[21] Noether E 1918 *Königl. Ges. Wiss. Göttingen Nachr. Math.-physik. Kl.* **2** 235
[22] Nucci M C (1996) in: *CRC Handbook of Lie Group Analysis of Differential Equations. Vol. 3: New Trends in Theoretical Developments and Computational Methods* ed N H Ibragimov (Boca Raton: CRC Press) p 415
[23] Nucci M C 2011 *Theor. Math. Phys.* **168** 997
[24] Nucci M C 2011 *Quantizing preserving Noether symmetries* (arXiv:1108.2300)
[25] Nucci M C and Leach P G L 2007 *J. Math. Phys.* **48** 123510
[26] Nucci M C and Leach P G L 2008 *Lie Groups and Quantum Mechanics* (arXiv:0812.1362v1)
[27] Nucci M C and Leach P G L 2011 *Phys. Scripta* **83** 035007
[28] Nucci M C and Tamizhmani K M 2010 *J. Nonlinear Math. Phys.* **17** 167
[29] Schuch D 2007 *J. Math. Phys.* **48** 122701
[30] Schuch D 2010 *J. Phys.: Conf. Ser.* **237** 012020
[31] Smith C E 2010 *J. Phys.: Conf. Ser.* **237** 012021
[32] Sonin N Ya 1886 *Warsaw Univ. Izvestiya* **1–2** 1(in Russian)
[33] Volterra V 1887 *Rend. Acc. Lincei ser.IV* **III** 274
[34] Whittaker E T 1988 *A Treatise on the Analytical Dynamics of Particles and Rigid Bodies* (Cambridge: Cambridge University Press)

Journal of Physics: Conference Series **380** (2012) 012009 doi:10.1088/1742-6596/380/1/012009

Complex Riccati equations as a link between different approaches for the description of dissipative and irreversible systems

Dieter Schuch

Institut für Theoretische Physik, J.W. Goethe-Universität Frankfurt am Main, Max-von-Laue-Str. 1, D-60438 Frankfurt am Main, Germany

E-mail: `Schuch@em.uni-frankfurt.de`

Abstract. Quantum mechanics is essentially described in terms of complex quantities like wave functions. The interesting point is that phase and amplitude of the complex wave function are not independent of each other, but coupled by some kind of conservation law. This coupling exists in time-independent quantum mechanics and has a counterpart in its time-dependent form. It can be traced back to a reformulation of quantum mechanics in terms of nonlinear real Ermakov equations or equivalent complex nonlinear Riccati equations, where the quadratic term in the latter equation explains the origin of the phase-amplitude coupling. Since realistic physical systems are always in contact with some kind of environment this aspect is also taken into account. In this context, different approaches for describing open quantum systems, particularly effective ones, are discussed and compared. Certain kinds of nonlinear modifications of the Schrödinger equation are discussed as well as their interrelations and their relations to linear approaches via non-unitary transformations. The modifications of the aforementioned Ermakov and Riccati equations when environmental effects are included can be determined in the time-dependent case. From formal similarities conclusions can be drawn how the equations of time-independent quantum mechanics can be modified to also incluce the enviromental aspects.

1. Introduction

In his talk "Square root of minus one, complex phase and Erwin Schrödinger" [1] presented 25 years ago at a conference in London celebrating the centenary of the birth of Schrödinger, the Nobel Laureate Chen Ning Yang claimed that with quantum mechanics, for the first time, the imaginary unit enters physics in a fundamental way and "complex numbers became a conceptual element of the very foundation of physics". He continued that the very meaning of the fundamental equations of matrix mechanics (Heisenberg's commutation relation) and of wave mechanics (the time-dependent Schrödinger equation (TDSE)) "would be totally destroyed if one tries to get rid of i by writing them in terms of real and imaginary parts".

I totally agree with this statement but want to go even a step further. The reason it is not sufficient to simply write these complex equations in terms of two real ones will be explained in section 2 using Madelung's hydrodynamic formulation [2] of Schrödinger's theory where the complex wave function is written in polar form. This shows that phase and amplitude (or real and imaginary parts), and hence the corresponding equations of motion, are not independent of each other but uniquely coupled. For the time-independent (TI)SE, it will be shown that this coupling is connected with a conservation law that leads to a nonlinear (NL) formulation of quantum mechanics in terms of a (real) so-called Ermakov equation [3]. This can be rewritten in the form a complex NL Riccati equation where the origin of the coupling becomes obvious. Formally, a similar situation is also found in the case of the TDSE, at least for quadratic Hamiltonians with Gaussian wave packet (WP) solutions.

Since all realistic physical systems are in contact with some kind of environment, which introduces phenomena like dissipation and irreversibility, it will be discussed in section 3 how these aspects can be included in our investigation. Several of the many different approaches to take the environment into account will be mentioned. Our main emphasis will be on models where only the effect of the environment on the system of interest, but not the details of the environmental degrees of freedom, is considered. In this context, modifications of the TDSE leading to explicitly TD Hamiltonians or NL modifications of the SE will be discussed; in particular, a NLSE with complex logarithmic nonlinearity that avoids the shortcomings of other approaches.

In section 4 this logarithmic NLSE will be used to show the modifications of the Ermakov and Riccati equations that were obtained in the TD case when a dissipative environment is taken into account. The formal similarities between the TD and TI situation in the conservative case give a hint as to how the TISE can be modified in order to also include the environmental effects. This leads to a deeper understanding of the connections between several of the NL approaches.

The relations between the NL approaches and those based on explicitly TD Hamiltonians will be shown in section 5. The transition between two of these explicitly TD models, via a unitary transformation, bears similarity with the relation between two of the NL approaches. The connections between the complex Riccati equations that exist in all these approaches will be specified and reduced to one equation that is relevant for a description in physical terms.

Finally, the results are summarized in section 6 where the interrelations between the approaches discussed are also presented schematically and further perspectives given.

2. Phases, amplitudes and complex Riccati equations in time-independent and time-dependent quantum mechanics

In classical mecanics the deterministic particle aspect is expressed by the equations of motion for the trajectory. In the most general, the Hamiltonian form, the integration of these equations becomes trivial, if the Hamiltonian function is canonically transformed into one that is identically zero. The corresponding generating function is the action $S(\mathbf{r}, t)$, leading to the Hamilton–Jacobi equation

$$\frac{\partial}{\partial t} S + H = \frac{\partial}{\partial t} S + \frac{1}{2m} (\nabla S)^2 + V = 0 \tag{1}$$

with the momentum $p = \nabla S$. This equation was Schrödinger's starting point for the "derivation" of his wave equation [4].

It should be mentioned that all the classical equations of motion (i.e., also the Newtonian and Lagrangian form) are (for conservative forces) invariant under time reversal.

On the classical level, also a probabilistic description of a mechanical system exists in terms of a continuity equation for a probability density $\varrho_\Gamma(\mathbf{q}, \mathbf{p}, t)$ in the 6N dimensional phase space (Γ space, for an N-particle system),

$$\frac{\partial}{\partial t} \varrho_\Gamma + \nabla_\Gamma (\varrho_\Gamma \, \mathbf{v}_\Gamma) = 0 \, . \tag{2}$$

with coordinate $\mathbf{q} = (q_1, ..., q_{3N})$, momentum $\mathbf{p} = (p_1, ..., p_{3N})$ and $\nabla_\Gamma = \left(\frac{\partial}{\partial \mathbf{q}}, \frac{\partial}{\partial \mathbf{p}} \right)$, $\mathbf{v}_\Gamma = (\dot{\mathbf{q}}, \dot{\mathbf{p}})$, where overdots denote time-derivatives. For Hamiltonian systems $\nabla_\Gamma \mathbf{v}_\Gamma = 0$ is valid and Eq.

(2) reduces to the Liouville equation

$$\frac{\partial}{\partial t}\varrho_\Gamma + \mathbf{v}_\Gamma \nabla(\varrho_\Gamma) = \frac{\partial}{\partial t}\varrho_\Gamma + \{H, \varrho_\Gamma\}_{PB} = 0 , \tag{3}$$

whith $\{..,..\}_{PB}$ representing the Poisson brackets. Also Eqs. (2) and (3) are invariant under time-reversal.

Equations (1) and (2) appear in modified form also in quantum mechanics where the modifications are characteristic for the complex nature of the quantum mechanical wave function and reflect the coupling of these two equations and hence the coupling of the deterministic particle-like aspect and the probabilistic wave-like aspect in quantum mechanics. This becomes obvious using Madelung's hydrodynamic formulation [2], where the wave function is written in polar form as

$$\Psi(\mathbf{r}, t) = \varrho^{1/2}(\mathbf{r}, t) \, \exp\left(\frac{i}{\hbar}S(\mathbf{r}, t)\right) \tag{4}$$

with the square root of the probability density $\varrho = \Psi^*\Psi$ as amplitude and $\frac{1}{\hbar}S$ as phase ($\mathbf{r}$ is the position vector in three dimensions).

Inserting this form into the TDSE

$$i\hbar\frac{\partial}{\partial t}\Psi(\mathbf{r}, t) = \left(-\frac{\hbar^2}{2m}\Delta + V\right)\Psi(\mathbf{r}, t) = H_L\Psi(\mathbf{r}, t) \tag{5}$$

(with Δ = Laplace operator) leads to a modified Hamilton–Jacobi equation for the phase,

$$\frac{\partial}{\partial t}S + \frac{1}{2m}(\nabla S)^2 + V - \frac{\hbar^2}{2m}\frac{\Delta\varrho^{1/2}}{\varrho^{1/2}} = 0 , \tag{6}$$

and a continuity equation (in position space),

$$\frac{\partial}{\partial t}\varrho + \frac{1}{m}\nabla(\varrho\,\nabla S) = 0 , \tag{7}$$

for the amplitude.

In Eq. (6) that determines the evolution of the phase occurs an additional term, the last one on the rhs, that depends on the amplitude $\varrho^{1/2}$ of the wave function and is often deceptively called "quantum potential" (V_{qu}) although it is essentially related to the kinetic energy of the quantum system (see below). This equation reflects the classical (deterministic) trajectory aspect of the system. However, via ϱ, also a probabilistic element enters.

The probability density fulfills Eq. (7) that determines the evolution of the amplitude of the wave function, but contains the (gradient of the) phase in the probability current $\mathbf{j} = \varrho(\frac{1}{m}\nabla S)$, where the gradient of the action S is, as usual, associated with the momentum of the system. So, obviously the Hamilton–Jacobi equation (6) and the continuity equation (7) (or, phase and amplitude of the wave function Ψ) are coupled. This coupling is not arbitrary but related to a conservation law which will be shown in the following for the TI case.

2.1. Nonlinear formulations and invariants in time-independent quantum mechanics

In 1994, G. Reinisch [5] presented a NL formulation of the TISE. Since in this case $\frac{\partial}{\partial t}\varrho = 0$ and $\frac{\partial}{\partial t}S = -E$ are valid, the continuity equation (6) (we now use the notation $\varrho^{1/2} = |\Psi| = a$) turns into

$$\nabla(a^2\nabla S) = 0 \tag{8}$$

and the modified Hamilton–Jacobi equation into

$$-\frac{\hbar^2}{2m}\Delta a + (V - E)\, a = -\frac{1}{2m}(\nabla S)^2 a \,. \tag{9}$$

Equation (8) is certainly fulfilled for $\nabla S = 0$, turning (9) into the usual TISE for the real wave function $a = |\Psi|$ with position-independent phase $S(\mathbf{r})$. (NB: the kinetic energy term divided by a, $-\frac{\hbar^2}{2m}\frac{\Delta a}{a}$, is just identical to V_{qu}!)

However, Eq. (8) can also be fulfilled for $\nabla S \neq 0$ if only the conservation law

$$\nabla S = \frac{C}{a^2} \tag{10}$$

with constant (or, at least, position-independent) C is fulfilled. This relation now shows explicitly the coupling between phase and amplitude of the wave function. Inserting (10) into the rhs of Eq. (9) changes this into the real NL Ermakov equation

$$\Delta a + \frac{2m}{\hbar^2}(E - V)\, a = (\frac{1}{\hbar}\nabla S)^2\, a = \left(\frac{C}{\hbar}\right)^2 \frac{1}{a^3} \tag{11}$$

with inverse cubic nonlinearity (this equation had been studied by V. Ermakov [3] long before quantum mechanics was developed and we will come back to it in the TD case). Knowing the solution of (11), the phase (up to an integration constant) can be determined via (10) and thus the complex wave function can be obtained.

For a better understanding of the origin of the coupling expressed by Eq. (10) it can be used that the real NL Ermakov equation (11) is equivalent to a complex NL Riccati equation. In our case, this quadratic NL equation reads

$$\nabla\left(\frac{\nabla\Psi}{\Psi}\right) + \left(\frac{\nabla\Psi}{\Psi}\right)^2 + \frac{2m}{\hbar^2}(E - V) = 0 \tag{12}$$

with the complex variable $\left(\frac{\nabla\Psi}{\Psi}\right) = \frac{\nabla a}{a} + i\frac{1}{\hbar}\nabla S$.

It is straightforward to show that Eq. (12) can be linearized to yield the usual TISE,

$$-\frac{\hbar^2}{2m}\Delta\Psi + V\,\Psi = E\,\Psi \,, \tag{13}$$

but inserting the definition of the complex variable into Eq.(12) and separating real and imaginary parts shows that the imaginary part leads to the conservation law (10), whereas the real part leads to the Ermakov equation (11). The origin of the coupling between phase and amplitude is obviously the quadratic nonlinearity since it introduces a term proportional to $(\nabla S)^2$ into the real part (Eq. (11)) and is responsible for the product term proportional to $(\nabla a)(\nabla S)$ in the imaginary part, leading to Eq. (10).

The same type of NL Riccati and Ermakov equations also occurs in the TD case, as will be shown in the following.

2.2. Nonlinear formulations and invariants in time-dependent quantum mechanics

The TDSE (5) always possesses for at most quadratic Hamiltonians exact analytic solutions in the form of Gaussian WPs that can be written (in the following in one dimension with position variable x) as

$$\Psi(x,t) = N(t)\exp\left\{i\left[y(t)\tilde{x}^2 + \frac{1}{\hbar} < p > \tilde{x} + K(t)\right]\right\} \,, \tag{14}$$

where the maximum of the WP is at $x = \eta(t)$ which is identical to the mean value of the position, $< x >= \int_{-\infty}^{+\infty} dx \Psi^* x \Psi = \eta$, $\tilde{x} = x - \eta$ and the width of the WP is related with the imaginary part of the complex quantity $y(t)$ via $y_I = \frac{1}{4 < \tilde{x}^2 >}$ with the position uncertainty $< \tilde{x}^2 >=< x^2 > - < x >^2$. ($N$ is a, possibly TD, normalization factor, $< p >= m\dot{\eta}$ the classical momentum and K a purely TD function that will not be relevant for the following.) Inserting WP (14) into the SE (5) leads to the equations of motion for $\eta(t)$ and $y(t)$. Particularly for the harmonic oscillator (HO) (i.e., $V = \frac{m}{2}\omega^2 x^2$ with possibly TD $\omega(t)$), one obtains the usual Newtonian equation of motion

$$\ddot{\eta} + \omega^2(t)\eta = 0 \tag{15}$$

for the maximum and a complex Riccati equation, but now for the TD variable $\left(\frac{2\hbar}{m}y\right)$,

$$\frac{2\hbar}{m}\dot{y} + \left(\frac{2\hbar}{m}y\right)^2 + \omega^2(t) \ = \ 0 \ . \tag{16}$$

Introducing a new variable $\alpha_L(t)$ via

$$\frac{2\hbar}{m}y_I = \frac{\hbar}{2m < \tilde{x}^2 >_L} = \frac{1}{\alpha_L^2} \tag{17}$$

allows one to determine the real part $\frac{2\hbar}{m}y_R$ from the imaginary part of Eq. (16) as

$$\frac{2\hbar}{m}y_R = \frac{\dot{\alpha}_L}{\alpha_L}. \tag{18}$$

Inserting this into the real part of (16) yields again an Ermakov equation,

$$\ddot{\alpha}_L + \omega^2(t)\alpha_L \ = \ \frac{1}{\alpha_L^3} \ . \tag{19}$$

Eliminating $\omega^2(t)$ from Eqs.(15) and (19) leads to the dynamical invariant

$$I_L \ = \ \frac{1}{2}\left[(\dot{\eta}\ \alpha_L - \eta\ \dot{\alpha}_L)^2 \ + \ \left(\frac{\eta}{\alpha_L}\right)^2\right] \ = \ \text{constant} \tag{20}$$

that also exists for TD $\omega(t)$, whereas the corresponding Hamiltonian is no longer an invariant (for further details see, e.g., [3] [6] [7] [8] [9]).

Also the NL Riccati equation (16) can be linearized via the ansatz $\left(\frac{2\hbar}{m}y\right) = \frac{\dot{\lambda}}{\lambda}$ with complex $\lambda(t)$ to yield the **complex** Newtonian equation

$$\ddot{\lambda} + \omega^2(t)\lambda = 0 \ . \tag{21}$$

With the polar form $\lambda = \alpha\ e^{i\varphi}$ the complex variable $\frac{2\hbar}{m}y$ can be written as $\frac{\dot{\lambda}}{\lambda} \ = \ \frac{\dot{\alpha}}{\alpha} + i\ \dot{\varphi}$. Inserting this into Eq. (16) shows that the imaginary part leads to the **conservation law**

$$\dot{\varphi} = \frac{1}{\alpha^2} \tag{22}$$

which represents a kind of conservation of angular momentum for the motion of $\lambda(t)$ in the complex plane [10]. From the real part of Eq. (16) one finally obtains the Ermakov equation (19) with $\alpha \equiv \alpha_L$.

In figure 1, the time-dependent and space-dependent cases are compared to show the formal similarity. The complex Riccati equations are given, as well as their linearized forms, leading to second-order differential equations for complex variables. These variables are given in polar form and the relations between phase and amplitude of these variables are expressed in terms of conservation laws. These laws are equivalent to the imaginary part of the respective Riccati equations if the polar form of the complex variables is inserted. The real part of the Riccati equations is then identical with the NL Ermakov equations. The essential difference between the two cases is that temporal and spatial variables and the respective derivatives are interchanged, as well as $\omega^2(t)$ and $\frac{2m}{\hbar^2}(E - V(x))$.

	Time	Space
NL complex	$\frac{d}{dt}\left(\frac{\dot{\lambda}}{\lambda}\right) + \left(\frac{\dot{\lambda}}{\lambda}\right)^2 + \omega^2(t) = 0$	$\nabla\left(\frac{\nabla\Psi}{\Psi}\right) + \left(\frac{\nabla\Psi}{\Psi}\right)^2 + \frac{2m}{\hbar^2}\left(E - V(\underline{r})\right) = 0$
linearized	$\ddot{\lambda} + \omega^2\lambda = 0$	$\Delta\Psi + \frac{2m}{\hbar^2}(E - V)\Psi = 0$
complex variable	$\lambda = \alpha e^{i\varphi}$	$\Psi = a e^{iS/\hbar}$
conservation law	$\dot{\varphi} = \frac{1}{\alpha^2}$	$\nabla S = \frac{C}{a^2}$
NL real	$\ddot{\alpha} + \omega^2 = \dot{\varphi}^2\alpha = \frac{1}{\alpha^3}$	$\Delta a + \frac{2m}{\hbar^2}(E - V) = \frac{1}{\hbar^2}(\nabla S)^2 \cdot a = \frac{(C/\hbar)^2}{a^3}$

contains info about phase on rhs

Figure 1: Comparison between the space-dependent and the time-dependent Riccati equations, their linearized forms and the corresponding Ermakov equations for the conservative case.

3. Irreversibility, dissipation and effective NLSEs

Since realistic physical systems are always in contact with some kind of environment and this coupling usually introduces the phenomena irreversibility and dissipation (but not necessarily always both simultaneously!), the questions arise of how this can be taken into account in the formalism of classical (Hamiltonian) mechanics and, especially, in a quantum mechanical context.

In the trajectory picture, dissipation can be included by adding a phenomenological friction force that is proportional to velocity (or momentum) to the Newtonian equation of motion, thus turning it into an irreversible evolution equation, i.e.,

$$m\ddot{x} = m\dot{v} = -m\gamma v - \frac{\partial}{\partial x}V \tag{23}$$

which is the Langevin equation without stochastic contribution (γ is the friction coefficient). This equation is used to describe Brownian motion from a trajectory point of view. An equivalent description of this phenomenon can also be given in terms of (classical) probability distributions ϱ_{cl} via the Fokker–Planck equations that contain irreversible diffusion terms. Particularly in position space, this can be written in the form of the Smoluchowski equation for $\varrho_{cl}(x,t)$,

$$\frac{\partial}{\partial t}\varrho_{\rm cl} + \frac{\partial}{\partial x}\left(\frac{F(x)}{m\gamma}\varrho_{\rm cl}\right) - \frac{\mathsf{kT}}{m\gamma}\frac{\partial^2}{\partial x^2}\varrho_{\rm cl} = 0\,, \tag{24}$$

with the conservative force $F(x)$, Boltzmann's constant k and temperature T. Comparison with Einstein's theory of Brownian motion [11] shows that the coefficient of the diffusion term fulfils the Einstein relation $D = \frac{\mathsf{kT}}{m\gamma}$.

However, Eqs. (23) and (24) do not fit consistently into the Lagrange/Hamilton formalism that is invariant under canonical transformations and provides a basis for quantization. So, these phenomenological equations do not yet provide the answer to the questions posed at the beginning of this Section.

There are numerous approaches in the literature to answer these questions in different ways; most of them with some problematic aspects. In the following, several of them and their interrelations will be discussed, particularly those which give rise to Riccati and Ermakov equations similar to the ones investigated in Section 2, so that a comparison between the conservative and dissipative case is possible.

The models use different approaches to reach their goal, namely:

1. Modification of the classical Lagrange/Hamilton formalism with subsequent canonical quantization (e.g., Caldirola–Kanai [12] [13], expanding coordinate system [14];

2. system-plus-reservoir approach (e.g., Bateman [15], Caldeira–Leggett [16] [17] [18], Kossakowski–Lindblad [19] [20];

3. non-unitary time-evolution (e.g., Gisin [21] [22], Beretta [23]) [24];

4. addition of friction terms to the SE according to Ehrenfest's equation for mean values, leading to NLSEs (e.g., Kostin [25], Cho [26], Albrecht [27], Süssmann [28], Hasse [29]);

5. addition of an irreversible term on the level of the probability density, e.g., via a Smoluchowski equation, also leading to NLSEs (e.g., Schuch–Chung–Hartmann [30] [31] [32], Doebner–Goldin [33] [34]).

3.1. Modified Lagrange/Hamilton formalism

The addition of a so-called "friction potential" to the classical Lagrangian or Hamiltonian in order to obtain the above-mentioned friction force is not consistent with the conventional Lagrange or Hamilton formalism. Therefore, Caldirola [12] and Kanai [13] used a different approach by multiplying the classical Lagrangian by an exponential factor, leading to

$$\hat{L}_{\mathrm{CK}} = \left[\frac{m}{2}\dot{x}^2 - V(x)\right] \mathrm{e}^{\gamma t} \tag{25}$$

from where, via the Euler–Lagrange equation, the friction force in the equation of motion arises

$$\ddot{x} + \gamma \dot{x} + \frac{1}{m}\frac{\partial}{\partial x} V = 0 \,. \tag{26}$$

Defining the canonical momentum in the usual way as

$$\frac{\partial}{\partial \dot{x}} \hat{L}_{\mathrm{CK}} = \hat{p} = m\dot{x}\, \mathrm{e}^{\gamma t} = p\, \mathrm{e}^{\gamma t} \tag{27}$$

allows to construct the corresponding Hamiltonian function

$$\hat{H}_{\mathrm{CK}} = \frac{1}{2m}\mathrm{e}^{-\gamma t}\, \hat{p}^2 + \mathrm{e}^{\gamma t} V(x) \tag{28}$$

that, via the Hamiltonian equations of motion, also provides the desired friction term. It is important to note that the transition from the physical variables (x, p) to the canonical ones, $\hat{x} = x, \hat{p} = \mathrm{e}^{\gamma t} p$, is a *non-canonical* transformation! Applying canonical quantization to

$\hat{H}_{CK}$, i.e., replacing $\hat{p}$ by $\frac{\hbar}{i}\frac{\partial}{\partial x}$, provides the corresponding linear, but explicitly TD Hamiltonian operator and consequently the modified TDSE

$$i\hbar\frac{\partial}{\partial t}\,\hat{\Psi}_{\text{CK}}(x,t) \;=\; \left\{ \mathrm{e}^{-\gamma t}\left(-\frac{\hbar^2}{2m}\frac{\partial^2}{\partial x^2}\right) + \mathrm{e}^{\gamma t}V \right\}\,\hat{\Psi}_{\text{CK}}(x,t)\ . \tag{29}$$

However, calculating the uncertainty product of position and (physical) momentum using the solutions of (29) leads to an exponential decay and therefore an apparent violation of the uncertainty product. The reason for this, and a possible solution of this problem, will be given in section 5.

Another approach that does not face this problem and also starts from a modified Lagrange/Hamilton formalism is likewise based on a non-canonical transformation of the variables; however, this time also changes the position variable by transition to an exponentially-expanding coordinate system. The new (canonical) variables (Q,P) are related to the physical ones (x,p) via

$$Q \;=\; e^{\gamma t/2}\,x\ \ ,\ \ P \;=\; m\dot{Q} \;=\; m e^{\gamma t/2}(\dot{x} + \frac{\gamma}{2}\,x)\ , \tag{30}$$

leading to the Hamiltonian function

$$\hat{H}_{\text{exp}} \;=\; \frac{1}{2m}P^2 + \frac{m}{2}\Omega^2 Q^2 \;=\; \text{const.} \tag{31}$$

with $\Omega^2 = \left(\omega^2 - \frac{\gamma^2}{4}\right)$. $\hat{H}_{exp}$ is a constant of motion, provides the correct equation of motion including the friction force (when expressed in the physical variables) and can be canonically quantized via $P \to \frac{\hbar}{i}\frac{\partial}{\partial Q}$ (for further details, see [14]). The relation between this approach and the one of Caldirola–Kanai will be elaborated in section 5.

3.2. *System-plus-reservoir approach*

A more conventional approach is to consider the system of interest together with the environment as a closed Hamiltonian system and to determine the evolution of the system while interacting with the environment. Albeit, using this method means there are still different ways of describing the environment: a) by a single other degree of freedom (see Bateman [15]), b) by a large number of degrees of freedom (e.g, Caldeira–Leggett [16] [17] [18]) c) using an effective description without considering any individual environmental degree of freedom (see, e.g., sections 3.3 - 3.5).

Method (a) was introduced by Bateman applying the Hamiltonian

$$H_{\text{B}} \;=\; \frac{p_x p_y}{m} + \frac{\gamma}{2}(y p_y - x p_x) + m\left(\omega^2 - \frac{\gamma^2}{4}\right)xy\ . \tag{32}$$

The system with variables x and p_x obeys the equation of motion

$$\ddot{x} + \gamma\dot{x} + \omega^2 x \;=\; 0 \tag{33}$$

with the dissipative damping force, whereas the environment is represented by one other system with variables y and p_y obeying

$$\ddot{y} - \gamma\dot{y} + \omega^2 y \;=\; 0 \tag{34}$$

with an accelerating force of the same magnitude as the friction force, but with a different sign. The energy lost by the first system is gained by the second and the Hamiltonian H_B remains constant. There are attempts in the literature to link this approach to the one by Caldirola–Kanai [35] and a recent achievement was made to connect it directly to the approach using the expanding coordinates [36].

The usual system-plus-reservoir approach, however, takes into account explicitly the interaction between the system and individual environmental degrees of freedom, representing them for instance by a large number of HOs. An ansatz of this kind, e.g., the one by Caldeira and Leggett, can be described by the Hamiltonian $H_{Tot} = H_S + H_{SR} + H_R +$ correction term where H_S and H_R depend only on the variables of system or reservoir, respectively, whereas H_{SR} depends on both types of variables. The correction term was added by hand to compensate for a negative shift of $V(x)$. By averaging over the environmental degrees of freedom, letting the number of environmental degrees go to infinity etc., one finally obtains an equation of motion for the system of interest that contains a dissipative friction force. It has been shown [37] [38] that the Caldirola–Kanai Hamiltonian can be derived from the Caldeira–Leggett ansatz!

However, there is also a problematic aspect of this approach. Hamiltonian H_{Tot} is usually applied to derive the corresponding master equation for the density operator ϱ_{op}. But the dissipative term in the equation for the system can lead to violation of the positivity of ϱ_{op}! This problem can be avoided if the dissipative term has a particular formal structure. This requirement is fulfilled by the Kossakowski–Lindblad generators [19] [20] but they provide only a formal mathematical requirement that must be satisfied but give no physical guideline for the explicit choice of the operators in the dissipative term.

A general problem with the approaches using master equations for the density operator is that the computational effort scales at least with N^2 if N is the number of relevant degrees of freedom. Therefore, it would be more desirable to have a description in terms of *pure states* since, in principle, a linear scaling would be possible. In the following, approaches will be presented that assume a quantum system can always be described by a pure state wave function whether it interacts with some kind of environment or not. In these approaches the individual degrees of the environment are not explicitly taken into account only the effect of the environment on the observable properties of the system.

3.3. Non-unitary time-evolution

In his approach Gisin [21] [22] assumed that the time-evolution of an open quantum system should be characterized by a direction of time and therefore it should be non-unitary. This can be reached by adding an imaginary term, e.g., iB with $B = B^+$ to the SE; but this in general destroys the normalizability of the wave function. A way of eliminating this shortcoming is to subtract the mean value of this imaginary term, thus regaining normalizability, but for the price of introducing nonlinearity via the mean value. Gisin used the ansatz

$$i\hbar \frac{\partial}{\partial t} \Psi(t) = \{ H_L - i (B - < B >) \} \Psi(t) , \tag{35}$$

wich is due to the appearance of $< B >$ a NLSE. In order to also describe dissipation, Gisin made the assumption $B = kH_L$; but this provides a degree of energy dissipation that contradicts the known contribution from the classical part, i.e., $\dot{E} \neq -2\gamma T$ with $T =$ kinetic energy.

In a similar approach, Beretta [23] [24] tried to describe non-equilibrium systems (without dissipation) where the imaginary term was supposed to be related to the entropy of the system that should increase during time-evolution. The particular form of his NLSE was

$$i\hbar \frac{\partial}{\partial t} \Psi = \{ H_L + \gamma \frac{\hbar}{2i} (\ln \varrho - < \ln \varrho >) \} \Psi , \tag{36}$$

a form that will appear again in another approach in section 3.5..

3.4. Nonlinear "friction potentials"

Whereas the two afore-mentioned approaches were essentially motivated by the irreversibility of the time-evolution of an open quantum system, the ones that will be presented below focus on the dissipative aspect. As mentioned earlier, the equation of motion including a dissipative friction force cannot be obtained by simply adding a term to the classical Lagrangian or Hamiltonian function. The idea of the following approaches was, however, to start already on the quantum mechanical level and add a term that would provide, according to Ehrenfest's theorem, the correct velocity- (or momentum)-dependent friction force in the equation of motion for the mean-values. A problem arises in this context in that a velocity or momentum defined via $p_{op}\Psi = \frac{\hbar}{i}\frac{\partial}{\partial x}\Psi$ is in general a complex quantity due to the fact that $\Psi(x,t)$ is generally complex. For the mean values, this does not really matter since the (imaginary) contribution from the amplitude of the wave function, $v_+ = \frac{\hbar}{2mi}\frac{\nabla\varrho}{\varrho}$, always vanishes on an average, $< v_+ >= 0$, whereas the (real) contribution from the phase, $v_- = \frac{\hbar}{2mi}\nabla \ln \frac{\Psi}{\Psi^*}$, (on an average) provides the classical velocity, $< v_- >= \dot{\eta}$, (and this is actually the definition of velocity used in the continuity equation for the probability density $\varrho = \Psi^*\Psi$).

So the question arises: which definition of velocity is relevant for the construction of a "friction potential" that should generally provide a force proportional to velocity? Is it v_+, v_- or $v_+ + v_-$?

In his approach, Kostin [25] used the definition of v_- that provides the classical velocity when averaged, leading to the friction term

$$W_{\text{Kos}} = \frac{\gamma}{2}\frac{\hbar}{i}\left(\ln\frac{\Psi_{NL}}{\Psi_{NL}^*} - < \ln\frac{\Psi_{NL}}{\Psi_{NL}^*} >\right) \tag{37}$$

where the subtraction of the mean-value guarantees that the mean-value of the NL Hamiltonian $H_{NL} = H_L + W_{\text{Kos}}$ is identical to the mean-value of the kinetic and potential energies, i.e., $< H_{NL} >=< H_L >=< T > + < V >$.

However, this approach suffers certain shortcomings:

a) if Ψ is real, as in the case of the eigenfunctions of the HO, $\Psi = \Psi^*$ and W_{Kos} vanishes, meaning that the solutions of the undamped HO are also solutions of the frictionally-damped problem;

b) the frequency occurring for the damped HO is not the correct reduced frequency $\Omega = (\omega^2 - \gamma^2/4)^{1/2}$ known from the classical counterpart;

c) since W_{Kos} is real, it does not contribute to the equation of motion for the probability density. Therefore this remains a reversible continuity equation, whereas the averaged equation of motion and the NLSE are irreversible evolution equations; a fact that needs explanation.

An attempt at solving these problems was made by Cho [19] who tried to take into account an additional "diffusion velocity"which is similar to the abovementioned imaginary contribution to the velocity definition in quantum mechanics, $v_{\text{diff}} = -\frac{\hbar}{2m}\frac{\nabla\varrho}{\varrho}$. However v_{diff} is real and, although getting rid of the first of the three shortcomings, the other two, the wrong frequency for the damped HO and the reversible continuity equation for $\varrho(x,t)$, remain.

A more pragmatic approach was taken in constructing a most general "friction potential" W_{Gen} as a product of position and momentum and their mean values, requiring that

$$< \frac{\partial}{\partial x}W > = \gamma < p > \tag{38}$$

is always fulfilled [29]. The corresponding "friction potential" can be given as

$$W_{\text{Gen}} = \gamma < p > (x- < x >) + \frac{\gamma}{2}\, C\, [(x- < x >), (p- < p >)]_+ \tag{39}$$

(with $[\ ,\ ]_+$ = anticommutator), that fulfils the requirement for any value of C.

There are several particular choices for C in the literature:

a) $C = 1$: Süssmann [28] used this ansatz to obtain

$$W_{\text{Süs}} = \frac{\gamma}{2}\,[(x- < x >), p]_+ \tag{40}$$

which, for complex Ψ due to the operator $p = \frac{\hbar}{i}\frac{\partial}{\partial x}$, is generally a complex quantity. This, however, still provides the wrong frequency for the damped HO. More concerning the contribution of $W_{\text{Süs}}$ to the continuity equation can be seen in the next section.

b) $C = 0$: Albrecht's choice [27] leads to

$$W_{\text{Alb}} = \gamma < p > (x- < x >)\ , \tag{41}$$

which is a real quantity, therefore still leading to a reversible continuity equation for ϱ and also providing an undamped frequency for the damped HO.

c) $C = \pm 1/2$: The correct frequency $\Omega = (\omega^2 - \gamma^2/4)^{1/2}$ for the damped HO can be obtained for $C = \pm 1/2$, a choice taken by Hasse [29] where $C = -1/2$ leads to unphysical results (wrong sign in the exponent of the Gaussian WP). So, for $C = +1/2$, Hasse obtained

$$W_{\text{Has}} = \frac{\gamma}{2}[(x- < x >), (p+ < p >)]_+ \tag{42}$$

which is a complex quantity, thus also contributing an (irreversible) term to the equation for $\varrho(x,t)$. Only the mean value of W_{Has} does not vanish, $< W_{\text{Has}} > \neq 0$, so that $< H_{NL} > \neq < T > + < V >$.

Comparing Hasse's approach with the two afore-mentioned ones, one finds

$$W_{\text{Has}} = \frac{1}{2}\,(W_{\text{Süs}} + W_{\text{Alb}})\ . \tag{43}$$

3.5. Broken time-symmetry for the probability density

Approaches via the averaged equation of motion and using condition (38) contain a lot of arbitrariness. Equation (38) only makes statements about the derivative of the "friction potential" (and even worse, only about the mean value of this quantity) and compares it with the mean-value of velocity (which is itself ambiguous, as shown above). We therefore tried (see [30] [31] [32]) an approach that is not based on the dissipative aspect of the trajectory picture but on the irreversible aspect of the probability picture. For this purpose, we add a time-reversal symmetry-breaking diffusion term to the continuity equation for ϱ, thus turning it, as previously discussed, into an irreversible Smoluchowski equation,

$$\frac{\partial}{\partial t}\varrho + \frac{\partial}{\partial x}(\varrho\, v_-) - D\frac{\partial^2}{\partial x^2}\varrho = 0\ , \tag{44}$$

but now for the quantum mechanical $\varrho(x,t)$ and with the v_- mentioned earlier. For the continuity equation, Madelung [39] and Mrowka [40] had shown that with a bilinear ansatz for the probability current density, $j = \varrho\left(\frac{1}{m}\frac{\partial}{\partial x}S\right) = \varrho v_- = \left(\frac{\hbar}{2mi}\right)\left(\Psi^*\frac{\partial}{\partial x}\Psi - \Psi\frac{\partial}{\partial x}\Psi^*\right)$, this can be separated into the TDSE and its complex conjugate where the separation "constant" is proportional to the potential V. However, due to the diffusion term in (44), Ψ- and Ψ^*-dependent terms can no longer be separated because of the last term on the rhs of

$$-D\frac{\partial^2}{\partial x^2}\varrho = -D\left\{\Psi^*\frac{\partial^2}{\partial x^2}\Psi + \Psi\frac{\partial^2}{\partial x^2}\Psi^* + 2\left(\frac{\partial}{\partial x}\Psi\right)\left(\frac{\partial}{\partial x}\Psi^*\right)\right\}\ . \tag{45}$$

For separation one needs that the condition

$$-D\frac{\frac{\partial^2}{\partial x^2}\varrho}{\varrho} = F_1(\Psi) + F_2(\Psi^*) \tag{46}$$

has to be fulfilled.

This can be reached by introducing the additional separation condition

$$-D\frac{\frac{\partial^2}{\partial x^2}\varrho}{\varrho} = \gamma(\ln\varrho - <\ln\varrho>) , \tag{47}$$

where $\ln\varrho = \ln\Psi + \ln\Psi^*$. This leads to an additional *complex* logarithmic term [23] in the SE for $\Psi_{NL}(x,t)$,

$$i\hbar\frac{\partial}{\partial t}\Psi_{NL}(x,t) = \{H_L + \gamma\frac{\hbar}{i}(\ln\Psi_{NL} - <\ln\Psi_{NL}>)\}\Psi_{NL}(x,t) = \{H_L + W_{\mathrm{SCH}}\}\Psi_{NL}(x,t) , \tag{48}$$

where H_L is the usual linear Hamiltonian. The additional NL term W_{SCH} can be written as real and imaginary contributions in the form

$$W_{\mathrm{SCH}} = W_R + iW_I = \frac{\gamma\hbar}{2i}\left(\ln\frac{\Psi_{NL}}{\Psi^*_{NL}} - <\ln\frac{\Psi_{NL}}{\Psi^*_{NL}}>\right) + \frac{\gamma\hbar}{2i}(\ln\varrho_{NL} - <\ln\varrho_{NL}>) \tag{49}$$

where the real part only depends on the phase of the wave function and provides the friction force in the averaged equation of motion. The imaginary part does not contribute to dissipation but introduces irreversibility into the evolution of the wave function. It corresponds to the diffusion term in the Smoluchowski equation, but still allows for normalizability due to the subtraction of the mean value of $\ln\varrho$. Comparison with the afore-mentioned approaches shows that the real part is just identical to Kostin's term and the imaginary part corresponds to Beretta's term (for pure states).

The imaginary part breaks the time-reversal symmetry on the level of the probability density, introduces a non-unitary time evolution and turns the Hamiltonian into a non-Hermitian one while still guaranteeing normalizable wave functions and real energy mean values since its mean value vanishes.

Also from the real part of W_{SCH} no additional term to the energy mean value occurs, so this is still given by the mean value of the operators of kinetic and potential energies. This real part is however not arbitrary but is uniquely determined by the separation condition and provides the correct dissipative friction forces in the equation of motion for the mean values. Besides, the ratio of energy dissipation (for the classical contribution) is in agreement with the classical counterpart and arises because the mean values are calculated with Ψ_{NL} instead of Ψ_L (the solution of Eq. (5)).

The real part, by itself, would provide dissipation but retain a unitary time-evolution of the wave function, whereas the imaginary part, on its own, would provide irreversibility via a non-unitary time-evolution but no dissipation. Consequently only the combination of real and imaginary parts provides all the desired properties of the quantum system under consideration. The reason for this is the coupling of phase and amplitude of the wave function since W_R depends on the phase and W_I on the amplitude.

We should mention another approach by Doebner and Goldin based on the same Smoluchowski equation for $\varrho(x,t)$ that confirms the additional diffusion term on group-theoretical grounds [24]. The authors do not try to separate this term but add half of it to the SE and the other half to its complex conjugate, leading to a NLSE where an arbitrary real, possibly NL, term $R[\Psi, \Psi^*]$ can be added, i.e.,

$$i\hbar \frac{\partial}{\partial t}\Psi \;=\; \{H_L \;+\; i\frac{\hbar}{2}\; D\frac{\frac{\partial^2}{\partial x^2}\varrho}{\varrho} \;+\; R[\Psi, \Psi^*]\;\}\Psi \;. \tag{50}$$

Without $R[\Psi, \Psi^*]$, the additional term leads to an irreversible non-unitary time-evolution, but not necessarily to dissipation; this depends essentially on R. For Gaussian WPs, the diffusion term is identical to the $\ln \varrho$-term of W_{SCH}. These authors also discuss the possibility of linearizing NLSEs via NL gauge transformations [25], what has been compared with an alternative method for linearization (see section 5) using non-unitary transformations [26].

In the following, the NLSE (48) with complex logarithmic nonlinearity will be used to show the influence of a dissipative environment on the TDSE and the corresponding Riccati and Ermakov equations. This will allow one to draw conclusions for corresponding modifications in the time-independent case which will show closer connections between the approaches presented in sections 3.4 and 3.5.

4. Modifications of the complex Riccati and nonlinear Ermakov equations

4.1. Modifications due to dissipation and irreversibility in the time-dependent case

Inserting WP (14) into the NLSE (48) yields, for $\eta(t) =< x >_{NL}$, and hence for the motion of the WP maximum, the Newtonian equation including the friction force,

$$\ddot{\eta} + \gamma\dot{\eta} + \omega^2\eta = 0 \tag{51}$$

and for the complex variable $(\frac{2\hbar}{m}y)_{NL}$ the modified Ricatti equation

$$\left(\frac{2\hbar}{m}\dot{y}\right)_{NL} + \gamma\left(\frac{2\hbar}{m}y\right)_{NL} + \left(\frac{2\hbar}{m}y\right)^2_{NL} + \omega^2(t) \;=\; 0 \tag{52}$$

with an additional linear term depending on the friction coefficient γ.

The relation between the imaginary part of this variable and the position uncertainty or Ermakov variable, respectively, remains unchanged as in the conservative case,

$$\left(\frac{2\hbar}{m}y_I\right)_{NL} \;=\; \frac{\hbar}{2m < \tilde{x}^2 >_{NL}} = \frac{1}{\alpha^2_{NL}} \;, \tag{53}$$

but the real part differs by a contribution from the friction coefficient, i.e.,

$$\left(\frac{2\hbar}{m}y_R\right)_{NL} \;=\; \frac{\dot{\alpha}_{NL}}{\alpha_{NL}} \;-\; \frac{\gamma}{2} \;. \tag{54}$$

Inserting (53) and (54) into the Riccati equation (52) turns this into the Ermakov equation

$$\ddot{\alpha}_{NL} + \left(\omega^2 - \frac{\gamma^2}{4}\right)\alpha_{NL} = \frac{1}{\alpha^3_{NL}} \;, \tag{55}$$

i.e., in comparison with the conservative case, only ω^2 has been replaced by $\Omega^2 = \left(\omega^2 - \frac{\gamma^2}{4}\right)$. Again, from Eqs. (51) and (55) a dynamical invariant can be obtained that now has the form [44] [45] [46]

$$I_{NL} \;=\; \frac{1}{2}\, e^{\gamma t}\left[\left(\dot{\eta}\,\alpha_{NL} - \left(\dot{\alpha}_{NL} - \frac{\gamma}{2}\alpha_{NL}\right)\eta\right)^2 + \left(\frac{\eta}{\alpha_{NL}}\right)^2\right] . \tag{56}$$

Riccati equation (52) can also be linearized, now using the ansatz $\left(\frac{2\hbar}{m}y\right)_{NL} = \frac{\dot{\tilde{\lambda}}}{\tilde{\lambda}} = \frac{\dot{\lambda}}{\lambda} - \frac{\gamma}{2}$ to yield the Newtonian equation with linear friction term

$$\ddot{\tilde{\lambda}} + \gamma\dot{\tilde{\lambda}} + \omega^2\tilde{\lambda} \;=\; 0 \;, \tag{57}$$

for the complex variable $\tilde{\lambda}(t) = \lambda e^{-\gamma t/2} = \alpha_{NL}e^{-\gamma t/2+i\varphi}$. Inserting the polar form into Eq. (52) leads to the unchanged conservation law

$$\dot{\varphi} = \frac{1}{\alpha_{NL}^2} \tag{58}$$

as in the conservative case and the modified Ermakov equation (55).

Now we compare Eqs. (51), (52) and (57) with the corresponding Eqs. (15), (16) and (21) in the conservative case. In the linear second-order differential equations for η and λ, a linear term with first derivative has been added while, in the first-order Riccati equation, an additional term linear in $\left(\frac{2\hbar}{m}y\right)$ appears. All the additional terms depend on the coefficient of the friction force. But now the question arises: is it possible to modify the corresponding equations in the time-independent case (see Fig. 1) in a similar way to include the effect of a dissipative environment into the TISE? The answer will be given in the next subsection.

4.2. Modifications due to dissipation and irreversibility in the space-dependent case

The space-dependent Riccati equation corresponding to Eq. (16) was Eq. (12) that could be linearized to yield the TISE (13). The modifications from Eqs. (16) and (21) to (52) and (57) in the time-dependent case would mean, in the space-dependent situation, the addition of a term $\Gamma\left(\frac{\nabla\Psi}{\Psi}\right)$ to Eq. (12) or a term $\Gamma\nabla\Psi$ to Eq. (13), respectively, where the coefficient Γ should somehow be related to the friction force but, in general, could be a complex function of $\mathbf{r}$ and t, $\Gamma(\mathbf{r},t)$.

Since Ψ is complex, a contribution from the imaginary part of the additional term $\Gamma\nabla\Psi$ to the continuity equation for $\varrho = \Psi^*\Psi$ should arise. What should be the specific form of this term? To answer this question, the following assumptions shall be made:

1. The SE (5) can be obtained from the continuity equation (7) (according to Madelung [39] and Mrowka [40]) via separation. The log NLSE (48) can be obtained from the Smoluchowski equation (44) with additional diffusion term (and condition (47), as shown in section 3.5) also via separation. Therefore, the additional term $\Gamma\nabla\Psi$ in the modified TISE should also originate from an additional term in the continuity equation plus separation. That is, a term

$$\Gamma\;\nabla\varrho \;=\; \Gamma\,(\Psi^*\nabla\Psi \;+\; \Psi\nabla\Psi^*) \tag{59}$$

should occur in the evolution equation for ϱ. The contribution from this term would enter the SE with an *imaginary* coefficient.

2. The additional term in the equation for ϱ, at least for cases with Gaussian WP solutions (i.e., $V = 0$ and HO) discussed in this paper, should have the same effect as the diffusion term or the ln ϱ-term, i.e.,

$$-D\Delta_x\varrho \;=\; \gamma(\ln\varrho - <\ln\varrho>)\varrho \;=\; \frac{\gamma}{2}\left(1 \;-\; \frac{\tilde{x}^2}{<\tilde{x}^2>}\right)\varrho \tag{60}$$

Comparison shows that

$$\nabla_x\left(\frac{\gamma}{2}\tilde{x}\varrho\right) \;=\; \frac{\gamma}{2}\left(1 \;-\; \frac{\tilde{x}^2}{<\tilde{x}^2>}\right)\varrho \tag{61}$$

yields the desired result! (In our one-dimensional case ∇_x means $\frac{\partial}{\partial x}$ and the subscript x will be omitted in the following.)

After separation this contributes an additional term $W_{\rm TI} = \frac{\gamma}{2}\frac{\hbar}{i}\left(\frac{1}{2} + \tilde{x}\frac{\nabla\Psi}{\Psi}\right)$ in the SE that already looks familiar when rewritten as

$$W_{\rm TI}\,\Psi = \left(\frac{\gamma}{4}\frac{\hbar}{i} + \frac{\gamma}{2}\tilde{x}\frac{\hbar}{i}\nabla\right)\Psi = \frac{\gamma}{4}\,[\tilde{x}, p_{op}]_+\,\Psi = \frac{1}{2}W_{\rm Süs}\,\Psi\,. \tag{62}$$

However:

a) $< W_{\rm TI} > = \frac{\gamma}{4} < [\tilde{x}, \tilde{p}]_+ > = < W_{\rm Has} > \neq 0$, but this could be compensated for by a normalization factor; and

b) $< -\nabla W_{\rm TI} > = -\frac{\gamma}{2} < p >$, i.e., half of the friction force is missing.

Considering for instance the TISE for the HO with real wave functions, this would not actually cause a problem since, in this case, $< p >= 0$. However, if one would also take into account cases where the phase of the wave function matters, a further real contribution $\frac{\gamma}{2}\tilde{x} < p > = \frac{1}{2}W_{\rm Alb}$ would be needed, in addition to $W_{\rm TI}$. Where could the missing terms originate from?

To answer this question, we need to revisit Madelung's hydrodynamic formulation. From the continuity equation (7), rewritten in the notation used in section 2.1., i.e., $\varrho = a^2$,

$$\frac{\partial}{\partial t}a^2 + \frac{1}{m}\nabla\left(a^2\nabla S\right) = 0 \tag{63}$$

in the TI case with $\frac{\partial}{\partial t}a^2 = 0$, the conservation law (10), $\nabla S = C/a^2$, was derived. Taking into account the additional term (61), this is obviously no longer valid. However, introducing a modified action function S' via

$$\begin{aligned}\frac{\partial}{\partial t}a^2 + \frac{1}{m}\nabla\left(a^2\left(\nabla S + m\frac{\gamma}{2}\tilde{x}\right)\right) &= \\ \frac{\partial}{\partial t}a^2 + \frac{1}{m}\nabla\left(a^2\nabla S'\right) &= 0\end{aligned} \tag{64}$$

with

$$S' = S + m\frac{\gamma}{4}\tilde{x}^2 + f(t) \tag{65}$$

allows us to find, for $\frac{\partial}{\partial t}a^2 = 0$, again a conservation law of the form

$$\nabla S' = \frac{C}{a^2}\,. \tag{66}$$

Separation of Eq. (64) introduces the term $W_{\rm TI}$ into the modified TISE, what leads to an additional contribution of the phase in the Hamilton–Jacobi-type equation (6),

$$\frac{\partial}{\partial t}S + \frac{1}{2m}(\nabla S)^2 + V - \frac{\hbar^2}{2m}\frac{\Delta a}{a} + \frac{\gamma}{2}\tilde{x}(\nabla S) = 0\,, \tag{67}$$

i.e., the last term on the lhs.

To obtain a form that is consistent with the modified action in Eq. (64), S will be replaced by S' according to (65). Specifically, the time-derivative of the action supplies the additional terms,

$$\frac{\partial}{\partial t}S = \frac{\partial}{\partial t}S' + \frac{\gamma}{2} < p > \tilde{x} - \frac{\partial}{\partial t}f(t)\,, \tag{68}$$

leading to Eq. (67) expressed in terms of S' as

$$\frac{\partial}{\partial t}S' + \frac{1}{2m}(\nabla S')^2 + \left[V - \frac{m}{2}\frac{\gamma}{4}\tilde{x}^2 + \frac{\gamma}{2} < p > \tilde{x} - \frac{\partial}{\partial t}f\right] - \frac{\hbar^2}{2m}\frac{\Delta a}{a} = 0\,. \tag{69}$$

With $\frac{\partial}{\partial t}S' = -E'$ and $\frac{\partial}{\partial t}f = \frac{\gamma}{4} < [\tilde{x},\tilde{p}]_+ >$ and applying (66), Eq. (69) can be rewritten as the Ermakov equation

$$\Delta a + \frac{2m}{\hbar^2}\left[\hat{E}'_R - \hat{V}\right]a = (\frac{1}{\hbar}\nabla S')^2\, a = \left(\frac{C}{\hbar}\right)^2\frac{1}{a^3} \tag{70}$$

with $\hat{E}'_R = E' - \frac{\gamma}{2} < p > \tilde{x} + \frac{\gamma}{4} < [\tilde{x},\tilde{p}]_+ >$ and $\hat{V} = V - \frac{m}{2}\frac{\gamma}{4}\tilde{x}^2$.

Including the two additional terms originating from the time-dependence of S expressed in terms of S', the full irreversible and dissipative term $\tilde{W}_{TI}$ now reads

$$\tilde{W}_{\rm TI} = \frac{\gamma}{2}\,[\tilde{x},p_{op}]_+ + \frac{\gamma}{2} < p > \tilde{x} - \frac{\gamma}{4} < [\tilde{x},\tilde{p}]_+ > = W_{\rm Has} - < W_{\rm Has} > = W_{\rm SCH}\,, \tag{71}$$

where the first additional (real) term just provides the missing contribution $\frac{1}{2}W_{\rm Alb}$ that is necessary to obtain the correct friction force in the equation of motion for the mean values and the second term is equal to the mean-value of Hasse's friction term and must be subtracted so that $< \tilde{W}_{\rm TI} >= 0$. For the Gaussian WPs $\tilde{W}_{\rm TI}\,\Psi_{WP} = W_{\rm SCH}\,\Psi_{WP}$ is valid. The modified TISE including these two additional terms can be rewritten as

$$E'\Psi = \left\{-\frac{\hbar^2}{2m}\Delta_x + \frac{\gamma}{2}\tilde{x}\,\frac{\hbar}{i}\nabla_x + \left(V + \frac{\gamma}{2} < p > - \frac{\gamma}{4} < [\tilde{x},\tilde{p}]_+ > + i\hbar\frac{\gamma}{4}\right)\right\}\Psi \tag{72}$$

or in the form of the corresponding modified complex Riccati equation

$$\nabla\left(\frac{\nabla\Psi}{\Psi}\right) + i\gamma\frac{m}{\hbar}\left(\frac{\nabla\Psi}{\Psi}\right) + \left(\frac{\nabla\Psi}{\Psi}\right)^2 + \frac{2m}{\hbar^2}(\hat{E} - V) = 0 \tag{73}$$

with the complex quantity

$$\hat{E} = E' - \frac{\gamma}{2} < p > \tilde{x} + \frac{\gamma}{4} < [\tilde{x},\tilde{p}]_+ > - i\hbar\frac{\gamma}{4}\,. \tag{74}$$

In figure 2, the time-dependent and space-dependent cases are again compared but now including the environmental effect. Whereas in the time-dependent situation the coefficient of the additional linear term in the Riccati equation and the first-derivative term in its linearized form do not explicitly depend on time t (only an implicit time-dependence via $\gamma = \gamma(t)$ is possible), in the space-dependent case the coefficient depends linearly on the spatial variable and is, in contrast with the TD case, purely imaginary. In the TD case, the conservation law remains unchanged while in the space-dependent case the phase is changed in a way that corresponds to a unitary transformation of the wave function. On the other hand, the amplitude of the complex variable in the space-dependent situation remains unchanged whereas in the TD case, the amplitude is multiplied with an exponential damping factor. Nevertheless, the resulting Ermakov equations have the same structure in both cases. So, it seems that a change of the phase, corresponding to a unitary transformation, can have the same effect as a change of the amplitude of a complex variable which would correspond to a non-unitary transformation. The role of non-unitary transformations in the context of our dissipative quantum systems will be discussed in the next section.

	Time	Space
NL complex	$\frac{d}{dt}\left(\frac{\dot{\tilde{\lambda}}}{\tilde{\lambda}}\right)+\gamma\left(\frac{\dot{\tilde{\lambda}}}{\tilde{\lambda}}\right)+\left(\frac{\dot{\tilde{\lambda}}}{\tilde{\lambda}}\right)^2+\omega^2=0$	$\nabla\left(\frac{\nabla\Psi}{\Psi}\right)+i\gamma\frac{m}{\hbar}\tilde{x}\left(\frac{\nabla\Psi}{\Psi}\right)+\left(\frac{\nabla\Psi}{\Psi}\right)^2+\frac{2m}{\hbar^2}\left(\hat{E}_R-V+i\hbar\frac{\gamma}{4}\right)=0$
linearized	$\ddot{\tilde{\lambda}}+\gamma\dot{\tilde{\lambda}}+\omega^2\tilde{\lambda}=0$	$\Delta\Psi+i\gamma\frac{m}{\hbar}\tilde{x}(\nabla\Psi)+\frac{2m}{\hbar^2}\left(\hat{E}_R-V+i\hbar\frac{\gamma}{4}\right)\Psi=0$
complex variable	$\tilde{\lambda}=\alpha e^{i\varphi-\frac{\gamma}{2}t}$	$\Psi=ae^{iS/\hbar}$
conservation law	$\dot{\varphi}=\frac{1}{\alpha^2}$	$\nabla S'=\nabla\left(S+m\frac{\gamma}{4}\tilde{x}^2\right)=\frac{C}{a^2}$
NL real	$\ddot{\alpha}+\left(\omega^2-\frac{\gamma^2}{4}\right)\alpha=\frac{1}{\alpha^3}$	$\Delta a+\frac{2m}{\hbar^2}\left[\hat{E}'_R-\left(V-\frac{m}{2}\frac{\gamma^2}{4}\tilde{x}^2\right)\right]a=\frac{(C/\hbar)^2}{a^3}$
Riccati variables	$\frac{\dot{\tilde{\lambda}}}{\tilde{\lambda}}=\frac{\dot{\alpha}}{\alpha}-\frac{\gamma}{2}+i\frac{1}{\alpha^2}$	$\frac{\nabla\Psi}{\Psi}=\frac{\nabla a}{a}+i\frac{1}{\hbar}\left(\frac{C}{a^2}-m\frac{\gamma}{2}\tilde{x}\right)$

Figure 2: Comparison between the space-dependent and the time-dependent Riccati equations, their linearized forms and the corresponding Ermakov equations for the dissipative case.

5. Connections between nonlinear and linear dissipative approaches via non-unitary transformations

The connection between the explicitly TD approach of Caldirola–Kanai and the logarithmic NLSE (48) can be established by referring to Schrödinger's first communication on wave mechanics [1] which is based on the Hamilton–Jacobi form of classical mechanics. Writing the Hamilton–Jacobi equation in terms of S_c as

$$\frac{\partial}{\partial t}S_c+H\left(x,\frac{\partial}{\partial x}S_c,t\right) \;=\; 0 \tag{75}$$

(with the momentum $p_c = \frac{\partial}{\partial x}S_c$) Schrödinger arrived at the Hamiltonian operator $H_L = -\frac{\hbar^2}{2m}\frac{\partial^2}{\partial x^2}+V(x)$ via a variational ansatz after he introduced the wave function $\Psi(x,t)$ via $S_c = \frac{\hbar}{i}\ln\Psi$, where the subscript c (added by us) indicates that this action is a *complex* quantity, since Ψ is generally a complex function (a fact that Schrödinger did not like at all in the beginning [47]).

Reversing Schrödinger's procedure, but starting now with Eq. (48) (divided by Ψ, which causes no problems for Gaussian WPs) and, using the definition of S_c, one arrives at

$$\left(\frac{\partial}{\partial t}+\gamma\right)S_c+H \;=\; -\gamma < S_c > \;. \tag{76}$$

This is, of course, as little rigorous as Schrödinger's first attempt was, however it follows his idea of connecting the classical Hamilton–Jacobi theory with a wave (mechanical) equation. The purely TD term $-\gamma < S_c >$ is necessary mainly for normalization purposes (and can therefore be absorbed by the normalization coefficient) and shall be neglected in the following.

Multiplying the remaining Eq. (76) by $e^{\gamma t}$ and using the definitions

$$\hat{S}_c \;=\; e^{\gamma t}S_c \quad\text{and}\quad \hat{H} \;=\; e^{\gamma t}H \tag{77}$$

it can be rewritten as a Hamilton–Jacobi equation

$$\frac{\partial}{\partial t}\hat{S}_c + \hat{H} = 0 \ . \tag{78}$$

From the definition of the action function, it follows that the wave function $\hat{\Psi}(x,t)$ in the transformed (canonical) system is connected with the wave function $\Psi(x,t)$ in the physical system via the *non-unitary* relation

$$\ln\hat{\Psi} = e^{\gamma t} \ln\Psi \ . \tag{79}$$

Consequently, the (complex) momenta in the two systems are connected via

$$\hat{p}_c = \frac{\hbar}{i}\frac{\partial}{\partial x}\ln\hat{\Psi} = e^{\gamma t}\frac{\hbar}{i}\frac{\partial}{\partial x}\ln\Psi_{NL} = e^{\gamma t}p_c \ , \tag{80}$$

which is equivalent to the connection between the canonical and the kinetic momentum in the Caldirola–Kanai approach. The *non−canonical* connection between the classical variables (x,p) and $(\hat{x},\hat{p})$ (see section 3.1) corresponds to the *non − unitary* transformation between Ψ and $\hat{\Psi}$.

Note: Although Ψ and $\hat{\Psi}$ both depend explicitly on the same variables, x and t, the two wave functions are analytically different functions of x and t and have different physical meanings due to the non-unitary transformation (79). Ignoring this fact leads to unphysical results like violation of the uncertainty principle (for further details, see [48]).

Expressing $\hat{H}$ in terms of the canonical momentum $\hat{p}_c$ and following Schrödinger's quantization procedure, finally yields the modified SE (29) of the Caldirola–Kanai approach.

In section 3.1 it was already mentioned that a canonical description of the damped HO is also possible in terms of an exponentially-expanding system where the transformation (30) between the canonical and the physical variables is also a non-canonical transformation. Applying this canonical formalism, the equations of motion, after back-transformation into the physical variables, also provide the equation of motion (26) including the friction force. Whereas this approach and the one by Caldirola–Kanai are both connected with the physical level via a non-canonical transformation, the connection between both of them on the canonical level is given by the *canonical* transformation

$$\begin{aligned} Q &= e^{\gamma t/2}\hat{x} = e^{\gamma t/2}x \\ P &= e^{-\gamma t/2}\hat{p} + m\,\frac{\gamma}{2}e^{\gamma t/2}\hat{x} \ . \end{aligned} \tag{81}$$

Since $\hat{H}_{\text{CK}}$ is explicitly TD, while $\hat{H}_{\text{exp}}$ is a constant of motion, the transformation between them must involve an explicitly TD generating function $\hat{F}_2$, i.e.,

$$\hat{H}_{\text{exp}} = \hat{H}_{\text{CK}} + \frac{\partial}{\partial t}\hat{F}_2 \tag{82}$$

with

$$\hat{F}_2(\hat{x},P,t) = \hat{x}Pe^{\gamma t/2} - m\frac{\gamma}{4}\hat{x}^2e^{\gamma t} \ . \tag{83}$$

Consequently, also the action functions in both approaches are different and related by

$$\hat{S}_{\text{exp}} = \hat{S}_{\text{CK}} + m\frac{\gamma}{4}\hat{x}^2e^{\gamma t} \tag{84}$$

or with $\hat{S}_{\text{CK}} = S\,e^{\gamma t}$ and $\hat{x} = x$

$$\hat{S}_{\text{exp}} = \left(S + m\frac{\gamma}{4}x^2\right)e^{\gamma t} = S'\,e^{\gamma t} \ . \tag{85}$$

IOP Publishing
doi:10.1088/1742-6596/380/1/012009

The change from S to S' is the same that occurred in section 4.2 when the modified conservation law (66), $\nabla S' = C/a^2$, was obtained after the irreversible term (61) had been added to the continuity equation.

With Schrödinger's definition $S = \frac{\hbar}{i}\ln\,\Psi$, the canonical transformation corresponds to a phase-factor $\frac{im}{\hbar}\frac{\gamma}{4}\tilde{x}^2$ in the wave function. So, the transition between the NL physical level and the formal (linear) canonical level is given classically by a non-canonical transformation (corresponding to the non-unitary transformation (79) of the wave function). Whilst, on the canonical level, the transition between the two approaches discussed is possible via a canonical transformation (corresponding to a unitary transformation of the wave functions involving the above-mentioned phase factor). These relations are shown again schematically in fig. 3.

Finally a short comparison will be made with the canonical approach of Bateman that was introduced at the beginning of section 3.2. Earlier attempts [25] were made to link this approach to the one of Caldirola–Kanai by imposing some constraints. A recent attempt [26] takes advantage of the fact that the Bateman Hamiltonian H_{B}, Eq. (32), as well as the Hamiltonian for the expanding system $\hat{H}_{\mathrm{exp}}$, Eq. (31), represent the constant initial energy of system plus reservoir. Therefore, it is possible to transform the Bateman Hamiltonian exactly into $\hat{H}_{\mathrm{exp}}$ by expressing the variables of H_{B} in terms of the physical variables x and $\dot{x}$. Since the (canonical) transformation between $\hat{H}_{\mathrm{exp}}$ and $\hat{H}_{\mathrm{CK}}$ is known (see above), the exact link between H_{B} and $\hat{H}_{\mathrm{CK}}$ can also be obtained (further details will be published elsewhere).

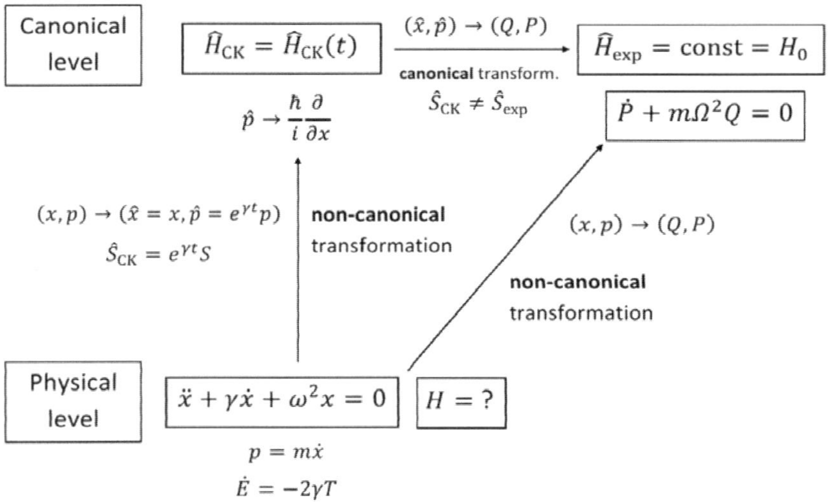

Figure 3: Connections between the physical and canonical descriptions of dissipative systems.

6. Conclusions

A fundamental difference between classical mechanics and quantum mechanics is the occurrence of complex quantities in quantum mechanics, not just for computational convenience but as a conceptual element of the foundations of physics. As a first example, Madelung's polar form of the complex wave function was used where the phase reflects the trajectory aspect of the system, whereas the amplitude takes into account the probabilistic aspect. Utilizing a NL formulation of the TISE, in terms of a Ermakov equation, it has been shown that phase and amplitude are not independent of each other but coupled via some kind of conservation law. The real NL Ermakov equation is equivalent to a complex NL Riccati equation where the quadratic term of this equation shows the origin of the afore-mentioned coupling. Formally, the same situation was found in the TDSE for cases where exact Gaussian WP solutions exist. A comparison of the respective quantities can be found in figure 4.

Several approaches were presented to effectively include the phenomena of irreversibility and dissipation into the systems mentioned above and their interrelations discussed. Specifically, a NL modification of the TDSE using a complex logarithmic nonlinearity was studied in detail and the changes in comparison with the corresponding conservative systems were analyzed. The modifications of the complex Riccati equation and its linearized form were of particular interest because, from there, the corresponding modifications for the TISE were concluded. This notably led to an additional first-derivate term with imaginary coefficient and a modified definition of the action in the phase of the wave function in order to obtain a similar conservation law and phase-amplitude coupling as in the conservative case (see also fig. 4).

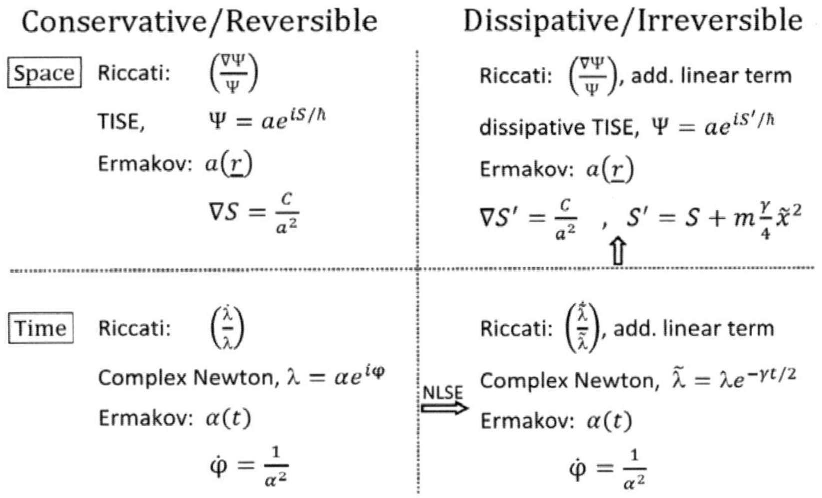

Figure 4: Interrelations between the space-dependent and time-dependent cases for conservative and dissipative systems.

This modification of the TISE allowed for a closer comparison between our logarithmic NLSE and the approaches by Süssmann, Albrecht and Hasse. On a level where operators and wave functions have the same physical meanings as in conventional quantum mechanics, our approach and that of Hasse's differ only by a phase factor. However, there are also linear approaches based on modified classical Hamiltonian functions (involving non-canonical transformations!) that attempt to describe the same dissipative systems but seem to provide, in certain cases, unphysical results. Analyzing the relation between our approach and the one of Caldirola–Kanai showed that this is mainly due to a non-unitary transformation between the log-NLSE using the conventional definitions of operators and wave functions, and the modified TDSEs using the canonical formalism but applying it to operators and wave functions that have different meanings from the conventional ones. Taking these facts into account, a consistent transition is possible between these two levels of description (see fig. 5).

Another canonical approach is related to the one of Caldirola–Kanai via a unitary transformation using an expanding coordinate system leading to a Hamiltonian that is a constant of motion. The corresponding change of the phase of the wave function is similar to the one that occurred when dissipation was introduced into the TISE ($S \to S'$). Finally, this constant Hamiltonian can be used to obtain an exact link between the NL and canonical approaches and the one by Bateman using one explicit, additional environmental system to mimic the surrounding heat bath.

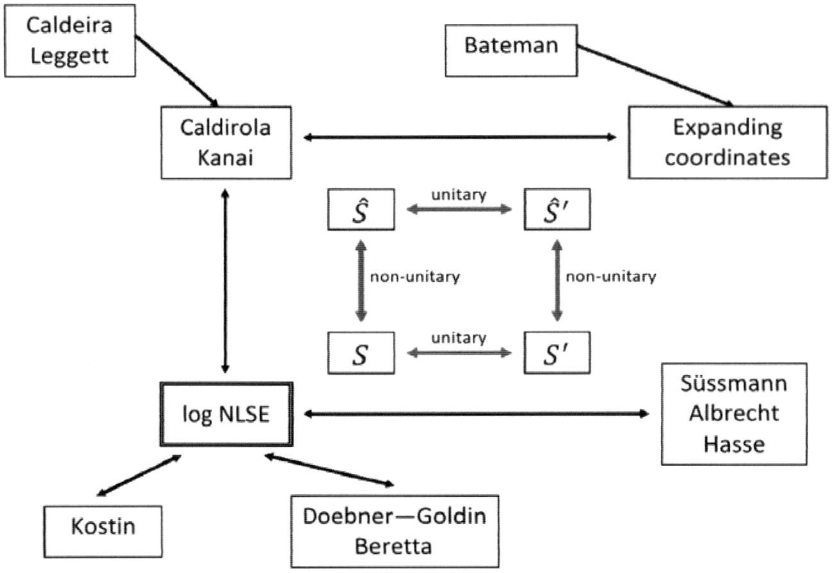

Figure 5: Interrelations between different physical and canonical descriptions of dissipative systems.

In conclusion, it was possible to establish connections between our logarithmic NL approach and the other approaches discussed in this paper (see fig. 5) based mainly on the complex Riccati equations that occur in this context.

Finally, it should be noted that the same type of Ermakov equation can be formally obtained either by changing the amplitude of the complex quantity that is used to linearize the Riccati equation (see lhs of fig. 2), or by changing the phase of the complex quantity (see rhs of fig. 2). In terms of quantum mechanical wave functions, this would mean that a non-unitary transformation could be replaced by a unitary one and vice-versa. This will be investigated in further detail elsewhere.

References

[1] Yang C N 1987 *Schrödinger – Centenary Celebration of a Polymath* ed C W Kilmister (Cambridge: Cambridge University Press) p 53
[2] Madelung E 1926 *Z. Physik* **40** 322
[3] Ermakov V P 1880 *Univ. Jzv. Kiev* **20** No 9, 1
[4] Schrödinger E 1926 *Ann. d. Phys.* **79** 361
[5] Reinisch G 1994 *Physica* A **206** 229
[6] Schuch D 2008 *SIGMA* **4** 043
[7] Schuch D and Moshinsky M 2006 *Phys. Rev.* A **73** 062111
[8] Schuch D 2005 *Phys. Lett,* A **338** 225
[9] Schuch D 2011 in: *Latin-American School of Physics XL ELAF; AIP Conf. Proceedings* **1334** 291
[10] Schuch D 1992 *Int. J. Quantum Chem.* **42** 663
[11] Einstein A 1905 *Ann. d. Phys.* **17** 549
[12] Caldirola P 1941 *Nuovo Cimento* **18** 393
[13] Kanai E 1948 *Progr. Theor. Phys.* **3** 440
[14] Schuch D 1999 *Int. J. Quantum Chem.* **72** 537
[15] Bateman H 1931 *Phys. Rev.* **38** 815
[16] Caldeira A O and Leggett A J 1981 *Phys. Rev. Lett.* **46** 211
[17] Caldeira A O and Leggett A J 1983 *Ann. Phys. (N.Y.)* **149** 374
[18] Caldeira A O and Leggett A J 1983 *Ann. Phys. (N.Y.)* **153** 445(E)
[19] Gorini V, Kossakowski A and Sudarshan E C G 1976 *J. Math. Phys.* **17** 821
[20] Lindblad G 1976 *Commun. Math. Phys.* **48** 119

[21] Gisin N 1981 *J. Phys.* A **14** 2259
[22] Gisin N 1982 *Physica* A **111** 364
[23] Beretta G P 1987 *Found. Phys.* **17** 365
[24] Beretta G P 2006 *Phys. Rev.* E **73** 026113
[25] Kostin M D 1972 *J. Chem. Phys.* **57** 3589
[26] Cho B R 1980 *Suhak Gwa Mulri (Korean) (Math. and Phys.)* **3** 37
[27] Albrecht K 1975 *Phys. Lett.* B **56** 127
[28] Süssmann G 1973 unpublished
[29] Hasse R W 1975 *J. Math. Phys.* **16** 2005
[30] Schuch D, Chung K-M and Hartmann H 1983 *J. Math. Phys.* **24** 1652
[31] Schuch D, Chung K-M and Hartmann H 1984 *J. Math. Phys.* **25** 3086
[32] Schuch D and Chung K-M 1986 *Int. J. Quantum Chem.* **29** 1561
[33] Doebner H-D and Goldin G A 1992 *Phys. Lett.* A **162** 397
[34] Doebner H-D and Goldin G A 1994 *J. Phys.* A **27** 1771
[35] Guerrero J, López-Ruiz F F, Aldaya V and Cossío F 2011 in: *GROUP 28: Physical and Mathematical Aspects of Symmetry; J. Phys.: Conf. Ser.* **284** 012062 (Bristol: IOP Publishing)
[36] Schuch D, Guerrero J, López-Ruiz F F and Aldaya V: in preparation
[37] Yu L H and Sun C P 1994 *Phys. Rev.* A **49** 592
[38] Sun C P and Yu L H 1995 *Phys. Rev.* A **51** 1845
[39] Madelung E 1950 *Die Mathematischen Hilfsmittel des Physikers* (Berlin: Springer) p 432
[40] Mrowka B 1951 *Z. Phys.* **130** 164
[41] Doebner H-D and Goldin G A 1996 *Phys. Rev.* A **54** 3764
[42] Goldin G A 1995 in; *Nonlinear, Deformed and Irreversible Quantum Systems* (Singapore: World Scientific) p 125
[43] Schuch D 2000 in: *Quantum Theory and Symmetries* (Singapore: World Scientific) p 152
[44] Nassar A B 1986 *J. Math. Phys.* **27** 755
[45] Nassar A B 1986 *J. Math. Phys.* **27** 2949
[46] Schuch D 2007 *J. Math. Phys.* **48** 122701
[47] Przibram K 1963 *Briefe zur Wellenmechanik* (Wien: Springer) 52
[48] Schuch D 1997 *Phys. Rev.* A **55** 935

Journal of Physics: Conference Series **380** (2012) 012010 doi:10.1088/1742-6596/380/1/012010

The Quantum Arnold Transformation for the damped harmonic oscillator: from the Caldirola-Kanai model toward the Bateman model

F F López-Ruiz[a,1], **J Guerrero**[b,a], **V Aldaya**[a] **and F Cossío**[a]

[a] Instituto de Astrofísica de Andalucía (IAA - CSIC), Apartado Postal 3004, 18080 Granada, Spain

[b] Departamento de Matemática Aplicada, Universidad de Murcia, Campus de Espinardo, 30100 Murcia, Spain

E-mail: `flopez@iaa.es`, `juguerre@um.es`, `valdaya@iaa.es`, `fcossiop@gmail.com`

Abstract. Using a quantum version of the Arnold transformation of classical mechanics, all quantum dynamical systems whose classical equations of motion are non-homogeneous linear second-order ordinary differential equations (LSODE), including systems with friction linear in velocity such as the damped harmonic oscillator, can be related to the quantum free-particle dynamical system. This implies that symmetries and simple computations in the free particle can be exported to the LSODE-system. The quantum Arnold transformation is given explicitly for the damped harmonic oscillator, and an algebraic connection between the Caldirola-Kanai model for the damped harmonic oscillator and the Bateman system will be sketched out.

1. Introduction

The interest in dissipative systems at the quantum level has remained constant since the early days of Quantum Mechanics. The difficulties in describing damping, which intuitively could be understood as a mesoscopic property, within the fundamental quantum framework, have motivated a huge amount of papers. In particular, the quantum damped harmonic oscillator, frequently described by the Caldirola-Kanai equation [1, 2], has attracted much attention, as it could be considered one of the simplest and paradigmatic examples of dissipative system.

The model for the quantum damped harmonic oscillator by Caldirola and Kanai (C-K), which includes a time-dependent Hamiltonian, has been considered to have some flaws. For instance, it is claimed that uncertainty relations are not preserved under time evolution and could eventually be violated [3, 4]. However, this inconsistency seems to be associated with a confusion between canonical momentum and "physical" momentum [5]. It should be understood that the C-K model accounts for a "canonical-level" description of the damped system, which in fact can be mapped into a "physical-level" one [6].

Coherent states for a generalized C-K model were calculated in [7] by finding creation and annihilation operators, built out of operators which commute with the Schrödinger equation. The corresponding number operator turns out to be an auxiliary, conserved operator, obviously different from the time-dependent Hamiltonian. This paper also defined the so-called loss energy

[1] Now at the Department of Applied Physics, University of Cádiz, Spain.

Journal of Physics: Conference Series **380** (2012) 012010 doi:10.1088/1742-6596/380/1/012010

states for the damped harmonic oscillator. The report by Dekker [8] provides a historical overview of some relevant results.

In a purely classical context, the symmetries of the equation of the damped harmonic oscillator with time-dependent parameters were found in [9]. Two comprehensive articles, [10, 11], are of special interest. In those papers the authors found, for the damped harmonic oscillator, finite-dimensional point symmetry groups for the corresponding Lagrangian (the un-extended Schrödinger group [12, 13, 14, 15]) and the equations of motion ($SL(3,\mathbb{R})$) respectively, and an infinite contact one for the set of trajectories of the classical equation. They singled out a "non-conventional" Hamiltonian from those generators of the symmetry, recovering some results from [7]. Then, they concluded that the damped harmonic oscillator should not be claimed to be dissipative at all at the quantum level, as this true, "non-conventional" Hamiltonian is conserved, and should be related to an oscillator with variable frequency.

There exists another interesting approach to the study of the classical damped harmonic oscillator, based on the observation that its classical equation of motion is a special case of the set of linear second-order ordinary differential equations (LSODE for short). In Classical Mechanics the family of solutions of a second-order differential equation corresponding to the motion of a given physical problem is sometimes related to that of a simpler system, considered as a toy model, in order to import from it simple general properties which could be hidden in the real problem. Both physical systems should share global properties of the solution manifold, such as topology and symplectic structure. The paradigmatic example is the transformation described by Arnold in [16], which brings any LSODE to the simplest form of the free Galilean particle equation. This transformation turns out to be extremely useful. In particular, it is possible to obtain the symmetry group of a particular instance of LSODE [9], in which the symmetries of the action of the corresponding system can be found as a subgroup [17].

Therefore, it is rather natural to generalize the Arnold transformation to the quantum level, to be denoted as Quantum Arnold Transformation (QAT), as much insight can be gained in the study of any system classically described by a LSODE and, in particular, the damped harmonic oscillator. This generalization was presented in [18] by the authors, and we shall provide here a short review on the subject[2].

Besides the Caldirola-Kanai model, the Bateman dual system appears as an alternative description of dissipation in the damped harmonic oscillator. In his original paper [19], Bateman looked for a variational principle for equations of motion with a friction term linear in velocity, but he allowed the presence of extra equations. This trick effectively doubles the number of degrees of freedom, introducing a time-reversed version of the original damped harmonic oscillator, which acts as an energy reservoir and could be considered as an effective description of a thermal bath. The Hamiltonian that describes this system was rediscovered by Feschbach and Tikochinsky [20, 21, 22, 8] and the corresponding quantum theory was immediately analyzed.

Some issues regarding Bateman's system arose. The Hamiltonian presents a set of complex eigenvalues of the energy (see [23] and references therein), and the vacuum of the theory seems to decay with time. This last feature was treated in [24], where Celeghini et al. suggested that the quantum theory of the dual system could find a more natural framework in quantum field theory. However, we understand that the ultimate reason is nevertheless the lack of a vacuum representation of the relevant group: there is no vacuum vanishing at all as there is no conventional vacuum [25, 26]: defining such a vacuum is an artifact of the usual Feschbach-Tikochinski construction [20, 21, 8]. On the other hand, in [23] the generalized eigenvectors corresponding to the complex eigenvalues are interpreted as resonant states, and the real spectrum is obtained (see also [25, 26]).

Bateman's dual system is still frequently discussed [27]. Many authors have considered this

[2] Several partial generalizations of the classical Arnold transformation to the quantum theory can in fact be found in the literature (see [18] for references).

model as a *good* starting point for the formulation of the quantum theory of dissipation. One of the aims of this paper will be to show that the study of the symmetries of the Caldirola-Kanai model leads to the Bateman dual system, thus to be considered as a *natural* starting point for the study of quantum dissipation.

The paper is organized as follows. In Section 2 we review the classical Arnold transformation (CAT) as well as its quantum counterpart (QAT), and provide the particular expressions for the case of the Caldirola-Kanai model for the damped harmonic oscillator. Section 3 is devoted to the algebraic derivation of the Bateman dual system closing a Lie algebra for the conserved operators obtained through the QAT for the Caldirola-Kanai model together with the (non-conserved) Hamiltonian.

2. The Arnold transformation for the damped harmonic oscillator

The Caldirola-Kanai equation [1, 2], which attempts to describe the damped harmonic oscillator in one spatial dimension (with classical equation $\ddot{x}(t)+\gamma\dot{x}(t)+\omega^2x(t)=0$), at the canonical level, can be written:

$$i\hbar\frac{\partial\phi}{\partial t}=\hat{H}_{DHO}\,\phi\equiv-\frac{\hbar^2}{2m}e^{-\gamma t}\frac{\partial^2\phi}{\partial x^2}+\frac{1}{2}m\omega^2x^2e^{\gamma t}\phi\,, \tag{1}$$

where ω is the given, constant frequency and γ is the friction constant. Let $\mathcal{H}_{\rm FP}$ be the Hilbert space of solutions of the free particle, time-dependent Schrödinger equation, and $\mathcal{H}_{\rm DHO}$ the one corresponding to the damped harmonic oscillator. We shall denote by $\psi(\kappa,\tau)\in\mathcal{H}_{\rm FP}$ the free particle solutions (κ and τ denote the spatial and time variable for the free particle, respectively) and by $\phi(x,t)\in\mathcal{H}_{\rm DHO}$ the damped harmonic oscillator ones. Then, the QAT [18], relating solutions of time-dependent Schrödinger equations for the free particle and the damped harmonic oscillator, or rather, its inverse, is given by:

$$\phi(x,t)=\hat{A}^{-1}\psi(\kappa,\tau)=\frac{1}{\sqrt{u_2(t)}}e^{\frac{i}{2}\frac{m}{\hbar}\frac{1}{W(t)}\frac{\dot{u}_2(t)}{u_2(t)}x^2}\psi(\frac{x}{u_2(t)},\frac{u_1(t)}{u_2(t)})\,. \tag{2}$$

This transformation is unitary, and its form is quite general [18]. For a general, homogeneous LSODE, the classical Arnold transformation (CAT) [16], present in (2), is explicitly

$$\begin{aligned}A:\ &\mathbb{R}\times T\longrightarrow\mathbb{R}\times\mathcal{T}\\ &(x,t)\longmapsto(\kappa,\tau)=A\big((x,t)\big)=\big(\tfrac{x}{u_2(t)},\tfrac{u_1(t)}{u_2(t)}\big)\,.\end{aligned} \tag{3}$$

$\mathcal{T}$ and T are open intervals of the real line containing $\tau=0$ and $t=0$, respectively, $u_1(t)$ and $u_2(t)$ are two independent solutions of the LSODE (here dots mean derivation with respect to t):

$$\ddot{x}+\dot{f}(t)\dot{x}+\omega(t)^2x=0\,, \tag{4}$$

and $W(t)$ is the Wronskian $W(t)=\dot{u}_1(t)u_2(t)-u_1(t)\dot{u}_2(t)$ of the two solutions. Applying the change of variables A to a solution $x(t)$ of (4) on a given interval of time T, a solution $\kappa(\tau)$ of the classical equation of motion $\ddot{\kappa}=0$ on a interval $\mathcal{T}$ is obtained, where now dots mean derivation with respect to τ.

We impose on u_1 and u_2 the condition that they preserve the identity of τ and κ, i.e., that (κ,τ) coincide with (x,t) at an initial point t_0, arbitrarily taken to be $t_0=0$:

$$u_1(0)=0,\ u_2(0)=1,\qquad \dot{u}_1(0)=1,\ \dot{u}_2(0)=0\,. \tag{5}$$

This fixes a unique form of A for a given "target" LSODE-type physical system. However, the quantum Arnold transformation is still valid if solutions u_1 and u_2 do not satisfy (5) (see [18] for details).

For the case of the damped harmonic oscillator $\dot{f} = \gamma$ and $\omega(t) = \omega$, and the two independent solutions can be chosen (see [18] for details) as

$$u_1(t) = \frac{1}{\Omega} e^{-\frac{\gamma}{2}t} \sin \Omega t, \qquad u_2(t) = e^{-\frac{\gamma}{2}t} \cos \Omega t + \frac{\gamma}{2\Omega} e^{-\frac{\gamma}{2}t} \sin \Omega t, \tag{6}$$

where, computing, $W(t) \equiv \dot{u}_1(t) u_2(t) - u_1(t) \dot{u}_2(t) = e^{-\gamma t}$, and

$$\Omega = \sqrt{\omega^2 - \frac{\gamma^2}{4}}\,. \tag{7}$$

Note that these solutions have good limit in the case of critical damping $\omega = \frac{\gamma}{2}$.

Then, with the change of variables given by (3), the classical equation of the free particle $\ddot{\kappa}(\tau) = 0$ transforms into the classical equation of the damped harmonic oscillator $\ddot{x}(t) + \gamma \dot{x}(t) + \omega^2 x(t) = 0$ up to a time-dependent global factor.

Coming back to the quantum level, the QAT can be diagrammatically represented as:

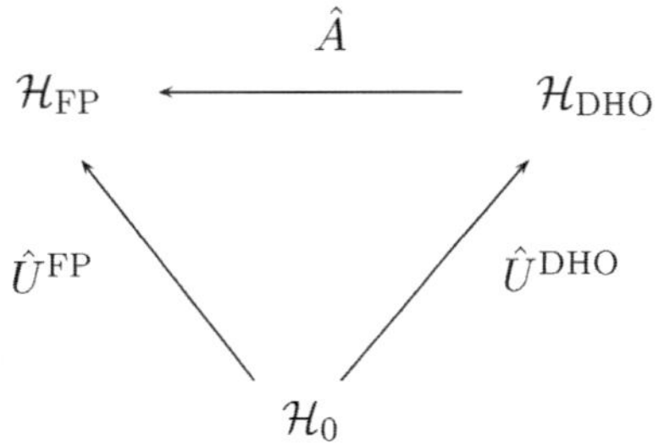

where $\hat{U}^{\mathrm{FP}}$ and $\hat{U}^{\mathrm{DHO}}$ stand for the evolution operators of the free particle and the damped harmonic oscillator, respectively, while $\mathcal{H}_0$ is the Hilbert space, either for the free particle or the damped harmonic oscillator, of solutions of their respective Schrödinger equations at $t = 0$. The fact that $\mathcal{H}_0$ is common to both systems is a consequence of imposing the conditions (5) (see [18] for the required modifications when these conditions are not satisfied).

Thanks to the commutativity of this diagram, and the unitarity of the operators appearing in it, we can map objects (wave functions, operators, expectation values, uncertainties) from one system to the other. In [18] we benefited from this fact for transporting the simplicity of the free particle to more involved systems, finding, for instance, an analytic expression for the evolution operator of the complicated systems (even with time-dependent Hamiltonians) in terms of the free particle evolution operator.

As already remarked, the basic symmetries of the free system are inherited by the LSODE-type system, as we are now able to transform the infinitesimal generators of translations (the Galilean momentum operator $\hat{\pi}$, corresponding to the classical conserved quantity 'momentum') and non-relativistic boosts (the position operator $\hat{\kappa}$, corresponding to the classical conserved quantity 'initial position'). They are, explicitly,

$$\hat{\pi} = -i\hbar \frac{\partial}{\partial \kappa} \tag{8}$$

$$\hat{\kappa} = \kappa + \frac{i\hbar}{m} \tau \frac{\partial}{\partial \kappa}\,, \tag{9}$$

that is, those basic, canonically commuting operators with constant expectation values, that respect the solutions of the free Schrödinger equation, have constant matrix elements (and constant expectation values in particular) and fall down to well defined, time-independent operators in the Hilbert space of the free particle $L^2(\mathbb{R})$. In general, these properties are satisfied whenever an operator $\hat{O}(t)$ can be written as

$$\hat{O}(t) = \hat{U}(t, t_0)\, \hat{O}\, \hat{U}^\dagger(t, t_0)\,, \tag{10}$$

where $\hat{O}$ is $\hat{O}(t_0)$ and $\hat{U}(t, t_0)$ is the evolution operator satisfying the Schrödinger equation[3]. If the Hamiltonian is time-independent, as in the free particle case, time-evolution is a one-parameter group and then $\hat{U}(\tau, \tau_0) = \hat{U}(\tau - \tau_0)$.

Let us apply the QAT (2) to (8) and (9). For the operator $\hat{\pi}$ acting on $\mathcal{H}_{\rm FP}$ we have the corresponding operator $\hat{P} = \hat{A}^{-1}\hat{\pi}\hat{A}$ on $\mathcal{H}_{\rm DHO}$. The action on functions $\phi(x,t)$ can then be obtained as follows:

$$\begin{aligned}
\hat{P}\phi(x,t) &= \hat{A}^{-1}\hat{\pi}\hat{A}\phi(x,t) = \hat{A}^{-1}\hat{\pi}A^*\big(\sqrt{u_2}\, e^{-\frac{i}{2}\frac{m}{\hbar}\frac{1}{W}\frac{\dot{u}_2}{u_2}x^2}\,\phi(x,t)\big) = \\
&= \frac{1}{\sqrt{u_2}} e^{+\frac{i}{2}\frac{m}{\hbar}\frac{1}{W}\frac{\dot{u}_2}{u_2}x^2} A^{*-1}\Big(-i\hbar\frac{\partial}{\partial\kappa}A^*\big(\sqrt{u_2}\, e^{-\frac{i}{2}\frac{m}{\hbar}\frac{1}{W}\frac{\dot{u}_2}{u_2}x^2}\,\phi(x,t)\big)\Big) = \\
&= \frac{1}{\sqrt{u_2}} e^{+\frac{i}{2}\frac{m}{\hbar}\frac{1}{W}\frac{\dot{u}_2}{u_2}x^2}\Big(-i\hbar u_2\frac{\partial}{\partial x}\big(\sqrt{u_2}\, e^{-\frac{i}{2}\frac{m}{\hbar}\frac{1}{W}\frac{\dot{u}_2}{u_2}x^2}\,\phi(x,t)\big)\Big) = \\
&= \big(-i\hbar u_2\frac{\partial}{\partial x} - m\frac{\dot{u}_2}{W}x\big)\phi(x,t)\,.
\end{aligned} \tag{11}$$

We can perform the same computation for the position operator and then we have, explicitly:

$$\hat{P} = -i\hbar e^{-\frac{\gamma t}{2}}(\cos\Omega t + \frac{\gamma}{2\Omega}\sin\Omega t)\frac{\partial}{\partial x} + m\,\frac{\omega^2}{\Omega}\,e^{\frac{\gamma t}{2}}\,\sin\Omega t\, x\,, \tag{12}$$

$$\hat{X} = e^{\frac{\gamma t}{2}}(\cos\Omega t - \frac{\gamma}{2\Omega}\sin\Omega t)\,x + i\hbar\frac{e^{-\frac{\gamma t}{2}}}{m\Omega}\,\sin\Omega t\frac{\partial}{\partial x}\,, \tag{13}$$

thus providing the generators of the realization of the (centrally-extended) Heisenberg-Weyl symmetry on the damped harmonic oscillator.

One key observation is that $\hat{H}_{DHO}$ does not make sense as an operator acting on the space of solutions of (1), while $\hat{P}$ and $\hat{X}$ do respect solutions. This can be proved by direct calculation or even obtaining the evolution operator $\hat{U}(t, t_0)$ and using it to de-evolve these operators to check that they are actually conserved: they fall to well defined operators on the space $\mathcal{H}_0$. The fact that $\hat{H}_{DHO}$ does not commute at different times $[\hat{H}_{DHO}(t_1), \hat{H}_{DHO}(t_2)] \neq 0$ makes its calculation trickier using conventional methods, but QAT can be used to obtain a simple, exact form [18]. The action of $\hat{U}(t)$ on $\hat{H}_{DHO}(t)$ shows that it does not fall down to the quotient by the time evolution generated by itself.

Let us stress that QAT can be useful to quickly perform some calculations [28, 29], avoiding tedious, direct evaluations which can become extremely involved in the system under study. For example, it can be used to compute the quantum propagator for any LSODE-type quantum system, following the idea of Takagi in [30] for the simple case of the harmonic oscillator, or even, as already mentioned, the evolution operator $\hat{U}(t)$, which becomes very difficult to evaluate exactly when the Hamiltonian is time-dependent and does not commute with itself at different times, which is the case of the C-K Hamiltonian.

Actually, the evolution operator of the C-K damped harmonic oscillator can be related to the free evolution operator. Having in mind the diagram above, we write:

$$\hat{A}\big(\hat{U}^{\rm DHO}(t)\phi(x)\big) = \hat{U}^{\rm FP}(\tau)\varphi(\kappa)\,. \tag{14}$$

[3] Note that $\hat{O}(t)$ is *not* the usual Heisenberg picture version $\hat{O}_H(t) = \hat{U}(t,t_0)^\dagger\, \hat{O}\, \hat{U}(t,t_0)$ of its associated operator in Schrödinger picture $\hat{O}$, although their relation is very simple when the Hamiltonian is time-independent.

Here ϕ and φ are *the same* function of only one argument (κ or x) and we shall denote it $\varphi = \phi = \psi$. Then,

$$\hat{U}^{\rm DHO}(t)\psi(x) = \hat{A}^{-1}\big(\hat{U}^{\rm FP}(\tau)\psi(\kappa)\big) = \frac{1}{\sqrt{u_2}} e^{\frac{i}{2}\frac{m}{\hbar}\frac{1}{W}\frac{\dot{u}_2}{u_2}x^2} A^{*-1}\big(\hat{U}^{\rm FP}(\tau)\psi(\kappa)\big) =$$
$$= \frac{1}{\sqrt{u_2}} e^{\frac{i}{2}\frac{m}{\hbar}\frac{1}{W}\frac{\dot{u}_2}{u_2}x^2} A^{*-1}\big(\hat{U}^{\rm FP}(\tau)\big) A^{*-1}\big(\psi(\kappa)\big) . \quad (15)$$

To factorize the function ψ and single out the general action of $\hat{U}^{\rm DHO}(t)$, we compute

$$A^{*-1}\big(\psi(\kappa)\big) = \psi(\tfrac{x}{u_2}) = e^{\log(1/u_2)x\frac{\partial}{\partial x}}\psi(x) , \quad (16)$$

where $e^{\log(1/u_2)x\frac{\partial}{\partial x}}$ is a dilation operator which is *not* unitary. To unitarize this operator, the generator must be shifted from $x\frac{\partial}{\partial x}$ to $x\frac{\partial}{\partial x}+\frac{1}{2}$, so that the true unitary operator is then

$$\hat{U}_D(\tfrac{1}{u_2}) = e^{\log(1/u_2)(x\frac{\partial}{\partial x}+\frac{1}{2})} = \frac{1}{\sqrt{u_2}} e^{\log(1/u_2)x\frac{\partial}{\partial x}} . \quad (17)$$

But the factor $\frac{1}{\sqrt{u_2}}$ is already present in the previous expression of $\hat{U}^{\rm DHO}(t)$. Therefore, it now reads

$$\hat{U}^{\rm DHO}(t) = e^{\frac{i}{2}\frac{m}{\hbar}\frac{1}{W}\frac{\dot{u}_2}{u_2}x^2} A^{*-1}\big(\hat{U}^{\rm FP}(\tau)\big)\hat{U}_D(\tfrac{1}{u_2}) =$$
$$= \frac{1}{\sqrt{u_2}} e^{\frac{i}{2}\frac{m}{\hbar}\frac{1}{W}\frac{\dot{u}_2}{u_2}x^2} e^{\frac{i\hbar}{2m}u_1u_2\frac{\partial^2}{\partial x^2}} e^{\log(1/u_2)x\frac{\partial}{\partial x}} . \quad (18)$$

Its inverse is given by

$$\hat{U}^{\rm DHO}(t)^{-1} = \hat{U}^{\rm DHO}(t)^{\dagger} = \sqrt{u_2} e^{\log(u_2)x\frac{\partial}{\partial x}} e^{\frac{-i\hbar}{2m}u_1u_2\frac{\partial^2}{\partial x^2}} e^{-\frac{i}{2}\frac{m}{\hbar}\frac{1}{W}\frac{\dot{u}_2}{u_2}x^2} . \quad (19)$$

Interestingly, we have been able to obtain an *exact* expression for the evolution operator as a product of operators. No perturbative approximation method, which could become cumbersome in some cases, is needed for *any* LSODE-related quantum system to obtain the evolution operator.

3. Deriving dissipative forces from a symmetry

Even though it is possible to set up a clear framework to deal with the quantum Caldirola-Kanai system for the DHO by employing the QAT, this does not provide by itself a well-defined operator (on solutions) associated with the actual time evolution. As mentioned previously, this is rooted in the fact that the conventional time evolution is not included in the symmetry group that can be imported from the free system: the Hamiltonian does not preserve the Hilbert space of solutions of the DHO Schrödinger equation. One may wonder what happens if time evolution symmetry is forced. We shall pursue this issue for the damped harmonic oscillator in this Section.

3.1. Including time symmetry

In the Caldirola-Kanai model of the damped harmonic oscillator, neither the operator $i\hbar\frac{\partial}{\partial t}$, nor $\hat{H}_{DHO}$ (which coincides with the former on solutions) close a Lie algebra under commutation with $\hat{X}$ and $\hat{P}$. We shall impose the condition for the time translation to be a symmetry. But will do it in an elegant way, trying to close an algebra of (constant, symmetry generating) observables, taking advantage of the expressions obtained for the basic operators. Implicitly, we are forcing the system to become conservative, but this is the only information we are going to provide. The implementation is as follows: we wonder whether it is possible to incorporate $i\hbar\frac{\partial}{\partial t}$ into the basic Lie algebra of operators, trying to close an enlarged Lie algebra acting on the (possibly enlarged) Hilbert space. The answer to this question is in the affirmative, but it requires a delicate analysis.

The resulting enlarged algebra will include $\hat{X}, \hat{P}, \hat{H} \equiv \hat{H}_{DHO} = i\hbar\frac{\partial}{\partial t}$ and four more generators (plus the central one $\hat{I}$), denoted by $\hat{Q}, \hat{\Pi}, \hat{G}_1$ and $\hat{G}_2$. Then, together with the generators $\hat{X}$ and $\hat{P}$ and the Hamiltonian, let us introduce the following operators:

$$\hat{\Pi} = i\hbar e^{-\frac{\gamma t}{2}}(\cos\Omega t - \frac{\gamma}{2\Omega}\sin\Omega t)\frac{\partial}{\partial x} - m\frac{\omega^2}{\Omega}e^{\frac{\gamma t}{2}}\sin\Omega t\, x \tag{20}$$

$$\hat{\hat{Q}} = e^{\frac{\gamma t}{2}}(\cos\Omega t - \frac{3\gamma}{2\Omega}\sin\Omega t)\, x + i\hbar\frac{e^{-\frac{\gamma t}{2}}}{m\Omega}\sin\Omega t\frac{\partial}{\partial x} \tag{21}$$

$$\hat{G}_1 = \frac{1}{4\Omega^2}\left(-4\omega^2 + \gamma^2\cos 2\Omega t + 2\gamma\Omega\sin 2\Omega t\right), \tag{22}$$

$$\hat{G}_2 = -\frac{\gamma}{\Omega^2}\sin^2\Omega t, \tag{23}$$

so that they close the seven-dimensional algebra:

$$\begin{aligned}
&\left[\hat{X}, \hat{P}\right] = i\hbar\hat{I} && \left[\hat{\hat{Q}}, \hat{\Pi}\right] = 2i\hbar\hat{G}_1 + i\hbar\hat{I} \\
&\left[\hat{X}, \hat{\hat{Q}}\right] = \frac{i\hbar}{m}\hat{G}_2 && \left[\hat{X}, \hat{\Pi}\right] = i\hbar\hat{G}_1 \\
&\left[\hat{\hat{Q}}, \hat{P}\right] = -i\hbar\hat{G}_1 + i\hbar\gamma\hat{G}_2 && \left[\hat{P}, \hat{\Pi}\right] = -i\hbar m\omega^2\hat{G}_2 \\
&\left[\hat{H}, \hat{X}\right] = \frac{i\hbar}{m}\hat{\Pi} && \left[\hat{H}, \hat{P}\right] = 2i\hbar m\omega^2\hat{X} - i\hbar m\omega^2\hat{\hat{Q}} \\
&\left[\hat{H}, \hat{\hat{Q}}\right] = -2i\hbar\gamma\hat{X} - \frac{i\hbar}{m}\hat{P} + i\hbar\gamma\hat{\hat{Q}} && \left[\hat{H}, \hat{\Pi}\right] = -3i\hbar m\omega^2\hat{X} + 2i\hbar m\omega^2\hat{\hat{Q}} - i\hbar\gamma\hat{\Pi} \\
&\left[\hat{H}, \hat{G}_1\right] = -i\hbar\gamma\hat{G}_1 + 2i\hbar\omega^2\hat{G}_2 && \left[\hat{H}, \hat{G}_2\right] = -2i\hbar\hat{G}_1 + i\hbar\gamma\hat{G}_2 - 2i\hbar\hat{I}\,.
\end{aligned} \tag{24}$$

The outcome of this process depends on the realization, which in fact contains crucial information: we have started with a reducible representation of the Heisenberg-Weyl algebra with details about the time evolution of the system and we have arrived at an algebra whose physical content, if any, must be unveiled.

We see that this algebra corresponds to a centrally extended algebra. The central extensions determine the actual basic conjugated pairs and classify possible quantizations. The operators $\hat{\hat{Q}}$ and $\hat{\Pi}$ (plus $\hat{I}$) expand a Heisenberg-Weyl subalgebra, and $\hat{H}, \hat{G}_1$ and $\hat{G}_2$ expand a 2-D affine algebra (with $\hat{H}$ acting as dilations). However, in this realization $\hat{\hat{Q}}$ and $\hat{\Pi}$ are not basic, while $\hat{H}$ and $\hat{G}_2$ are basic, resulting in time being a canonical variable. This result is puzzling: time is not a coordinate or momentum of any degree of freedom and therefore the corresponding generator should not appear as an element of a conjugate pair. Is it possible to take advantage of the information encoded in this algebra to describe a physical system in which we have an evolution with respect to an external time variable? In fact, it is.

Our strategy here is to consider other possible quantizations of the un-extended algebra. A detailed study of the (projective) representations of the enlarged (7+1) dimensional Lie algebra (that is, the possible central extensions) is going to show that there are three relevant kinds of representations, describing systems with different degrees of freedom. The underlying "classical symmetries" shared by this class of physical realizations are the same, and contain the information about the time evolution of the system. And we shall see that there is in fact a unique quantization in which time is a non-canonical variable. This is the reason why the whole process is unambiguously defined.

Thinking of the algebra above as an abstract Lie algebra, it can be shown that a parameter k controls the central extensions which are allowed by the Jacobi identity of Lie algebras:

$$\begin{aligned}
&\left[\hat{X},\hat{P}\right] = i\hbar\hat{I} && \left[\hat{\hat{Q}},\hat{\Pi}\right] = 2i\hbar\hat{G}_1 + i\hbar k\hat{I} \\
&\left[\hat{X},\hat{\hat{Q}}\right] = \frac{i\hbar}{m}\hat{G}_2 && \left[\hat{X},\hat{\Pi}\right] = i\hbar\hat{G}_1 \\
&\left[\hat{\hat{Q}},\hat{P}\right] = -i\hbar\hat{G}_1 + i\hbar\gamma\hat{G}_2 + i\hbar(1-k)\hat{I} && \left[\hat{P},\hat{\Pi}\right] = -i\hbar m\omega^2\hat{G}_2 \\
&\left[\hat{H},\hat{X}\right] = \frac{i\hbar}{m}\hat{\Pi} && \left[\hat{H},\hat{P}\right] = 2i\hbar m\omega^2\hat{X} - i\hbar m\omega^2\hat{\hat{Q}} \\
&\left[\hat{H},\hat{\hat{Q}}\right] = -2i\hbar\gamma\hat{X} - \frac{i\hbar}{m}\hat{P} + i\hbar\gamma\hat{\hat{Q}} && \left[\hat{H},\hat{\Pi}\right] = -3i\hbar m\omega^2\hat{X} + 2i\hbar m\omega^2\hat{\hat{Q}} - i\hbar\gamma\hat{\Pi} \\
&\left[\hat{H},\hat{G}_1\right] = -i\hbar\gamma\hat{G}_1 + 2i\hbar\omega^2\hat{G}_2 && \left[\hat{H},\hat{G}_2\right] = -2i\hbar\hat{G}_1 + i\hbar\gamma\hat{G}_2 - i\hbar(1+k)\hat{I}\,.
\end{aligned} \tag{25}$$

It is convenient to perform the linear shift:

$$\hat{Q} \equiv -\hat{\hat{Q}} + (1-k)\hat{X}\,, \tag{26}$$

so that the actual degrees of freedom diagonalize:

$$\begin{aligned}
&\left[\hat{X},\hat{P}\right] = i\hbar\hat{I} && \left[\hat{Q},\hat{\Pi}\right] = -i\hbar(k+1)\hat{G}_1 - i\hbar k\hat{I} \\
&\left[\hat{X},\hat{Q}\right] = -\frac{i\hbar}{m}\hat{G}_2 && \left[\hat{X},\hat{\Pi}\right] = i\hbar\hat{G}_1 \\
&\left[\hat{Q},\hat{P}\right] = i\hbar\hat{G}_1 - i\hbar\gamma\hat{G}_2 && \left[\hat{P},\hat{\Pi}\right] = -i\hbar m\omega^2\hat{G}_2 \\
&\left[\hat{H},\hat{X}\right] = \frac{i\hbar}{m}\hat{\Pi} && \left[\hat{H},\hat{P}\right] = i\hbar m\omega^2(1+k)\hat{X} + i\hbar m\omega^2\hat{Q} \\
&\left[\hat{H},\hat{Q}\right] = i\hbar\gamma(1+k)\hat{X} + \frac{i\hbar}{m}\hat{P} && \left[\hat{H},\hat{\Pi}\right] = -i\hbar m\omega^2(2k+1)\hat{X} \\
&\qquad + i\hbar\gamma\hat{Q} + \frac{i\hbar}{m}(1-k)\hat{\Pi} && \qquad - 2i\hbar m\omega^2\hat{Q} - i\hbar\gamma\hat{\Pi} \\
&\left[\hat{H},\hat{G}_1\right] = -i\hbar\gamma\hat{G}_1 + 2i\hbar\omega^2\hat{G}_2 && \left[\hat{H},\hat{G}_2\right] = -2i\hbar\hat{G}_1 + i\hbar\gamma\hat{G}_2 - i\hbar(1+k)\hat{I}\,.
\end{aligned} \tag{27}$$

Noticing the appearance of the central generator $\hat{I}$ on the right hand side of these commutation relations, we can see that the representations of this algebra include:

- For arbitrary k, a generic family with 3 degrees of freedom: $(\hat{X},\hat{P})$, $(\hat{Q},\hat{\Pi})$ and $(\hat{H},\hat{G}_2)$, then time being a canonical variable.
- For $k=1$, already described, an anomalous family with 2 degrees of freedom: $(\hat{X},\hat{P})$ and $(\hat{H},\hat{G}_2)$, then time being a canonical variable.
- For $k=-1$, a family with 2 degrees of freedom: $(\hat{X},\hat{P})$ and $(\hat{Q},\hat{\Pi})$.

The third case is what we are looking for: it contains two degrees of freedom and time is *not* a canonical variable. Explicitly, for $k=-1$ we have:

$$
\begin{aligned}
&\left[\hat{X},\hat{P}\right] = i\hbar\hat{I} && \left[\hat{Q},\hat{\Pi}\right] = i\hbar\hat{I} \\
&\left[\hat{X},\hat{Q}\right] = -\frac{i\hbar}{m}\hat{G}_2 && \left[\hat{X},\hat{\Pi}\right] = i\hbar\hat{G}_1 \\
&\left[\hat{Q},\hat{P}\right] = i\hbar\hat{G}_1 - i\hbar\gamma\hat{G}_2 && \left[\hat{P},\hat{\Pi}\right] = -i\hbar m\omega^2\hat{G}_2 \\
&\left[\hat{H},\hat{X}\right] = \frac{i\hbar}{m}\hat{\Pi} && \left[\hat{H},\hat{P}\right] = i\hbar m\omega^2\hat{Q} \\
&\left[\hat{H},\hat{Q}\right] = \frac{i\hbar}{m}(\hat{P}+2\hat{\Pi}) + i\hbar\gamma\hat{Q} && \left[\hat{H},\hat{\Pi}\right] = i\hbar m\omega^2(\hat{X}-2\hat{Q}) - i\hbar\gamma\hat{\Pi} \\
&\left[\hat{H},\hat{G}_1\right] = -i\hbar\gamma\hat{G}_1 + 2i\hbar\omega^2\hat{G}_2 && \left[\hat{H},\hat{G}_2\right] = -2i\hbar\hat{G}_1 + i\hbar\gamma\hat{G}_2\,.
\end{aligned}
\tag{28}
$$

In this case we notice that the operators $\hat{G}_1$ and $\hat{G}_2$ commute with the basic couples $(\hat{X},\hat{P})$ and $(\hat{Q},\hat{\Pi})$), and therefore will be represented as constants (times the identity operator) in any irreducible representation. Even more, their commutation rules with $\hat{H}$ determine that the value of these constants must be zero ($[\hat{H},\hat{G}_1] = [\hat{H},\hat{G}_1] = 0$, as $\hat{G}_1$ and $\hat{G}_2$ are proportional to the identity $\hat{I}$; then, solve for $\hat{G}_1$ and $\hat{G}$ using the expressions above). Technically, the operators $\hat{G}_1$ and $\hat{G}_2$ are *gauge*, in the sense that, in the resulting physical system, their symmetry transformation does not produce any change in the parameters associated with basic operators (to be intuitive, they do not change coordinate or momentum of any degree of freedom of the system, so that their action is not physically observable, in analogy with a gauge transformation in electrodynamics). In consequence, the effective dimension of the algebra is $5+1$: $(\hat{X},\hat{P})$, $(\hat{Q},\hat{\Pi})$, $\hat{H}$ and $\hat{I}$.

$$
\begin{aligned}
&\left[\hat{X},\hat{P}\right] = i\hbar\hat{I} && \left[\hat{Q},\hat{\Pi}\right] = i\hbar\hat{I} \\
&\left[\hat{X},\hat{Q}\right] = 0 && \left[\hat{X},\hat{\Pi}\right] = 0 \\
&\left[\hat{Q},\hat{P}\right] = 0 && \left[\hat{P},\hat{\Pi}\right] = 0 \\
&\left[\hat{H},\hat{X}\right] = \frac{i\hbar}{m}\hat{\Pi} && \left[\hat{H},\hat{P}\right] = i\hbar m\omega^2\hat{Q} \\
&\left[\hat{H},\hat{Q}\right] = \frac{i\hbar}{m}(\hat{P}+2\hat{\Pi}) + i\hbar\gamma\hat{Q} && \left[\hat{H},\hat{\Pi}\right] = i\hbar m\omega^2(\hat{X}-2\hat{Q}) - i\hbar\gamma\hat{\Pi}\,.
\end{aligned}
\tag{29}
$$

This way, we end up with two pairs of independent, canonical operators, plus a Hamiltonian. As we have selected the representation in which $\hat{H}$ is not a basic operator, it can be written in terms of the basic ones in an irreducible representation:

$$\hat{H} = -\frac{1}{m}\hat{\Pi}\hat{P} - \frac{\gamma}{2}(\hat{Q}\hat{\Pi} + \hat{\Pi}\hat{Q}) - \frac{\hat{\Pi}^2}{m} + m\omega^2\hat{X}\hat{Q} - m\omega^2\hat{Q}^2\,. \tag{30}$$

Note that the normal ordering has been chosen to solve the ordering ambiguity, as customary in canonical quantization. The corresponding classical version of the Hamiltonian is[4]:

$$H = -\frac{1}{m}\Pi P - \gamma Q\Pi - \frac{\Pi^2}{m} + m\omega^2 XQ - m\omega^2 Q^2\,. \tag{31}$$

[4] This expression can be obtained directly regarding the previous algebra as a Poisson subalgebra and determining what function satisfies those Poisson brackets.

The physical content of this Hamiltonian can be analyzed classically by solving Hamilton's equations of motion. After that, a canonical quantization would also be possible. However, it is interesting to establish the relationship of the physical system at which we have arrived with a known one: the new system with two degrees of freedom is, actually, Bateman's dual system.

3.2. The Bateman dual system
The classical Hamiltonian (31) can be transformed, using the linear, constant, canonical transformation:

$$\begin{aligned} X &= y + \frac{1}{m\gamma}p_y + \frac{1}{2}x \\ P &= p_x - m\frac{\gamma}{2}y - m\frac{\omega^2}{\gamma}x \\ Q &= y + \frac{1}{m\gamma}p_y - \frac{1}{2}x \\ \Pi &= -p_x - m\frac{\gamma}{2}y + m\frac{\omega^2}{\gamma}x\,, \end{aligned} \tag{32}$$

into Bateman's dual Hamiltonian [19]

$$H_B = \frac{p_x p_y}{m} + \frac{\gamma}{2}(yp_y - xp_x) + m\Omega^2 xy\,. \tag{33}$$

Let us remark the fact that this canonical transformation is time-independent and, therefore, it is just a change of variables for H, so that $H = H_B$. Then, it can be checked that H (or H_B) describes a damped particle (x, p_x) and its time reversal (y, p_y) computing the second order classical equations out of Hamilton's equations:

$$\ddot{x} + \gamma\dot{x} + \omega^2 x = 0\,, \qquad \ddot{y} - \gamma\dot{y} + \omega^2 y = 0\,. \tag{34}$$

The system is conservative, so our objective of including time evolution among the symmetries has been accomplished, at the cost of including a new degree of freedom.

Performing the usual canonical quantization, the quantum Bateman Hamiltonian can be written:

$$\hat{H}_B = \frac{\hat{p}_x\hat{p}_y}{m} + \frac{\gamma}{2}(\hat{y}\hat{p}_y - \hat{x}\hat{p}_x) + m\Omega^2\hat{x}\hat{y}\,, \tag{35}$$

and the Schrödinger equation for the Bateman system is given by:

$$i\hbar\frac{\partial\phi(x,y,t)}{\partial t} = \left[-\frac{\hbar^2}{m}\frac{\partial^2}{\partial x\partial y} - i\hbar\frac{\gamma}{2}(y\frac{\partial}{\partial y} - x\frac{\partial}{\partial x}) + m\Omega^2 xy\right]\phi(x,y,t)\,. \tag{36}$$

However, it has been argued that the quantum Bateman system possesses inconsistencies, such as complex eigenvalues and non-normalizable eigenstates. Chruściński & Jurkowski [23] showed that $\hat{H}_B$ has real, continuous spectrum, and that the complex eigenvalues are associated with resonances, which in last instance are the responsible of dissipation.

The Bateman system also admits an equivalent description in terms of a first-order Schrödinger equation, which can be derived from the symmetry group associated with the Bateman model. In this framework, the questions of unitarity and the lack of a proper definition of a vacuum, which are poorly treated in the literature, are much better addressed. This matter is under study [25, 31] and will be published elsewhere [26].

Acknowledgments
Work partially supported by the Spanish MCYT, Junta de Andalucía and Fundación Séneca under projects FIS2008-06078-C03-01, P06-FQM-01951 and 08816/PI/08.

References

[1] Caldirola P 1941 *Nuovo Cimento* **18** 393
[2] Kanai E 1948 *Prog. Theor. Phys.* **3** 440
[3] Brittin W E 1950 *Phys. Rev.* **77** 396
[4] Razavy M 2005 *Classical and Quantum Dissipative Systems* (Imperial College Press)
[5] Schuch D 1997 *Phys. Rev. A* **55** 935–940
[6] Schuch D 2007 *J. Math. Phys.* **48** 122701
[7] Dodonov V V and Man'ko V I 1979 *Phys. Rev. A* **20** 550
[8] Dekker H 1981 *Phys. Rep.* **80** 1
[9] Martini R and Kersten P H M 1983 *J. Phys. A* **16** 455
[10] Cerveró J and Villarroel J 1984 *J. Phys. A* **17** 1777
[11] Cerveró J M and Villarroel J 1984 *J. Phys. A* **17** 2963
[12] Hagen C 1972 *Phys. Rev. D* **5** 377
[13] Niederer U 1972 *Helv. Phys. Acta* **45** 802
[14] Niederer U 1973 *Helv. Phys. Acta* **46** 191
[15] Niederer U 1974 *Helv. Phys. Acta* **47** 167
[16] Arnold V 1998 *Geometrical methods in the theoy of ordinary differential equations* (Springer)
[17] Lutzky M 1978 *J. Phys. A* **11** 249
[18] Aldaya V, Cossío F, Guerrero J and López-Ruiz F 2011 *J. Phys. A* **44** 065302 arXiv:1010.5521
[19] Bateman H 1931 *Phys. Rev.* **38** 815
[20] Tikochinsky Y 1978 *J. Math. Phys.* **19** 888
[21] Feshbach H and Tikochinsky Y 1977 *A Festschrift for I.I. Rabi* vol 38 (Trans. New York Ac. Sc. Ser. 2.) p 44
[22] Morse P and Feshbach H 1953 *Methods of Theoretical Physics* vol I (New York: McGraw-Hill) p 298
[23] Chruściński D and Jurkowski J 2006 *Ann. Phys.* **321** 854
[24] Celeghini E, Rasetti M and Vitielo G 1992 *Ann. Phys.* **215** 156
[25] Aldaya V, Cossío F, Guerrero J and López-Ruiz F F A symmetry trip from Caldirola to Bateman damped systems arXiv:1102.0990
[26] Guerrero J, López-Ruiz F F, Aldaya V and Cossío F Symmetries of the quantum damped harmonic oscillator in preparation
[27] Blasone M and Jizba P 2004 *Ann. Phys.* **312** 354–397
[28] Guerrero J, López-Ruiz F F, Aldaya V and Cossío F 2011 *J. Phys. A* **44** 445307 arXiv:1010.5525v3
[29] López-Ruiz F F, Guerrero J and Aldaya V 2011 *J. Russ. Laser Res.* **32** 372 arXiv:1109.1124
[30] Takagi S 1990 *Prog. Theor. Phys.* **84** 1019
[31] Guerrero J, Aldaya V, Cossío F and López-Ruiz F F 2010 *Proceedings of the 28th International Colloquium on Group-Theoretical Methods in Physics* (Newcastle, UK: Journal of Physics Conference Series (IOP))

Journal of Physics: Conference Series **380** (2012) 012011 doi:10.1088/1742-6596/380/1/012011

Quantum theory with arrow of time: symmetry breaking and non-local spinor realization with non-commuting operators of energy and decay

Vadim V Asadov[1], **Oleg V Kechkin**[1,2]

[1] Neur OK–III, Lomonosov Moscow State University, Vorob'jovy Gory, 119899 Moscow, Russia
[2] Skobeltsyn Institute of Nuclear Physics, Lomonosov Moscow State University, Vorob'jovy Gory, 119899 Moscow, Russia

E-mail: asadov@neurok.com,kechkin@gmail.com

Abstract.
We present a modified quantum dynamics with a non-Hermitian Hamiltonian $\hat{H} \neq \hat{H}^+$, and a complex parameter of evolution. The theory is developed with $\left[\hat{H}^+, \hat{H}\right] = \kappa^2$, where κ is the new fundamental constant. It is shown that this theory loses its symmetry in the case of $\kappa \neq 0$. We demonstrate that this theory allows a thermodynamical interpretation and possesses an 'arrow of time' for isothermal and adiabatic regimes of evolution. A realization of this theory is a modified four-dimensional Dirac spinor system, which appears non-local in the physical three-dimensional space.

1. Introduction

General principles of quantum theory in Dirac's understanding provide a universal framework for the construction of any fundamental theory [1–3]. Thermodynamics is another 'supertheory', which also imposes restrictions on a realistical theoretical scheme [4, 5]. Both of these theories deal with statistical ensembles in their description of reality, and both have other remarkable similarities. It seems possible to construct a theory, which will unify quantum and thermodynamic principles and which will provide a dynamical realization of the thermodynamic laws in a consequent and precise form.

We think that this unified theory can be constructed by 'minimal' extensions of the standard quantum mechanics formalism. This extensions do not destroy the general and fundamental principles of quantum theory. In fact, we try to realize the idea of a total 'analytic continuation' of the conventional quantum theoretical scheme. We consider theories with a complex parameter of evolution and a non-Hermitian Hamiltonian and study the dynamics of the resulting system with respect to its reversibility and other properties that are interesting in the thermodynamic framework.

In our previous report, given at 'Symmetries in Science XIV', we have presented results obtained for theories with

$$\left[\hat{H}^+, \hat{H}\right] = 0, \tag{1}$$

where $\hat{H}$ and $\hat{H}^+$ are the Hamiltonian and its Hermitian conjugate, respectively [6]. The presence of an arrow of time in the dynamics of this quantum system was established. We have also constructed explicitly several realizations of above mentioned quantum theory and have presented their results. This activity was highly motivated by the study of hidden symmetries of modified quantum systems (see [7–9] for the general symmetry context). Particularly, $\hat{H}$ and $\hat{H}^+$ are the symmetry operators in the dynamics under consideration [10–14]. The problem of time reversal invariance violation was also studied in [15–19].

In this article we discuss results obtained by a modified quantum theory with

$$\left[\hat{H}^+, \hat{H}\right] = \kappa^2, \tag{2}$$

where $\kappa \neq 0$ is the new fundamental constant of the theory. We prove the presence of an arrow of time in the modified dynamics and study general properties of this new fundamental quantum scheme, including the possibility to explicitly solve the Cauchy problem. Our interest to this version of the theory is related to the closure of the set of fundamental constants in relativistic quantum theories of this type: taking κ into account as a third fundamental parameter besides the reduced Planck constant $\hbar$ and the speed of light in vacuum c, one fixes the 'physical scales' in natural way (we use the notation where $\hbar = 1$ and $c = 1$). It is important to note that in this theory only the Hamiltonian $\hat{H}$ is the symmetry operator – the symmetry that corresponds to the operator $\hat{H}^+$ becomes broken if $\kappa \neq 0$. We discuss this in detail below.

2. General formalism

2.1. Quantum theory in bra and ket vector terms

Conventional quantum theory is based on principles that must be preserved in all possible realistic modifications (superposition principle, relativistic invariance, etc.). We modify the general quantum theory scheme using an analytic continuation of all its 'building blocks'. The resulting 'minimally extended' quantum theory possesses an arrow of time and gives the dynamical base for non-equilibrium statistical mechanics.

2.1.1. Standard elements A state vector $|\Psi\rangle$ depends on the evolutionary parameter τ according to the Hamiltonian $\hat{H}$ in quantum theory. The corresponding dynamics is covered by the linear and homogenous Schrödinger equation

$$i\left|\Psi\right\rangle_{,\tau} = \hat{H}\left|\Psi\right\rangle, \tag{3}$$

which defines the state vector $|\Psi\rangle = |\Psi(\tau)\rangle$ completely with initial data $|\Psi(\tau_0)\rangle = |\Psi_0\rangle$. In fact, quantum theory deals with a Cauchy problem of first order in τ. The most interesting and actually fundamental quantum systems have a Hamiltonian independent of τ, i.e. they yield the additional stationarity condition

$$\hat{H}_{,\tau} = 0. \tag{4}$$

For such dynamical systems the solution of the Cauchy problem reads:

$$|\Psi(\tau)\rangle = \exp\left[-i\hat{H}(\tau - \tau_0)\right]|\Psi_0\rangle. \tag{5}$$

The expectation value $\bar{A}$ of the observable $\hat{A} = \hat{A}^+$ corresponding to the (non-normalized) state vector $|\Psi\rangle$ is defined by the relation

$$\bar{A} = \frac{A}{Z}, \tag{6}$$

where

$$A = \langle\Psi|\,\hat{A}\,|\Psi\rangle \tag{7}$$

and

$$Z = \langle\Psi\,|\Psi\rangle\,. \tag{8}$$

2.1.2. Analytic extension Conventional quantum theory deals with some complex linear space of state-vectors, i.e. $|\Psi\rangle \neq |\Psi\rangle^*$. We propose to extend it to the theory with a complex parameter of evolution $\tau \neq \tau^*$ and a non-Hermitian Hamiltonian $\hat{H} \neq \hat{H}^+$. It can be performed in the 'minimal' analytic form, which is defined by the additional relations

$$|\Psi\rangle_{,\tau^*} = 0 \text{ and } \hat{H}_{,\tau^*} = 0\,. \tag{9}$$

Note that the solution of the initial problem is given by Eq. (5). Using the parametrization

$$\tau = t - \frac{i}{2}\beta \text{ and } \hat{H} = \hat{E} - \frac{i}{2}\hat{\Gamma} \tag{10}$$

with the real variables t and β and the Hermitian operators $\hat{E}$ and $\hat{\Gamma}$, one obtains the following results for the expectation values of the corresponding observables:

$$\bar{E} = -\left(\log Z\right)_{,\beta} \text{ and } \bar{\Gamma} = -\left(\log Z\right)_{,t}\,. \tag{11}$$

This leads to the interpretation of $\hat{E}$ and β as the energy operator and the inverse temperature of a system, whereas t and $\hat{\Gamma}$ corresponds to the 'usual' time and the operator of decay, respectively. The quantity Z is not only the norm of the state vector $|\Psi\rangle$, but also the statistical sum of the quantum ensemble under consideration in the framework of this interpretation. We work with systems with $Z > 0$ in this article.

2.1.3. Definition of energy and decay operators We consider quantum systems with the operators $\hat{E}$ and $\hat{\Gamma}$ that are canonically conjugated up to the constant

$$\left[\hat{E}, \hat{\Gamma}\right] = ik, \tag{12}$$

where k = const (note that $\hbar = 1$ in the our notation). Specifically, we extend the standard quantum system with time parameter t and Hermitian Hamiltonian $\hat{E}$ using the modified one by the substitutions $\hat{E} \to \hat{H}$ and $t \to \tau$ according to Eq. (10). Here, Eq. (12) plays the role of the defining relation for the construction of the decay operator $\hat{\Gamma}$ in the problem with the given operator $\hat{E}$. This equation is linear and non-homogenous. We construct its solution in a form that guarantees a 'minimal generalization' of the starting conventional quantum system.

It is important to note that $\hat{H}$ is the symmetry operator of the system for arbitrary values of the parameter k, whereas $\hat{H}^+$ has this property only for $k = 0$. This means that the last symmetry breaks if the constant k becomes nontrivial. Actually, let us suppose that the state vector $|\Psi\rangle$ satisfies the Schrödinger equation. Then the map

$$|\Psi\rangle \to |\Psi'\rangle = \hat{H}^+\,|\Psi\rangle \tag{13}$$

defines the transformed state vector. This state vector satisfies the equation

$$i\,|\Psi'\rangle_{,\tau} = \hat{H}\,|\Psi'\rangle + k\,|\Psi\rangle\,, \tag{14}$$

which coincides with Eq. (3) in the case of $k = 0$ only.

2.1.4. Arrow of time It is clear that the quantities Z, $\bar{E}$, and $\bar{\Gamma}$ are the functions of evolutionary parameters t and β on the dynamics of the quantum system. Performing the calculations one obtains:

$$\bar{\Gamma}_{,t} = -k - \overline{\Gamma \circ \Gamma}, \quad \bar{E}_{,\beta} = -\frac{k}{4} - \overline{E \circ E}, \tag{15}$$

$$\bar{\Gamma}_{,\beta} = \bar{E}_{,t} = -\overline{E \circ \Gamma}.$$

Here $E \circ \Gamma$ means a correlation between the observables $\bar{E}$ and $\bar{\Gamma}$; it is defined as

$$\overline{E \circ \Gamma} = \overline{\Gamma \circ E} = \frac{1}{2}\left[\overline{(\hat{E}-\bar{E})(\hat{\Gamma}-\bar{\Gamma})} + \overline{(\hat{\Gamma}-\bar{\Gamma})(\hat{E}-\bar{E})}\right], \tag{16}$$

whereas

$$\overline{E \circ E} = D_E^2, \qquad \overline{\Gamma \circ \Gamma} = D_\Gamma^2 \tag{17}$$

are the corresponding dispersions.

Let us consider two temperature regimes, i.e. a dynamics with one additional relation $\beta = \beta(t)$ imposed. We study the isothermal and the adiabatic evolution of the system.

(i) In the isothermal case one has $\beta = \text{const}$, and for the total time derivative of the function $\bar{\Gamma}[t] = \bar{\Gamma}[t, \beta(t)]$ one has:

$$\frac{d\bar{\Gamma}}{dt} = -k - \overline{\Gamma \circ \Gamma}. \tag{18}$$

(ii) For the adiabatic evolution we put $\bar{E} = \bar{E}[t, \beta(t)] = \text{const}$, and for the total time derivative of the same quantity $\bar{\Gamma}$ one obtains the following result:

$$\frac{d\bar{\Gamma}}{dt} = -\frac{\frac{k^2}{4} + k(\overline{E \circ E} + \frac{1}{4}\overline{\Gamma \circ \Gamma}) + \overline{E \circ E} \cdot \overline{\Gamma \circ \Gamma} - (\overline{E \circ \Gamma})^2}{\frac{k}{4} + \overline{E \circ E}}. \tag{19}$$

¿From Eqs. (18) and (19) it follows that

$$\frac{d\bar{\Gamma}}{dt} \leq 0 \tag{20}$$

if $k \geq 0$ for an arbitrary initial state $|\Psi_0\rangle$ that specifies the solution of a Cauchy problem completely.

Thus, the quantity $\bar{\Gamma}[t]$ detects the presence of an arrow of time in dynamics of the quantum system in these two temperature regimes. We fix the non-negative sign of the constant k and put

$$k = \kappa^2, \tag{21}$$

where κ is the real non-negative constant.

2.2. Ladder operators analysis

Let us put

$$\hat{H} = \kappa\, \hat{a}^{+}, \tag{22}$$

then

$$[\hat{a},\, \hat{a}^{+}] = 1. \tag{23}$$

From the formal point of view $\hat{a}$ and $\hat{a}^{+}$ are lowering and raising operators, respectively.

2.2.1. 'Oscillatory' basis Let us define a linear subspace $|0\rangle$ by the relation

$$\hat{a}\,|0\rangle = 0. \tag{24}$$

Let the number $\mathcal{K}$ denote the dimensionality of this subspace and $\{|0,\,k\rangle\}$ its orthonormal basis ($k = 1,\,\ldots,\,\mathcal{K}$). Thus, we have the corresponding scalar products

$$\langle 0,\,k_1\,|0,\,k_2\rangle = \delta_{k_1 k_2}, \tag{25}$$

and the decomposition

$$|0\rangle = \sum_k C_{0\,k}\,|0,\,k\rangle \tag{26}$$

for an arbitrary vector $|0\rangle$ with the coefficients $C_{0\,k}$. Let us introduce the infinite set of vectors

$$|n,\,k\rangle = \frac{a^{+\,n}}{\sqrt{n!}}\,|0,\,k\rangle\,, \tag{27}$$

where $n = 1, \ldots$ is an arbitrary integer number. It is possible to show that these vectors are orthonormal too for any $k = 1,\,\ldots,\,\mathcal{K}$ and $n \in \mathrm{N}$:

$$\langle n_1,\,k_1\,|n_2,\,k_2\rangle = \delta_{n_1 n_2}\delta_{k_1 k_2}\,. \tag{28}$$

The statement is that these vectors with the starting subset $|0,\,k\rangle$ form the general solution of the eigenproblem

$$\hat{n}\,|n,\,k\rangle = n\,|n,\,k\rangle \tag{29}$$

for a Hermitian operator

$$\hat{n} = \hat{a}^{+}\hat{a} \tag{30}$$

with the eigenvalues $n = 0, 1, \ldots$. Finally, this set of orthonormal vectors is a basis in the linear space of thestate vectors of quantum theory in accordance to Steklov's theorem with the decomposition formula:

$$|\Psi\,(\tau)\rangle = \sum_{n=0}^{+\infty}\sum_{k=1}^{\mathcal{K}} C_{n,\,k}\,(\tau)\,|n,\,k\rangle\,. \tag{31}$$

The expectation values of the energy and decay operators calculated over the state vector $\{|n,\,k\rangle\}$ are obtained as

$$\bar{E} = \bar{\Gamma} = 0 \tag{32}$$

and the dispersions as

$$\overline{E^2} = \frac{1}{4}\overline{\Gamma^2} = \frac{\kappa^2}{2}\left(n + \frac{1}{2}\right). \tag{33}$$

It is also interesting to note that

$$\overline{E \circ \Gamma} = 0. \tag{34}$$

To prove these statements one must calculate averaged values of the operators $\hat{H}$, $\hat{H}\hat{H}$ and $\hat{H}\hat{H}^{+}$ and take into account the underlying commutation relation (12) and Eq. (21).

2.2.2. Exact solution: 'Big Bang of vacuum state' In terms of the coefficient functions $C_{n,\,k} = C_{n,\,k}\,(\tau)$ from the decomposition formula (31), Schrödinger's equation (3) reads:

$$\frac{dC_{0,\,k}}{d\tau} = 0, \qquad \frac{dC_{n+1,\,k}}{d\tau} = -i\kappa\,\sqrt{n+1}\,C_{n,\,k}. \tag{35}$$

Let us construct a special exact solution of the problem with initial data specified by the relations

$$C_{n,\,k}\,(0) = \delta^0_n c_k \text{ and } \sum_{k=1}^{\mathcal{K}} |c_k|^2 = 1, \tag{36}$$

where $c_k = \text{const}$. This solution describes the dynamics of the quantum system with only 'vacua modes' $|0,\,k\rangle$ excited at the initial moment of evolution $\tau = 0$. This moment corresponds physically to an initial time $t_0 = 0$ and an absolute temperature $T_0 = 1/\beta_0 = \infty$. The solution of the Cauchy problem (35)–(36) reads:

$$C_{n,\,k}\,(\tau) = c_k \frac{(-i\kappa\tau)^n}{\sqrt{n!}}. \tag{37}$$

For the statistical sum, which is also the norm of the constructed solution, one has:

$$\mathcal{Z}\,(t,\,\beta) = \exp\left(\kappa^2\,|\tau|^2\right) = \exp\left[\kappa^2\left(t^2 + \frac{\beta^2}{4}\right)\right], \tag{38}$$

so

$$\bar{\Gamma}_{,t} = -2\kappa^2 < 0 \tag{39}$$

and

$$\bar{E}_{,T} = \frac{\kappa^2}{2T^2} > 0. \tag{40}$$

It is seen that this solution describes the dynamics of the consecutive excitation of the originally non-excited modes $|n,\,k\rangle$ of an explosion type. The system demonstrates an irreversibility (see Eq. (39)) and has a positive heat capacity – as it follows from Eq. (40).

2.3. Analysis in energy representation
Dirac's theory of quantum representations is based on the use of the complete set of commuting Hermitian variables $\hat{Q}_\alpha = \hat{Q}^+_\alpha$,

$$[\hat{Q}_\alpha,\,\hat{Q}_\beta] = 0, \tag{41}$$

and the related basis of eigenvectors $|Q\,\rangle$:

$$\hat{Q}_\alpha|Q\,\rangle = Q_\alpha|Q\,\rangle. \tag{42}$$

The wave function in the Q-representation is defined by the relation

$$\Psi(\tau,\,Q) = \langle\,Q|\Psi(\tau)\,\rangle. \tag{43}$$

2.3.1. Energy representation The energy operator $\hat{E}$ in the energy representation is included in the set $\{\hat{Q}_\alpha\}$. Let us put $\hat{Q}_1 = \hat{E}$; then $Q_1 = E$ and for the wave function we have the notation $\Psi(\tau, Q) = \Psi(\tau, E, \ldots)$. In this representation

$$\hat{E} = E \text{ and } \hat{\Gamma} = -i\kappa^2 \frac{\partial}{\partial E}. \tag{44}$$

Actually, the second relation in Eq. (44) defines the Hermitian decay operator that yields the underlying condition (12).

2.3.2. Solution of initial problem Schrödinger's equation in the energy representation reads:

$$\Psi_{,\tau} - \frac{i\kappa^2}{2}\Psi_{,E} = -iE\Psi. \tag{45}$$

It has the simplest form

$$\Phi_{,v} = \frac{u+v}{2\kappa^2}\Phi \tag{46}$$

in terms of the new variables

$$u = E + \frac{i\kappa^2}{2}\tau \text{ and } v = E - \frac{i\kappa^2}{2}\tau \tag{47}$$

and the new function $\Phi(u, v, \ldots) = \Psi(\tau, E, \ldots)$. Solving Eq. (46) yields

$$\Phi(u, v, \ldots) = \phi(u, \ldots) \exp\left[(u+v)^2/4\kappa^2\right], \tag{48}$$

where ϕ is an arbitrary function of all variables excepting v. Thus, the general solution of Eq. (45) is

$$\Psi(\tau, E, \ldots) = \phi\left(E + \frac{i\kappa^2}{2}\tau, \ldots\right) \exp\left(E^2/\kappa^2\right). \tag{49}$$

Taking into account the initial data $\Psi_0(E, \ldots) = \Psi(0, E, \ldots)$, where we put $\tau_0 = 0$ again, one can fix the introduced function ψ:

$$\phi(E, \ldots) = \Psi_0(E, \ldots) \exp\left(-E^2/\kappa^2\right). \tag{50}$$

Finally, for the solution of the quantum Cauchy problem we obtain:

$$\Psi(\tau, E, \ldots) = \Psi_0\left(E + \frac{i\kappa^2}{2}\tau, \ldots\right) \exp\left(-iE\tau + \kappa^2\tau^2/4\right). \tag{51}$$

3. Two specific realizations

In this section we give two examples of the modified quantum theory considered above. These systems have 'truncations' (defined by the map $\tau \to t$, $\hat{H} \to \hat{E}$) to the conventional relativistic quantum mechanics.

3.1. Modified transport equation
This model is intended for $(1+1)$–space-time and its energy operator is equal to the momentum defined by a single component:

$$\hat{E} = \hat{p}. \tag{52}$$

We introduce the decay operator as

$$\hat{\Gamma} = -\kappa^2 \hat{x}, \tag{53}$$

where $\hat{x}$ is the operator of the coordinate that is canonically conjugated to $\hat{p}$. In fact, the momentum representation coincides with the energy one for this theory, so the solution of Cauchy problem using the wave function $\Psi(\tau, p)$ can be found in the previous section.

3.2. Modified Dirac equation
Our second example is related to conventional Dirac's spinor theory in four dimensions. We show that its modified version with an arrow of time is non-local in physical three-dimensional space.

3.2.1. 4D spinor theory The energy operator for Dirac's spinor theory is defined as

$$\hat{E} = \gamma^0 \left(m + \gamma^k \hat{p}^k \right), \tag{54}$$

where m is the mass parameter, $k = 1, 2, 3$, and γ^μ are the Dirac matrices ($\mu = 0, \ldots, 3$). The wave function Ψ is the 4-column in the representation of the lowest matrix dimensionality.

3.2.2. Construction of decay operator An analysis shows that the introduction of the decay operator to this quantum system, i.e. the map

$$\hat{E} \to \hat{H}, \tag{55}$$

is realized by the substitution

$$m \to \hat{M} = m\left(1 + \frac{i\kappa^2}{2}\hat{\alpha}\right) \text{ and } \hat{p}^k \to \hat{P}^k = \hat{p}^k + \frac{i\kappa^2}{4}\hat{\pi}^k, \tag{56}$$

with

$$\hat{\alpha} = \frac{1}{6}\left[\left(\hat{p}^l\right)^{-1}\hat{x}^l + \hat{x}^l\left(\hat{p}^l\right)^{-1}\right] \text{ and } \hat{\pi}^k = \left(\hat{\alpha}\hat{p}^k + \hat{p}^k\hat{\alpha}\right). \tag{57}$$

For the multiplicity number (see Eq. (31)) one obtains $\mathcal{K} = 4$ in this approach.

3.2.3. Non-locality in coordinate representation The momentum representation (where $\hat{p}^k = p^k$, $\hat{x}^k = i\partial/\partial p^k$, and $\tilde{\Psi} = \Psi\left(t, p^k\right)$) for the mass- and momentum-like operators (56) gives:

$$\hat{M} = m\left\{1 + \frac{i\kappa^2}{12}\left[\left(\frac{1}{p^l}\right)^2 - \frac{2}{p^l}\frac{\partial}{\partial p^l}\right]\right\}, \quad \hat{P}^k = p^k - \frac{i\kappa^2}{12}\left[\frac{1}{p^k} - p^k\left(\frac{1}{p^l}\right)^2 + 2\frac{p^k}{p^l}\frac{\partial}{\partial p^l}\right]. \tag{58}$$

Subjecting the standard modified Dirac equation to the map (56), (58), we obtain

$$i\Psi_{,t} = \gamma^0\left\{\left(\gamma^k p^k + m\right)\left[1 + \frac{\kappa^2}{12}\left(\frac{1}{(p^l)^2}\right) - \frac{2}{p^l}\frac{\partial}{\partial p^l}\right] - \frac{\kappa^2}{12}\frac{\gamma^k}{p^k}\right\}\Psi. \tag{59}$$

This equation must be supplemented by the Cauchy-Riemann condition

$$\Psi_{,\beta} = -\frac{i}{2}\Psi_{,t} \tag{60}$$

instead of the first relation in Eq. (9). Eq. (59) is of the same type as the conventional Dirac equation in the coordinate representation. Actually, it is of first order with respect to the derivatives, but its coefficients are no longer constant. It is important to note that Eq. (59) corresponds to the action

$$S = \int dt d^3p\, L \tag{61}$$

with the Lagrangian

$$L=\frac{i}{2}\left(\bar{\Psi}\gamma^0\Psi_{,t}-\bar{\Psi}_{,t}\gamma^0\Psi\right)-\frac{1}{2}\left[\bar{\Psi}\gamma^k\left(\hat{P}^k\Psi\right)+\left(\hat{P}^{k*}\bar{\Psi}\right)\gamma^k\Psi+\bar{\Psi}\left(\hat{M}\Psi\right)+\left(\hat{M}^*\bar{\Psi}\right)\Psi\right]. \tag{62}$$

Thus, the modified Dirac theory in the $(t,\, p^k)$-representation possesses a Lagrangian formulation despite it has an arrow of time.

The modified Dirac equation in the coordinate representation, where $\hat{x}^k = x^k$, $\hat{p}^k = -i\partial/\partial x^k$, and $\tilde{\Psi} = \Psi\left(t,\, x^k\right) = \Psi\left(x^\mu\right)$, has the following form:

$$\left(i\gamma^\mu\partial_{x^\mu} - m\right)\Psi = \frac{\kappa^2}{24}\left(m - i\gamma^k\partial_{x^k}\right)\int_0^{+\infty} d\xi \sum_{l=1}^{3}\left[\xi\Psi_{l+} - \left(2x^l + i\gamma^l\right)\Psi_{l-}\right]. \tag{63}$$

Here we put $\Psi_{\pm k} = \Psi_{\pm k}\left(t,\, x^k,\, \xi\right)$, where

$$\begin{aligned}
\Psi_{\pm 1} &= \Psi\left(t,\, x^1 - \xi,\, x^2,\, x^3\right) \pm \Psi\left(t,\, x^1 + \xi,\, x^2,\, x^3\right),\\
\Psi_{\pm 2} &= \Psi\left(t,\, x^1,\, x^2 - \xi,\, x^3\right) \pm \Psi\left(t,\, x^1,\, x^2 + \xi,\, x^3\right),\\
\Psi_{\pm 3} &= \Psi\left(t,\, x^1,\, x^2,\, x^3 - \xi\right) \pm \Psi\left(t,\, x^1,\, x^2,\, x^3 + \xi\right).
\end{aligned} \tag{64}$$

To obtain Eq. (63) from Eq. (59) one must Fourier transform and take into account the convolution theorem. It is seen that Eq. (63) is essentially an integro-differential equation that is non-local with respect to the physical three-dimensional space.

4. Conclusions

The main results of this article are:

(i) A quantum theory with a complex parameter of evolution and a non-Hermitian Hamiltonian does provide a dynamical framework for a non-equilibrium statistical mechanics and thermodynamics.

(ii) This dynamical framework possesses an arrow of time for the quantum theory with operators of energy and decay conjugated canonically up to some positive constant.

(iii) This nontrivial constant is an additional fundamental constant in the theory, which leads to some hidden dynamical symmetry breaking.

(iv) A realization of this theory can be constructed on the base of the four-dimensional Dirac theory. This realization is a Lagrangian one in momentum space.

(v) The constructed modified Dirac theory is essentially non-local in the physical three-dimensional space. This fact seems really interesting for a potential resolution of the Einstein-Podolsky-Rosen paradox [20, 21].

Acknowledgments
The authors would like to thank Profs. B. Ishkhanov and Yu. P. Rybakov for fruitful discussions. The authors are really grateful to the organizers of the International Symposium 'Symmetries in Science XV' for the possibility to present and discuss these results.

References

[1] Dirac P A M 1957 *The principles of quantum mechanics* (Oxford: Clarendon Press)
[2] Green H S 1965 *Matrix mechanics* (Groningen: P Nordhoff)
[3] Sudbery E 1986 *Quantum mechanics and particles of Nature* (Cambridge: Cambridge University Press)
[4] Fermi E 1937 *Thermodynamics* (New York: Dover publications)
[5] Kubo R 1968 *Thermodynamics* (Amsterdam: North-Holland Publishing Company)
[6] Asadov V V and Kechkin O V 2010 *J. Phys. Conf. Ser.* **237** 012002
[7] Stephani H 1989 *Differential Equations. Their Solution Using Symmetries* (Cambridge: Cambridge
[8] Olver P J 1986 *Applications of Lie groups to differential equations* (New York: Springer-Verlag)
[9] Olver P J 1995 *Equivalence, invariance and symmetry* (Cambridge: Cambridge University Press)
[10] Asadov V V and Kechkin O V 2011 *Proc. of the Conference in honour of Murray Gell-Mann's 80th birthday "Quantum mechanics, elementary particles, quantum cosmology and complexity"* (Singapore: World Scientific Publishing Co) 575
[11] Asadov V V and Kechkin O V 2008 *Moscow University Physics Bulletin* **2** 105
[12] Asadov V V and Kechkin O V 2009 *Bulletin of the Peoples Friendship University of Russia (Mathematics. Informatics. Physics)* **4** 80
[13] Asadov V V and Kechkin O V 2009 *Gravit. Cosmol.* **15** 295
[14] Asadov V V and Kechkin O V 2009 *Bulletin of the Peoples Friendship University of Russia (Mathematics. Informatics. Physics)* **3** 99
[15] Mackey M C 1991 *Time's arrow: the origins of thermodynamic behavior Springer* (New York: Springer-Verlag)
[16] Dekker H 1981 *Phys. Rep.* **80** 1
[17] Tarasov V E 2001 *Phys. Lett.* A **288** 173
[18] Celeghini E, Rasetti M and Vitiello G 1992 *Annals Phys.* **215** 156
[19] Feshbach H and Tikochinsky Y 1977 *Trans. New York Acad. Sci.* **38** 44
[20] Walborn S P, Salles A, Gomes R M, Toscano F and Souto Ribeiro P H 2011 *Phys. Rev. Lett.* **106** 130402
[21] Reid M D, Drummond P D, Bowen W P, Cavalcanti E G, Lam P K, Bachor H A, Andersen U L and Leuchs G 2011 *Rev. Mod. Phys.* **81** 1727

Symmetries in Science XV IOP Publishing
Journal of Physics: Conference Series **380** (2012) 012012 doi:10.1088/1742-6596/380/1/012012

Entanglement Dissipation: Unitary and Non-unitary Processes

Allan I. Solomon

Department of Physics and Astronomy, The Open University, MK7 6AA, UK
and
LPTMC, Université de Paris VI, France

Abstract. Dissipative processes in physics are usually associated with non-unitary actions. However, the important resource of entanglement is not invariant under general unitary transformations, and is thus susceptible to unitary "dissipation". In this note we discuss both unitary and non-unitary dissipative processes, showing that the former is ultimately of value, since reversible and thus not a true dissipative process, which enables the production of entanglement; while even in the presence of the latter, more conventional non-unitary and non-reversible, process there exist nonetheless invariant entangled states.

1. Introduction to Bipartite Entanglement

1.1. Definition and Measure (Concurrence)

Bipartite entanglement involves the direct product space V of two (complex) vector spaces, V_1 and V_2, of dimension m and n respectively. If V_1 has basis $\{e_i,\ i = 1 \dots m\}$ and V_2 has basis $\{f_j,\ j = 1, \dots n\}$ then $\{e_i \otimes f_j,\quad i = 1 \dots m, j = 1 \dots n\}$ is a basis for $V = V_1 \otimes V_2$.

In this note we specialize to the specific and more familiar case of two two-qubit spaces, $m = n = 2$, with the same standard bases $\{e_i(= f_i), i = 1, 2\}$ for V_1 and V_2. We shall also use the matrix forms $\{e_1, e_2\} = \{[1, 0], [0, 1]\}$ ‡ as well as the ket notation $e_1 = |0\rangle$ and $e_1 \otimes e_1 = |0, 0\rangle$ etc. Note that in the context of quantum mechanics we refer to vectors as *pure* states.

Definition 1 (Entangled pure state) *Every vector in $V_1 \otimes V_2$ is a sum of products; but not every vector is a product. If it is a product, then it is said to be* non-entangled *or* separable.

It is a straightforward matter to determine whether a vector $v \in V$ is entangled or not.

If $v \in V$ is non-entangled, i.e. separable, then

$$\begin{aligned} v \in V_1 \otimes V_2 = & \sum_{i,j=1}^{2} c_{ij} e_i \otimes e_j \\ & = (x_1 e_1 + x_2 e_2) \otimes (y_1 e_1 + y_2 e_2) \\ \Rightarrow \qquad & c_{ij} = x_i y_j \quad \{i, j = 1, 2\} \end{aligned} \tag{1}$$

from which we deduce that the matrix c of coefficients c_{ij} has determinant zero, $\det c = 0$.

‡ For typographical simplicity we write all our (column) vectors as row vectors.

Journal of Physics: Conference Series **380** (2012) 012012 doi:10.1088/1742-6596/380/1/012012

Example 1 (Separable pure state) *Consider the bipartite pure state*

$$(1/\sqrt{50})(|0,0\rangle + 3|0,1\rangle + 2|1,0\rangle + 6|1,1\rangle). \tag{2}$$

The matrix c of coefficients is given by

$$c = 1/\sqrt{50}\begin{bmatrix} 1 & 3 \\ 2 & 6 \end{bmatrix} \tag{3}$$

for which $\det c = 0$ *and so the state is separable.*

Example 2 (Bell state) *An example of a maximally entangled two-qubit state is given by the Bell state* $\frac{1}{\sqrt{2}}(|0,0\rangle + |1,1\rangle)$ *for which*

$$c = \frac{1}{\sqrt{2}}\begin{bmatrix} 1 & 0 \\ 0 & 1 \end{bmatrix} \tag{4}$$

and so $\det c = 1/2$.

This simple criterion for pure state separability in fact gives a *measure* of entanglement for pure states. To obtain this measure, we normalize so that a Bell state, such as that of Example 2, has maximal measure of entanglement equal to 1, and we arrive at the definition of the *concurrence* applicable to pure states:

Definition 2 (Pure state Concurrence) *A measure of entanglement for pure bipartite states (belonging to two two-qubit spaces) is given by the* concurrence $\mathcal{C} = 2|\det c|$.

The concurrence $\mathcal{C}$ is essentially equivalent to the measure $\mathcal{E}$ called *Entanglement of Formation*, based on the Von Neumann entropy of the partial trace[1]§.

The entanglement measure $\mathcal{C}$ may be extended to *general*, or mixed, states (density matrices). We describe this extension in Section 4.3.

1.2. General states

We are initially concerned in this note with *pure* states; i.e. represented by vectors v in V. However, we can equally represent our vector v by the matrix $\rho(v) = vv^\dagger$. Of course the overall irrelevant phase information is lost in this form. It is easily verified that ρ is a hermitian matrix of rank one, that is, all sub-matrices of order 2 or more have determinant zero. And it has a sole non-zero eigenvalue which is equal to 1. It has trace equal to one, assuming that v is normalized. A hermitian matrix all of whose eigenvalues are ≥ 0 is called a *positive matrix* (more accurately, semi-positive). We may extend this description of the matrix associated with a pure state to give the following definition of a *state* in general (*mixed state* or *density matrix*):

Definition 3 (State) *A state ρ (acting on a space V) is a positive matrix of trace 1.*

Equivalently,

Definition 4 (State as convex sum) *A state ρ is a convex sum $\sum_i \lambda_i \rho_i$ $(\lambda_i \geq 0 \quad \sum_i \lambda_i = 1)$ of pure states ρ_i.*

We simply note here the definition of separability for (general) states:

Definition 5 (Separable state) *The state ρ acting on $V_1 \otimes V_2$ is said to be* separable *if is given by a convex sum $\sum_i \lambda_i \rho_i^1 \otimes \rho_i^2$ $(\lambda_i \geq 0 \quad \sum_i \lambda_i = 1)$ where ρ_i^α acts on V_α .*

§ Writing $f = -xlog_2x - (1-x)log_2(1-x)$ then $\mathcal{E} = f((1+\sqrt{1-\mathcal{C}^2})/2)$.

When $\rho = \rho^1 \otimes \rho^2$ it is said to be *simply separable*. The above definition extends immediately to *multipartite* states.

If we have a measure of entanglement $\mathcal{E}$ for pure states (such as that given in Definition 2) we may extend it to general states by

Definition 6 (Entanglement of general state) *The entanglement* $\mathcal{E}(\rho)$ *of a mixed bipartite state* ρ *acting on* $V_1 \otimes V_2$ *is given by* $\mathcal{E}(\rho) = \min\{\sum_i \mu_i \mathcal{E}(\psi_i) | \rho = \sum_i \mu_i \rho(\psi_i)\}$ *where the* ψ_i *are pure states in* $V_1 \otimes V_2$.

1.3. Unitary and Local Unitary Transformations

Since every (normed) vector $v \in V$ can be transformed to the (non-entangled) state $|0, 0\rangle$ by a unitary transformation, it is clear that entanglement is *not* invariant under unitary transformations. However, under a local unitary transformation, defined by $U = U_1 \otimes U_2$, one can see that the concurrence, for example, is invariant:

Theorem 1 *The concurrence* $\mathcal{C}$ *is invariant under local unitary transformations.*
Let $v = \sum_{i,j=1\ldots 2} a_{ij} e_i \otimes e_j \in V = V_1 \otimes V_2$, *and the unitary matrix* $U = U_1 \otimes U_2$ *be a local unitary matrix; then*

$$\begin{aligned} Uv &= \sum a_{ij} U_1 e_i \otimes U_2 e_j \\ &= \sum a_{ij} (U_1)_{ik} e_k \otimes (U_2)_{jr} e_r \\ &= \sum c_{kr} e_k \otimes e_r \end{aligned}$$

where $c_{kr} = \sum_{ij} a_{ij} (U_1)_{ik} (U_2)_{jr}$ *so that* $c = \tilde{U_1} a U_2$ *whence*

$$\begin{aligned} |\det c| &= |\det(\tilde{U_1} a U_2)| \\ &= |\det \tilde{U_1}|\ |\det a|\ |\det U_2| \\ &= |\det a| \quad \text{since} \quad |\det U_i| = 1. \end{aligned}$$

We may see rather more immediately from Definition 5 that the property of *being* separable is invariant under local unitary transformations; and this extends to the multipartite case. However, an extension of Theorem 1 to multipartite systems, namely that such local transformations preserve the *measure* of entanglement, would depend on a definition of measure (or measures) of entanglement for such systems, which is currently unavailable. For general multipartite states, local unitary equivalence does not preserve all the relevant (state and substate) entanglement properties [2].

2. Unitary "Dissipation" of entanglement

Although the notion of dissipation is more usually associated with a non-unitary process, from the preceding we see that *entanglement* is subject to unitary degradation, since unitary evolution associated with a (hermitian) hamiltonian does not necessarily preserve entanglement. This is of course *not* a true dissipative process, and the good news is the other side of this coin; that is, entanglement may be produced by the evolution induced by a quantum control hamiltonian. Quantum control applied to multipartite systems has been well treated, see for example [3]. We choose a simple example to illustrate the bipartite case.

Example 3 (Entanglement production and decay) *Consider the unitary evolution $U(t)$ induced by the hamiltonian H given by*

$$H = \begin{bmatrix} x_1 & 0 & 0 & y \\ 0 & x_2 & 0 & 0 \\ 0 & 0 & x_3 & 0 \\ y & 0 & 0 & x_1 \end{bmatrix}. \tag{5}$$

This is essentially a free hamiltonian with the addition of an off-diagonal time-independent control term $y\|$. Note that without this latter term H would not change the entanglement since H would then be a local transformation.

We act by the unitary evolution matrix $U(t) = \exp(itH)$ induced by Eq.(5) on the base vector $v_0 \equiv [1, 0, 0, 0]$. (Note that for such calculations it is important to choose a fixed basis - here the standard basis.)

$$v(t) = \exp(itH)v_0 \tag{6}$$
$$= e^{itx_1}\left[\cos(t\,y), 0, 0, i\sin(t\,y)\right]. \tag{7}$$

Apart from an overall phase factor, only the control term y plays a rôle in the entanglement production.

Using the measure of entanglement for pure states given in Definition 2, the concurrence for $v(t)$ is given by $|\sin(2ty)|$ (See Figure 1).

2.1. Unitary Entanglement Loss

Example 4 (Unitary decay of entanglement) *Referring to the previous example and Figure 1, we see immediately that at $t = \pi/4$ (in units of $1/y$ where y is the control frequency) we have the maximally entangled (Bell) state $\frac{1}{\sqrt{2}}[1, 0, 0, i]$ (up to an overall phase factor). The unitary action $U(t)$ destroys the entanglement, completely at $t = \pi/2$.*

3. Environmental Dissipation

For the usual description of dissipative processes we must use the standard definition of a *general* quantum state ρ given by Definition 3. Thus ρ is an $N \times N$ positive matrix (and for our two-qubit examples, $N = 4$). For a non-dissipative process, the basic equation which determines the evolution of a hamiltonian quantum system may be written in the form of a differential equation for the quantum state, the Liouville-Von Neumann equation[4] (choosing units in which $\hbar = 1$):

$$i\frac{d}{dt}\rho(t) = [H, \rho(t)] \equiv H\rho(t) - \rho(t)H \tag{8}$$

where H is the total hamiltonian of the system. The standard form of a general *dissipative* process in Quantum Mechanics is governed by the Liouville equation obtained by adding a dissipation (super-)operator $L_D[\rho(t)]$ to Eq.(8):

$$i\dot{\rho}(t) = [H, \rho(t)] + iL_D[\rho(t)]. \tag{9}$$

$\|$ For calculational simplicity we have chosen a degeneracy between the first and last energy levels, $x_1 = x_4$.

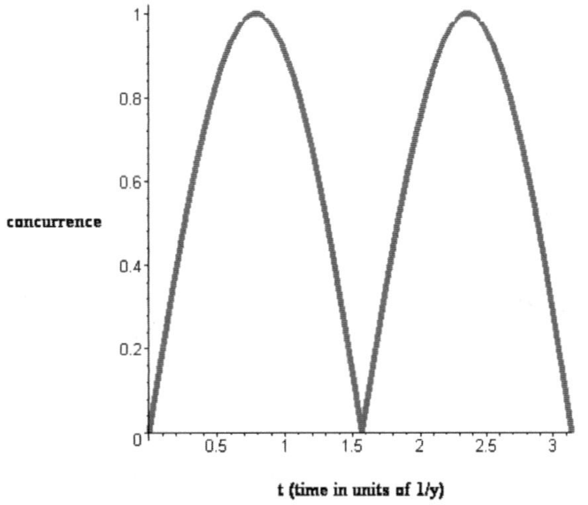

Figure 1. Concurrence *v* Time t (units of 1/control frequency)

3.1. Liouville Dissipation

In general, uncontrollable interactions of the system with its environment lead to two types of dissipation: phase decoherence (dephasing) and population relaxation.

Phase decoherence occurs when the interaction with the environment destroys the phase correlations between states, which leads to changes in the off-diagonal elements of the density matrix:

$$\dot{\rho}_{kn}(t) = -i([H, \rho(t)])_{kn} - \Gamma_{kn}\rho_{kn}(t) \tag{10}$$

where Γ_{kn} (for $k \neq n$) is the dephasing rate between $|k\rangle$ and $|n\rangle$.

Population relaxation occurs, for instance, when a quantum particle in state $|n\rangle$ spontaneously emits a photon and moves to another quantum state $|k\rangle$, which changes the populations according to¶

$$\dot{\rho}_{nn}(t) = -i([H, \rho(t)])_{nn} + \sum_{k \neq n} [\gamma_{nk}\rho_{kk}(t) - \gamma_{kn}\rho_{nn}(t)] \tag{11}$$

where $\gamma_{kn}\rho_{nn}$ is the population loss for level $|n\rangle$ due to transitions $|n\rangle \rightarrow |k\rangle$, and $\gamma_{nk}\rho_{kk}$ is the population gain caused by transitions $|k\rangle \rightarrow |n\rangle$. The population relaxation rate γ_{kn} is determined by the lifetime of the state $|n\rangle$, and for multiple decay pathways, the relative probability for the transition $|n\rangle \rightarrow |k\rangle$.

Phase decoherence and population relaxation lead to a dissipation superoperator (represented by an $N^2 \times N^2$ matrix) whose non-zero elements are [5]

$$\begin{array}{lll} (L_D)_{[k;n],[k;n]} & = -\Gamma_{kn} & k \neq n \\ (L_D)_{[n;n],[k;k]} & = +\gamma_{nk} & k \neq n \\ (L_D)_{[n;n],[n;n]} & = -\sum_{n \neq k} \gamma_{kn} & \end{array} \tag{12}$$

¶ also known as the Pauli Master Equation.

where Γ_{kn} and γ_{kn} are taken to be positive numbers, with Γ_{kn} symmetric in its indices.

In Eq.(12) we introduced the convenient notation

$$[m; n] = (m-1)N + n. \tag{13}$$

The $N^2 \times N^2$ matrix superoperator L_D may be thought of as acting on the N^2-vector $\mathbf{r}$ obtained from ρ by

$$\mathbf{r}_{[m;n]} \equiv \rho_{mn}. \tag{14}$$

The resulting vector equation is

$$\dot{\mathbf{r}} = L\mathbf{r} = (L_H + L_D)\mathbf{r} \tag{15}$$

where L_H is the anti-hermitian matrix corresponding to the hamiltonian H. We obtain L_H explicitly by using the standard algebraic trick applied in evaluating Liouville equations (see, for example [6]). The correspondence between ρ and $\mathbf{r}$ as given in Eq. (14) tells us, after some manipulation of indices, that

$$\rho \to \mathbf{r} \Rightarrow A\rho B \to A \otimes \tilde{B}\ \mathbf{r} \tag{16}$$

using the direct (Kronecker) product of matrices. From Eq.(16), the superoperator corresponding to the hamiltonian term in Eq. (8) is

$$L_H = -i(H \otimes I - I \otimes \tilde{H}). \tag{17}$$

4. Physical Processes

The quantum Liouville equation (9) is very formal; it covers both physical and non-physical processes and may tell us little about an actual physical dissipation process. For example, the *values* of the dissipation parameters Γ_{kn} and γ_{kn} are not determined and a general choice will not lead to a physical process - that is, one under which the state $\rho(t)$ remains a physical state - unless the parameters satisfy various constraints[5]. This (completely) positive evolution and the appropriate constraints emerge from physical stochastic dissipation equations such as those given by Lindblad and others in differential form [7], as well as in global form [8]. Nevertheless, the virtue of Eq.(9) is that essentially every dissipation process will have to satisfy it and so results derived from its use will have great generality.

Since in our examples we wish to restrict ourselves to *physical* processes, we obtain our dissipation parameters Γ and γ by use of the Lindblad equation.

4.1. Lindblad Equation

Completely positive evolution of the system is guaranteed by the Lindblad form of the dissipation superoperator L_D

$$L_D[\rho(t)] = \frac{1}{2}\sum_{s=1}^{N^2}\left\{[V_s\rho(t), V_s^\dagger] + [V_s, \rho(t)V_s^\dagger]\right\} \tag{18}$$

where the matrices V_s are arbitrary. The standard basis for $N \times N$ matrices is given by

$$(E_{ij})_{mn} = \delta_{im}\delta_{jn} \qquad (i, j, m, n = 1 \dots N) \tag{19}$$

Relabelling, using the notation $[m; n] = (m-1)N + n$, we choose

$$V_{[i;j]} = a_{[i;j]}E_{[i;j]}. \tag{20}$$

All the Γ's and γ's are determined by the (absolute values of) the N^2 (=16 here) parameters $a_{[i;j]}$.

4.2. Pure decoherence only

So far we have discussed the most general case, when in principle all relaxation and decoherence parameters may be present in the dissipation matrix. However, experimentally, the relaxation time T_1 for most systems is much longer than the dephasing time T_2 so that we may effectively neglect the relaxation rates γ. In the *pure decoherence (dephasing)* case, comparison of Eq.(18) and Eq.(20) with Eq.(12) tells us that the γ terms vanish if we choose $a_{[i;j]} = 0$ for $i \neq j$. The decoherence parameters Γ_{ij} are then given by

$$\Gamma_{ij} = \frac{1}{2}(|a_{[i;i]}|^2 + |a_{[j;j]}|^2) \quad (i, j = 1 \ldots N \quad i \neq j). \tag{21}$$

This leads to a mathematically very simple situation, as the dissipation matrix L_{D0} is then diagonal. For the $N = 4$ system, this gives 6 pure dephasing parameters ($\Gamma_{ij} = \Gamma_{ji}$ and $\Gamma_{ii} = 0$), determined by 4 constants, so there are two relations between the Γ's - see Eq.(23) below.

Explicitly for the two-qubit, 4-level case,

$$\begin{aligned} L_{D0} = \operatorname{diag}\{&0, -\Gamma_{12}, -\Gamma_{13}, -\Gamma_{14}, -\Gamma_{21}, 0, -\Gamma_{23}, -\Gamma_{24}, \\ &- \Gamma_{31}, -\Gamma_{32}, 0, -\Gamma_{34}, -\Gamma_{41}, -\Gamma_{42}, -\Gamma_{43}, 0\} \end{aligned} \tag{22}$$

with $\Gamma_{ij} = \Gamma_{ji}$.

The constraints imposed by the physical process are

$$\Gamma_{12} + \Gamma_{34} = \Gamma_{14} + \Gamma_{23} = \Gamma_{13} + \Gamma_{24}. \tag{23}$$

4.3. Concurrence

In Sections 3 and 4 we are perforce dealing with general states, and so we must use the extended definition of concurrence for such (mixed) states[9]:

Definition 7 (Concurrence: General definition) *The* concurrence $\mathcal{C}$ *of a (mixed) two-qubit state* ρ *is given by*

$$\mathcal{C} = \max\{\lambda_1 - \lambda_2 - \lambda_3 - \lambda_4, 0\}, \tag{24}$$

where the quantities λ_i *are the square roots of the eigenvalues of the* 4×4 *matrix*

$$\rho(\sigma_2 \otimes \sigma_2)\rho^*(\sigma_2 \otimes \sigma_2) \tag{25}$$

in descending order, where $\sigma_2 = \begin{bmatrix} 0 & -i \\ i & 0 \end{bmatrix}$.

This applies whether the density matrix ρ is either pure or mixed; for pure states it reduces to the form given in Definition 2. As already implied in the footnote given after Definition 2, the entanglement of formation is a monotonic function of the concurrence $\mathcal{C}$, varying between a minimum of zero for $\mathcal{C} = 0$, and a maximum of 1 for $\mathcal{C} = 1$.

4.4. Decoherence of Bell State

We now give an example of a standard dephasing process acting on a maximally entangled state.

Example 5 *Consider the Bell state* $v_B = 1/\sqrt{2}[1, 0, 0, 1]$. *Using Eq.(14), the Liouville vector* $\mathbf{r}$ *corresponding to the equivalent density matrix* $v_B v_B^\dagger$ *is* $1/2[1, 0, 0, 1, 0, 0, 0, 0, 0, 0, 0, 0, 1, 0, 0, 1]$. *The action of the dephasing operator* L_{D0} *of Eq.(22) is given, as in Eq.(15), by*

$$\dot{\mathbf{r}} = L\mathbf{r} = (L_{D0})\mathbf{r} \tag{26}$$

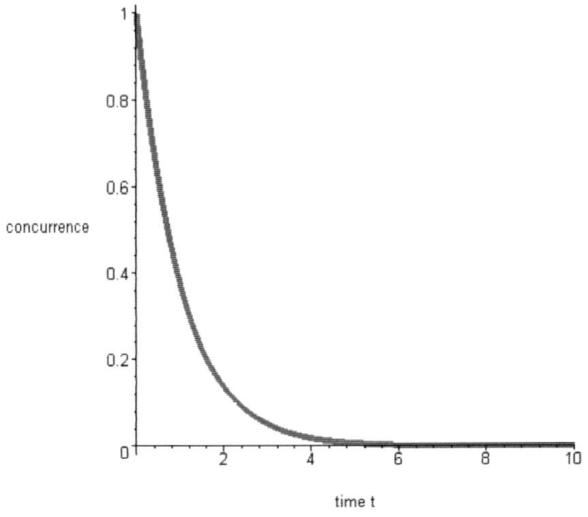

Figure 2. Concurrence *v* Time t (units of 1/Decoherence Γ_{14})

which may be immediately integrated to give

$$\mathbf{r}(t) = 1/2[1, 0, 0, e^{-\Gamma_{14}t}, 0, 0, 0, 0, 0, 0, 0, 0, e^{-\Gamma_{14}t}, 0, 0, 1] \tag{27}$$

corresponding to the density matrix

$$\rho(t) = 1/2 \begin{bmatrix} 1 & 0 & 0 & e^{-\Gamma_{14}t} \\ 0 & 0 & 0 & 0 \\ 0 & 0 & 0 & 0 \\ e^{-\Gamma_{14}t} & 0 & 0 & 1 \end{bmatrix}. \tag{28}$$

Note that this does not represent a pure state except at $t = 0$*. The concurrence as defined in Definition 7 evaluates to* $\exp(-\Gamma_{14}t)$*. (See Figure 2.)*

The results of Example 5 are essentially unchanged in the presence of an additional *free* hamiltonian. From Eq.(17), the superoperator corresponding to the diagonal matrix $H_0 = \text{diag}\{h_{(1)}, \ldots h_{(N)}\}$ is

$$(L_{H_0})_{[u;v],[r;s]} = -i \ \ (h_{(u)} - h_{(v)})\delta_{ur}\delta_{vs} \tag{29}$$

using the notation of Eq. (13) for which

$$(A \otimes B)_{[u;v],[r;s]} = A_{ur}B_{vs}. \tag{30}$$

Therefore we have

$$[L_D, L_{H_0}]_{[k;n],[r;s]} = -i(h_{(k)} - h_{(n)} - h_{(r)} + h_{(s)})(L_D)_{[k;n],[r;s]}. \tag{31}$$

Consulting Eq. (12) we note that all the non-zero elements of L_D have vanishing coefficients in Eq. (31), whence $[L_D, L_{H_0}] = 0$. Therefore the superoperator corresponding to a *diagonal* hamiltonian commutes with the dissipation superoperator L_{D0}, and indeed commutes with the terms Eq.(12) of L_D in general [10], and thus the additional free Hamiltonian only introduces a phase factor.

Journal of Physics: Conference Series **380** (2012) 012012 doi:10.1088/1742-6596/380/1/012012

4.5. Stable Bell state

In general, entanglement will decay under the type of dissipative processes noted here. However, as is clear from the last example, under special values of the decoherence parameters, entanglement will be preserved. In the case of Example 5 when $\Gamma_{14} = 0(= \Gamma_{41})$ then the maximal entanglement of the state does not decay. Of course, in general we are not able to specify the values of the dephasing Γ's, but one may predict theoretically which types of state will remain invariant under the appropriate decoherence parameters.

References

[1] Nielsen M A and Chuang I L 2000 Quantum Computation and Quantum Information (Cambridge: Cambridge University Press)
[2] Solomon A I and Ho C-L 2010 *Links and Quantum Entanglement* in *Gell-Mann's 80th Birthday Conference* (Singapore:World Scientific) 646
Ho C-L, Solomon A I and Oh C-H 2010 Europhysics Letters 92 30002
[3] Schirmer S G, Pullen I C H and Solomon A I 2005 J. Opt. B: Quantum Semiclass. Opt.**7** S293
[4] Breuer H-P and Petruccione F 2006 The Theory of Open Quantum Systems (Oxford: Clarendon)
[5] Schirmer S G and Solomon A I 2004 Phys. Rev. **A 70**,022107
[6] Havel T F 2003 J. Math. Phys. **44**,534
[7] Lindblad G 1976 Comm. Math. Phys. **48**, 119
Lindblad G 1975 Comm. Math. Phys. **40**, 147
Gorini V, Kossakowski A, and Sudarshan E C G 1976 J. Math. Phys. **17**, 821
[8] Sudarshan E C G, Matthews P M, and Rau J 1961 Phys. Rev.**121**, 920
Kraus K 1971 *Ann.Phys.***64** 311
[9] Wootters W K 1998 Phys. Rev. Lett. **80**, 2245-2248
[10] Solomon A I 2008 European Physical Journal (Special Topics) **160** 391

Breaking time-inversion invariance through decoherence — Energetic consequences for attosecond neutron scattering

C A Chatzidimitriou-Dreismann[1], E MacA Gray[2], and T P Blach[2]

[1] Institute of Chemistry, Technical University of Berlin, D-10623 Berlin, Germany
[2] Queensland Micro- and Nanotechnology Centre, Griffith University, Brisbane 4111, Australia

E-mail: dreismann@chem.tu-berlin.de

Abstract.
Nuclei and electrons in condensed matter and/or molecules are usually entangled, due to the prevailing (mainly electromagnetic) interactions. However, the "environment" of a microscopic scattering system (e.g. a proton) causes ultrafast decoherence, thus making atomic and/or nuclear entanglement effects not directly accessible to experiments. However, our neutron Compton scattering experiments from protons (H-atoms) in condensed systems and molecules have a characteristic collisional time about 100—1000 attoseconds. The quantum dynamics of an atom in this ultrashort, but finite, time window is governed by non-unitary time evolution due to the aforementioned decoherence. Unexpectedly, recent theoretical investigations have shown that decoherence can also have the following energetic consequences. Disentangling two subsystems A and B of a quantum system AB is tantamount to erasure of quantum phase relations between A and B. This erasure is widely believed to be an innocuous process, which e.g. does not affect the energies of A and B. However, two independent groups proved recently that disentangling two systems, within a sufficiently short time interval, causes increase of their energies. This is also derivable by the simplest Lindblad-type master equation of one particle being subject to pure decoherence.

Our neutron-proton scattering experiments with H_2 molecules provide for the first time experimental evidence of this effect. Our results reveal that the neutron-proton collision, leading to the cleavage of the H-H bond in the attosecond timescale, is accompanied by larger energy transfer (by about 2—3%) than conventional theory predicts. Preliminary results from current investigations show qualitatively the same effect in the neutron-deuteron Compton scattering from D_2 molecules. We interpret the experimental findings by treating the neutron-proton (or neutron-deuteron) collisional system as an entangled open quantum system being subject to fast decoherence caused by its "environment" (i.e., two electrons plus second nucleus of H_2 or D_2). The presented results seem to be of generic nature, and may have considerable consequences for various processes in condensed matter and molecules, e.g. in elementary chemical reactions.

1. Introduction

Quantum entanglement (QE) has emerged as the most emblematic feature of quantum mechanics. It provides novel insights into the fundamental structure of physical reality [1, 2, 3, 4], and has raised widespread interest in various branches of science, and recently also in quantum information technology [5].

The environment surrounding a quantum system interacts with it and can, in effect, monitor (or "measure") some of the system's observables. As a result, the phenomenon of environment-induced decoherence [6, 7, 8, 9] emerges, and QE is destroyed. Other, supposedly more general models of decoherence, e.g. those appearing in fundamental modifications of the Schrödinger time evolution, play no role in the present paper. A deeper understanding of decoherence is also expected to lead to new insights into quantum fundamentals, particularly quantum measurement and the quantum-classical transition.

Owing to the prevailing electromagnetic interactions, nuclei and electrons in condensed matter and/or molecules are also in principle entangled. Under ambient conditions, the "environment" of microscopic systems (e.g. a neutron colliding with a proton) causes ultrafast decoherence, thus making atomic and/or nuclear entanglement effects not directly accessible to experiments. In this context, it may also be noted that decoherence is recognized as one of the main obstacles to realizing quantum computing and quantum information processing [5].

Decoherence (and, in particular, disentanglement of quantum-entangled systems) is an omnipresent phenomenon of dynamics of quantum open systems. It is usually an extremely fast dynamical process; for various examples, see e.g. [10] and the textbooks cited above. However, *neutron Compton scattering* (NCS, also called "deep inelastic neutron scattering", DINS) [11, 12] is an ultrafast scattering technique that can probe quantum dynamics of nuclei in the attosecond time window, and so it may open up a way for investigating short-lived QE even in condensed matter at ambient conditions. NCS was first suggested in the pioneering work of Hohenberg and Platzman [11] as a means of measuring nuclear momentum distributions in condensed matter. Conceptually based on the well-known Compton effect [13] the NCS method is analogous to the measurement of electronic momentum distributions in atoms and solids by Compton scattering of high-energy photons from electrons, or measurement of nucleon momenta by quasi-elastic scattering of high-energy electrons from nuclei.

For neutron—proton scattering, and for the experimental setup reported below, the effectuated momentum and energy transfers are large,

$$\Delta E \sim 1 - 30 \text{ eV} , \quad |\mathbf{q}| \sim 20 - 120 \text{ Å}^{-1} ,$$

and the "collisional time window" or "scattering time" τ_{sc} is very short,

$$\tau_{sc} < 10^{-15} \text{ s}$$

[12, 14]. This is of the order of the timescale of "electronic motion", e.g. the duration of electronic rearrangements accompanying the formation or the breaking of chemical bonds. Thus one may expect that NCS can reveal unknown features of quantum dynamics of atoms participating in chemical bonds [15]. Such features are revealed by the "anomalous" decrease of NCS-intensity from protons (in short, H-intensity deficit) of several materials; cf. [16, 17, 18, 19]. Furthermore, this attosecond effect was confirmed by Cooper et al. applying an independent experimental method, i.e. electron—atom Compton scattering [17, 20]; see also the detailed theoretical analysis by Bonham et al. in the frame of the first Born and Born-Oppenheimer approximations [21].

However, this paper does not deal with the usual objectives of NCS, i.e. widths of Compton profiles, or anomalous scattering intensities. Instead, here we consider only the simplest property of the NCS collisional process which is directly accessible to the experiment: i.e. the mean energy transfer from the neutron to the struck nucleus, called "recoil energy" E_{rec}. This quantity is completely determined by the position of the recoil peak-maximum (or center-of gravity) in the measured time-of-flight(TOF) interval.

Our experimental results presented below demonstrate experimentally that decoherence in neutron—proton Compton scattering from isolated H_2 molecules can also have unexpected energetic consequences. The theoretical basis of new phenomenon under consideration may be described as follows. Disentangling two subsystems A and B of a quantum system AB is tantamount to erasure of quantum phase relations (i.e. decoherence) between A and B. It is widely believed that this erasure is an innocuous process, which e.g. does not affect the energies of A and B. Surprisingly, recent theoretical investigations by two independent groups showed that disentangling two systems, within a sufficiently short time interval, can cause increase of their energies. Moreover, this intrinsic energy increase is also shown to be derivable by the simplest Lindblad-type master equation, which describes one particle being subject to pure decoherence (in the Markovian limit).

Our neutron—proton scattering experiments with H_2 molecules provide for the first time experimental evidence of this effect [22]. The presented results reveal that the neutron-proton collision leading to the cleavage of the H—H bond on an attosecond timescale is accompanied by a larger energy transfer (by about 2—3%) than conventional theory predicts. Moreover, preliminary results from current investigations reveal qualitatively the same effect in the neutron—deuteron Compton scattering from D_2 molecules. We interpret the experimental findings by treating the neutron—proton (or neutron—deuteron) collisional system as an entangled open quantum system being subject to fast decoherence caused by its "environment" (i.e., two electrons plus second nucleus of H_2 or D_2).

The outline of the paper is as follows. After some short introductory remarks on quantum entanglement and nonlocality in Sec. 2, we discuss in Sec. 3 the recent theoretical models exhibiting the energetic consequences of decoherence. In addition, we show here the presence of the same effect in the well-known formalism of Lindblad-type master equations describing pure decoherence. Sec. 4 contains the experimental method and the main experimental results. The proposed theoretical interpretation of the observations in Sec. 5, which is only qualitative, appears to confirm the new decoherence effect. In Sec. 6, among additional remarks and conclusions, we present current (preliminary) results of scattering from molecular D_2 which give further confirmation of the phenomenon under consideration. For readers interested in experimental details, the Appendices provide a concise (but rather self-contained and fully referenced) description of the experimental method and its specific features, of data analysis procedure, and of associated technical and/or instrumental details.

2. Remarks on quantum entanglement

The notion of entanglement (in short QE) was introduced by Schrödinger. It describes the physical consequences of the mathematical fact that the tensor product $\mathsf{H}_A \otimes \mathsf{H}_B$ of two vector spaces H_A and H_B is larger than their Cartesian product $\mathsf{H}_A \times \mathsf{H}_B$ [23]. Namely, not every element of the tensor product can be factorized as $\psi_A \otimes \psi_B$, with $\psi_i \in \mathsf{H}_i$ $(i = A, B)$.

The crucial physical consequence of this fact becomes obvious by taking H_A and H_B to be the state spaces of two quantum objects A and B. A pure state $|\Psi\rangle \in \mathsf{H}_A \otimes \mathsf{H}_B$ of the combined quantum system is called separable, or not entangled, if it can be factorized as $|\Psi\rangle = |\psi_A\rangle|\psi_B\rangle$. But due to the superposition principle of quantum mechanics there are also states of the form

$$|\Psi\rangle = c_1|\psi_A\rangle|\psi_B\rangle + c_2|\phi_A\rangle|\phi_B\rangle + \dots \ , \tag{1}$$

$(\psi_i, \phi_i \dots \in \mathsf{H}_i,\ i = A, B)$. which are not separable in general. In that case, the individual quantum objects cannot be assigned independent pure states, and one says that the two objects are entangled with each other, or simply entangled.

A well known example of an entangled quantum state is the so-called singlet state

$$|\Psi^-\rangle = \frac{1}{\sqrt{2}}\left(|0\rangle_A|1\rangle_B - |1\rangle_A|0\rangle_B\right) \tag{2}$$

of two (distinguishable) qubits (or two spin-1/2 particles, etc.). In a singlet each individual particle A and B has no definite value 0 or 1 (or: spin direction). For example, the state of the particle A is not a pure state but the mixed state

$$\rho_A = \mathbf{Tr}_B|\Psi^-\rangle|\langle\Psi^-| = \frac{1}{2}(|0\rangle_A\langle 0|_A + |1\rangle_A\langle 1|_A) \ . \tag{3}$$

QE has various counter-intuitive consequences, one of them being nonlocality. Einstein, Podolsky and Rosen (EPR) argued that spatially separated objects must have separable physical descriptions and thus concluded that the quantum mechanical description of objects in an entangle state must be incomplete, i.e. so-called "hidden variables" must exist [1]. This constitutes the main point of the famous EPR argument.

Three decades later, Bell, in looking more thoroughly at the measurements that can be made on two entangled but spatially well-separated, objects, showed that the probabilistic predictions of quantum mechanics can be incompatible with the existence of such hidden variables [2]. In particular, Bell showed this for specific entangled states. Twenty-five years later, Gisin succeeded in proving that this is true for all entangled states [3]. Thus [23]

$$\text{Entanglement} \Rightarrow \text{Nonlocality} \tag{4}$$

Today one can say that entanglement and nonlocality are fundamental properties of physical reality.

The context in which the notion of entanglement was first introduced and discussed was purely theoretical and/or philosophical (see e.g. [4]). Only in the last 15—20 years was it recognized that entanglement and decoherence should have practical applications: quantum computing, quantum coding, quantum cryptography, etc. [5].

2.1. Dynamical creation of entanglement.

Quantum entanglement (QE) in molecules and/or condensed matter is the consequence of many-body interactions. Let Ψ_A and Ψ_B be the state vectors of two systems A and B. The two systems, with "free" Hamiltonians H_A and H_B, may interact, the corresponding Hamiltonian being V_{AB}. Let the state of the composite system 'A and B', in short AB, at time $t = 0$ be separable, $\Psi_{AB}(0) = \Psi_A \cdot \Psi_B$. With the exception of pathological cases, V_{AB} does not commute with H_A and H_B. Then it holds for $t > 0$ that the complete evolution operator of the composite system AB,

$$U_{AB}(t) = \mathrm{e}^{-\frac{i}{\hbar}(H_A+H_B+V_{AB})t} \tag{5}$$

does not factorize into a product of two "free" evolution operators, U_A and U_B. Consequently, the wave function $\Psi_{AB}(t)$ at time $t > 0$, $\Psi_{AB}(t) = U_{AB}(t)\Psi_{AB}(0)$, does not factorize into a product of two wave functions (say, Ψ'_A and Ψ'_B), each of them representing the state of one of the two systems A and B; i.e.

$$\Psi_{AB}(t) = U_{AB}(t)\Psi_{AB}(0) \neq \Psi'_A(t) \cdot \Psi'_B(t) \ . \tag{6}$$

That is, $\Psi_{AB}(t)$ is an entangled state being caused by V_{AB} which couples the specific degrees of freedom appearing into it. For example, Coulombic interactions, which depend on the spatial degrees of freedom of the particles (i.e., nuclei and electrons) lead to spatial entanglement; magnetic interactions between spins lead to spin entanglement being of central importance for the emerging quantum technology of "spintronics" [5].

For closed quantum systems the time evolution is t-inversion invariant, and the process of Eq. (6) represents a unitary transformation, because the operator U_{AB} is unitary, i.e. $U_{AB}(t)\, U^{\dagger}_{AB}(t) = \hat{1}$.

Journal of Physics: Conference Series **380** (2012) 012013 doi:10.1088/1742-6596/380/1/012013

3. Energetic consequences of decoherence

Decoherence is the typical quantum dynamical phenomenon which destroys quantum state superpositions (quantum interference, quantum phases), thus leading to the "appearance of a classical world in quantum theory"; cf. the textbooks and review articles [6, 7, 8, 9].

Decoherence, similarly to many other irreversible mechanisms, is a time-oriented process, i.e. it breaks the well-known time-inversion invariance of the Schrödinger equation. A considerable number of models of decoherence, based on a variety of physical motivations, have been proposed and investigated in the scientific literature; cf. [6, 7, 8, 9]. A particularly illuminating comparison of some models and their motivations has been provided by Diósi, who also demonstrated that a possible intrinsic time uncertainty of quantum mechanics can cause decoherence [24].

Quite unexpectedly, it turns out that "pure" decoherence, which is known to be much faster than friction [6, 7, 8], may have a specific peculiar consequence. Namely, as shown below in the frame of three well-known theoretical models, pure decoherence turns out to be intrinsically connected with an unavoidable increase of mean energy of the system — in the absence of any direct interaction with other systems. This is certainly unexpected, since erasure of quantum phases is usually not acknowledged as a source of energy.

3.1. Lindblad equation and energy increase

To describe the dynamics of open quantum systems and decoherence, various generalizations of the Schrödinger equation have been proposed; see the textbooks [6, 7, 8]. Among these theoretical approaches, the Born-Markov master equation plays an enormously important role. Master equations of the so-called Lindblad form refer to a particular (albeit quite general) class of Markovian master equations, which ensure positivity of the reduced density operator $\rho(t)$ describing the system, i.e. $\langle\rho(t)\rangle \geq 0$, for any pure state $|\psi\rangle$ of the system and for all t. The most general mathematical form of such equations was derived by Gorini, Kossakowski and Sudarshan [25] and Lindblad [26].

Here we consider the simplest Lindblad-type ansatz for the master equation for the statistical operator ρ of an open quantum system, which includes only one Lindblad operator R; in a real system we would have a multitude of such dynamical variables. We set

$$\frac{\partial \rho}{\partial t} = -\frac{i}{\hbar}[H, \rho] - \Gamma[R, [R, \rho]] \tag{7}$$

where $\Gamma > 0$ is a positive constant and H is the Hamiltonian. The first term on the right-hand side (rhs) describes the usual unitary time evolution of the state. The double commutator term describes decoherence (and/or dephasing). It may be noted that this equation describes "pure" decoherence, i.e. it does not contain any term describing friction, the latter being included in Caldeira-Leggett-type equations [27].

Considering Eq. (7) in the R-representation, the double commutator term takes the form

$$-\Gamma(r - r')^2 \langle r|\rho|r'\rangle$$

which, for positive Γ, leads to an exponential decay of the nondiagonal elements of the density matrix:

$$\langle r|\rho(t)|r'\rangle = \langle r|\rho(0)|r'\rangle \exp(-\Gamma(r - r')^2 t) \qquad \text{for } \ 0 \leq t \tag{8}$$

where $t = 0$ is the initial time [6, 7, 8]. Obviously, this result does not hold for $t \leq 0$. In other words, coherent superpositions of different states $|r\rangle$ and $|r'\rangle$ will be suppressed over time. The most well-known specialization of Eq. (7) is the case of an one-particle system that interacts with an external (thermal) environment, in which case R is taken to be the position operator, q. This corresponds to the Joos-Zeh master equation [28].

It is well known that Eq. (7) preserves the normalization $Tr\rho = 1$, which is very satisfactory. However, as Ballentine [29] pointed out, the expectation value of the energy, $\langle H\rangle = Tr(H\rho)$ is in general not conserved, as the following short calculation proves:

$$\frac{d\langle H\rangle}{dt} = \frac{d}{dt}Tr(H\rho) = Tr\left(H\frac{\partial\rho}{\partial t}\right) = -\Gamma\, Tr([H,R][R,\rho]) \tag{9}$$

Here the operator identity $Tr(A[B,C]) = Tr([A,B]C)$ was used. Obviously, if the Hamiltonian and the Lindblad operator R do not commute, $\langle H\rangle$ is not conserved for every state ρ.

Of particular importance for illustration, as well as for the following discussions, is the special case of a "free" particle in one dimension with momentum operator p and Hamiltonian $H = p^2/2m$. The decoherence in position owing to interactions with the "environment" is described by taking R to be the particle's position operator q [28]. Then, since $[p,q] = \hbar/i$ and

$$[p^2,q] = p^2q - qp^2 = (p^2q - pqp) + (pqp - qp^2) = p[p,q] + [p,q]p = \frac{2\hbar}{i}p$$

one obtains from the last equation

$$\begin{aligned} \frac{d\langle H\rangle}{dt} &= -\Gamma\, Tr([p^2/2m,q][q,\rho]) = -\frac{\Gamma}{2m}\frac{2\hbar}{i}Tr(p[q,\rho]) = -\frac{\Gamma}{2m}\frac{2\hbar}{i}Tr([p,q]\rho) \qquad (10)\\ &= +\frac{\Gamma\hbar^2}{m} > 0 \qquad (11)\end{aligned}$$

This surprising result holds also in three dimensions, and for inclusion of scalar and/or vector potentials. It also remains valid for more general cases, e.g. even in cases with the decoherence factor Γ becoming a function of the "distance" $|r - r'|$; see [29, 30].

This result is quite disturbing, as it contradicts conventional expectations about what the consequences of decoherence should be (besides the destruction of coherent superpositions of quantum states). Namely, theorem (11), which follows from the Lindblad equation (7) for a "free" particle, seems to represent a serious weakness of the theory, because it implies that the system's mean energy should always increase; or, in other terms, the environment would act as an inexhaustible energy source. As discussed by Ballentine, this continuous gain of energy is clearly incompatible with the system's attainment of equilibrium [29]. For a more thorough discussion of these difficulties, see Refs. [29, 30].

3.2. The theoretical model by Schulman and Gaveau

Surprisingly, a recent independent and more general theoretical analysis by Schulman and Gaveau [31, 32] has derived essentially the same result. We begin with a short description of the general model.

A quantum system A, e.g. a quantum oscillator, with free Hamiltonian H_A makes an elastic collision with a second system B with free Hamiltonian H_B. Let the interaction Hamiltonian (potential) be V_{AB}. The total Hamiltonian is then

$$H = H_A + H_B + V_{AB}\ . \tag{12}$$

Initially, i.e. before the collision, the two particles are not entangled and so one takes the complete density matrix $\rho(0)$ to be a product, $\rho(0) = \rho_A(0)\otimes\rho_B(0)$. Subsequent to their collision they become entangled and the exact density operator

$$\rho(t) = U(t)\,\rho(0)\,U^\dagger(t)$$

($U(t)$: time evolution operator) is not representable by the product of the individual density operators $\rho_A(t) = \mathrm{Tr}_B\rho(t)$ and $\rho_B(t) = \mathrm{Tr}_A\rho(t)$. ($Tr_i$ denotes the partial trace over the variables of system i.

However, it is widely believed that once the particles are separated the quantum correlations associated with the entanglement can be dropped (provided one does not perform an experiment of Einstein-Podolsky-Rosen type [1]), simply because measurements of physical quantities of each of the two particles cannot depend on their QE. Thus the replacement

$$\rho(t) \rightarrow \rho_A(t) \otimes \rho_B(t) \ , \tag{13}$$

i.e. the erasure of entanglement, is assumed to be innocuous, as it should e.g. not affect the energies of the systems.

The striking result of Schulman and Gaveau is that this does not hold in general [31]. Putting

$$\Delta\rho(t) = \rho_A(t) \otimes \rho_B(t) - \rho(t)$$

and for a particular form of the interaction Hamiltonian, they show that for sufficiently *short* times the following relation holds:

$$\Delta E \equiv Tr(\Delta\rho(t)H) = Tr(\Delta\rho(t)\, V_{AB}) > 0 \ . \tag{14}$$

In other words, the replacement of the entangled $\rho(t)$ by the non-entangled state $\rho_A(t) \otimes \rho_B(t)$, i.e. the decoherence of the AB system, necessarily increases the systems's energy, for sufficiently short times [31, 32]. This appears highly paradoxical since, as Schulman and Gaveau put it: "...losing quantum correlations should not heat the gas. You do not burn your finger because of a partial trace over a density matrix" [31].

A detailed derivation of this inequality is presented in [32]. Moreover, this result was shown to be valid for a large class of potentials, e.g. for two-body interactions, although it does not hold universally [32]. This is an interesting point, and therefore the following example should be mentioned. The spin-boson model Hamiltonian

$$H = H_A + H_B + V_{SB} = \omega_a a^\dagger a + \omega_b b^\dagger b + g(a^\dagger + a)(b^\dagger + b)$$

was shown to exhibit the considered effect, in clear contrast to the related Jaynes-Cummings model

$$H = H_A + H_B + V_{JC} = \omega_a a^\dagger a + \omega_b b^\dagger b + g(a^\dagger b + b^\dagger a) \ ,$$

in which the effect is absent [31].

One may object that the above result, Eq. (14), is unphysical since it violates energy conservation. This, however, is not the case, as the detailed discussions of Ref. [31] showed. It was stressed that, in the situation contemplated, the coupling Hamiltonian must be considered as time dependent, because the physical approach and separation of the particles leads to a time-dependent coupling coefficient. Thus, energy conservation need not apply. Moreover, it was pointed out that the translational degrees of freedom of A and B supply this "additional" energy $\Delta E > 0$ [31].

To make qualitative contact with our NCS experiment under consideration, the impinging neutron n may be associated with system A, and the struck proton p together with its effective Born-Oppenheimer (BO) potential with system B. The neutron-proton interaction is associated with V_{AB}; see also below. In this context it should be noted that the phenomena of entanglement and decoherence play no role in conventional neutron scattering theory [12, 39, 40, 41]. However, the existence of quantum entanglement follows from first principles within quantum theory, and is essentially a consequence of the non-commutativity of the interaction Hamiltonian with the free Hamiltonian.

3.2.1. Digression: On van Kampen's "repeated randomness assumption". In [32] it was mentioned that some physical topics like the Boltzmann H-theorem and Kampen's criticism of Green-Kubo formulas may be related to the aforementioned erasure of quantum correlations and its effects. Hence it may be interesting to consider in some detail the possible conceptual interrelation and/or similarity between

(*a*) the Schulman-Gaveau approach of applying the decoherence assumption (13) before each new collision of system A with B, and

(*b*) van Kampen's "repeated randomness assumption", which was shown to underly the fundamental basis of statistical mechanics of irreversible processes [33, 34, 35, 36].

The mentioned "randomness assumption" refers to erasure (or coarse graining, averaging out) of quantum phases being represented with the non-diagonal elements of the exact density operator $\rho(t)$. This corresponds to decoherence or dephasing, in present-day nomenclature. Scrutinizing linear response theory and the validity of the well-known Green-Kubo formulas, van Kampen concluded the following: the assumption of the Kubo theory [37, 38] that linearity of the macroscopic response to an external force F means that the equation of motion for $\rho(t)$ should also be solved to first order in F, during a short time interval Δt, constitutes a "fatal flaw" (see [36], p. 280). Moreover, this microscopic linearity was shown to be equivalent to a tacitly assumed randomness (i.e. decoherence) not only at the initial time $t = 0$, as the Kubo theory claims, but also continuously during the whole Δt [36].

For illustration, let us consider the case of electric conductivity in a metal. Van Kampen's result implies here that the well-known appearance of resistivity, and associated energy transfer (dissipation) from the electrons to the lattice, can only be proven within Kubo's theory by assuming that decoherence takes place continuously during the whole Δt, thus keeping the density operator at equilibrium (and hence diagonal in the energy representation), $\rho(t) = \rho_{eq}$, during Δt. Or, in simpler terms: if one assumes $\rho(0) = \rho_{eq}$ only, as standardly done in theoretical treatments, and if linearization of $\rho(t)$ owing to a small disturbing electric field $F(t)$ is carried out correctly and without applying further assumptions, then one obtains zero resistance and zero energy dissipation!

The analogy to the Schulman-Gaveau results of calculations presented in [31] is given by the repeated disentanglement $\rho(t) \rightarrow \rho_A(t) \otimes \rho_B(t)$ after each collision and before the next one takes place. Without erasing quantum correlations there is no heating up of the systems A and B.

3.3. The model by Erez et al. and Gordon et al.

It may be noted that the model of Ref. [31] assumes the appearance of decoherence, as the replacement $\rho(t) \rightarrow \rho_A(t) \otimes \rho_B(t)$ shows, but does not describe the disentanglement process explicitly. Thus, the environment and the couplings of the two systems to it do not appear explicitly in the model.

Very recently, theoretical investigations that address these issues have been presented by Erez et al. [42] and Gordon et al. [43], which explore explicitly the aforementioned decoherence process in a different physical context. The focus here is the cooling of two-level model systems (quantum bits) on ultrashort time scales. It is shown that these anomalous non-Markov cooling processes stem from the hitherto unfamiliar coherent quantum dynamics of the system-bath interaction well within the bath memory time. This is an interesting point, since the significance of the bath's dynamics becomes apparent.

In the course of their analysis the authors derive a result analogous to Eq. (14), first for brief quantum non-demolition (QND) measurements [42] and later for the general case of any abrupt disturbances, e.g. phase changes [43].

In Ref. [42] the "measurement" (or interaction) of a quantum system S by a "detector" D (i.e a second quantum system) is considered, represented by the interaction $H_{SD}(t)$. The authors study the total Hamiltonian

$$H_{tot} = H_S + H_B + H_{SB} \tag{15}$$

of the system (with Hamiltonian H_S) that interacts with the bath (with Hamiltonian H_B) and is intermittently perturbed by the coupling of the system to the detector:

$$H(t) = H_{tot} + H_{SD}(t) \ . \tag{16}$$

It is essential that the interaction Hamiltonians H_{SB} and H_{SD} do not invoke the rotating-wave approximation; i.e. energy conservation between the system and the bath or the detector is not imposed, on the ultrashort time scales considered [42].

It was proved [42] that a nearly impulsive (projective) disturbance of the system always produces an energy increase (called "heating") of the equilibrium state immediately thereafter. This general conclusion was supported by numerical results, for a disturbance of finite, albeit brief, duration. For an ultrashort time interval $0 \leq t \leq \tau$, the authors show the important results

$$\begin{aligned} \langle H_{SB}(0)\rangle \leq 0 \quad &\rightarrow \quad \langle H_{SB}(\tau)\rangle = 0 \\ \langle H_{SD}(t)\rangle \quad &= \quad -\langle H_{SB}(t)\rangle \end{aligned} \tag{17}$$

where $\langle H_D\rangle$ remains unchanged. This transfer of energy, associated with a change in the entanglement between the system and the bath, triggers the quantum dynamics that redistributes their mean energy (and entropy).

At time $t = \tau + \Delta t$ after the measurement (because $H_{SD}(t \geq \tau) = 0$), and during sufficiently short time intervals Δt, the rotating wave approximation breaks down and $\langle H_S + H_B\rangle$ is not conserved; only $\langle H_{tot}\rangle$ is conserved, until the next measurement. But now $\langle H_{SB}\rangle$ decreases because the correction to the ground-state energy of H_{tot} should be negative for a weakly perturbing interaction H_{SB}. As a result, one obtains the relations

$$\begin{aligned} \frac{d}{dt}(\langle H_S + H_B\rangle)\rfloor_{\tau+\Delta t} \ &> \ 0 \\ \frac{d}{dt}\langle H_{SB}\rangle\rfloor_{\tau+\Delta t} \ &< \ 0 \end{aligned} \tag{18}$$

which implies a post-disturbance "heating" of the system and the bath combined [42]. These results were recently extended to the general case of any abrupt disturbance [43].

To make a qualitative contact with the basic features of our NCS experiment under consideration, one may associate the proton p with the quantum system S, the proton's environment (denoted by $\mathcal{E}$ below) with the bath B, and the neutron with D. The aforementioned "additional" energy ΔE immediately after the disturbance —in Ref. [43] denoted by $\delta\langle H_{tot}\rangle^M$ for projective measurements and by $\delta\langle H_{tot}\rangle^\phi$ for an abrupt phase shift— was proved to be positive, $\Delta E > 0$; i.e., a "heating" of the equilibrium state appears immediately after the external abrupt disturbance [43]. Recall that the NCS process is not a QND process, and thus one may expect that ΔE can contain a contribution from the translation energy of the impinging neutron.

However, the "environment" of the struck proton hardly resembles a thermal bath and the coupling H_{SB} is very strong. Therefore the above theoretical results cannot be directly related with our experimental findings on H_2 presented below. However, a relation to NCS experimental results from liquid or solid ^{4}He, in the normal or superfluid state, could be appropriate, because the struck He (i.e. the system) interacts weakly with the atoms of its environment.

4. Experimental results

We now proceed to our new experimental results [22] and their physical interpretation in the theoretical context of the models of the preceding section. It will be shown that, although one usually thinks of the destruction of quantum correlations as purely an information issue, here it appears to have measureable energetic consequences.

We reiterate that issues regarding the "anomalous" scattering intensity effect [16, 17, 18, 19], or the widths of Compton profiles [12], are not addressed in the present paper.

4.1. NCS measurements

In this section we present very recent experimental NCS results from gaseous molecular H_2 and D_2 and associated current discussions [22, 44, 45], which provide for the first time evidence for the novel decoherence effect predicted by the theoretical models of the preceding section.

To facilitate the presentation, we first outline some essential elements of data recording and analysis; full details are given in the Appendices and in [22].

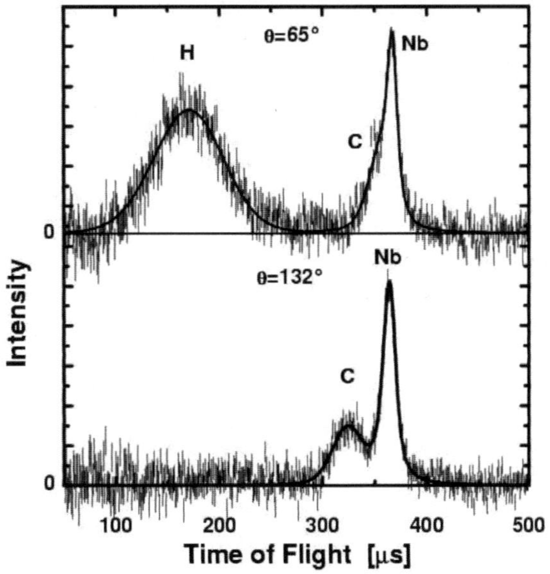

Figure 1. Examples of measured TOF spectra. The data are from liquid benzene (C_6H_6) in an Nb cell, at the specified scattering angles [48]. Solid lines are fits to the data. The H recoil peak is well separated from those of C and Nb. As expected, there is no H recoil peak in the backward scattering spectrum. In backward scattering regime, the C- and Nb-recoil peaks are resolved owing to the larger energy transfers involved.

From the measured time-of-flight (TOF) spectra (see Figure 1 for two examples), momentum ($\hbar\mathbf{q}$) and energy ($\hbar\omega$) transfers from the neutron to the stuck particle are determined by standard methods; see Appendix B. The main object of interest is here the position of the recoil peak of a struck nucleus (proton or deuteron) in TOF space. According to conventional theory, this immediately translates to the energy transfer from the neutron to the nucleus in an elementary collisional two-body process, and therefore it is considered to be well understood.

Within conventional NCS theory which assumes validity of the impulse approximation (IA) [12, 46, 47], the elastic collision of a neutron with an approximately free atom with mass M and initial momentum $\mathbf{p}$ results in the neutron's lost energy $\hbar\omega$ being transferred to the struck atom (see Appendix B):

$$\hbar\omega = \frac{(\hbar\mathbf{q}+\mathbf{p})^2}{2M} - \frac{p^2}{2M} = \frac{(\hbar q)^2}{2M} + \frac{\hbar\mathbf{q}\cdot\mathbf{p}}{M} = \hbar\omega_{rec} + \frac{\hbar\mathbf{q}\cdot\mathbf{p}}{M}\,. \quad (19)$$

($\hbar\mathbf{q}$: momentum transfer from the neutron to the struck atom.) This equation represents energy conservation. The recoil energy $\hbar\omega_{rec} = \hbar^2 q^2/2M$ represents the energy of recoil of a stationary nucleus owing to the collision. Thus scattering from a gaseous sample of such atoms leads to a recoil peak centered at $\hbar\omega_{rec}$, exhibiting a peak width caused by the term $\hbar\mathbf{q}\cdot\mathbf{p}/M$ which represents Doppler broadening. The above energy balance holds exactly in the IA, which however is not completely fulfilled at moderate momentum and energy transfers. In such cases, so-called final state effects (FSE) become apparent. They are caused by environmental forces on the struck particle. It was shown that FSE cause a shift of the maximum of the recoil peak to *lower* energy than the recoil energy [12, 46, 22].

In the context under consideration, this point is crucial. In the framework of conventional NCS theory, a shift to higher energy transfers than $\hbar\omega_{rec}$ would imply that there is a higher energy loss in the scattering process than the IA predicts. Physically this means that the struck atom would be "more free" than even an isolated atom floating in space, which of course is meaningless. Hence, deviations from the IA must give peak shifts to less than $\hbar\omega_{rec}$, since they are always caused by the atom not being free, owing to its interaction with other atoms. Thus there is an additional resistance to motion of the struck atom, which manifests as a slightly lower energy transfer than the IA predicts. Summarizing, a peak-maximum shift to higher energies than $\hbar\omega_{rec}$ is impossible within conventional NCS theory.

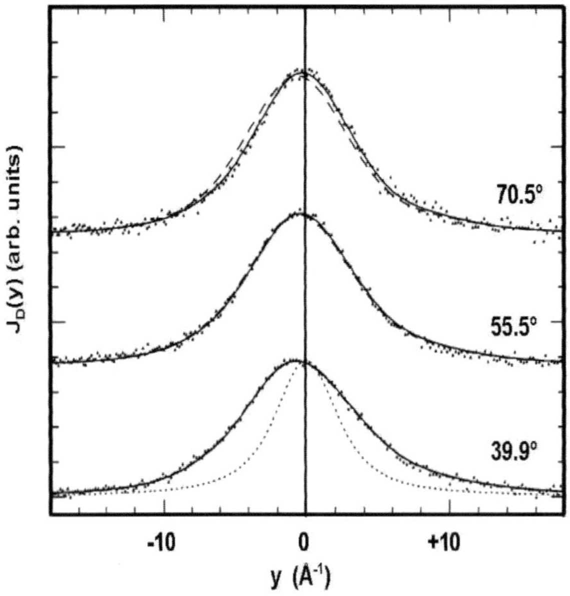

Figure 2. Compton profiles of ortho-D_2 at 20 K, measured at ISIS (parallel shifted for visibility). Adapted from Fig. 26 of [50]. Points are experimental data (after data processing). The lines represent various theoretical models discussed in that reference.

The measured TOF spectra may be straightforwardly transformed to obtain the associated experimental Compton profiles [12, 46, 47] of the recoiling atoms, commonly denoted $J(y)$; see Appendix C. This function represents the distribution of the one-dimensional projection $\hbar y$ of atomic momentum $\mathbf{p}$ (before collision) along the $\mathbf{q}$ direction represented by the unit vector $\mathbf{e_q}$. According to the IA, [12, 46, 47]

$$J(y) = \int n(\mathbf{p})\, \delta(\mathbf{p} \cdot \mathbf{e_q} - \hbar y)\, d\mathbf{p} \ . \tag{20}$$

Here, $n(\mathbf{p})$ is the momentum distribution of the struck particle with mass M before collision, which is the quantity of interest for theoretical investigations. For isotropic systems, the specific direction $\mathbf{e_q}$ becomes immaterial.

In the IA, $J(y)$ must be centered at $y = 0$ [12, 44, 46, 47]; for further details, see Appendix C. The aforementioned conclusion about peak shifts implies that a shift of the peak maximum (or center of gravity) of $J(y)$ to $y_{max} > 0$ values is impossible within conventional theory, since deviations from the IA must displace $J(y)$ to $y_{max} < 0$. This theoretical expectation is nicely confirmed with the experimental NCS results from earlier liquid ortho-D_2 investigations reproduced in Fig. 2. For the low scattering angle shown, the Compton profile $J_D(y)$ is clearly shifted to $y_{max} < 0$, whereas the peak is centered at $y_{max} = 0$ for the highest scattering angle, for which the IA appears to be valid, as expected by conventional theory.

4.2. Main results in forward scattering

Figs. 3-4 show the Compton profiles $J(y)$ of D and H obtained from measurements of gaseous H_2 and D_2 on the Vesuvio (formerly eVS) NCS spectrometer at the ISIS neutron spallation source. The data are from detectors positioned at scattering angles $\theta < 90°$, i.e. in the "forward" scattering regime.

These experimental data were obtained by transforming to y space the TOF spectra measured as described above and in Appendix B. For our data analysis we used calibration parameters measured directly with steel rules and a protractor [58] by the ISIS Metrology Group; cf. Appendix C.

We have analyzed the sensitivity [49] of the peak positions of $J(y)$ in y-space to calibration uncertainties and found that the results in y-space are reliable to better than 0.3 Å^{-1}, for each single detector. Of course, this value improves for an average $\overline{J}(y)$ obtained from several detectors. This estimate is supported by the small scatter in the positions of the individual-detector $J(y)$ curves shown in all four Figs. 3(a,b) and 4(a,b). The most important findings are:

(A) The peak maxima (or the centers-of-gravity) of the D Compton profiles are found to be at $y_{max} \leq 0$ as conventionally expected, for all momentum transfers investigated (Fig. 3a,b). The IA appears to be well obeyed for high momentum transfers. In these experiments, the D—D bond is not broken (see below). The Compton profiles (a) and (b) are in full agreement with the well-established results of Andreani et al. [51]. Therefore they also represent a confirmation of both the correct working of the present configuration of the eVS/Vesuvio spectrometer and our data analysis, and a baseline for comparison with our results for H, measured and analyzed in the same way. (Data measured with backward scattering detectors, during the same time, are reported in subsection *6.1.* below.)

(B) The H Compton profiles obtained at low scattering angles are centered at $y_{max} \leq 0$ as conventionally expected and experimentally observed earlier [51, 52]; see Fig. 4a. In contrast, the H Compton profiles measured at θ_{high} are centered at momentum values $y_{max} > 0$, thus contradicting conventional theory; see Fig. 4b. Equivalently, the actual energy transferred in the collision is larger (by about 2-3%) than the IA predicts.

The magnitude of this effect may be estimated as follows: From Eq. (20), the excess energy

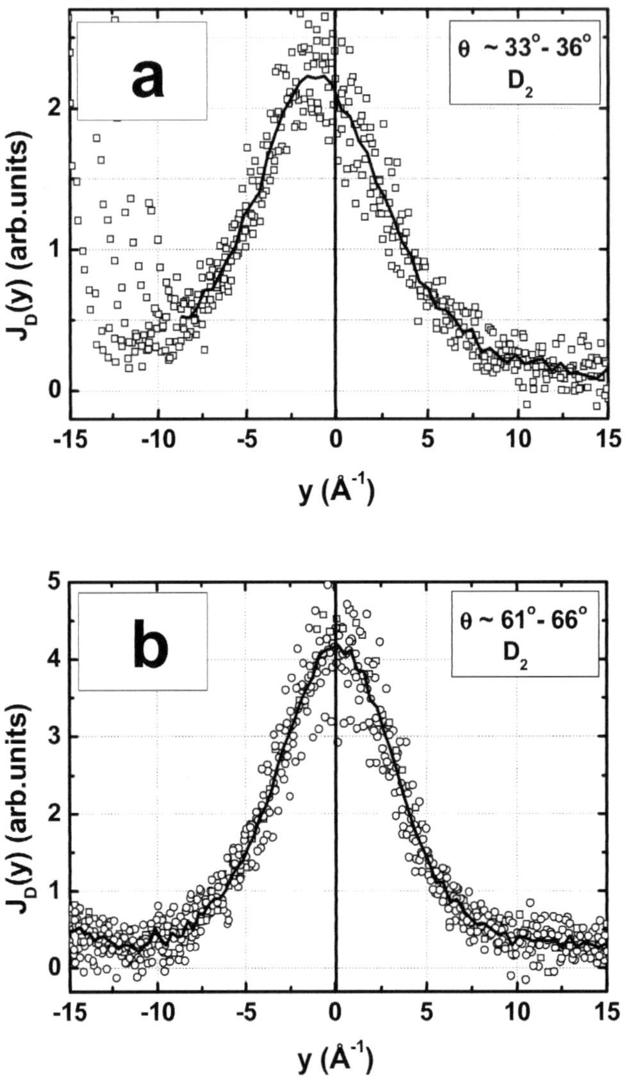

Figure 3. Experimental (not normalized, with no corrections included) Compton profiles $J_D(y)$ of D in D_2 gas at 44 K and 10 bar in an Al cell. This figure is identical with Fig. 1 of Ref. [22]. **(a)** $J_D(y)$ measured with one block of 8 detectors, at the lowest ($\theta_{low} \sim 33° - 36°$) scattering angles available. The signal contribution on the left is due to scattering from the Al cell. The results clearly exhibit conventional FSE for low momentum transfers. **(b)** $J_D(y)$ measured with one block of 8 detectors, at the highest ($\theta_{high} \sim 61° - 66°$) scattering angles available presently. These Compton profiles are in agreement with earlier well-established results [51, 52], like e.g. those in Fig. 2.

transfer is

$$\delta E = \hbar^2 q\, y_{max}/M \tag{21}$$

and thus the ratio R of δE to the recoil energy is $R \equiv \delta E/\hbar\omega_{rec} = 2y_{max}/q$. For H, $m \approx M$ and $q \approx k_1 \tan\theta$, where k_1 is the wavevector of the scattered neutron. From the observed shift of $J_H(y)$ at high q we estimate $y_{max} \approx 1.5\,\text{Å}^{-1}$, which amounts to an excess energy $\delta E \approx 590$

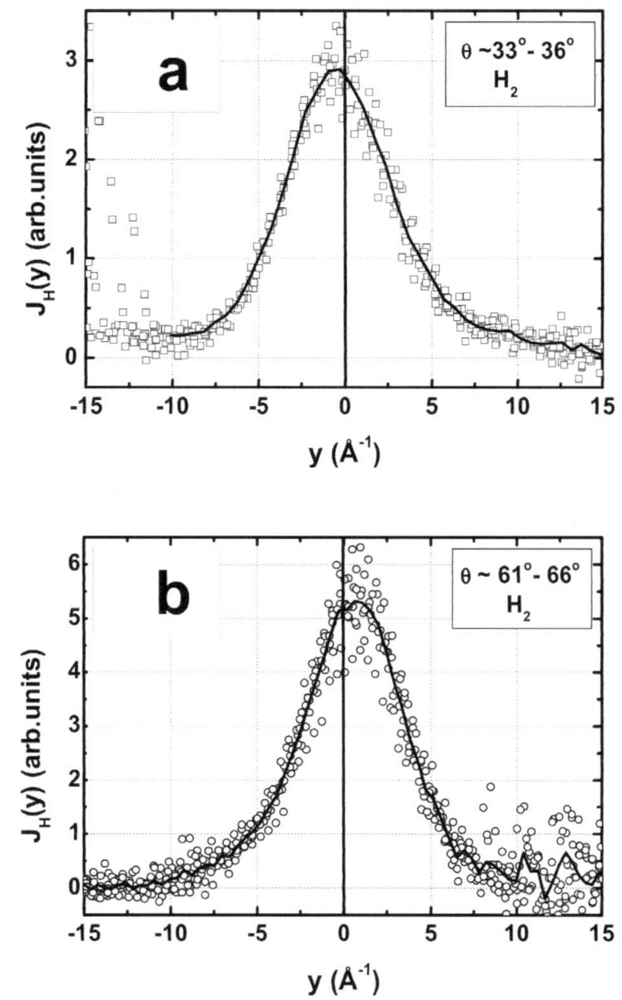

Figure 4. Experimental (not normalized, with no corrections included) Compton profiles $J_H(y)$ of H in H_2 gas at 41 K and 10 bar. This figure is identical with Fig. 2 of Ref. [22]. The instrumental setup was the same as for the experiment with D_2; see the text and Fig. 1. The results **(a)** at θ_{low} are in agreement with the corresponding results of Andreani et al. [51], in which conventional FSE are visible in the experimental Compton profiles $J_H(y)$. The H—H bonds are not broken after scattering. In contrast, the results **(b)** at θ_{high} (with $q \sim 95\,\text{Å}^{-1}$ and $\hbar\omega \sim 19$ eV) are centered at momentum values $y_{max} > 0$. Scattering brakes the H-H bonds. We note that results from detectors at $\theta > 50°$ were *not* reported in Refs. [51, 52].

meV. Therefore one obtains $R \approx (2 \cdot 1.5)/(48.6 \cdot \tan 63°) = 0.03$.

This is interpreted to represent the quantity ΔE of Eq. (14), i.e. $\Delta E = \delta E$.

Some additional comments related to the working conditions of the spectrometer and/or reliability of the above main results are in order here. The agreement between the D results and conventional theory necessarily means that H exhibits an "anomalous" behavior. Namely, the H recoil peak is shifted to positive y values at high energy transfer, which is impossible within any conventional theory. However, by artificially changing the instrument's calibration parameters, the position of the H recoil peak in y space could be "shifted back" to become centered at $y = 0$ and in agreement with conventional theory. But then the D recoil peak would be also shifted to negative y values, thus contradicting numerous literature reports [51, 52, 50] that D obeys conventional theory well as to the position of its recoil peak and thereby refuting those reports [44]. See also Appendix C.

5. Discussion

Observation (B) stands in strict contrast to conventional NCS theory for the reasons described above. It may, however, find a plausible explanation within the framework of open quantum system dynamics, as we propose in the following.

Before collision, the proton is assumed to be bound in an effective BO potential which, in the usual theoretical picture [12], is related to the "environment" $\mathcal{E}$ (the two electrons and the second proton of H_2); the three quantum systems n, p and $\mathcal{E}$ are not entangled. The $n-p$ collision creates QE between the two particles, indicated by $\rho_{np}(t)$, cf. Sect. 2. Furthermore, within the time window τ_{sc}, the struck proton collides and/or strongly interacts with its environment $\mathcal{E}$, causing a complex many-body quantum dynamical process. These interactions create new entanglement between the involved charged quantum systems p and $\mathcal{E}$, which then naturally leads to decoherence of the neutron-proton system under investigation. Thus one may write

$$\rho_n(0) \otimes \rho_p(0) \otimes \rho_{\mathcal{E}}(0) \to \rho_{np}(t) \otimes \rho_{\mathcal{E}}(t) \to \rho_n(t') \otimes \rho_{p\mathcal{E}}(t') \to ... \tag{22}$$

for $0 \leq t < t' \leq \tau_{sc}$. The third step of this dynamical chain of events is due to the "measurement" of p by the environment $\mathcal{E}$ [6, 7, 8]. (Later on, at time t'', and after equilibration of the highly excited system "p plus $\mathcal{E}$", a non-entangled state $\rho_p(t'') \otimes \rho_{\mathcal{E}}(t'')$ may occur again.)

For our purposes, it is important to clarify the physical cause of decoherence in the actual experimental context. To do so, we note that the $n-p$ scattering time related to our NCS experiment turns out to be of the *same order* as the characteristic time of electronic motion and/or charge redistribution, $\tau_{el.motion}$, following the violent excitation and/or electronic "shake-up" of $\mathcal{E}$, both lying in the attosecond time range [53]

$$\tau_{sc} \sim \tau_{el.motion} \ . \tag{23}$$

These excitations are caused by the sudden movement of the struck proton which, in our experiments, has a mean kinetic energy and an energy spread of several electron volts after the collision, which is of the order of the separation of the Born-Oppenheimer molecular electronic levels. Under such conditions, neither the adiabatic nor the sudden approximation [54] can be applied, and thus the "environment" $\mathcal{E}$ appears to play an active role in the quantum dynamics of the $p-n$ system. Additionally, neither can $\mathcal{E}$ be considered here as a memory-less (Markov) bath, nor is it completely "frozen". Using common illustrative terms, one can say that $\mathcal{E}$ continuously monitors, observes, or measures the $p-n$ system. This leads to disentanglement and decoherence, which may be expected to have a sub-femtosecond characteristic time [53].

Obviously, the relation (23) holds only for a restricted range of neutron energies (e.g. those of our experiment), and thus is certainly not valid in conventional thermal and/or cold neutron scattering.

That $\tau_{el.motion}$ lies in the sub-femtosecond time regime has been confirmed by attosecond laser spectroscopy. For example, Goulielmakis et al. [55] reported the real-time observation of valence electron motion in Kr atoms. After ionization, the electron-ion pair was demonstrated to constitute an entangled system and its attosecond decoherence was probed.

As mentioned above, this physical picture is not in line with conventional NCS (and general neutron scattering) theory, in which the entanglement between the neutron and the struck particle is ignored. However, as shown in Sect. 2, denying this entanglement contradicts the principles of quantum mechanics.

Furthermore, conventional NCS theory assumes that single-particle properties are measured, that the environment has no dynamical degrees of freedom (since it is represented by an effective Born-Oppenheimer potential), and that stationary scattering states are fully sufficient for the theoretical description; see also below.

That the effect discussed here is observed for H_2 but not for D_2 (at forward scattering angles) indicates that molecular dissociation might play an important role here. At the recoil energy, H_2 dissociation occurs at approximately $q > 68\,\text{Å}^{-1}$, a condition readily met at θ_{high} where $q_{rec} \sim 95\,\text{Å}^{-1}$ (and corresponding $\hbar\omega_{rec} \sim 19$ eV). In contrast, dissociation in D_2 would occur at about $q > 96\,\text{Å}^{-1}$, which is much higher than the available $q_{rec} \sim 60 - 65\,\text{Å}^{-1}$ at θ_{high} [52]. Consequently, at θ_{high} the neutron—H scattering probes final unbound states in the continuum, whereas neutron—D scattering takes place in the physical space of bound states only. The irreversible electronic shake-up caused by H—H bond breaking should exhibit quite different dynamics from the excitation of bound states of H_2 or D_2, thus leading to potentially different decoherence dynamics. Thus it is tempting to assume that the specific decoherence mechanisms in these two cases be the cause for the qualitatively different scattering behavior of H_2 and D_2.

Needless to say, the complex five-body dynamical problem (i.e. neutron plus two nuclei plus two electrons) has no analytic solution, and a proper quantitative theoretical model of it is not available thus far. However, first theoretical-numerical investigations of the proton momentum distribution $n_H(\mathbf{p})$ in condensed systems, in relation with NCS processes, have been recently initiated [56].

6. Additional remarks and conclusions

6.1. Evidence of decoherence effect in D_2 in backward scattering

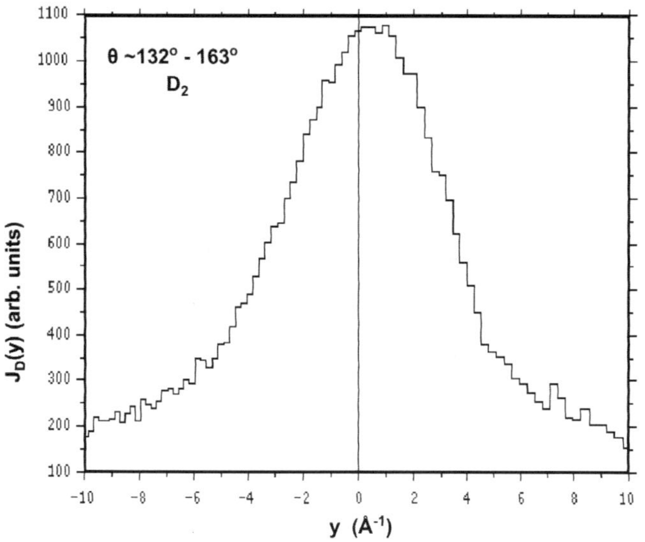

Figure 5. Compton profiles of gaseous D_2 at 41 K, preliminary results. Backward scattering, $\overline{J_D}(y)$ obtained from 87 detectors. The data were taken during the same experiment as that reported above; see Fig. 3.

The above results obtained from H_2, together with the proposed theoretical interpretation, suggest that NCS from D_2 at *backward scattering* angles, $\theta_{back} > 90°$, may also show the decoherence effect under consideration. This is because the energy and momentum transfers on D are, in this regime, considerably higher than in the "forward" scattering regime. The preliminary findings of our current investigations are as follows.

Fig. 5 shows the accumulated D Compton profile (not normalized, as measured) measured by 87 detectors positioned in the "backward" scattering regime of Vesuvio at the specified angular range. The shift of $J_D(y)$ to positive y-values is again clearly visible in the data, even without any data processing or corrections. We emphasize again that this stands in blatant contradiction of conventional theory. Qualitatively the same shift was observed in various groupings containing 10 detectors each. Therefore we can confidently state that the effect under consideration is reproducible, significant and, within experimental error, constant over the available backward angular range of $132° - 163°$.

For the determination of this $J_D(y)$ we used the "standard" instrument calibration data (i.e., the associated file IP0002) provided to the users of Vesuvio, since direct calibration data are not available for the backward-scattering detectors. The standard procedure of the instrumental calibration was very recently published [58]. Since there is no H recoil peak in backward scattering to compare with, as done in the analysis of measurements in the forward scattering regime, we determined the position of the Al and Pb recoil peaks originating from the cell, and found that they are centered at the conventionally expected value $y = 0$.

6.2. On conventional theory of neutron scattering and decoherence

It is sometimes claimed that the phenomena of entanglement and decoherence are irrelevant and/or nonexistent in the physical context of general neutron scattering and NCS [59]. Therefore the following explanations and clarifications should be helpful.

Conventional NCS theory [12], as well as general neutron scattering theory [39, 40, 41], are based on quantum theory of systems with well defined Hamiltonians, which implies that these are *closed* quantum systems, and the associated *unitary* time evolution. The basic formalism is that of time-dependent first-order perturbation theory, and in particular Fermi's Golden Rule; cf. [40], Eq. (2.2).

To be explicit, let us refer here to the basic formula for the partial differential cross-section, which is the measured quantity in a neutron scattering experiment.

$$\left(\frac{d^2\sigma}{d\Omega d\omega}\right)_{\nu_0\to\nu_1} = \frac{k_1}{k_0}\left(\frac{m}{2\pi\hbar^2}\right)^2 \left|\langle\nu_1\mathbf{k}_1|V(\mathbf{r})|\mathbf{k}_0\nu_0\rangle\right|^2 \delta(E_{\nu_0} - E_{\nu_1} + E_0 - E_1) \; ; \tag{24}$$

cf. [40], Eq. (2.15). The right-hand side of this formula contains Fermi's Golden Rule [40, 41] with the delta-function explicitly included in order to guarantee strict energy conservation. V is the Fermi pseudopotential; b is the bound scattering length of atom; $|\nu_i\rangle$, with $i = 0, 1$ are stationary states (with energies E_{ν_i}) of the complete closed system before and after scattering, respectively; the remaining symbols are defined above.

The assumption that $|\nu_i\rangle$ are stationary states of the assumed closed system, i.e. eigenfunctions of the complete Hamiltonian H_c of the N-body system, is essential. This is because the further derivations leading to the well known van Hove results, e.g. the intermediate correlation function

$$F(\mathbf{q}, t) = \frac{1}{N}\sum_{j,k}^{N}\langle\exp(-i\mathbf{q}\cdot\mathbf{r}_j(0))\exp(i\mathbf{q}\cdot\mathbf{r}_k(t))\rangle \tag{25}$$

and the dynamic structure factor

$$S(\mathbf{q}, \omega) = \frac{1}{2\pi}\int_{-\infty}^{\infty}\exp(-i\omega t)F(\mathbf{q}, t)\, dt \tag{26}$$

are based on the introduction of the Heisenberg operators

$$\mathbf{r}(t) = U^{\dagger}(t)\mathbf{r}U(t) \tag{27}$$

where $U(t)$ the *unitary* time evolution operator

$$U(t) = \exp(-\frac{i}{\hbar}H_c t) \ , \tag{28}$$

i.e., $U^{\dagger}(t) = U^{-1}(t)$ holds. Recall that H_c is the Hamiltonian of the closed N-body system — not only of the single nucleus in the two-body neutron—nucleus collision [40, 41]. Note also that neutron–nucleus entanglement, caused by their interaction during the scattering process, plays no role in this formalism.

As an approximation in conventional NCS theory, the complete Hamiltonian is replaced "by hand" with a Hamiltonian of one particle in some effective Born-Oppenheimer potential [12, 50]. In this theoretical framework, of course, any non-unitary decoherence processes in the system play no role "by definition".

However, the scattering nucleus in the aforementioned experiments represents an open quantum system strongly interacting with particles in its environment. Hence the assumed N-body stationary states of the complete Hamiltonian are a means to perform calculations and lack direct physical significance.

The aforementioned impossibility to incorporate decoherence in the standard formalism of neutron scattering, which is governed by t-invariant evolution, leads sometimes to bizzare assumptions. For example, the criticism of Ref. [59] against the theoretical interpretation [53] of the "anomalous" intensity decrease effect [16, 17, 18, 19] is based on the unphysical form of a decohering density matrix introduced into the calculations by

$$\rho_{i,j}(r - r') \rightarrow \rho_{i,j}(r - r') \exp(-\Lambda(r - r')^2|t|) \ , \tag{29}$$

see [59], Eq. (36). Note the incorrect assumption represented by the absolute value $|t|$. This makes the functional form of the assumed density matrix t-inversion invariant, thus standing in blatant contrast to the spirit and physical nature of decoherence.

6.3. Further remarks

(A) Since the observed new effect appears to have a fundamental theoretical basis given by three independent models (see Section 3), we believe our results are of generic nature.

(B) Generally speaking, the results presented above suggest that the relevant theoretical framework of "fast" collisional processes is that of non-unitary dynamics of open quantum systems [6, 7, 8, 9] Moreover, decoherence seems to play a more prominent role here which has not been recognized so far. The quantitative theoretical analysis of the new results remains a challenge to modern theory.

The decoherence appearing in the complex dynamics summarized in relations (22) merits further comments. Decoherence is the focus of very active research on the foundations of quantum theory. Since 1970, various pioneering theoretical investigations of related processes showed that complete information about a particle (here: struck proton) is carried away into the "environment" by interacting microscopic objects and/or a thermal bath. The corresponding information transfer should be considered irreversible, at least in the sense that the "lost" information cannot practically be retrieved. The reader interested in the historical development of the fundamental theory of decoherence may also refer to the original references [60, 61, 62, 27, 28].

(C) The above NCS results and the suggested qualitative interpretation may lead to the following proposition: Not a wavefunction (state vector) Ψ, but a (reduced) density operator

ρ is the physical entity of basic importance. In other terms, ρ is not a mere "approximation" related to "missing knowledge" of an "exact state" Ψ, but rather the opposite holds true: Ψ ought to be an approximation for the description of real physical states.

This viewpoint is in line with the theoretical work of Beretta [63]. This work deals with model evolution equations (applicable not only to open systems but also to closed isolated systems) capable of describing – simultaneously and in competition with the usual Hamiltonian unitary evolution – the natural tendency of any initial non-equilibrium state to relax towards thermodynamic equilibrium. The author also makes the following interesting point:

> "The central conceptual difference between the proposed approach, and the approaches based on attempting to derive the KSGL [Kossakowski-Sudarshan-Gorini-Lindblad] equation, is that this approach regards a non-pure density matrix as representing a real ontological object, the actual state of the world, and is not understood as just an epistemic ignorance of which particular pure state the world is 'really' in." (See [63], p. 142.)

Furthermore, the characteristic time parameter τ_D in Beretta's theory (see [63], Eq. (35)) called "intrinsic dissipative time", seems to be conceptually connected with the characteristic decoherence time of the scattering process under consideration.

Moreover, Beretta's geometrically motivated construction effectively suppresses the term corresponding to

$$R^{\dagger}\rho R \tag{30}$$

in the Lindblad equation (7). This difference between the two theoretical models can be subjected to experimental test. Namely, in view of the novel experimental NCS findings presented above, it would be very interesting to investigate the predictions of Beretta's theory with respect to the experimental observations, and compare them with those of the Lindblad equation. This may shed new light on the fundamentals of decoherence, microscopic irreversibility, and other non-unitary dynamical processes.

(D) In this context it should be pointed out that the description of open quantum systems is possible not only in terms of a reduced density matrix, but also in terms of wave functions, but these are then usually solutions of effective nonlinear Schrödinger equations (see, e.g. [64]).

(E) One may wonder why the dynamics of open quantum systems should be relevant for the system "neutron plus proton of H_2", although "neutron plus H_2 molecule" (in gas phase) is a closed system (in the theoretical frame of non-relativistic quantum mechanics.) Here it is instructive to refer to the aforementioned investigation of krypton atoms by Goulielmakis et al. [55], applying attosecond laser-spectroscopic techniques. As a result, even this monoatomic system was found to exhibit short-lived entanglement and decoherence, which are typical phenomena of open quantum systems (see above).

(F) Owing to the fact that the timescale of the considered scattering process and of electronic motion are of the same order, both being in the attosecond regime, we anticipate the above phenomenon to be a starting point for more sophisticated time-dependent studies in atoms (like e.g. ion—electron decoherence accompanying electron extraction from an atom with an attosecond laser pulse) and in molecules (e.g. chemical bond cleavage caused by collisions of other probe particles with molecules).

(G) A few remarks on the "anomalous" intensity effect [16, 17, 20, 53, 19] in molecular hydrogen are in order. Previous NCS experiments on liquid H_2 and D_2 [57] performed in 2003/04 with the earlier instrumental setup eVS showed that the "anomalous" ratio of the scattering intensities I_H/I_D is independent of θ in the whole scattering angle range $30° < \theta < 70°$. However, the instrumental energy resolution at that time was significantly inferior than that of the present setup Vesuvio (see [58] for details) and thus possible peak-overlap effects might have masked an angular dependence in the measured ratio I_H/I_D. Additionally, the full details of the associated calibration parameters (cf. Appendix C) are not available to us, and thus a possible positive y-shift of the H-recoil peak in the earlier data [57] cannot be tested reliably. (Note that the detectors of eVS were freely movable and their positions were often changed after each experiment.) Decoherence has been proposed to be the common theoretical basis (*i*) of the "anomalous" intensity [53, 18] and (*ii*) of the y-shift of Compton profile [22], but a quantitative treatment is still the subject of current theoretical explorations. Further experimental investigations on H_2–D_2 mixtures and HD are planned.

(H) The general character of the decoherence effect of Refs. [31, 32, 42, 43] leads one to speculate that the new findings presented in this paper might be of relevance for quite different scattering processes (e.g., electron—molecule scattering, proton—nucleus scattering in high energy physics, etc.).

Additionally, as decoherence represents the main barrier to the realization of a quantum computer [5], the above results may also be of relevance for the performance of read-write processes and/or quantum-gate operations in quantum information devices. Undoubtedly a deeper understanding of the relation between decoherence and disentanglement, and the underlying non-unitary quantum dynamics, will be of importance for both the foundation of quantum mechanics and practical quantum information applications.

6.4. Note added

Very recently (March 2012), the preliminary results of Fig. 5, section 6.1, have been confirmed and presented in detail in Ref. [80].

Acknowledgments

We thank J. Mayers and M. A. Adams (ISIS) for assistance with the measurements and discussions about instrumental details, A. Dreismann for discussions about ultrafast scattering, and ISIS Facility (Rutherford Appleton Laboratory) for providing beamtime. EMG and TPB acknowledge financial support from the Australian Reseach Council and the Access to Major Research Facilities Program.

Appendix A. Neutron Compton scattering

This appendix aims at presenting a complete, although concise, account of all essential details of the applied experimental NCS method, instrumental details, and data analysis leading to the results presented in the main text.

We begin with a short description of the basics of NCS and its characteristic time window, following Watson [12].

NCS differs from other neutron scattering methods by the high energy and momentum transfers involved, typically in excess of several eV and 20 Å^{-1}, respectively. Neutron scattering is renowned as a sensitive probe of collective properties in condensed matter, but in this extreme range of energies and momenta – the so-called impulse limit – it is widely believed that single-particle properties are probed. The physical picture advocated is that the neutron-nucleus (or neutron-atom) scattering process occurs so rapidly, compared with the time-scales of atomic motion in the sample, that the measured response is rather simply related to the equilibrium momentum distribution of the atoms.

In the impulse limit, the scattered intensity, as a function of energy measured by a detector at a fixed scattering angle, consists of a peak centred at the "recoil energy",

$$E_{rec} = \hbar\omega_{rec} = \hbar^2 q^2 / 2M$$

which represents the energy of recoil of a stationary nucleus with mass M owing to the collision. The dependence of the peak position on the mass M of the struck nucleus (atom) implies that the scattering from different atoms appears at different energies. This ability to separate contributions from different atomic species (and isotopes) is a very useful feature of the NCS technique. Due to the smallness of the nucleus compared with the neutron's de Broglie wavelength (of the order of 0.1 Å), the scattering is isotropic (i.e., s-wave scattering) in the centre-of-mass system.

Doppler broadening results in a recoil peak whose width is proportional to the mean kinetic energy of the atoms, and whose detailed shape depends on the distribution of atomic momenta. The principal tool in interpreting experiments is the so-called impulse approximation, IA [12, 46]. The IA treats the scattering essentially as a billiard-like collision of two mass points and predicts a precise relationship between the scattering data, in the form of the Compton profile (denoted as $J(y)$, see below), and the atomic momentum distribution. This makes it possible to compare results of experiments with predictions of theoretical models of microscopic properties, e.g. effective Born-Oppenheimer potentials of the struck atoms as being calculated by quantum-chemistry methods.

NCS is closely related to techniques of other fields of physics [65]. Compton scattering using x-rays, gamma rays, or electrons in the 10—100 keV regime provides information on electron momentum distributions [66, 67, 68]. In nuclear physics, one probes the momentum distribution of nucleons in nuclei with deep inelastic scattering (DIS) [69] of protons or electrons with energies of 100—1000 MeV. At still higher energies (e.g., 10 GeV and above), DIS of electrons, muons or neutrinos [70, 71, 72] probes the internal structure of nucleons and has been a key experiment in the confirmation of the existence of quarks.

Appendix A.1. Scattering time

We now consider the characteristic timescale of an elementary scattering process. As mentioned in the Introduction, this "scattering time" is very short; in our NCS experiments, for the neutron-proton collision it is of the order $\tau_{sc} \sim 10^{-16} - 10^{-15}$ s.

The quantity τ_{sc} was first introduced by Sears [46, 12] as

$$\tau_{sc} \approx \frac{M}{q\sqrt{\langle p_q^2 \rangle}} , \tag{A.1}$$

where p_q is the projection of particle's momentum $\mathbf{p}$ along $\mathbf{q}$ and the expectation value $\langle ... \rangle$ is taken over the particle's ground state. It is also found that $\tau_{sc} \approx \hbar/\Delta E_q$, where ΔE_q is the width of the dynamic structure factor at $q =$ const. A similar estimate of τ_{sc} can be obtained with the relation

$$\Delta_E \tau_{sc} \geq \pi\hbar/2 \tag{A.2}$$

derived by Ballentine [73], where τ_{sc} may be interpreted as the shortest time required for the state vector $|\psi(t)\rangle$ to become orthogonal to the initial state $|\psi(0)\rangle$, and Δ_E represents a standard measure of the statistical spread of energy in the state.

The terminology "scattering time" can be misleading in some circumstances. The physical meaning of τ_{sc} is as follows. First, the time interval during which the neutron and the nucleus are in physical proximity is much shorter, as the following estimate shows. For example, a neutron with kinetic energy $E_0 \approx 10$ eV will pass a distance of 10^{-5} Å, which is about the range of the strong nuclear force, in a much shorter time, i.e. $10^{-19} - 10^{-20}$ s. However, this is not in conflict with the above estimate, for the following important reason: τ_{sc} gives a measure of the length of the time interval during which the neutron-proton collision may occur [22]. Therefore, in view of the "attosecond double-slit" experiment by Paulus and co-workers [74], the scattering process under consideration may be viewed as a quantum interference process over the time interval τ_{sc}. This conceptual viewpoint (i.e., interference-in-time) is also in line with the results obtained from a temporally diffracted beam of slow Cesium atoms, using a Young-slit-type interferometer, in the time domain [75]. The first theoretical study of diffraction by a time slit was presented by Moshinski [76].

Appendix B. Experimental procedure

Our experiments applying the NCS method were done at the electron volt spectrometer (presently called "Vesuvio") or the neutron spallation source ISIS, Rutherford Appleton Laboratory, UK. As originally proposed by Hohenberg and Platzman, neutron scattering at high energy and momentum transfers (say, $\hbar\omega > 1$ eV, $q > 20$ Å^{-1}) can be used to directly measure the distribution of momentum $\hbar\mathbf{p}$ of light atoms in condensed matter systems [11]. Here it is assumed that the scattering is essentially incoherent and the so-called Impulse Approximation (IA) applies, and thus single-particle properties are probed [11, 46, 12].

The Vesuvio spectrometer applies the standard time-of-flight (TOF) technique to determine the energy and momentum transfers to the neutron during each scattering event. The TOF for each detected neutron is given by

$$t = \frac{L_0}{v_0} + \frac{L_1}{v_1} + t_0 \tag{B.1}$$

Here L_0 (equal to 11.005 m [58]) is the distance from the source to the sample, L_1 (typically $0.50 - 0.65$ m) that from sample to detector. v_0 and v_1 are the velocities of the incident and scattered neutron, respectively. The detector is at scattering angle θ (about $30° - 67°$ for forward scattering). t_0 is a small time offset due largely to electronic delays. For the data analysis of our experiment [22, 44] we used values for L_1 and θ for each individual detector measured directly with steel rules and a purpose-built protractor [58].

Vesuvio is a so-called "inverse geometry" instrument meaning that the final velocity v_1 of the neutrons contributing to the TOF spectra is fixed, whereas v_0 varies; for details see e.g. [77]. The final neutron energy is presently fixed at $E_1 = 4906$ meV (the resonance energy of Au-197 used as analyzer foil [22]), corresponding to velocity $v_1 = 3.063 \times 10^4$ m/s and wavevector of scattered neutrons $k_1 = 48.66\,\text{Å}^{-1}$.

It follows from Eq. (B.1) that to each t corresponds to an initial velocity v_0 and thus to an energy transfer

$$\hbar\omega = \frac{(\hbar k_0)^2}{2m} - \frac{(\hbar k_1)^2}{2m} \tag{B.2}$$

For the corresponding momentum transfer $\hbar q$ from the neutron to the struck atom, $\hbar\mathbf{q} = \hbar\mathbf{k}_0 - \hbar\mathbf{k}_1$,

$$q = \sqrt{k_0^2 + k_1^2 - 2k_0 k_1 \cos\theta}\ . \tag{B.3}$$

Hence, for each value of t in an experimental TOF spectrum recorded at a particular detector, the associated momentum ($\hbar\mathbf{q}$) and energy ($\hbar\omega$) transfers from the neutron to the struck particle are *uniquely* determined. For scattering from H, q and ω vary considerably over the range of the recoil peak, especially for high scattering angles.

From the measured time-of-flight (TOF) spectra, momentum ($\hbar\mathbf{q}$) and energy ($\hbar\omega$) transfers from the neutron to the stuck particle are determined by standard methods. Within the IA, the elastic collision of a neutron with an approximately free atom with mass M and initial momentum $\mathbf{p}$ results in the neutron's lost energy $\hbar\omega$ being transferred to the struck atom:

$$\hbar\omega = \frac{(\hbar\mathbf{q} + \mathbf{p})^2}{2M} - \frac{p^2}{2M} = \hbar\omega_{rec} + \hbar\mathbf{q}\cdot\mathbf{p}/M \tag{B.4}$$

The recoil energy $\hbar\omega_{rec} = \hbar^2 q^2/2M$ represents the energy of recoil of a stationary nucleus owing to the collision. ($\hbar q$: momentum transfer from neutron to the struck nucleus/atom.)

Thus scattering from a gaseous sample of such atoms leads to a recoil peak centered at $\hbar\omega_{rec}$, exhibiting a peak width caused by the term $\hbar\mathbf{q}\cdot\mathbf{p}/M$ which represents Doppler broadening.

The above energy balance holds exactly in the IA If however final state effects (FSE) [46, 78] become apparent, as in the above experiments at lower scattering angles, the maximum of the recoil peak is shifted to *lower* energy than the recoil energy. This shift is also analytically proven for the three-dimensional quantum harmonic oscillator [79].

As discussed in the main text, a peak-maximum shift to higher energies than $\hbar\omega_{rec}$ is impossible in conventional NCS theory.

For scattering from *free* atoms with *zero* initial momentum, $p = 0$, conservation of kinetic energy and momentum in an elastic neutron-atom collision yield the kinematic relation

$$\frac{v_1}{v_0} = \frac{k_1}{k_0} = \frac{\cos\theta + \sqrt{(M/m)^2 - \sin^2\theta}}{M/m + 1} \tag{B.5}$$

which holds in both quantum and classica l mechanics. Here m and M are the masses of the neutron and struck nucleus, respectively. This physically corresponds to neutrons detected at the centre of the measured recoil peak.

Appendix C. Compton profiles and their measurement

From the measured time-of-flight (TOF) spectra one determines the dynamic structure factor [40], which in the case of the IA takes the simple form [12]

$$S(\mathbf{q},\omega) = \int n(\mathbf{p})\ \delta(\hbar\omega - \hbar\omega_{rec} - \hbar\mathbf{q}\cdot\mathbf{p}/M)\, d\mathbf{p}\ . \tag{C.1}$$

$n(\mathbf{p})$ is the momentum distribution of the struck particle before collision. The delta function incorporates energy conservation.

To compare results obtained with different detectors (and/or to improve counting statistics), one usually applies the so-called West scaling (or y scaling) [46, 47], which, for isotropic samples, amounts to the following. In the IA can be shown that, for each kind of struck particles of mass M, the two variables ω and q are uniquely coupled through a new scaling variable with the physical dimensions of momentum, commonly denoted by $\hbar y$ (or simply y) and defined by

$$\hbar y = \frac{M}{\hbar q}\left(\hbar\omega - \hbar\omega_{rec}\right)\ , \tag{C.2}$$

which simplifies the dynamic structure factor to [12]

$$S(q,\omega) = \frac{M}{\hbar q} J(y), \quad \text{with} \quad J(y) = \int n(\mathbf{p})\ \delta(\mathbf{p}\cdot\mathbf{e_q} - \hbar y)\, d\mathbf{p} \tag{C.3}$$

$J(y)$ is the Compton profile, which gives the distribution of the one-dimensional projection $\hbar y$ of atomic momentum $\mathbf{p}$ (before collision) along the $\mathbf{q}$ direction represented by $\mathbf{e_q}$, i.e. $|\mathbf{e_q}| = \mathbf{1}$. This scaling property of the IA implies that all detectors (at various scattering angles) should yield the same Compton profile $J(y)$ [12, 46, 47]. This quantity, and the associated $n(\mathbf{p})$, are widely believed to depend on the effective Born-Oppenheimer potential $V(r)$ of the struck particle [12]. For a harmonic $V(r)$ the associated $J(y)$ is Gaussian.

The Compton profiles shown in Figs. 3-5 were obtained by transforming to y space the TOF spectra measured as described above.

We have analyzed the sensitivity of the peak positions of $J(y)$ in y space to calibration uncertainties and found that the results in y space are reliable to better than 0.3 Å^{-1} for each single detector [49]. This estimate is supported by the small scattering of the positions of the individual-detector $J(y)$ curves shown in all four Figs. 3a,b and 4a,b.

Our measurements on H_2 and D_2 gas were made using an all-aluminium pressure cell with parallel internal surfaces perpendicular to the incoming beam. The mounting of the cell within the cryostat ensured reproducibility of the sample position between experiments to better than about 2 mm, with negligible effect on the comparison of results between H_2 and D_2 [49]. The spectrometer configuration was identical in the two experiments. Exactly the same analysis was applied to each data set.

As mentioned above and detailed in our papers [22, 44], we used calibration parameters measured directly with steel rules and a protractor [58] by the ISIS Metrology Group, for our data analysis. In contrast, the standardly available Vesuvio calibration parameters [45] (listed in instrument parameter files) are obtained by an iterative sequential procedure involving inelastic and diffraction measurements [58]. Using the standard calibration parameters, we demonstrated [44] that, if the experimental results for H in gaseous H_2 are in agreement with conventional theory (as claimed in [45]), then those for D in gaseous D_2 obtained in the same way cannot be, and vice-versa. Crucial to this alternative, however, the standardly available calibration parameters may be "adjusted" to ensure that the position of the H peak conforms to conventional theory; see Ref. [10] cited in [44].

[1] Einstein A, Podolsky B and Rosen N 1935 *Phys. Rev.* **47**, 777
[2] Bell J S 1964 *Physics* **1** 195
[3] Gisin N 1991 *Phys. Lett.* A **154** 201
[4] Jammer M 1974 *The Philosophy of Quantum Mechanics* (New York: Wiley)
[5] Nielsen M A and Chuang I L 2000 *Quantum Computation and Quantum Information* (Cambridge: Cambridge University Press)
[6] Giulini D, Joos E, Kiefer C, Kupsch J, Stamatescu I O, and Zeh H D 1996 *Decoherence and the Apperance of a Classical World in Quantum Theory* (Berlin: Springer)
[7] Breuer H P and Petruccione F 2002 *The Theory of Open Quantum Systems* (Oxford: Oxford University Press)
[8] Schlosshauer M 2007 *Decoherence and the Quantum-to-Classical Transition* (Berlin: Springer)
[9] Akulin V M, Sarfati A, Kurizki G, and Pellegrin S (ed) 2005 *Decoherence, Entanglement and Information Protection in Complex Quantum Systems* NATO Science Series II, Vol 189 (Dordrecht: Springer)
[10] Tegmark M 1993 *Found. Phys. Lett.* **6** 571
[11] Hohenberg P C and Platzman P M 1966 *Phys. Rev.* **152** 198
[12] Watson G I 1996 *J. Phys.: Condens. Matter* **8** 5955
[13] Compton A H 1923 *Phys. Rev.* **21** 483
[14] Karlsson E B, Chatzidimitriou-Dreismann C A, Abdul-Redah T, Streffer R M, Hjörvarsson B, Öhrmalm J, and Mayers J 1999 *Europhys. Lett.* **46** 617
[15] Chatzidimitriou-Dreismann C A, Abdul-Redah T, and Kolarić B 2001 *J. Am. Chem. Soc.* **123** 11945
[16] Chatzidimitriou-Dreismann C A, Abdul-Redah T, Streffer R M F, and Mayers J 1994 *Phys. Rev. Lett.* **79** 2839
[17] Chatzidimitriou-Dreismann C A, Vos M, Kleiner C, and Abdul-Redah T 2003 *Phys. Rev. Lett.* **91** 057403
[18] Chatzidimitriou-Dreismann C A and Krzystyniak M 2010 *Laser Physics* **20** 990
[19] Karlsson E B 2012 *Int. J. Quant. Chem.* **112** 587
[20] Cooper G, Hitchcock A P, and Chatzidimitriou-Dreismann C A 2008 *Phys. Rev. Lett.* **100** 043204
[21] Bonham R A, Cooper G, and Hitchcock A P 2009 *J. Chem. Phys.* **130** 144303
[22] Chatzidimitriou-Dreismann C A, Gray E MacA, and Blach T P 2011 *AIP Advances* **1** 022118
[23] Sudbery A 2006, in: *Quantum Theory and Symmetries QTS-4* Vol. 1, ed V K Dobrev (Sofia: Heron Press) pp 10-21
[24] Diósi L 2005 *Brazil. J. Phys.* **35** 260
[25] Gorini V, Kossakowski A, and Sudarshan E C G 1976 *J. Math. Phys.* **17** 821
[26] Lindblad G 1976 *Commun. Math. Phys.* **48** 119
[27] Caldeira A O and Leggett A J 1983 *Physica* A **121** 587
[28] Joos E and Zeh H D 1985 *Z. Phys.* B **59** 223
[29] Ballentine L E 1991 *Phys. Rev.* A **43** 9
[30] Gallis M R and Fleming G N 1991 *Phys. Rev.* A **43** 5778

[31] Schulman L S and Gaveau B 2006 *Phys. Rev. Lett.* **97** 240405
[32] Gaveau B and Schulman L S 2010 *J. Phys. A: Math. Theor.* **43** 055308
[33] van Kampen N G 1954 *Physica* **20** 603
[34] van Kampen N G 1956 *Fortschritte der Physik* **4** 405
[35] van Kampen N G 1962 *Fundamental problems in statistical mechanics* ed E G D Cohen (Amsterdam: North-Holland) pp 173-202
[36] van Kampen N G 1971 *Physica Norvegica* **5** 279
[37] Kubo R 1957 *J. Phys. Soc. Japan* **12** 570
[38] Kubo R 1966 *Rep. Prog. Phys.* **29** 255
[39] van Hove L 1954 *Phys. Rev.* **95** 249
[40] Squires G L 1996 *Introduction to the Theory of Thermal Neutron Scattering* (Mineola: Dover)
[41] Lovesey S W 1984 *Theory of Neutron Scattering form Condensed Matter* (Oxford: Oxford University Press)
[42] Erez N, Gordon G, Nest M, and Kurizki G 2008 *Nature* **452** 724
[43] Gordon G, Bensky G, Gelbwaser-Klimovsky D, Bhaktavatsala Rao D D, Erez N, and Kurizki G 2009 *New J. Phys.* **11** 123025
[44] Chatzidimitriou-Dreismann C A, Gray E MacA, and Blach T P 2011 *AIP Advances* **1** 049102
[45] Mayers J and Reiter G 2011 *AIP Advances* **1** 049101
[46] Sears V F 1984 *Phys. Rev.* B **30** 44
[47] West G B 1975 *Phys. Rep.* **18** 263
[48] Chatzidimitriou-Dreismann C. A., Abdul-Redah T. Streffer R. M. F. and Mayers J. *J. Chem. Phys.* 2002 **116** 1511
[49] Gray E MacA, Chatzidimitriou-Dreismann C A, and Blach T P 2012 *Nucl. Instr. Meth.* A **661** 64
[50] Andreani C, Colognesi D, Mayers J , Reiter G F and Senesi R 2005 *Adv. Phys.* **54** 377
[51] C. Andreani, A. Filabozzi and E. Pace, *Phys. Rev.* B 1995 **51** 8854
[52] C. Andreani, D. Colognesi, and E. Pace, 1999 *Phys. Rev.* B **60** 10008
[53] Chatzidimitriou-Dreismann C A and Stenholm S 2005 in: Ref. [9], pp 555-562; also available at arXiv:quant-ph/0702038v1.
[54] A. Messiah 1965 *Quantum Mechanics*, Vol. II (North Holland, Amsterdam)
[55] E. Goulielmakis et al. 2010 *Nature* **466** 739
[56] Lin L, Morrone J A, Car R, and Parrinello M, 2010 *Phys. Rev. Lett.* **105** 110602
[57] Chatzidimitriou-Dreismann C A, Abdul-Redah T and Krzystyniak M 2005 *Phys. Rev.* B **72** 054123
[58] Mayers J and Adams M A 2009 *Calibration of the electron volt spectrometer VESUVIO at ISIS*, Technical Report RAL-TR-2009-022; main parts of it published in: 2011 *Nucl. Instr. Meth.* A **625** 47
[59] Colognesi D 2007 *Physica* B **398** 89
[60] Zeh H D 1970 *Found. Phys.* **1** 69
[61] Mensky M B 1979 *Phys. Rev. D* **20** 384
[62] Zurek W H 1981 *Phys. Rev. D* **24** 1516
[63] Beretta G P 2009 *Rep. Math. Phys.* **64** 139
[64] Strunz W T 2002, in: *Coherent Evolution in Noisy Environments* ed A Buchleitner and K Hornberger (Berlin: Springer) paragraph 5.3.5
[65] Silver R N and Sokol P E (ed) 1989 *Momentum Distributions* (New York: Plenum)
[66] Williams B (ed) 1977 *Compton Scattering* (New York: McGraw-Hill)
[67] Cooper M J 1985 *Rep. Prog. Phys.* **48** 415
[68] Kane P P 1992 *Phys. Rep.* **218** 67
[69] Antonov A N, Hodgson P E, and Petkov I Zh 1988 *Nucleon Momentum and Density Distributions in Nuclei* (Oxford: Oxford University Press)
[70] Taylor R E 1991 *Rev. Mod. Phys.* **63** 573
[71] Kendall H W 1991 *Rev. Mod. Phys.* **63** 597
[72] Friedman J I 1991 *Rev. Mod. Phys.* **63** 615
[73] Ballentine L E 1998 *Quantum Mechanics - A Modern Development* (Singapore: World Scientific)
[74] Lindner F, et al. 2005 *Phys. Rev. Lett.* **95** 040401
[75] Szriftgiser P, Gury-Odelin D, Arndt M, and Dalibard J 1996 *Phys. Rev. Lett.* **77** 4
[76] Moshinski M 1952 *Phys. Rev.* **88** 625
[77] Mayers J and Abdul-Redah T 2004 *J. Phys: Condens. Matter* **16**, 4811
[78] Glyde H R 1994 *Phys. Rev.* B **50** 6726
[79] Schimmel H G, Montfrooli W, Verhoeven V W J, and de Schepper I M 2002 *Europhys. Lett.* **60** 868
[80] Chatzidimitriou-Dreismann C A, Gray E MacA, and Blach T P 2012 *Nucl. Instr. Meth.* A **676** 120

Neutron Compton scattering — critical analysis of some basic theoretical assumptions

I C Tietje

Institutes of Physics, Technical University of Berlin, D-10623 Berlin, Germany

E-mail: ingmari.c.tietje@campus.tu-berlin.de

Abstract. A detailed overview of the conventional (van Hove) neutron scattering theory and its specification to the *Impulse Approximation* (IA) are given. The IA constitutes the basis of all interpretations of results from neutron scattering experiments in the Compton regime, in which is widely believed that single-particle properties are probed. Here the validity of this approximation is carefully scrutinized and it is shown that there are several different steps in the derivation, whose validity can be questioned: (1) the scattering system is a closed system (in the quantum mechanical sense); (2) all entanglement in the scattering system is absent (N-body operators are replaced by single particle operators); (3) consequently the one-body momentum distribution $n(\mathbf{p})$ is introduced ad hoc; (4) the δ-function in energy assumed in the derivation is not valid when the interaction times underlying the IA are so short that the uncertainty relation allows a noticeable spread in the energy balance. Additionally, (5) the entanglement due to the direct interaction of the neutron with a nucleus is completely absent in the general formalism. The concrete experimental context of recent neutron Compton scattering (NCS) experiments at the ISIS neutron spallation source is considered and emphasis is put on the ultrafast (attosecond) scattering time for the neutron-proton scattering. Recent modern theoretical developments are shortly mentioned in the discussion, which take into account the neutron's coherence length being larger than the de Broglie wavelength, or decoherence phenomena during the short but finite scattering time.

1. Neutrons, NCS and the eVs/Vesuvio instrument

In this paper the standard theory of neutron scattering [1, 2, 3] is presented in considerable detail as well as its specification to neutron Compton scattering (NCS), also called deep inelastic neutron scattering (DINS) [4]. This technique involves epithermal neutrons. All interpretations of results from neutron scattering in the Compton regime are based on the so-called *impulse approximation* (IA) [4]. The short scattering times as well as the intuitively simple theoretical "one-body picture" provided by the IA are commonly taken to represent the essential physical features of NCS. In more illustrative terms, the main contents of this article may be considered to refer to an 'ideal machine' where single nuclei are probed.

In this paper the validity of the standard (or conventional) theory and in particular the IA is carefully scrutinized. Modern developments, e.g. questions concerning the neutron's coherence length λ_{coh} (see [5]) are shortly considered in section 6.

The VESUVIO spectrometer of the ISIS neutron spallation source (Rutherford Appleton Laboratory, U.K.) provides neutrons in the epithermal regime; see [5]. The resulting momentum and energy transfers from the probing neutron to the scattering system lie in the range of

$$\Delta E_{NCS} \approx 1 - 50 \text{ eV} , \quad |\mathbf{q}| \sim 20 - 150 \text{ Å}^{-1} . \tag{1}$$

The measured peaks in the TOF-spectra (time-of-flight) are well-resolved due to the high momentum transfers, cf. fig. 1. Each recoil peak X corresponds to scattering from one kind of nuclei, each having the mass M_X. The neutron flux provided by the Vesuvio instrument is low and therefore scattering of two or more neutrons from one nucleus at the same time is practically impossible.

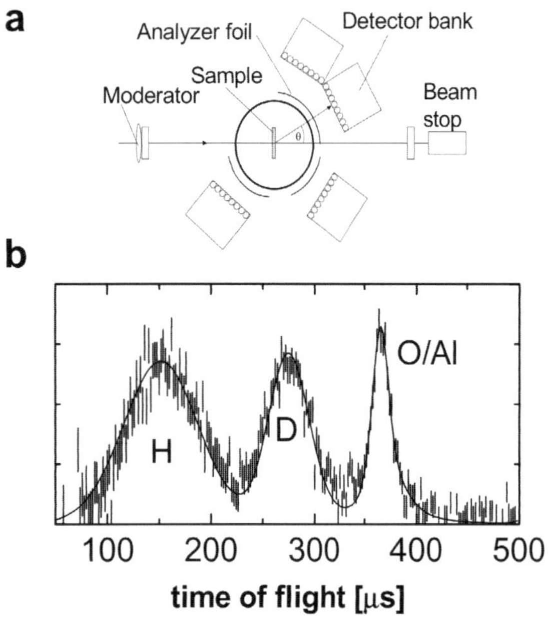

Figure 1. (a) Schematic representation of time-of-flight (TOF) spectrometer eVS/Vesuvio of ISIS. (b) A TOF-spectrum of a 20:80 H_2O-D_2O mixture in an Al can together with the corresponding fit (full line). The H and D recoil peaks are well separated from each other and from the joint O/Al peak. The peak area of atom X (i.e., H, D, etc.) gives the neutron-X scattering intensity. All figures adapted from [6].

In sections 2 and 3 the conventional formalism [1, 2] is derived as well as its specification to NCS - the IA [4]. Section 4 gives an insight into the characteristic scattering time τ_{sc} of the neutron Compton scattering regime [7, 4, 5]. In sections 5 and 6 several particularly significant steps in the derivation of sections 2 and 3 are closely examined:

(i) that the scattering system is assumed to be closed (in the quantum mechanical sense);
(ii) the physical justification of the closure relation for specific experimental prerequisites;
(iii) that the δ-function assuring strict energy conservation, as assumed in the derivation, is not valid when the interaction times are so short that the uncertainty relation allows a noticeable spread in the energy balance;
(iv) that all particle-particle entanglement of the scattering system in the IA is assumed to be "broken" (N-body wavefunctions and operators are replaced by single-particle wavefunctions and operators) and the one-body momentum distribution $n(\mathbf{p})$ is introduced ad hoc.

All these points are discussed qualitatively. A quantitative evaluation would require a large amount of numerical calculations that are beyond the scope of this paper. An additional estimate of the characteristic 'scattering time' is presented which implies that the systems involved in NCS are open quantum systems [8, 9].

In the context of the discussions of section 6 it will furthermore become obvious that the coherence length λ_{coh} of the neutron appears to be related with new phenomena (see latest developments [5]). Moreover, it will be mentioned that the phenomena of quantum entanglement and decoherence between the particles of the scattering system may play a significant role in the NCS dynamical process [10, 11].

2. Conventional neutron scattering theory

The general neutron scattering theory is given by the fundamental van Hove formalism [3], which is based on first order perturbation theory. The latter corresponds, for scattering processes, to the first Born approximation [12].

One considers a closed N-body system described by the Hamiltonian H_c (the subscript "c" indicating "closed" and/or "complete"). The system is perturbed by the impinging neutron.

The starting point in the theory [1, 3, 2] is Fermi's golden rule [12] which is derived within first order perturbation theory, assuming that the perturbation is sufficiently weak. In the domain of neutron scattering, however, this is not justified in a strict mathematical sense as the perturbation is very strong and being represented by Fermi's pseudo-potential well localized in space (see Eq. (6) below). Still, according to Squires ([1] p. 16), the golden rule and the pseudo-potential together hold as a good approximation[1]:

> "The justification for the use of the golden rule in these circumstances is that, when combined with the pseudopotential, it gives the required result of isotropic scattering for a single fixed nucleus."

In the first Born approximation the incident and scattered wave functions of the neutron are assumed to be plane waves with wave vectors $\mathbf{k}_0$ and $\mathbf{k}_1$, respectively. (This is an approximation, since s-wave scattering means that the two-particle wave-function has spherical symmetry in the center-of-mass system.)

To assure that the total energy is conserved and to obtain an expression for the double differential cross section, a corresponding δ-function

$$\delta(E_\nu - E_{\nu'} + E_0 - E_1) \tag{2}$$

is added to the time-independent expression for the partial differential scattering cross section [1]

$$\left(\frac{d\sigma}{d\Omega}\right)_{\nu\to\nu'} = \frac{k_1}{k_0}\left(\frac{m_n}{2\pi\hbar^2}\right)^2 \left|\langle\nu'\mathbf{k}_1|V(\mathbf{r})|\nu\mathbf{k}_0\rangle\right|^2 \tag{3}$$

Thus Eq. (2), denoting that the overall system "neutron & N-body scattering system" is closed, joined together with Eq. (3) give the basic equation of neutron scattering (see [1] Eq. 2.15])

$$\left(\frac{d^2\sigma}{d\Omega d\omega}\right)_{\nu\to\nu'} = \frac{k_1}{k_0}\left(\frac{m_n}{2\pi\hbar^2}\right)^2 \left|\langle\nu'\mathbf{k}_1|V(\mathbf{r})|\nu\mathbf{k}_0\rangle\right|^2 \delta(E_\nu - E_{\nu'} + \hbar\omega)\ , \tag{4}$$

where $d\Omega$ is the small solid angle in which the scattered neutron with wave-vector $\mathbf{k}_1$ is detected and the final state of the neutron together with the system $|\nu'\mathbf{k}_1\rangle$ is assumed to fulfil $|\nu'\rangle \otimes |\mathbf{k}_1\rangle$. Initial and final eigenstates of the unperturbed and closed Hamiltonian H_c (with energies E_ν and $E_{\nu'}$, respectively) refer to the collective states $\{\nu\}$ and $\{\nu'\}$ of the overall N-body scattering system and correspond to the single transition $|\nu\rangle \to |\nu'\rangle$. $\hbar\mathbf{q}$ and $\hbar\omega$ denote the momentum and energy transfers from the neutron to the probe, respectively; i.e.

$$\hbar\mathbf{q} = \hbar\mathbf{k}_0 - \hbar\mathbf{k}_1\ , \qquad \hbar\omega = E_0 - E_1\ . \tag{5}$$

[1] To quote Squires in full ([1] p. 16): "The justification for the use of the golden rule in these circumstances is that, when combined with the pseudopotential, it gives the required result of isotropic scattering for a single fixed nucleus. It may be noted that the pseudopotential does not correspond even approximately to the actual potential"

The subscript '0' labels the initial, the subscript '1' the final quantities of the neutron involved in the scattering process.

Introducing Fermi's pseudo-potential

$$V(\mathbf{r}) = \frac{2\pi\hbar^2}{m_n}\, b\, \delta(\mathbf{r}) \;, \tag{6}$$

where b denotes the bound scattering length of the nucleus and m_n denotes the mass of the neutron, into the basic equation (Eq. (4)) and taking the sum over all scattering centres (denoted by the subscript 'j') one obtains straightforwardly (see [1] Eq. 2.40)

$$\left(\frac{d^2\sigma}{d\Omega d\omega}\right)_{\nu\to\nu'} = \frac{k_1}{k_0}\Big|\sum_j^N b_j \langle\nu'|\exp(i\mathbf{q}\cdot\mathbf{r}_j|\nu\rangle\Big|^2 \delta(E_\nu - E_{\nu'} + \hbar\omega) \;. \tag{7}$$

Please note that in Eq. (7) the variables representing the neutron are c-numbers k_0, k_1 as well as the vector momentum transfer $\mathbf{q} = \mathbf{k}_0 - \mathbf{k}_1$. As mentioned above, The states $|\nu'\mathbf{k}_1\rangle$ are assumed to factorize $|\nu'\rangle \otimes |\mathbf{k}_1\rangle$ which practically means that the quantum dynamical variables of the neutron are not taken into account. In other words, possible entanglement of the neutron with the system under consideration is presumed to be absent.

Following van Hove [3], one substitutes the δ-function in Eq. (7) by its spectral decomposition

$$\delta(E_\nu - E_{\nu'} + \hbar\omega) = \frac{1}{2\pi\hbar}\int_{-\infty}^{\infty} dt\, \exp[i(E_{\nu'} - E_\nu)t/\hbar]\exp(-i\omega\, t) \;, \tag{8}$$

and one obtains

$$\begin{aligned}\left(\frac{d^2\sigma}{d\Omega d\omega}\right)_{\nu\to\nu'} &= \frac{k_1}{k_0}\frac{1}{2\pi\hbar}\sum_{j,j'}^N b_j b_{j'} \int_{-\infty}^{\infty} \exp(-i\omega\, t)\langle\nu|\exp(-i\mathbf{q}\cdot\mathbf{r}_{j'}|\nu'\rangle \\ &\quad \langle\nu'|\exp(iE_{\nu'}t/\hbar)\,\exp(i\mathbf{q}\cdot\mathbf{r}_j)\,\exp(-iE_\nu t/\hbar)|\nu\rangle\, dt \;.\end{aligned} \tag{9}$$

In Eq. (10) the time-independent matrix element originating from the golden rule (cf. [1], Eq. (2.2)) is put inside the time integral of the spectral decomposition, Eq. (8), and the time-depending energy terms $\exp(-iE_{\nu'}t/\hbar)$ and $\exp(-iE_\nu t/\hbar)$ of the spectral decomposition Eq. (8) are inserted into the second matrix element.

The second, now time-depending matrix element of Eq. (10) is transformed by using time unitarity

$$\begin{aligned}&\langle\nu'|\exp(iE_{\nu'}t/\hbar)\,\exp(i\mathbf{q}\cdot\mathbf{r}_j)\,\exp(-iE_\nu t/\hbar)|\nu\rangle \\ &= \langle\nu'|\exp(iH_c t/\hbar)\,\exp(i\mathbf{q}\cdot\mathbf{r}_j)\,\exp(-iH_c t/\hbar)|\nu\rangle \\ &= \langle\nu'|U^\dagger(t)\,\exp(i\mathbf{q}\cdot\mathbf{r}_j)\,U(t)|\nu\rangle \;,\end{aligned} \tag{10}$$

where

$$U(t)|\nu\rangle = \exp(-iE_\nu t/\hbar)|\nu\rangle \tag{11}$$

with

$$U(t) = \exp(-\frac{i}{\hbar}H_c t) \tag{12}$$

denoting the unitary time evolution operator. After this transition one yields

$$\left(\frac{d^2\sigma}{d\Omega d\omega}\right)_{\nu\to\nu'} = \frac{k_1}{k_0}\frac{1}{2\pi\hbar}\sum_{j,j'}^{N} b_j b_{j'}\int_{-\infty}^{\infty}\exp(-i\omega\, t) \times\langle\nu|\exp\left(-i\mathbf{q}\cdot\mathbf{r}_{j'}(0)\right)|\nu'\rangle\langle\nu'|\exp(i\mathbf{q}\cdot\mathbf{r}_j(t))|\nu\rangle\, dt\ , \tag{13}$$

where Eq. (14) refers to the Heisenberg representation of the position operators

$$\mathbf{r}_j(t) \equiv U^{\dagger}(t)\mathbf{r}_j U(t)\ . \tag{14}$$

In order to obtain a measurable expression for the double differential cross-section, Eq. (13) (which still includes all initial and final states of the system), as the first step one sums over all final states ν', keeping the initial state ν fixed, and then, as the second step, averages over all ν [1]. The elimination of the final states is done through the mathematical procedure of closure[2]

$$\sum_{\nu'}|\nu'\rangle\langle\nu'| = \widehat{\mathbf{1}}\ , \tag{15}$$

or for a pair of operators $\hat{\mathbf{A}}$ and $\hat{\mathbf{B}}$

$$\sum_{\nu'}\langle\nu|\hat{\mathbf{A}}|\nu'\rangle\langle\nu'|\hat{\mathbf{B}}|\nu\rangle = \langle\nu|\hat{\mathbf{A}}\hat{\mathbf{B}}|\nu\rangle\ . \tag{16}$$

In order to take the initial states ν into account probabilities W_ν of the N-body scattering system are introduced and

$$\sum_{\nu} W_\nu\langle\nu|\hat{\mathbf{O}}|\nu\rangle\ = \langle\hat{\mathbf{O}}\rangle \tag{17}$$

($\hat{\mathbf{O}}$: arbitrary operator) denotes a thermodynamic average over the initial states, which for thermal neutron scattering belong to thermal equilibrium. $W_\nu = \frac{1}{Z}\exp(-E_\nu\beta)$ is given by the Boltzmann distribution, $Z = \sum_\nu \exp(-E_\nu\beta)$ is known as the partition function which ensures $\sum_\nu W_\nu = 1$, and $\beta = \frac{1}{k_B T}$.

These operations lead to the result

$$\frac{d^2\sigma}{d\Omega d\omega} = b^2\frac{k_1}{k_0}\frac{1}{2\pi}\int_{-\infty}^{\infty} dt\ \exp(-i\omega t)\sum_{j,j'}^{N}\langle\exp(-i\mathbf{q}\cdot\mathbf{r}_{j'}(0))\exp(i\mathbf{q}\cdot\mathbf{r}_j(t))\rangle\ , \tag{18}$$

where, for simplicity, all b_j are assumed to be equal. The intermediate correlation function [1, 2]

$$F(\mathbf{q},t) = \frac{1}{N}\sum_{j,j'}^{N}\langle\exp(-i\mathbf{q}\cdot\mathbf{r}_{j'}(0))\exp(i\mathbf{q}\cdot\mathbf{r}_j(t))\rangle \tag{19}$$

and the dynamical structure factor

$$S(\mathbf{q},\omega) = \frac{1}{2\pi}\int_{-\infty}^{\infty}\exp(-i\omega\, t)F(\mathbf{q},t)\, dt \tag{20}$$

give the well-known, final result [1, 2]

$$\frac{d^2\sigma}{d\Omega d\omega} = Nb^2\frac{k_1}{k_0}S(\mathbf{q},\omega)\ . \tag{21}$$

[2] From a mathematical viewpoint using the closure relation is a distinct and useful procedure. But are there any physical limitations to this mathematical proceeding in experimental set-ups? (see discussion below)

It should be noted that the quantity we measure in an experiment is not the cross-section itself but the number of neutrons (I_{sc}) with fixed initial energy E_0, scattered per second into a small solid angle $d\Omega$ (in a given direction) with final energy between E_1 and $E_1 + dE_1$ (and thus $\hbar d\omega = dE_1$). It is proportional to the partial differential cross-section, i.e.

$$I_{sc}(\Omega, \omega) \propto \frac{d^2\sigma}{d\Omega d\omega} \, . \tag{22}$$

3. Neutron Compton Scattering (NCS)

The wavevector transfer in NCS is much larger in extend than the inverse of the nearest-neighbour distance of the nuclei

$$q = |\mathbf{q}| \gg \frac{2\pi}{d} \, . \tag{23}$$

As a consequence we will not detect details of the coherent superposition of scattering amplitudes from different nuclei as it will diminish over a small solid angle — the detector opening and becomes effectively incoherent.[3] According to this common view, the intermediate correlation function in the incoherent approximation [4, 7, 13] takes a much simpler form

$$F(\mathbf{q}, t) = \langle \exp(-i\mathbf{q} \cdot \mathbf{r}(0)) \exp(+i\mathbf{q} \cdot \mathbf{r}(t)) \rangle \, , \tag{24}$$

where the index j of the jth particle can be obviously dropped and $\mathbf{r}(t)$ denotes the Heisenberg position operators.

The intermediate correlation function $F(\mathbf{q}, t)$ belongs to the class of stationary time correlation functions which are time translation invariant

$$C(t_{ini}, t_{end}) = C(t_{end} - t_{ini})$$

and therefore fulfil

$$C(t_{ini}, t_{end}) = C(t_{ini} + s, t_{end} + s) \, ,$$

where C is an arbitrary stationary time correlation function describing thermal equilibrium, cf. Eq. (19), s is an arbitrary translation in time and t_{ini} and t_{end} are two time points with $t_{ini} \le t_{end}$. The time points $'0'$ and $'t'$ of Eq. (24) correspond to t_{ini} and t_{end}, respectively; i.e. $F(\mathbf{q}, t)$ depends on the time interval $t_{end} - t_{ini} = t$ only and not on any individual time points $'0'$ and $'t'$. The argument t in the Heisenberg position operator $\mathbf{r}(t)$, therefore, ought to be regarded as a time interval $t_{end} - t_{ini}$. The same conclusion also holds for $\mathbf{r}(0)$, since $'0'$ corresponds to t_{ini}.

Eq. (24) is often misinterpreted to "prove" that only one-body dynamics is effective in NCS. Since the Heisenberg position operators $\mathbf{r}(t)$ are N-body position operators (and do not merely account for "the single particle j") several particles of the system under consideration can become entangled.

The pioneers of NCS Hohenberg and Platzmann [14] proposed the possibility of obtaining information about the single-particle momentum distribution of helium with neutrons of sufficiently high energies (i.e., for helium atoms in the range of 1 eV). Their original idea can be described along the following lines: The neutron scattering acts as a microscopic probe [14] (NCS $\sim$ 0.1 Å) so the scattering consists of two-body collisions. The 'time passage' of the neutron through the sample must be short compared to the characteristic time scales of the

[3] In other terms, for large enough wave vector transfers (as e.g. in NCS) the many contributions with $j \neq j'$ in the intermediate correlation function, Eq. (19), correspond to uncorrelated phase differences causing fine-oscillating (i.e. constructive and destructive) interference patterns which tend to average to zero, and therefore the scattering becomes effectively incoherent.

dynamic response of the system (see also [2], p. 73). In terms of the uncertainty relation the 'time passage' or the 'time window' happens to be short if the recoil energy of the particle is much greater than the characteristic energy of its bond (e.g. for atoms of a molecule) or kinetic energy (e.g. for helium atoms in the normal or superfluid state). Under these circumstances, one conventionally believes that the dynamic structure factor contains information about the single-particle or single-nucleus momentum distribution.

This interpretation of NCS in terms of the *impulse approximation* (IA) is not entirely correct, because as described above particle-particle entanglement contained in the Heisenberg position operators $\mathbf{r}(t)$ of Eq. (24) is not included, and by the assumption of a mere single-particle or single-nucleus momentum distribution broken 'by hand'. Therefore also particle-particle decoherence phenomena are left out entirely.

Let us describe the further derivation based on the IA in some detail. For sufficiently short times the Heisenberg position operators are expanded and take the form [4, 13]

$$\mathbf{r}\,(t) = \mathbf{r} + \frac{t}{M}\mathbf{p}\ , \tag{25}$$

where M denotes the jth nucleus mass and $\mathbf{p}$ the momentum conjugate to $\mathbf{r}$. As noted above, the operator entity $\mathbf{r}(t)$ (which may be written as $\mathbf{r}(t_{ini}, t_{end})$) on the left-hand side of Eq. (25) is a N-body operator, as well as the operators on the right-hand side $\mathbf{r}$ and $\mathbf{p}$ (which correspond to $\mathbf{r}(t_{ini})$, $\mathbf{p}(t_{ini})$). This interpretation directly follows from Eq. (24) as $F(\mathbf{q}, t)$ is a stationary time correlation function which has the property of time translation. In other words, neither time $'0'$ nor time $'t'$ in the expansion of the Heisenberg position operator is a preferred instant in time. Only the time window $t_{end} - t_{ini} = t$ is of importance for the evaluation of $F\,(\mathbf{q}, t)$. Accordingly, we write

$$\mathbf{r}\,(t_{ini}, t_{end}) = \mathbf{r}(t_{ini}) + \frac{t_{end} - t_{ini}}{M}\mathbf{p}(t_{ini})\ .$$

Note also that the linear expansion of $\mathbf{r}(t)$ is only possible for velocity independent interatomic forces (cf. [7], p.45).

Sears introduces the *impulse approximation* in a very elegant way [7], which is as follows. For simplicity we may assume that the scattering system consists of identical particles (atoms). In a very short but finite time interval, the quantum velocity operators $\mathbf{v}(t) = \mathbf{p}(t)/M$ of a particle with mass M may not commute for different times. Therefore the intermediate correlation function involves the time-ordering operator $\mathcal{T}$

$$F(\mathbf{q}, t) = \exp(i\omega_r t)\left\langle \mathcal{T}\exp\left[i\mathbf{q}\cdot\int_0^t \mathbf{v}(t')dt'\right]\right\rangle \tag{26}$$

where

$$\hbar\omega_r = \hbar^2 q^2/2M \tag{27}$$

is the recoil energy of a stationary nucleus in the collision with the neutron. For sufficiently large q, formally $q \to \infty$, the right-hand side of Eq. (26) is appreciably different from zero only if $t \to 0$, and thus $\mathbf{v}(t') \approx \mathbf{v}(0) \equiv \mathbf{v}$. Thus, in the limit of $q \to \infty$ one gets

$$F(\mathbf{q}, t) \to \langle \exp[i(\omega_r + \mathbf{q}\cdot\mathbf{v})t)]\rangle \equiv F_{IA}(\mathbf{q}, t) \tag{28}$$

and going to ω space by Fourier transform we obtain an expression for the dynamic structure factor in the short-time limit or, equivalently, in the limit of high energies and high impulse transfers

$$S(\mathbf{q}, \omega) \to \langle \delta(\hbar\omega - \hbar\omega_r + \hbar\mathbf{q}\cdot\mathbf{p}/M)\rangle \equiv S_{IA}(\mathbf{q}, \omega) \tag{29}$$

with $\mathbf{v} = \mathbf{p}/M$). The δ-function expresses conservation of energy and momentum for the collision of an epithermal neutron with an atom of initial velocity $\mathbf{v}$. Eq. (29) represents the IA and can be written in the following equivalent form that explicitly expresses the scattering from single nuclei [4, 7, 13]

$$\begin{aligned} S_{IA}(\mathbf{q},\omega) &= \int n(\mathbf{p})\, \delta\left(\hbar\omega + \frac{\hbar^2 p^2}{2M} - \frac{\hbar^2(\mathbf{p}+\mathbf{q})^2}{2M}\right) d\mathbf{p} \\ &= \int n(\mathbf{p})\, \delta(\hbar\omega - \hbar\omega_r - \hbar\mathbf{q}\cdot\mathbf{p}/M)\, d\mathbf{p}\ . \end{aligned} \tag{30}$$

where $n(\mathbf{p})$ is the (one-body) momentum distribution of the scattering nucleus before collision, i.e. that of the unperturbed system. The well-known partial cross section in the IA writes [1, 3, 2]

$$\frac{d^2\sigma}{d\Omega d\omega} = Nb^2 \frac{k_1}{k_0} S_{IA}(\mathbf{q},\omega)\ . \tag{31}$$

The short scattering times as well as the intuitively simple theoretical "one-body picture" provided by the IA, as described above, are regarded to represent the essential physical features of NCS [4, 7, 13].

While for sufficiently short times the N-body position operators $\mathbf{r}(t)$, according to the derivations above, become one-body operators the N-body wavefunction of the system is not said to factorize into single-particle wavefunctions. Nevertheless, in Eq. (30) the possible entanglement by this N-body wavefunction is tacitly assumed to be negligible entanglement, or in other terms, plays no role by definition.

.

4. Scattering and collision time in NCS

Exploring the magnitude of the scattering time window in NCS of the elementary neutron-atom (especially, neutron-proton) scattering process a result is obtained that is found in the literature as the outcome of a couple of different estimates (a good overview can be found in [6] with references therein). Recollecting that the momentum and energy transfers in NCS are large (see Eq. (1)) the characteristic time scale — often termed "scattering time" τ_{sc} [4] — of the neutron-proton scattering process lies in the attosecond time range [7, 4, 14].

$$\tau_{sc} \sim 100 - 1000 \text{ as} \tag{32}$$

(as: attosecond). The estimate of the scattering time corresponds to $\hbar/\Delta E$, where ΔE is the width of the recoil peaks (more precisely, as measured at constant q). It also can be taken from the theoretical result valid in the IA [7, 15, 16]

$$\tau_{sc}\, |\mathbf{q}|\, v_0 \approx 1\ , \tag{33}$$

where $v_0 = \frac{\Delta\mathbf{p}}{M}$ ($\Delta\mathbf{p}$: width of the momentum distribution) denotes the root-mean-square (rms) velocity of the nucleus and $\hbar\mathbf{q}$ is the momentum transfer. This formula, introduced by Sears [7] represents the decay time of the intermediate correlation function $F(q,t)$ and is connected to the dynamic structure factor $S(\mathbf{q},\omega)$, Eq. 20, by Fourier transform. The intuitively expected 'actual duration' of one single scattering event (neutron with proton) may be even much shorter [6]. A neutron with a kinetic energy of $E_0 \sim 10$ eV (and velocity $v_0 \sim 44$km/s) will go over a distance equal to the range of the strong interaction, $R_{strong} \sim 10^{-5}$ Å, in a time about three orders of magnitude shorter than τ_{sc}, i.e.

$$t_{strong} = R_{strong}/v_0 \sim 2 \times 10^{-20} \text{ s.} \tag{34}$$

The comparison of these two characteristic times leads us to the following clarification [17](b):

The scattering time τ_{sc} gives a statistical measure of the length of the time interval during which an elementary neutron-proton collision may occur, in the same way that the spatial extent of a particle wavefunction (or wavepacket) gives a statistical measure of the extent of the region in which the particle may be found.

For further considerations of these two characteristic times Eq. (32) and Eq. (34) see section 6 below.

5. Critical Analysis

Some specific points of the derivation of sections 2 and 3 are reconsidered/examined in more detail.

5.1. Applicability of the closure relation

The application of closure, Eq. (15), is the crucial step for the transition from Eq. (13) to the final expression for the double differential cross section Eq. (18) of the general neutron scattering theory. Needless to say, the application of closure is mathematically justified. However, the justification for applying it to an experiment in some cases lacks physical prerequisites as is shown in the following.

E.g., let us think of a standard (direct geometry) instrument where the initial energy of the probing particle E_0^n is fixed. Then E_0^n gives a natural upper bound for the available energy transfers, which, in terms of section 2 is expressed as

$$E_{\nu'} - E_\nu \leq E_0^n \ .$$

Thus, for any initial energy of the system E_ν there exists a maximal accessible final energy which we denote by $E_{\nu'}^{max}$.

Is E_0^n large enough, so that any possible state of the N-body system can in principle be excited, then the applicability of the closure relation is also physically justified. If instead E_0^n is smaller than a set of energy transfers $\{E_{\nu'} - E_\nu\}$, then the closure does not correspond to the physical situation under consideration as the more energetic final states cannot be excited. In this sense, let us put

$$\hat{\mathbf{1}} = \sum_{0}^{\nu'_{max}} |\nu'\rangle\langle\nu'| + \sum_{\nu'_{max}}^{\infty} |\nu'\rangle\langle\nu'| \ . \tag{35}$$

Formally, the introduction of the closure Eq. (35) into Eq. (13), by using the linearity of the integral, yields

$$\begin{aligned}\int_{-\infty}^{\infty} \exp(-i\omega\, t)\langle\nu|\hat{\mathbf{A}}\hat{\mathbf{1}}\hat{\mathbf{B}}|\nu\rangle\, dt &= \int_{-\infty}^{\infty} \exp(-i\omega\, t) \sum_{0}^{\nu'_{max}} \langle\nu|\hat{\mathbf{A}}|\nu'\rangle\langle\nu'|\hat{\mathbf{B}}|\nu\rangle\, dt \\ &+ \int_{-\infty}^{\infty} \exp(-i\omega\, t) \sum_{\nu'_{max}}^{\infty} \langle\nu|\hat{\mathbf{A}}|\nu'\rangle\langle\nu'|\hat{\mathbf{B}}|\nu\rangle\, dt \ .\end{aligned} \tag{36}$$

The operators $\hat{\mathbf{A}}$ and $\hat{\mathbf{B}}$ are defined as $\hat{\mathbf{A}} = \exp(-i\mathbf{q}\cdot\mathbf{r}_{j'}(0))$ and $\hat{\mathbf{B}} = \exp(i\mathbf{q}\cdot\mathbf{r}_j(t))$, the sum over the j and j' terms in Eq. (13) as well as the constants are left out for simplicity.

Note that it is erroneous to believe that the first term on the right-hand side of Eq. (36) contains only the states $\{\nu' \leq \nu'_{max}\}$. All the states ν' are intrinsically incorporated in the Heisenberg operator, namely $\hat{\mathbf{B}}$, and therefore all states ν' exist in the first term of the right-hand side.

Correspondingly, the second term on the right-hand side of Eq. (36) contains all states ν' and not only $\{\nu' \geq \nu'_{max}\}$.

Journal of Physics: Conference Series **380** (2012) 012014 doi:10.1088/1742-6596/380/1/012014

5.2. Physical constraints of the δ-function

As pointed out in section 2, the δ-function appearing in Eq. (4) has been introduced "by hand" in order to guarantee strict energy conservation (see [1], p.14). In the conventional derivation this introduction is justified in the limit $t \to \infty$ of the expression Eq. (37), see below, and [18], Eq. 42.4. Therefore, the question arises whether this introduction of the delta function is physically justified for the short scattering times in the neutron Compton scattering regime.

Let us proceed according to Landau and Lifshitz [18]. Essentially, we follow the presentation of §41 to §44: According to §44 [18] the transition rate for the system under the influence of a time independent perturbation from a state of energy E to a state of energy E' is proportional to

$$\frac{\sin^2 \frac{E'-E}{2\hbar}t}{\pi(E'-E)^2 t} \; . \tag{37}$$

In the following let us consider a system of two interacting particles [18]. The energies E and ε of these two particles at a particular point of time, in general, deviate from the energies of the system E' and ε' measured after an interval of time Δt. The most probable value of the difference $E' - E$ is of the order $\frac{\hbar}{\Delta t}$. For the system under consideration an estimate is obtained

$$|E + \varepsilon - E' - \varepsilon'|\Delta t \sim \hbar \; , \tag{38}$$

which does not depend on the magnitude of the perturbation (cf. Eq. (44.1), [18]).

In the favourable case that ε and ε' can be exactly determined ($\Delta\varepsilon = \Delta\varepsilon' = 0$) one obtains

$$\Delta(E' - E) \sim \frac{\hbar}{\Delta t} \; . \tag{39}$$

In the context of NCS the uncertainty in energy $\Delta(E' - E)$ corresponds to the width of the dynamic structure factor $S(\mathbf{q}, \omega)$. The mean energy transfer ΔE_{NCS} (usually called recoil energy E_{rec}) happens to be of similar magnitude as the width of the structure factor. (This is an experimental fact.) Thus:

$$\Delta E_{NCS} \sim \Delta(E' - E) \; .$$

Therefore Eq. (39) gives us an estimate of the time interval Δt through the energy transfer ΔE_{NCS} for neutron-proton scattering

$$\Delta t \sim \frac{\hbar}{\Delta E_{NCS}} \leq 1 \text{ fs} \; .$$

In NCS, the scattering time fulfils the relation $\tau_{sc} \sim \frac{\hbar}{\Delta(E'-E)}$. Therefore we have

$$\tau_{sc} \sim \Delta t \; .$$

Recall that the scattering process is accomplished during a time interval of the order τ_{sc} and therefore large times, $t \to \infty$, play no role in the dynamics of the neutron-proton Compton scattering process.

However, as is well-known and pointed out above, the transition of Eq. (37) into the delta function $\delta(E' - E)$ is possible only for $t \gg \frac{\hbar}{(E'-E)}$, or formally, in the limit of $\Delta t \to \infty$ (see [18], Eq. 42.4).

Recalling that the intermediate correlation function Eq. (19) is based upon the basic formula, Eq. (4), where the delta distribution is supposed to assure energy conservation, we conclude that the approximation of short times done in the IA is in blatant contradiction to these basic assumptions of the conventional formalism.

5.3. The transition from the N-body to the effective one-body picture in NCS
The scattering in the Compton regime, as pointed out in section 3

$$q = |\mathbf{q}| \gg \frac{2\pi}{d}$$

is incoherent and the intermediate correlation function becomes

$$F(\mathbf{q}, t) = \langle \exp(-i\mathbf{q} \cdot \mathbf{r}(0)) \exp(+i\mathbf{q} \cdot \mathbf{r}(t)) \rangle , \tag{40}$$

as the index j of the jth particle can be dropped. But still, Eq. (40) corresponds to the whole N-body system (cf. section 2) as is the N-body Heisenberg operator $\mathbf{r}(t) \equiv U^\dagger(t)\, \mathbf{r}\, U(t)$.

In order to arrive at the final result of the IA, Eq. (28), the following key approximation (i.e. a short-time expansion) is introduced (see: section 2, Eq. (25); [4], Eq. (6); [13], Eq. (7))

$$\mathbf{r}(t) = \mathbf{r} + \frac{t}{M}\mathbf{p} . \tag{41}$$

At this point, it should be emphasized that whereas $\mathbf{r}(t)$ is an N-body operator, all operators on the right-hand side, and especially $\mathbf{p}$, are one-body operators. In other words, this key approximation claims to represent a mathematically strict transition from N-body to one-body operators in Eq. (41), yielding the intermediate correlation function of the IA $F_{IA}(\mathbf{q}, t) \equiv \langle \exp[i(\omega_r + \mathbf{q} \cdot \mathbf{v})t)] \rangle$, Eq. (28). Yet, the physical context requires a different conclusion; see section 6(B). Possible entanglement and decoherence phenomena between different particles are therefore manually excluded from having any significance in Eq. (28).

Summing up, the system under consideration in the above approximation is a single nucleus in an effective Born-Oppenheimer potential. Together, they define an effective single particle Hamiltonian. Therefore, the momentum distribution $n(\mathbf{p})$ of Eq. (30) describes a single particle in a Born-Oppenheimer potential. This also implies that the effective wave-functions appearing in the main formula Eq. (28) are one-body quantities (see [4], p. 5957).

6. Discussion

In the preceding sections several steps of the conventional theory of neutron scattering were analysed in some detail with particular focus on the Compton regime of the theory. Here we summarize the main results and provide several additional comments and explanations.

(A) The standard, conventional theory as found in the textbooks [1, 2], also called the van Hove formalism [3], is based on the theoretical framework of first order perturbation theory, specifically, on so-called Fermi's golden rule. Within this theoretical framework, all dynamical quantities (operators) of the impinging particle, i.e. the neutron, are missing in the main formula of the theory, see e.g. Eq. (19). This presumes that the neutron-nucleus entanglement plays no role in the derivation and final results although, in general, the entanglement is caused by the same interaction of the two particles that results in the scattering. E.g., note that the final Eq. for $F(\mathbf{q}, t)$ comprises only position operators of the scattering nuclei, but not of the neutron. The same holds for the wavefunctions included in the thermodynamic average $\langle ... \rangle$ of Eq. (19). Quantities referring to the neutron appear exclusively in the form of mere c-numbers — the momentum transfer $\hbar\mathbf{q} = \hbar\mathbf{k}_0 - \hbar\mathbf{k}_1$ and the magnitudes of the wavevectors of incident and scattered neutron k_0 and k_1.

(B) The transition from the N-body Heisenberg position operators $\mathbf{r}(t)$ to single-particle operators $\mathbf{r}$ and $\mathbf{p}$ in Eq. (25) is achieved "by hand", and not by a mathematical proof. The expansion of Eq. (25) for small t, i.e. in the limit $t \to 0$, seems to clearly show that the operators on the rhs ought to be Schrödinger one-body operators. Yet, this is erroneous, since the quantity $\mathbf{r}(t)$ appears in the equilibrium correlation function $F(\mathbf{q}, t)$, which is by definition t−translation invariant, and thus **r** of the rhs of Eq. (25) must represent a position operator at any time point t (and not only at $t = 0$). This, in consequence, means that **r** ought to be the N-body Heisenberg-operator of Eq. (14) at time t_{ini}. (For t_{ini} see section 3; Eq. (25) of section 3 refers to Eq. (41) in section 5.)

(C) A further consequence of (B) is that the introduction of the momentum distribution $n(\mathbf{p})$ (cf. Eq. (30)), which is assumed to be a single-particle quantity, is ad hoc. A similar conclusion holds for the wavefunctions appearing in Eq. (28) as they are just assumed to be single-particle quantities. This is a tacit assumption, claiming that the many-body wavefunctions factorize into effective single-particle wavefunctions, which is uncertain for condensed matter systems or liquids frequently studied in NCS. Summing up, entanglement and decoherence phenomena between two or more different particles in the systems studied is presumed to be absent and manually excluded.

(D) Due to the ultrashort scattering time τ_{sc} in NCS one has to consider a significant spread in energy and therefore the δ-function is a questionable oversimplification. It cannot be overemphasized that if the δ-function cannot be introduced into the formalism, Eq. (8), there is no way to arrive at the Heisenberg position operators. If there are no Heisenberg position operators the closure procedure leading from Eq. (13) to Eq. (18) cannot be carried out [1]. As Squires puts it (see [1], pp. 20-21. I replaced the λ' terms in the quotation by ν' terms as in the derivations above, and also properly renumbered the equations):

> We can see at this stage the purpose of expressing the δ-function for conservation of energy as an integral with respect to time in Eq. (8). The $E_{\nu'}$ term in the argument of the δ-function prevents us from using Eq. (15) to sum over ν'. By means of Eq. (8) we are able to bring the $E_{\nu'}$ term inside one of the matrix elements, where it reappears as an operator depending on the Hamiltonian. There is now no term in ν' outside the two matrix elements, and the sum over ν' can be carried out immediately.

Summing up, if the δ-function cannot be introduced there is no possibility to arrive at the main result of the conventional theory Eq. (19) and consequently the main result of the IA Eq. (28) cannot be obtained.

(E) In the light of Relativity Theory, the results of section 4 show further crucial consequences in the Compton regime of neutron scattering. For illustration, let us present a numerical example in the following. Let us consider a typical epithermal neutron with a mean root velocity $v_0 \sim 44$ km/s corresponding to $E_0 \sim 10$ eV. For a scattering system where single nuclei can be resolved, a semiclassical estimate is carried out regarding each scattering event to be independent of the others. In this semiclassical picture it takes $\tau_{col} \sim 2 \times 10^{-20}$ s for the neutron to go past the range of the nucleus given by the strong force 10^{-5} Å.

The effective scattering system (e. g. a proton and parts of its surroundings) therefore must have a linear dimension smaller than [6]

$$\Delta s_{caus} = c \cdot \Delta t_{strong} \sim 0.1 \text{ Å}$$

where c is the velocity of light, since the neutron-nucleus scattering dynamics during Δt_{strong} cannot be causally influenced by other particles being more than Δs_{caus} apart from the colliding nucleus. The dimension of a typical covalent bond H-X (with X= C, O, N, etc.) is roughly 1 Å and exceeds the event-horizon defined by Δs_{caus}. For the H_2 molecule with an internuclear distance of approximately 0.75 Å, consequently, only one proton and a part of the electrons' degrees-of-freedom causally participate in the collision process. In other terms, the usual view of the H_2 molecule to represent a *closed* system appears to contradict the principle of causality. In consequence, the notion of a *closed* "effective one-body" scattering system has no physical significance in this example. Therefore, the above line of argumentation implies that the systems studied in NCS experiments ought to be treated — for fundamental reasons — as open quantum systems.

(F) Last but not least, some modern theoretical developments may be mentioned here. In the conventional theory [1, 4] the neutron's *coherence length* plays no role. In contrast, it is of great significance in the theoretical model of Ref. [5], in which the observed "anomalous" NCS intensity effect from protons [20] is explained as a result of interference when the scattering particle (i.e. neutron) interacts with more than one hydrogen nucleus. The coherence volume of the actual setup, which limits the number of interfering particles, is therefore an important parameter. In particular, the author of [5] provides a *quantitative* description of the neutron's coherence volume for the actual VESUVIO spectrometer and of the probability for the scattering neutron to interact coherently with two or more protons or deuterons.

Another theoretical model for the interpretation of the aforementioned "anomalous" intensity effect [20, 15, 21] is proposed in Ref. [22]. Central to this qualitative model is the decoherence process of the scattering proton and its adjacent particles during the time-window of scattering time τ_{sc}.

Note that quantum coherence and entanglement is preserved over the sub-fs scattering times relevant for the NCS process, even in the condensed systems studied [5, 20, 15, 21]. Compton scattering opens, therefore, a time-window for studies of coherence and decoherence not easily accessible by other methods.

Acknowledgments
I am indebted to Prof. C A Dreismann for illuminating discussions. I am very thankful for the invitation of the organizers of the symposium. Partial financial support by the 'Gesellschaft von Freunden der TU-Berlin e.V.' and the 'Vereinigung von Freunden und Förderern der Goethe Universität Frankfurt' is greatfully acknowledged.

References

[1] Squires G L 1996 *Introduction to the Theory of Thermal Neutron Scattering* (Mineola: Dover)
[2] Lovesey S W 1984 *Theory of Thermal Neutron Scattering from Condensed Matter* (Oxford: Oxford University Press)
[3] van Hove L 1954 *Phys. Rev.* **95** 249
[4] Watson G I 1996 *J. Phys.: Condens. Matter* **8** 5955
[5] Karlsson E B 2012 *Int. J. Quantum Chemistry* **112** 587-602
[6] Chatzidimitriou-Dreismann C A and Tietje I C 2010 *J. Phys.: Conf. Ser.* **237** 012010
[7] Sears V F 1984 *Phys. Rev.* B **30** 44
[8] Joos E, Zeh H D, Kiefer C, Guilini D J W, Kupsch J, Stamatescu I-O 1996 2003 *Decoherence and the Appearance of a Classical World in Quantum Theory* (Springer-Verlag Berlin Heidelberg New York)
[9] Breuer H P and Petruccione F 2002 *The Theory of Open Quantum Systems* (Oxford: Oxford University Press)
[10] Schulman L S and Gaveau B 2006 *Phys. Rev. Lett.* **97** 240405
[11] Chatzidimitriou-Dreismann C A, Gray E MacA, and Blach T P 2011 *AIP Advances* **1** 022118
[12] Ballentine L E 1998 *Quantum Mechanics – A Modern Development* (Singapore: World Scientific)
[13] Andreani C, Colognesi D, Mayers J , Reiter G F and Senesi R 2005 *Adv. Phys.* **54** 377
[14] Hohenberg P C and Platzman P M 1966 *Phys. Rev.* **152** 198
[15] Chatzidimitriou-Dreismann C A, Vos M, Kleiner C and Abdul-Redah T 2003 *Phys. Rev. Lett.* **91** 057403
[16] Karlsson E B, Chatzidimitriou-Dreismann C A, Abdul-Redah T, Streffer R M , Hjörvarsson B, Öhrmalm J and Mayers J *Europhys. Lett.* 1999 **46** 617.
[17] Gidopoulos N I (a) *Phys. Rev.* B 2005 **71** 054106; (b) *Non-adiabatic electronic excitations in neutron Compton scattering*, ISIS Science Highlights, 2004.
[18] Landau L D, Lifschitz E M (a) 1958 *course of theoretical physics 3: quantum mechanics* (Pergamon Press London - Paris) (b) 1979 *Lehrbuch der theoretischen Physik III: Quantenmechanik* (Akademie-Verlag Berlin)

Journal of Physics: Conference Series **380** (2012) 012014 doi:10.1088/1742-6596/380/1/012014

[19] Marshall W and Lovesey S W 1971 *Theory of Thermal Neutron Scattering*, (Oxford: Oxford University Press)

[20] Chatzidimitriou-Dreismann C A, Abdul-Redah T, Streffer R M F and Mayers J 1997 *Phys. Rev. Lett.* **79** 2839

[21] Cooper G, Hitchcock A P and Chatzidimitriou-Dreismann C A 2008 *Phys. Rev. Lett.* **100** 043204

[22] Chatzidimitriou-Dreismann C A and Stenholm S, in: V M Akulin, A Sarfati, G Kurizki, and S Pellegrin, *Decoherence, Entanglement and Information Protection in Complex Quantum Systems*, NATO Science Series II – Vol. **189** (Springer, Dordrecht, 2005); pp. 555-562; also available at arXiv:quant-ph/0702038v1.

[23] Merzbacher E 1961 *Quantum mechanics* (John Wiley & Sons, Inc.)

Journal of Physics: Conference Series **380** (2012) 012015 doi:10.1088/1742-6596/380/1/012015

Comparison of the non-equilibrium predictions of Intrinsic Quantum Thermodynamics at the atomistic level with experimental evidence.

Charles E Smith and Michael R von Spakovsky

Center for Energy Systems Research, Mechanical Engineering Department
Virginia Polytechnic Institute and State University, Blacksburg, VA 24061
E-mail: Michael.von.Spakovsky@vt.edu, cesmith@vt.edu

Abstract. Theoretical predictions from Intrinsic Quantum Thermodynamics (IQT) are compared to published experimental findings in order to examine whether or not calculations using IQT are consistent with experimental data. IQT makes the assertion that entropy is an intrinsic property of matter in the same way that inertial mass, energy, and momentum are and must, thus, exist even for single particles. Entropy as defined by IQT is a measure of the distribution of a system's internal energy at any given instant of time amongst the available internal degrees of freedom, i.e., the energy eigenlevels of the system. In this paper, it is shown that IQT models the internal relaxation of a 5-level rubidium system in a way that is consistent with the experimental data available in the literature. In addition, other experimental data found in the literature is compared with results obtained from the equation of motion of IQT that allows for heat interactions. It is shown that the decay of so-called "cat states" for single ions that are contained in Paul traps and that interact with a heat reservoir is also consistent with the calculations made using the theory of IQT.

1. Introduction and background

The foundations of Intrinsic Quantum Thermodynamics (IQT) were first presented thirty-five years ago by Hatsopoulos and Gyftopoulos [1] with important preliminary work contributed by Park and Simmons [2,3]. A fundamental dynamical contribution was introduced in 1981 by Beretta [4], and important subsequent work includes Refs. [5-22]. The theory of IQT is based on the assertion that the second law of thermodynamics, with its implications of irreversibility, applies at all physical levels of description from the macroscopic and classical to the atomistic and quantum [1]. Central to the foundations of IQT is the recovery of the concept of "state of a system", a bedrock of physical thought and a concept lost in quantum statistical mechanics (QSM) in which the state necessarily refers only to the state of an ensemble, which consists of an infinite number of identical systems not identically prepared [1,2]. Also central to IQT is the fact that the dynamics of any change in state is at all times consistent with the laws of physics and thermodynamics [1-7]. To satisfy both the requirements of thermodynamics, particularly that of the second law, as well as those of quantum mechanics (QM), IQT relies on two fundamental insights. The first is that the ontological entity representing *any* state of a quantum system[1] is not the density operator (i.e., wave function projector) of QM[2] but instead a density or "state" operator based on an ensemble consisting of an infinite number of identical systems identically prepared [1]. The second is that the Schrödinger equation of motion of QM, though correct, is incomplete since it is unable to describe the dynamics of a system with entropies greater than zero [1,4-7].

The rationale behind the concept that the density operator is synonymous with the state of a system is based on the idea that in QM the density operator contains all the information necessary to characterize the state (i.e., the so-called pure state) of a quantum system at a any given instant of time. Thus, for example, the expectation value $\langle S\rangle$ for the entropy, as defined by von Neumann, can be written in terms of the density or "state" operator ρ via

$$\langle S\rangle = -k_b Tr(\rho \ln \rho)\,. \tag{1}$$

Here k_b is Boltzmann's constant. In QM, the value of $\langle S\rangle$ is necessarily zero, i.e. is zero for all pure states. For values greater than zero, i.e. for non-pure or mixed states, the density operator of QM can be replaced with that of QSM. However, the problem is that not only does this introduce the difficulty mentioned above about the loss of the concept of "state of the system" but it as well leads to an entropy, which is not the entropy of thermodynamics [12]. In fact, QSM leads to the so-called 'irreversibility paradox", the resolution of which requires a simultaneous consideration of questions that specifically go to the heart of issues surrounding the second law of thermodynamics, namely, i) what the physical roots of "*entropy*" and "*irreversibility*" are, ii) whether or not "*entropy generation*" due to irreversibility is merely a *statistical illusion*[3], and iii) what a general description of *non-equilibrium* is [21]. Such a general description is not possible with QSM without the addition of a non-thermodynamic principle (microscopic reversibility), assumption (e.g., small perturbations to ensure linear behaviour), or approximation (e.g., sufficiently small deviations from stable equilibrium) [21]. In contrast, IQT and its density operator exhibit none of these drawbacks and instead lead to an entropy that is *physical* and exists for all states (pure and non-pure) and all systems regardless of size. Moreover, the entropy at a fundamental level of description is seen as a measure of how the system energy $\langle E\rangle$ is distributed amongst the system's available degrees of freedom, i.e., its energy eigenlevels, while the entropy generation is a measure of how the energy $\langle E\rangle$ is redistributed in a change of state [21].

The second insight mentioned above that the Schrödinger equation of motion is incomplete has prompted the search for an equation able to describe irreversible processes. In Quantum Thermodynamics (QT), formerly known generally as Quantum Dissipative Dynamics, this has led to master equations of various types [23, 24]. These equations have been developed to model the so-called "open quantum systems"[4] introduced by Lindblad and Kossakowski and others [25-27] where the increase in the entropy of a system is brought about through interactions with an external reservoir. In IQT, the generalization to irreversible or "dissipative"[5] processes is provided by an equation of motion originally developed by Beretta [4-7], which assumes a priori a tendency for an intrinsic increase in system entropy, i.e., the entropy of thermodynamics. Unlike the master equations of QT, which are often second order approximations [28], the Beretta equation represents the full nonlinear dynamics, which describes the irreversible evolutions in state of systems that are arbitrarily far from stable equilibrium.

Thus, the outlook provided by IQT avoids the inconsistencies pointed out by Loschmidt [29], which arise when trying to force irreversible, non-unitary behaviour from dynamics that is intrinsically reversible and unitary. Avoided as well is the need for the exogenous statistics found in QSM that destroy the concept of state by requiring that a system possessing entropy be described with ensembles

[1] All systems are in the end quantum systems; but as the size of a system increases, the importance of quantum effects decreases, since the dispersions representing observables approach delta functions.
[2] The density operator of QM only represents a limited class of states and is, thus, a special case of the density operator of QT.
[3] In fact, this is the conclusion drawn from statistical mechanics that entropy generation due to irreversibility does not result from the endogenous dynamics but instead from temporal changes of some exogenous statistical description.
[4] This is the closed-system-plus-reservoir model used as the basis for describing the concept of "dissipation" in QT, which in QT is understood to be a loss of information between the system and the reservoir. This "dissipation" is, thus, not physical.
[5] In contrast to QT, the concept of "dissipation" in IQT is that of a physical entity internal to the system and is not dependent on the closed-system-plus-reservoir model as used in QT.

of "pure" states which themselves have no entropy. Statistical mixes of pure states have the additional inconsistency which arises when work can be extracted from subsets of the ensemble even though none can be extracted from the ensemble as a whole. Clearly, this violates the second law [21].

IQT also avoids the violations of the second law that are inherent with the QT approach. In contrast to QSM, QT is not based on statistical mixes of pure states and, thus, the definition of "state of the system" is preserved. However, because the entropy for the "open system" (i.e., closed system plus reservoir) cannot increase, the potential for extracting energy to do work remains unchanged (i.e., is not degraded) regardless of whether gradients of thermodynamic potentials between the closed system and reservoir exist and change over time. This cannot be since it suggests that regardless at what point in time energy is extracted that the potential to do work remains unchanged even when all thermodynamic potentials have ceased to exist. This is inconsistent with what is observed in nature and, thus, violates the second law. IQT obviates such difficulties because the increase in entropy is a process that is intrinsic to the closed (isolated) system, leading to a degradation in the potential to do work.

The equation of motion of IQT governs how the diagonal and off-diagonal elements of the thermodynamic state or density operator (or matrix) ρ [6] evolve in time. The formulation is based on the hypothesis that physical systems naturally seek the path of local steepest entropy increase on their way to stable equilibrium [4]. For an isolated or non-isolated (experiencing a work interaction), single constituent (i.e., a single particle, a single assembly of indistinguishable particles, or a single field) closed system, this equation is given by

$$\frac{d\rho}{dt} = -\frac{i}{\hbar}[H,\rho] - \frac{1}{\tau_D} D \,. \tag{2}$$

Here, H is the Hamiltonian operator, τ_D a scalar time constant or functional[7], and $-D/\tau_D$ the so-called dissipation term, which is a function of ρ, $\ln\rho$, and H, and which pulls the state operator in the direction of the projection of the gradient of the entropy functional $\langle S\rangle$ onto the hyper-plane of the constant energy $\langle E\rangle$ of the system. Forms of this equation for closed systems composed of distinguishable particles (or distinguishable assemblies of particles) and/or experiencing a non-work interaction (i.e., a heat interaction) are found in Beretta [7,19,20] as well as in Section 2.3 below. Both the first and the second laws of thermodynamics are implied by this equation and its other forms[8].

The first term on the right hand side of Eq. (2) is the Schrödinger term which governs the reversible (linear) dynamics for the system, and it along with the time-derivative term on the left are equivalent to the time-dependent Schrödinger equation. This part of Eq. (2) governs the relative phases between system energy eigenlevels and quantum interference effects. Equation (2) also includes a dissipation operator D scaled by an associated time functional τ_D. This term governs the dissipation of a system's adiabatic availability [31] as its state relaxes to one of maximal entropy. In general, due to its high degree of non-linearity, this equation must be solved numerically except in the simplest of cases, e,g., a two-energy-eigenlevel system [8].

Since its development, there has been a limited effort to verify Eq. (2) experimentally and as a consequence the theory of IQT. The first by Beretta [8] uses IQT to model the effect of irreversibilities on the resonance fluorescence, absorption and stimulation emission of a two-level atom driven near resonance by a monochromatic laser beam. Beretta's results suggest that the discrepancies between experiment and the theoretical predictions of QM are due to an irreversible atomic relaxation to stable equilibrium at constant energy, which can be predicted using the IQT equation of motion.

[6] The state operator is a linear, self-adjoint, non-negative definite, unit-trace operator (i.e., an operator whose diagonal elements sum to one) on Hilbert space $\mathcal{H}$.
[7] Note, that a lower bound for τ_D and, thus, an upper bound on D is based on the time-energy Heisenberg uncertainty relation [20].
[8] This equation implies the *first law* because as is proven in Beretta et al. [6], each of the generators (e.g., the identity and Hamiltonian operators) of the motion (i.e., the evolution or change in state of the system) is also a constant of the motion of the system. Thus, $\langle E\rangle = Tr(\rho H)$ is conserved and the ΔE in any adiabatic process is uniquely related to the amount of work involved in the process. These equations also imply the *second law* since as proven in Beretta et al. [6], a system admits of one and only one stable equilibrium state for given finite mean values of the generators of the motion. This, of course, is simply a generalization of the Hatsopoulos-Keenan statement of the second law [1,30].

Nonetheless, further high-resolution experimental work, which has yet to be done, is needed to verify Beretta's predicted results [8].

In a second attempt, Çubukçu [11] argues that unlike QM, IQT can predict the experimental results developed by Kukolich [32] for the damped-oscillatory behavior seen in the absorption rate of transmitted light by rubidium atoms in the vapor phase in a magnetic field. However, to support this contention, he uses the analytical solution of the IQT equation of motion for a 2-level, spin-½ system given by Beretta [33] even though the experimental rubidium spin system actually represents a 5-level system. Thus, the match between his solution and the actual experimental results is not perfect. Nonetheless, the implication is that the damping effects predicted by the IQT equation are due to the irreversible relaxation of the spin states of the rubidium atom to stable equilibrium.

To improve on Çubukçu's results, Kukolich's experiment [32] is revisited in this paper and the actual five-level system of Kukolich's experiment numerically (as opposed to analytically) modeled using the IQT equation of motion, Eq. (2). In addition, a new comparison is made with the experimental data of Turchette et al. [34] for the relaxation of superposition states for a single ion in a Paul trap using the IQT equation of motion initially developed by Beretta [19] to include a heat interaction term and subsequently modified by the present authors as outlined in Section 2.3 below and in Smith [35].

2. Comparison of IQT with experiment

2.1 The Kukolich experiment

Kukolich [32] uses the spin states of rubidium to demonstrate the time dependence of the state operator of QM (i.e., the projector of the wave function). A diagram of the experimental apparatus is shown in Figure 1.

The rubidium (Rb) atoms and a helium (He) buffer gas are contained in a closed glass vessel with an applied external magnetic field that is oriented along the z-axis. The technique of optical pumping with circularly polarized light is used to prepare the Rb atoms in the desired spin state. The He is used to slow the relaxation of the Rb due to collisions with other Rb atoms as well as collisions with the walls of the container. Once the Rb atoms are in the upper most spin energy eigenlevel, the direction of the magnetic field is changed suddenly to the x-direction, perpendicular to the z-axis. A photoelectric detector opposite the light source along the z-axis is used to measure the intensity of the light beam as a function of time. The intensity of the light that is transmitted through the Rb vapor is found to oscillate and can be used to determine the quantum eigenstates of the Rb atoms.

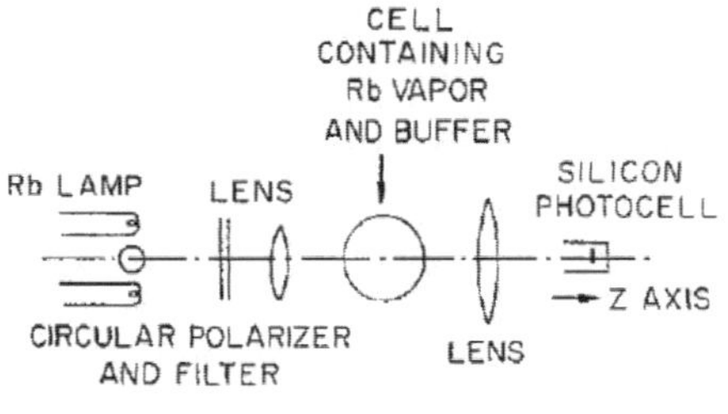

Figure 1. Diagram of the experimental apparatus of Kukolich [32].

The spin states for Rb occur due to the coupling of the $3/2$ spin of the nucleus with the $1/2$ spin of the electrons. An energy level diagram is shown in Figure 2. The external magnetic field splits the energy eigenlevels E_0 and E_1 into sets of 5-level spin energy eigenlevels denoted as F=1 and F=2, respectively. The energies between each spin energy eigenlevel are multiples of Planck's modified constant $\hbar$ as shown. The external circularly polarized laser light, which has a wavelength of 7948 Å, pumps the majority of the atoms into the highest energy eigenlevel (i.e., F=2, $|+2\rangle_z$), which leaves the Rb atoms in a spin-2 eigenstate.

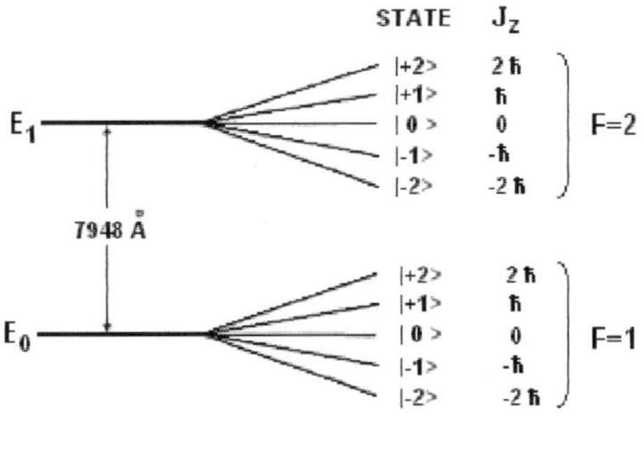

Figure 2. Energy level diagram for the state of ^{87}Rb [32].

The switching of the magnetic field to the x-direction changes the thermodynamic state of the Rb atoms. In QM, the new state or density operator (matrix) ρ_{new}^{QM} as seen along the z-axis is calculated by means of the time-dependent von Neumann (or equivalent Schrödinger) equation or equivalently by the unitary (non-dissipative) transformation given by

$$\rho_{new}^{QM} = |\Psi_z\rangle_{new}\langle\Psi_z|_{new} = U^{-1}|\Psi_z\rangle_{old}\langle\Psi_z|_{old}U \tag{3}$$

where $|\Psi_z\rangle_{new}$ (or its bra $\langle\Psi_z|_{new}$) is the new state vector and the unitary transformation matrix U and its inverse U^{-1} are Hermitian, i.e.,

$$U^{-1} = U^{\dagger}. \tag{4}$$

As done in Kukolich [32], the unitary transformation matrix U can be derived using the method outlined in Feynman [36]. The QM solution results in phases of state (i.e., $(J_z\omega_p/\hbar)t = (-\varepsilon_i/\hbar)t$ for i = -2,-1,0,1,2) that change periodically in time t. The precession frequency of the phase, ω_p, is proportional to the absolute magnitude of the spin energy eigenvalues ε_i. As a consequence, the resulting *non-dissipative* state operator ρ_{new}^{QM} of the system must contain elements that oscillate with the precession frequency of the Rb atoms as well as twice this precession frequency, with the precession frequency being directly proportional to the strength of the applied magnetic field in the x-direction. After the transformation, the new state vector as seen from along the z-axis is expressed as

$$|\Psi_z\rangle_{new} = \begin{bmatrix} \frac{1}{8}\cos 2\omega_p t + \frac{1}{2}\cos\omega_p t + \frac{3}{8} \\ \frac{1}{4} i\sin 2\omega_p t + \frac{1}{2} i\sin\omega_p t \\ \frac{1}{2}\sqrt{\frac{3}{8}}(\cos 2\omega_p t - 1) \\ \frac{1}{4} i\sin 2\omega_p t - \frac{1}{2} i\sin\omega_p t \\ \frac{1}{8}\cos 2\omega_p t - \frac{1}{2}\cos\omega_p t + \frac{3}{8} \end{bmatrix} \tag{5}$$

where the transformed state vector $|\Psi_z\rangle_{new}$ (or state operator ρ_{new}^{QM}) is that of a pure state (i.e., a thermodynamic state with zero entropy).

After the field direction is switched, the amount of transmitted light from the same light source that is used for the optical pumping is employed to produce a signal whereby the thermodynamic state of the system can be determined. This is deduced from the intensity of the light transmitted through the cell containing the Rb atoms given by [32]

$$I = I_0 - A = I_0 - \tfrac{7}{2}C + 2C\cos\omega_p t + \tfrac{3}{2}C\cos 2\omega_p t \tag{6}$$

where I_0 is the intensity of the light source and A the magnitude of the light absorbed by the atoms expressed as

$$A = \tfrac{7}{2}C - 2C\cos\omega_p t - \tfrac{3}{2}C\cos 2\omega_p t\,. \tag{7}$$

As seen in Eq. (6), the intensity contains two precession frequencies: ω_p and $2\omega_p$.

A diagram of I as a function of time is shown in Figure 3 and clearly shows the oscillatory behaviour of the thermodynamic state of the Rb atoms also seen in the experimental results given in Figure 4. As expected, the damping seen in the experimental results does not appear in Figure 3 since the unitary transformation of QM is reversible and damping is an inherently irreversible phenomenon. The damping, in fact, indicates that there is a relaxation from the initial non-equilibrium state in which the Rb atoms are placed to one of stable equilibrium. In addition, as noted in Kukolich [32], the relative amplitude between the two frequency components in Figure 4 differ from the expected theoretical value in part because the photocell attenuates the fundamental frequency more than the double frequency component. Finally, the braces under some of the oscillations in part (a) of Figure 4 indicate a "stair step" aspect of the data not attributable to a simple exponential decay of the signal. As will be seen below, both this aspect and the damping are captured by the IQT simulation.

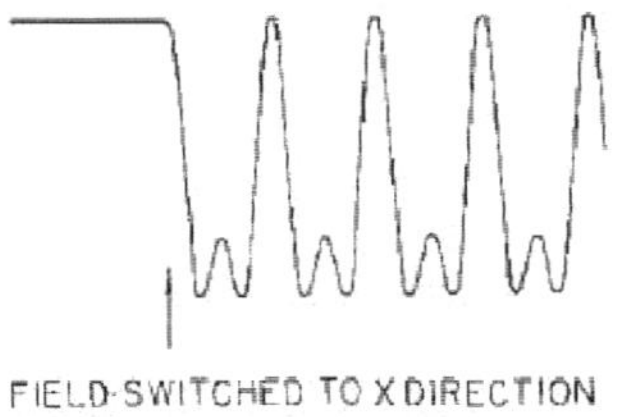

Figure 3. Time-dependent photocell current as calculated for a spin-2 system without dissipation [32].

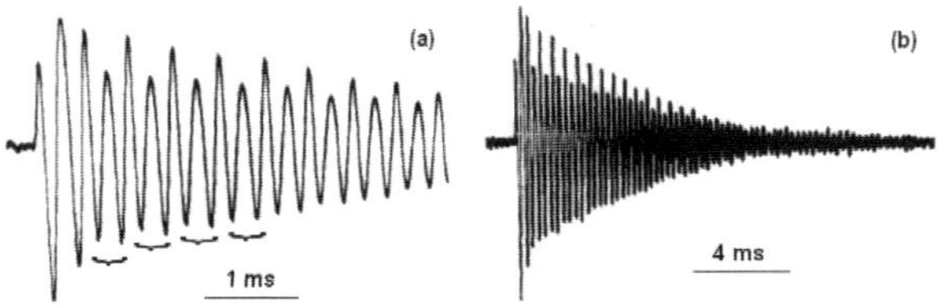

Figure 4. Experimental trace of the photocell current for two precession frequencies ((a) ω_p and (b) $2\omega_p$) after the field is switched to the x-direction [32].

In an experiment similar to that of Kukolich [32], Nagel and Haworth [37] use the magnetic field of the earth, which is oriented in the x-direction, instead of an applied external field to produce the oscillations. Their results shown in Figure 5 indicate an initial spin attenuation that is more gently curved than would be indicated by a simple exponential decay. In fact, these relaxation curves look more like the gradual, S-shaped curves that are predicted by IQT where sparsely occupied energy eigenlevels for a system gradually increase their occupation probabilities at the expense of those with larger initial occupation probabilities. Clearly, the same gradual, S-shaped drop off is not seen in the Kukolich experiment. Although one can only speculate as to why, since Kukolich makes no comment, it is reasonable to assume that only the latter part of the signal, which corresponds in the Nagel-Haworth experiment to the bottom half of the S of the S-shaped decrease, is reported in Kukolich.

To initiate the solution of the IQT equation of motion for the Rb relaxation, an initial density or state operator ρ, which represents the thermodynamic state of the system after the magnetic field is switched to the x-direction, is chosen. Thus, the initial state operator, ρ_{init}, is taken as ρ_{new}^{QM} (Eq. (3)) slightly perturbed (see the discussion below following Eq. (9)) and evaluated at t = 0. The IQT solution assumes that each Rb atom can be treated as a separate, isolated system. Çubukçu [11] sug-

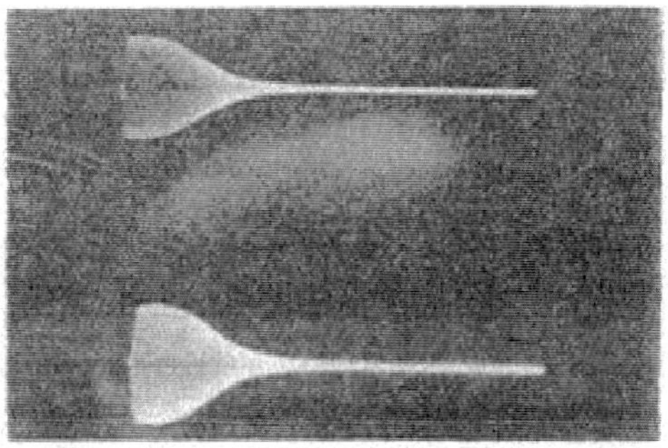

Figure 5. The results of the field switching experiment for Rb atoms by Nagle and Haworth [37]; the sweep time for the signal is 10 ms.

gests this assumption is valid at sufficiently low pressures based on Franzen's [38] measurements of relaxation time as a function of Rb vapour pressure which indicate that the time to relax increases with decreasing pressure until the latter is sufficiently low after which time it plateaus out, i.e., collisions as the driving mechanism for relaxation become negligible. Çubukçu concludes that the gradual relaxation to stable equilibrium is attributed to each atom's own internal relaxation from a state of non-equilibrium to one of stable equilibrium and not due to collisions with other atoms (i.e., Rb or those of the buffer gas or of the wall)[9]. However, the recent experiments of Balabas et al. [39] show with refined techniques, which further isolate the Rb atoms from collisions harmful to the spin state, that relaxation times are extended by 2 to 3 orders of magnitude over those viewed in the Franzen [38] experiment. Thus, redoing the latter with the techniques of Balabas et al. would be needed to see if the plateauing at low vapour pressures observed by Franzen indeed still occurs. Nonetheless, for now, the assumption of Çubukçu is made and as seen below, the characteristics of the data of Kukolich and Nagel and Haworth are modelled well using it and the IQT formalism.

To model this relaxation, the system of equations which result from Eq. (2) for this 5-level, spin-2 system is written as

$$\frac{d\rho_{ij}}{dt} = -\frac{i}{\hbar}[H,\rho]_{ij} - \frac{1}{\tau_D}D_{ij} \tag{8}$$

where the ρ_{ij} are the elements of the density matrix (state operator) that are found as functions of time and i and j each vary from 1 to 5. The first term on the right hand side of Eq. (8) are the commutator elements that represent the decay of the relative phases between the spin energy eigenlevels of each Rb atom, while the second term are the elements of the dissipation operator of IQT expressed as[10]

$$D_{ij}=\frac{\begin{vmatrix} (\rho\ln\rho)_{ij} & \rho_{ij} & E_{ij}\rho_{ij} \\ \sum(\rho\ln\rho)_{ii} & 1 & \sum E_{ii}\rho_{ii} \\ \sum E_{ii}(\rho\ln\rho)_{ii} & \sum E_{ii}\rho_{ii} & \sum E_{ii}^2\rho_{ii} \end{vmatrix}}{\begin{vmatrix} 1 & \sum E_{ii}\rho_{ii} \\ \sum E_{ii}\rho_{ii} & \sum E_{ii}^2\rho_{ii} \end{vmatrix}}. \tag{9}$$

[9] Note that 'current theory' explains the plateauing in the following way: when the number of particles is sufficiently small so that collisions with the walls are still much more frequent than collisions with other particles, the relaxation time plateaus because each particle behaves as if the others were not present. To conclusively rule out that the effect being seen is not due to particle-wall interactions, one would have to redo the experiment with different wall materials to see if the relaxation time plateaus at different values or always at the same 'intrinsic' value.

[10] Note, that the logarithm of the matrix representation of the density operator ρ is found by computing its Jordan decomposition and then taking the logarithm of the Jordan blocks.

Because the state involves the decay of the relative phases between the spin energy eigenlevels of each Rb atom, the operators H and ρ do not commute so both diagonal and off-diagonal elements of ρ are included in the solution.

The results from the IQT equation of motion are seen in Figure 6, which shows the evolution of the density operator for the 5-level, spin-2 Rb system with internal dissipation. The initial state operator, ρ_{init}, is constructed from the outer product of the state vector shown in Eqs. (3) and (5). At $t = 0$, this product, ρ_{new}^{QM}, represents a pure state having no entropy. In order for the state of the system to evolve in time according to Eq. (8), the elements of ρ_{new}^{QM} are perturbed slightly from their starting values; and it is this perturbed density operator, ρ_{init}, which is used as the initial state operator in Eq. (8). The upper blue traces seen in Figure 6 result from the diagonal terms of the state operator, while the traces in red and green represent evolutions due to the off-diagonal terms. The time constant τ is chosen to qualitatively match the rate of dissipation shown in the Kukolich data. The solution is used in Eq. (6) to calculate the amount of transmitted laser light.

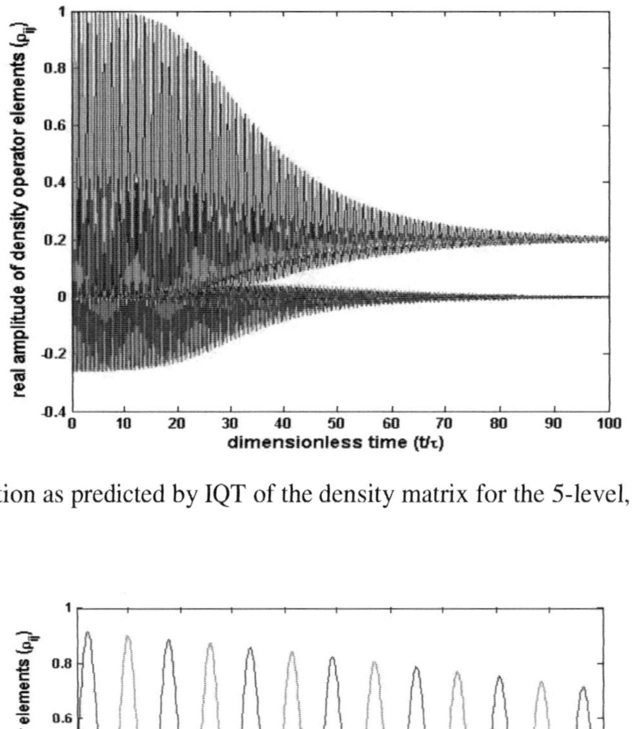

Figure 6. Evolution as predicted by IQT of the density matrix for the 5-level, spin-2 Rb system.

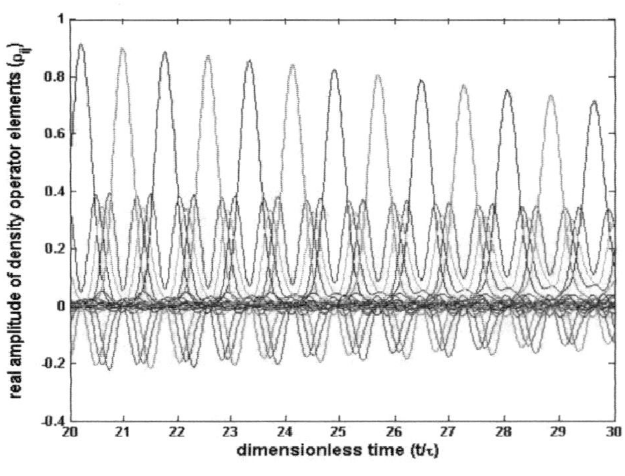

Figure 7. A detailed view of the IQT result for the density matrix evolution of the 5-level, spin-2 Rb system.

A close-up view of the IQT solution is seen in Figure 7 where it is evident that the two different frequencies contained in the wave form play a role in the time evolution of the state operator. In other similar experiments described in Balabas et al. [39] and Happer, Jau, and Walker [40], the light frequency unlike in the Kukolich experiment is detuned far from resonance so as to reduce any disturbance by the probe beam. As a result, the double frequency component is suppressed and the signal is dominated by just the precession frequency.

Since the relaxation to stable equilibrium is due to dissipation (i.e., the destruction of exergy), which is modelled by the second term on the right hand side of the IQT equation of motion (Eq. (8)), the entropy changes and evolves according to the results presented in Figure 8. As is evident from this figure, the entropy evolves somewhat slowly initially and then quickly increases, reaching its maximum or stable equilibrium value after about 80 dimensionless time units.

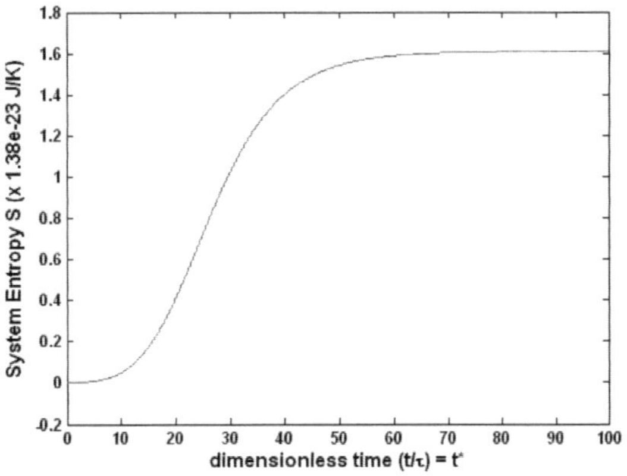

Figure 8. The IQT result for the entropy as a function of time for a 5-level, spin-2 Rb system.

Finally, as is done in Kukolich [32] with Eq. (6), the IQT solution is used to calculate the amount of transmitted light. The relative probabilities of photon absorption for the four lowest eigenenergies are given in Kukolich [32] as derived from Messiah [41,42]. Only the four lowest spin energies are able to absorb photons since there are no transitions above the $|+2\rangle$ energy state. These absorption probabilities are multiplied by their respective lowest four diagonal elements of ρ which are then summed together to represent the IQT version of the absorption function *A* of Eq. (6). The result is

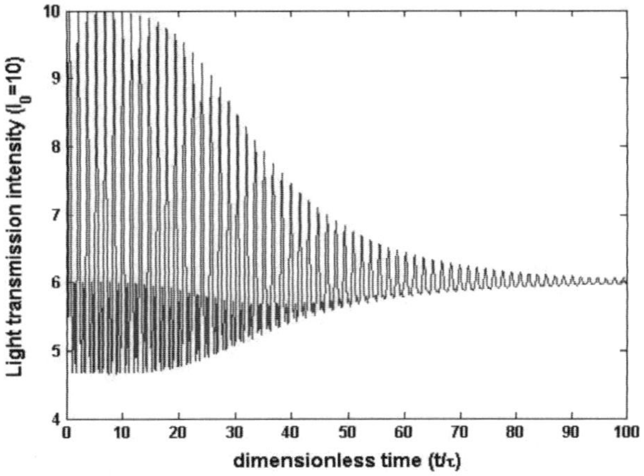

Figure 9. The transmission curve as predicted by IQT for the 5-level, spin-2 Rb spin system.

presented in Figure 9 and follows the same trends as the experimental data of Kukolich and Nagel and Haworth presented in Figures 4 and 5. The initial period of relaxation starts out relatively gradually, much like the result seen in Figure 5, an effect which is commonly seen in IQT simulations. The reason for this is that it takes some time for energy sharing between the occupied and unoccupied energy eigenlevels of the system to grow. For the 5-level, spin-2 Rb system, the lower four system energy eigenlevels start out with very low occupation probabilities due to theoretical pumping of the Rb atom into the fifth and highest level. The IQT simulation predicts a lag before the internal energy begins to be extensively shared with the lower four spin levels.

A close-up view of the predicted IQT transmission curve is shown in Figure 10. The double frequency component of the experiment is captured but is seen to damp out more rapidly than for the Kukolich data since the simulation leaves out the selective damping effects of the primary frequency of the signal by the optical filters and photocell that occur in the experiment. Furthermore, note the uneven stair-step feature of the wave minima in Figure 10, a feature seen in the experiment and one which contrasts with the smooth envelope that would be produced if one simply used decaying exponentials to describe the data (see Figure 11). The capture of these two key features by IQT lends weight to the physical basis for its predictions.

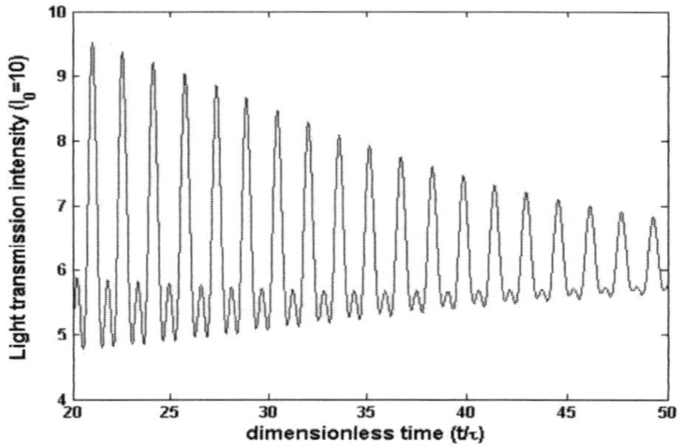

Figure 10. A close-up view of the transmission amplitude as predicted by IQT.

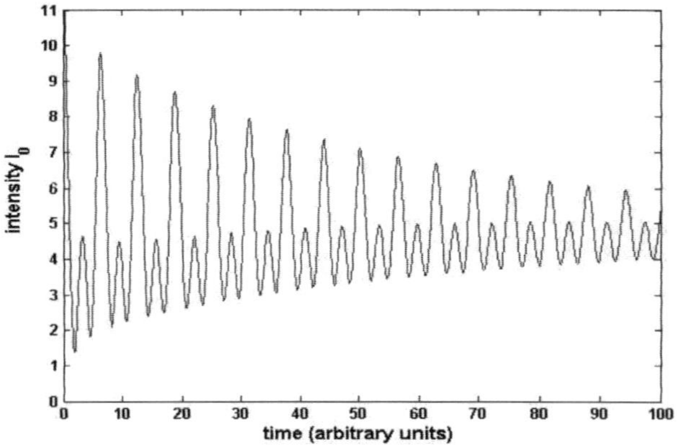

Figure 11. The transmission amplitude as calculated by QM and multiplied by a decaying exponential function with the extrema of the curve following the exponential envelope.

2.3 The Turchette et al. experiment

In the experiment of Turchette et al. [34], a thermodynamic system consisting of single trapped Be^+ ion contained in a Paul trap is put into various quantum superposition states much like in the "particle in a box" models that are used in physics texts. The decay of the initial state is observed and measured after the ion trap is put into contact with a range of engineered external heat sources. A schematic of the Paul trap is shown in Figure 12.

The edges of the slots serve as electrodes through which the RF (radio frequency) fields are produced to trap the ion as well as to input the noise signals used to serve as an external heat source. The strength of the fields is quadratic, so the particle behaves as a quantum harmonic oscillator within the trap. The harmonic superposition or "cat" or "motional" states that are produced in the experiments are also known as Fock states, and density matrices describing these states contain only diagonal elements [34].

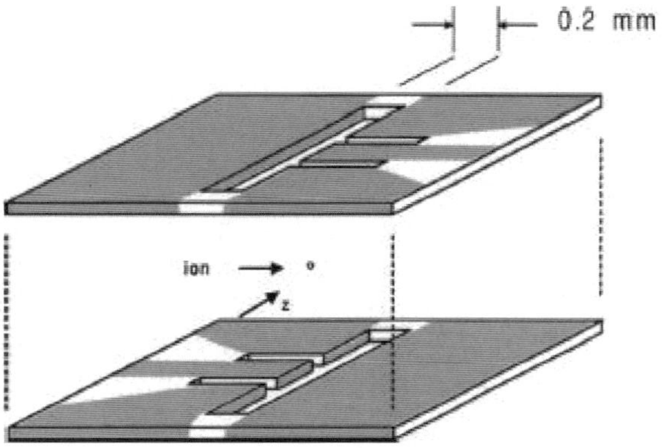

Figure 12. Schematic of the Paul trap used in the experiment of Turchette et al. [34].

The measurement of the amount of decoherence (i.e., in the view of IQT, the amount of dissipation) over the time interval is done through interferometry techniques. Nuclear spin states (up and down) are excited in the Be^+ ion and combined by means of optical pumping and laser cooling methods with the superpositions of the motional eigenstates of interest. The spins constitute a "carrier" signal that enables the degree of decoherence of the cat states to be readily measured. The resulting quantum (thermodynamic) state after preparation can be written in terms of the state vector such that

$$|\psi\rangle = \frac{1}{\sqrt{2}}(|\downarrow\rangle|\alpha_\downarrow\rangle + |\uparrow\rangle|\alpha_\uparrow\rangle) \tag{10}$$

where the arrows signify spin up or spin down and α is the particular cat state superposition between the ground state $|0\rangle$ and some higher energy oscillator state $|n\rangle$. Because the spin states are correlated with the energy eigenstates of the harmonic oscillator, any changes or degradation of the cat state will result in proportional changes between the phases of the up and down ($|\uparrow\rangle$ and $|\downarrow\rangle$) eigenstates.

During the experimental procedure, a state as shown in Eq. (10) is created and immediately coupled to a heat bath. After a given delay, typically 3 μs, a measurement is made. The phase shift between the spin components is seen as a loss of signal contrast, from which the magnitude of decoherence of the cat state can be calculated. The heat bath consists of a noise spectrum of a given mean frequency and power that is applied to the fields containing the ion in the Paul trap. The mean frequency of the noise corresponds to the temperature of the heat bath while the power (in volts^2) corresponds to the strength of the heat source coupled to the system. Numerous measurements are conducted to produce ensemble average values that make up each experimental data point. The time intervals between the application of the heat source and when the measurements are taken are held constant. The variation of the power turns out to give results that are close to being the same as keeping the strength of the heat source constant and varying the measurement time intervals and is evidently easier to do experimentally.

IQT simulations are compared with the relaxation data for the ion and use a modified form of Eq. (2), which includes an additional term to account for the heat interaction that occurs between the heat source and the system. This form of the IQT equation of motion was initially introduced by Beretta [19] and subsequently modified by Smith [35]. The simulations use 100 equally spaced energy eigenlevels to represent the lowest eigenlevels of the trap. The superposition eigenstates that are studied are the $|1\rangle$, $|2\rangle$ and $|3\rangle$ cat states from Turchette et al. [34] where $|1\rangle$ is the state associated with the energy eigenlevel one level above the ground energy level, $|2\rangle$ two levels above and $|3\rangle$ three levels above. In the experiments, the power applied to the heat source $\langle V^2\rangle$ is used to represent the relaxation time.

The initial state and relaxation time constants are varied to fit the IQT simulation results with the experimental probability distribution versus time data of Turchette et al. [34] as well as with the energy versus entropy plots created by Smith [35] using the experimental data of Turchette et al. [34]. The temperatures of the heat reservoirs of the experiment are estimated by noting the tightness of the probability distribution for the data as stable equilibrium is approached. Determining the initial state of the system for the IQT simulations is somewhat problematic since it is not specified in Turchette et al. [34] but is again estimated with reasonable success based on the experimental data.

Comparisons between the IQT results and the experimental data for the lowest 5 energy eigenlevels of the cat state are shown in Figures 13 to 15. The experimental data is indicated by the symbols. The solid lines in part a) of these figures are the theoretical probabilities predicted from the QT master equation used in Turchette et al. [34], while those in part b) are the probabilities predicted by IQT using an equation of motion which includes the heat interaction term as formulated by Smith [35]. The time constants used for the IQT simulation in Figure 13 are τ_D=20.0 and τ_Q=27.0 for the dissipation term and heat interaction term, respectively. In Figures 14 and 15, τ_D is still 20 but τ_Q changes to 20.0 and 25.0, respectively. The scaled reservoir temperature in each figure for a Boltzmann constant set to 1.0 is estimated to be 0.055, 0.11, and 0.15, respectively.

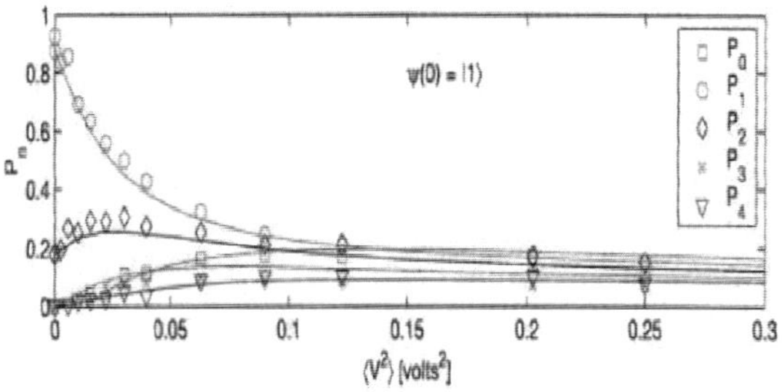

a)

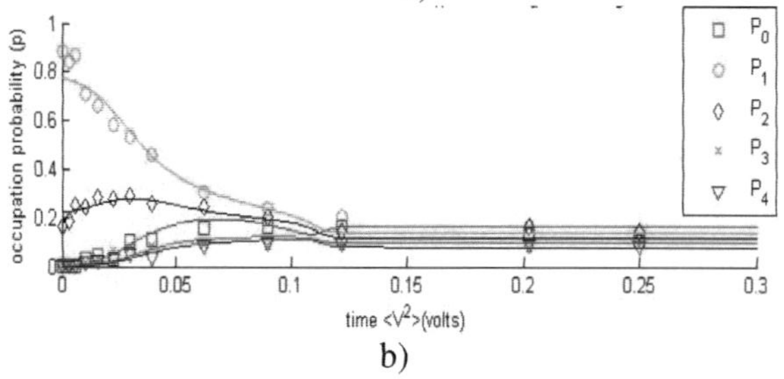

b)

Figure 13. Comparison of the experimentally measured dissipative decay of cat state $|1\rangle$ from Turchette et al. [34] for the lowest 5 energy eigenlevels with a) that predicted by Turchette et al. [34] using a QT master equation and with b) that predicted by the IQT equation of motion that includes a heat interaction term as formulated by Smith [35].

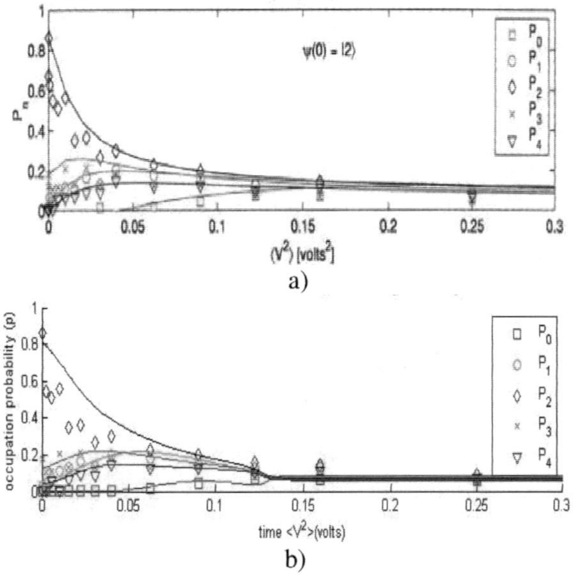

Figure 14. Comparison of the experimentally measured dissipative decay of cat state $|2\rangle$ from Turchette et al. [34] for the lowest 5 energy eigenlevels with a) that predicted by Turchette et al. [34] using a QT master equation and with b) that predicted by the IQT equation of motion that includes a heat interaction term as formulated by Smith [35].

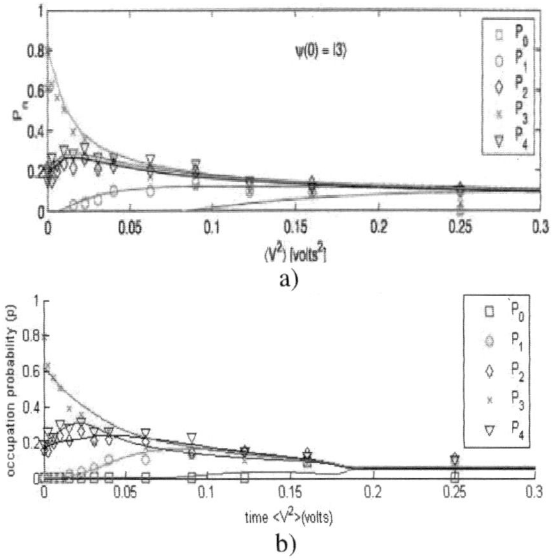

Figure 15. Comparison of the experimentally measured dissipative decay of cat state $|3\rangle$ from Turchette et al. [34] for the lowest 5 energy eigenlevels with a) that predicted by Turchette et al. [34] using a QT master equation and with b) that predicted by the IQT equation of motion that includes a heat interaction term as formulated by Smith [35].

Although both the QT and IQT simulations fit the data well, QT views the "dissipation" predicted by its master equation as a loss of coherence, resulting from the loss of correlations (information) between the system and the heat reservoir, which cyclically build-up and then are lost (i.e., dissipated) as the system relaxes to stable equilibrium. Such a loss is not physical since it is due solely to the exogenous statistics, i.e., to a loss of information about the state of the system. In contrast, IQT views the "dissipation" as the internal destruction of exergy, i.e., available energy [31]. Thus, the "dissipation" is due to the endogenous measurement statistics of QM alone and is, therefore, physical.

2.4 IQT equation of motion with heat interaction

The IQT equation of motion used for the simulations in the previous section is based on Smith [35] and is a modified version of the equation introduced by Beretta [19]. The latter is expressed as

$$\frac{d\rho}{dt} = -\frac{1}{\tau_D}\{\Delta G_D, \rho\} - \frac{1}{\tau_Q}\{\Delta G_Q, \rho\} \tag{11}$$

where the Schrödinger term representing the linear dynamics does not appear since for this case the state operator ρ of the trapped ion and the Hamiltonian H commute. The curly brackets in this equation represent the anti-commutator. The first term t the right hand side of Eq. (11) is the internal dissipation operator, while the second is the heat interaction operator. Both terms can be expressed in a fashion similar to what appears in Eqs. (2), (8), and (9) but have been put into an equivalent, more compact form here as developed in Beretta [19]. As already indicated in the previous section, τ_D and τ_Q are relaxation times that are adjusted to change the relative strengths of the two effects.

The quantities G_D and G_Q are "non-equilibrium Massieu free energy operators" for internal dissipation and a heat interaction with a reservoir, respectively, i.e.,

$$G_D(H,S) = S - \frac{H}{\theta_H} \tag{12}$$

and

$$G_Q(H,S) = S - \frac{H}{\theta_Q} \tag{13}$$

where θ_H and θ_Q are non-equilibrium temperatures. Here, S is the entropy operator written as

$$S = -k_B \ln \rho. \tag{14}$$

Furthermore, the symbol Δ in Eq. (11) signifies the deviation of an operator from its mean value. Thus, for any operator A,

$$\Delta A = A - \langle A \rangle I\,. \tag{15}$$

where I is the identity operator. For further details, the reader is referred to Beretta [19].

To address a drawback to the heat interaction term of Eq. (11), Smith [35] has developed a new term that drives the system to mutual stable equilibrium with the heat reservoir with which the system interacts along a path that not only is unique but is the path of steepest entropy ascent and steepest energy descent/ascent. The term results from a rotational transformation of the internal dissipation operator. This leads to a new non-equilibrium Massieu free energy heat interaction operator given by

$$\tilde{G}_Q(\tilde{H},\tilde{S}) = \tilde{S} - \frac{\tilde{H}}{\tilde{\theta}_H} \tag{16}$$

where $\tilde{\theta}_H$ is a non-equilibrium temperature expressed as

$$\tilde{\theta}_H = \frac{\langle \Delta\tilde{H}\Delta\tilde{H} \rangle}{\langle \Delta\tilde{H}\Delta\tilde{S} \rangle} \tag{17}$$

and the new $\tilde{S}$ and $\tilde{H}$ operators result from a rotation of the original S and H operators, i.e.,

$$\begin{bmatrix} \tilde{S} \\ \tilde{H} \end{bmatrix} = \begin{bmatrix} 1 & 0 \\ 0 & T^* \end{bmatrix} \begin{bmatrix} \cos\varphi & -\sin\varphi \\ \sin\varphi & \cos\varphi \end{bmatrix} \begin{bmatrix} 1 & 0 \\ 0 & 1/T^* \end{bmatrix} \begin{bmatrix} S \\ H \end{bmatrix}. \quad (18)$$

The angle of rotation $\varphi$is a function of the slope of the heat interaction trajectory and is expressed as

$$\varphi = \tan^{-1}\left(T_Q / T^*\right). \quad (19)$$

The quantity T^* is a constant with units of temperature and a value of one, while T_Q is yet another non-equilibrium temperature that corresponds to the slope of the line in the energy versus entropy operator plane, which connects the current state of the system and a state in mutual stable equilibrium with the heat reservoir.

By using the rotation matrix as shown in Figure 16, the path for a dissipation process is transformed from being horizontal to one at an angle of φ. Thus, instead of the *S* operator following a steepest path with the *H* operator held constant, the system now follows a steepest path with respect to the transformed $\tilde{S}$ operator while holding the transformed $\tilde{H}$ operator constant. Thus, the dissipation operator in the transformed coordinate space appears as a heat interaction operator in the original coordinate space.

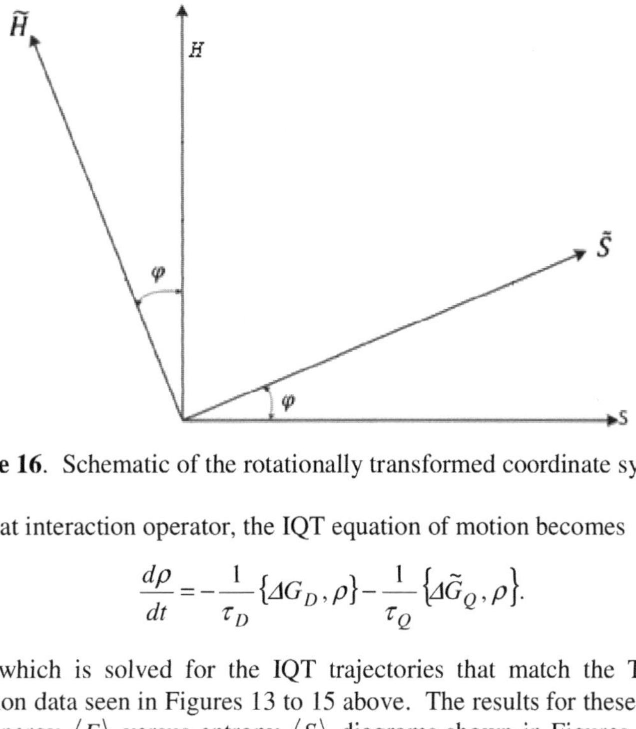

Figure 16. Schematic of the rotationally transformed coordinate system.

With this new heat interaction operator, the IQT equation of motion becomes

$$\frac{d\rho}{dt} = -\frac{1}{\tau_D}\{\Delta G_D, \rho\} - \frac{1}{\tau_Q}\{\Delta\tilde{G}_Q, \rho\}. \quad (20)$$

It is this equation, which is solved for the IQT trajectories that match the Turchette et al. [34] experimental relaxation data seen in Figures 13 to 15 above. The results for these IQT simulations are also plotted on the energy $\langle E \rangle$ versus entropy $\langle S \rangle$ diagrams shown in Figures 17 to 19. For the 5 lowest eigenlevels, the experimental data is given in dark blue, the values predicted by the QT master equation are in light blue, and the curve predicted from IQT is in magenta. Note that the fact that the experimental data as well as the QT and IQT trajectories curve back on themselves is, of course, physically impossible, i.e., violates the second law. However, this occurs here solely due to the fact that these trajectories are only based on the lowest 5 energy eigenlevels. When all 100 eigenlevels are considered, the result is the magenta curves, which show the evolution of the 100 energy eigenlevel model as predicted by IQT from the initial state designated by the cross in magenta to a state of mutual stable equilibrium with the heat reservoir. Clearly, for all three cat state relaxations, the IQT simulations do a good job of matching the experimental data.

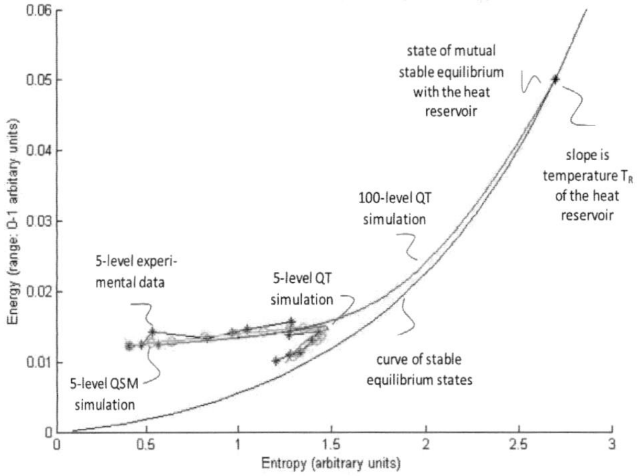

Figure 17. Non-equilibrium evolutions in thermodynamic state of cat state $|1\rangle$ on an energy versus entropy diagram when only considering the lowest 5 energy eigenlevels and in the case of the IQT simulation 100 eigenlevels as well.

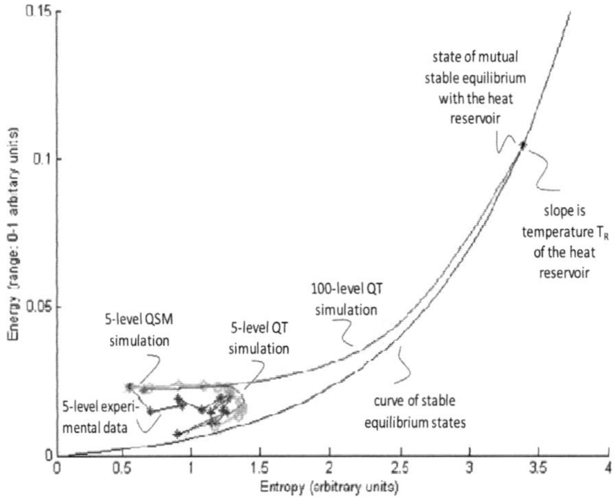

Figure 18. Non-equilibrium evolutions in thermodynamic state of cat state $|2\rangle$ on an energy versus entropy diagram when only considering the lowest 5 energy eigenlevels and in the case of the IQT simulation 100 eigenlevels as well.

3. Conclusions

Simulations using the equation of motion of IQT are seen to correspond closely to data from two experiments that show non-equilibrium relaxations at atomistic scales. Thus, there appears to be support for the idea as proposed by IQT that dissipation due to internal irreversibilities occur for systems even at these small scales, i.e., that there is no "irreversibility paradox". The findings though not conclusive nonetheless warrant the development of additional experiments specifically designed to verify the claims of IQT that the entropy is indeed a fundamental property of matter in the same way that inertial mass, energy, and momentum are.

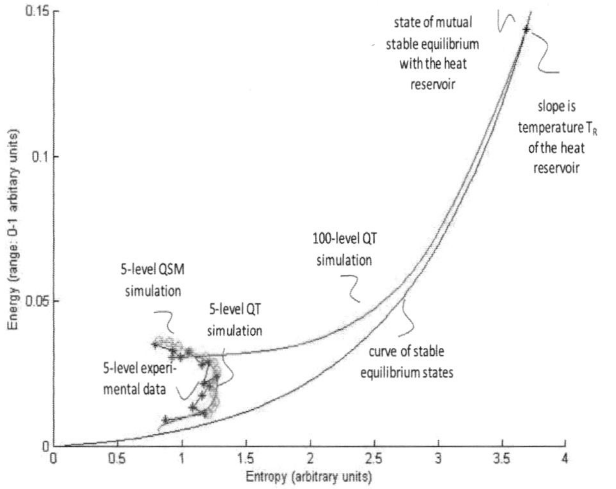

Figure 19. Non-equilibrium evolutions in thermodynamic state of cat state $|3\rangle$ on an energy versus entropy diagram when only considering the lowest 5 energy eigenlevels and in the case of the IQT simulation 100 eigenlevels as well.

Towards this end, a possible variation of the Turchette et al. experiment which could be used to test for the presence of internal dissipation is to employ an interaction time with the reservoir that is much shorter than the measurement time interval. The short interaction would serve as a perturbation after which the system would be allowed to evolve on its own without further influence from the reservoir. The system could then be examined to see whether the evolution is consistent with the expected magnitude of internal dissipation. Of course, care must be taken to differentiate the results from the effects of *1/f* noise from the electrodes as well as ambient vacuum noise, which, though relatively weak, is always present. It is also suggested that the time length of the measurement interval be varied instead of varying the strength of the heat reservoir as was done in Turchette et al. By doing so, there is greater assurance that the data represent a single trajectory and not portions of different trajectories that are associated with reservoirs of varying strengths.

Finally, another experiment involving a trapped ion would be to use the same procedure as Turchette et al. but with relatively weak reservoirs to study whether the signature of a gradual initial relaxation evolution can be detected. An initial shape of the relaxation curve that tends to be concave down would support the notion of a gradual increase in energy spreading among the energy eigenlevels of the system as is seen in IQT simulations.

Acknowledgements

The authors would like to acknowledge the ASME as original publisher of a version of this paper in the proceedings of the 2011 International Mechanical Engineering Congress and Exhibition (IMECE 2011). This paper was submitted for publication to the *Journal of Physics* with permission of the ASME.

Appendix

Be^{+}	beryllium ion
C	constant
D	IQT dissipation operator or matrix
D_{ij}	element of IQT dissipation operator or matrix
$\langle E\rangle$	expectation energy
E_{ii}	energy of an element in the density matrix

E_0, E_1	Rb spin states without the external magnetic field
F	Rb spin states after splitting by external magnetic field
H	Hamiltonian operator
I	identity matrix, operator
I_0	intensity of the laser beam probe
$\hbar$	modified Planck constant
J	Rb angular momentum operator
J_+	raising operator for the angular momentum
k_b	Boltzmann's constant
M	Rb spin operator
Q	energy exchange due to a heat interaction
IQT	Intrinsic Quantum Thermodynamics
QSM	Quantum Statistical Mechanics
Rb	rubidium
$\langle S \rangle$	expectation entropy
t	time
T_R	temperature of the heat reservoir
Tr	trace of a matrix or operator
U	unitary transformation matrix
$U^{\dagger}$	Hermitian conjugate of the unitary transformation matrix
$\langle V^2 \rangle$	average power of the external heat source

Greek

α	generic index for a harmonic superposition state
$\lvert \Psi \rangle_z$	state vector of the Rb spin state as seen along the z-axis
ρ	density matrix or state operator
τ	scalar time constant or functional for the IQT equation of motion
τ_D	time constant or functional for the dissipation term in the IQT equation of motion
τ_Q	time constant for the strength of heat interaction term in the IQT equation of motion
ω_p	precession frequency of a Rb spin state

Miscellaneous

$\lvert\uparrow\rangle, \lvert\downarrow\rangle$	spin up and spin down states for the Be^+ ion
$\lvert n \rangle$	a harmonic cat state between the ground energy level and the nth energy level

References[11]

[1] Hatsopoulos, G N and Gyftopoulos, E P 1976a,b,c,d A unified quantum theory of mechanics and thermo-dynamics – Part I: Postulates, Part IIa: Available energy, Part IIb: Stable equilibrium states, Part III: Irreducible quantal dispersions *Foundations of Physics* **6** 15-31, **6** 127-141, **6** 439-455, **6** 561-570
[2] Park J L 1968 *Am. J. Phys.* **36** 211
[3] Simmons Jr. R F and Park J L 1981 *Found. Phys.* **11** 297
[4] Beretta G P 1981 On the general equation of motion of Quantum Thermodynamics and the distinction between quantal and nonquantal uncertainties *Dr. Sc. thesis, MIT, Cambridge, MA.*
[5] Park J L and Simmons Jr. R F 1983 The knots of thermodynamics *Old and New Questions in Physics, Cosmology, Philosophy, and Theoretical Biology* ed A van der Merwe (New York: Plenum)
[6] Beretta G P, Gyftopoulos E P, Park, J L and Hatsopoulos G N 1984 A new equation of motion for a single constituent of matter *Il Nuovo Cimento B* **82** 169-191

[11] In references [4], [7], [10], [20] and [22] the term "Quantum Thermodynamics" is used to refer to what in the present paper is referred to as "Intrinsic Quantum Thermodynamics."

[7] Beretta G P, Gyftopoulos E P, and Park J L 1985 Quantum Thermodynamics - A new equation of motion for a general quantum system *Nuovo Cimento B* **87** 77

[8] Beretta G P 1985 Effect of irreversible atomic relaxation on resonance fluorescence, absorption and stimulated emission *Int. J. Theor. Phys.* **12** 1233

[9] Beretta G P 1986 *Frontiers of Nonequilibrium Statistical Physics*: *Proc of the NATO Advanced Study Institute, (Santa Fe, 1984, Series B: Physics, Vol. 135)* ed G.T. Moore and M.O. Scully (New York: Plenum) p 205

[10] Beretta G P 1987 Quantum thermodynamics of nonequilbrium. Onsager reciprocity and dispersion-dissipation relations *Foundations of Phys.* **17** 365

[11] Çubukçu E 1993 Thermodynamics as a nonstatistical theory, *Sc.D. dissertation, M.I.T.*

[12] Gyftopoulos E P and Cubukçu E 1997 The entropy of thermodynamics *Physical Review E* **55** 3851-3858

[13] Gheorghiu-Svirchevski S 2001 and 2001 Nonlinear quantum evolution with maximal entropy production *Phys Rev. A* **63** 022105 and Addendum *Phys Rev. A* **63** 054102

[14] Gyftopoulos E P and von Spakovsky M R 2003 Quantum theoretic shapes of constituents of systems in various states *Journal of Energy Resources Technology*, *(ASME transactions, March, vol. 125, no. 1)* pp 1-8

[15] Beretta G P and Gyftopoulos E P 2004 Thermodynamic derivations of conditions for chemical equilibrium and of Onsager reciprocal relations for chemical reactors *J. Chem. Phys.* **121** 2718

[16] Beretta G P 2005 Nonlinear extensions of Schrodinger-von Neumann quantum dynamics: A set of necessary conditions for compatibility with thermodynamics *Mod. Phys. Letters A* **20** 977

[17] Beretta G P 2006 Nonlinear model dynamics for closed-system, constrained, maximal-entropy-generation relaxation by energy redistribution *Phys. Rev. E* **73** 026113

[18] Beretta G P 2008 Modeling non-equilibrium dynamics of a discrete probability distribution: General rate equation for maximal entropy generation in a maximum-entropy landscape with time-dependent constraints *Entropy* **10** 160

[19] Beretta G P 2009 Nonlinear quantum evolution equations to model irreversible adiabatic relaxation with maximal entropy production and other nonunitary processes *Reports on Mathematical Physics* **64** 139

[20] Beretta G P 2010 Maximum entropy production rate in quantum thermodynamics *J. Phys.: Conf. Ser.* **237** 012004

[21] von Spakovsky M R 2008 The second law: A unified approach to thermodynamics applicable to all systems and all states *The Entropy Challenge: Proc. Am. Inst. Phys.*

[22] Smith C E, Sciacovelli A, von Spakovsky M R and Verda V 2010 3D quantum thermodynamic description of the non-equilibrium behavior of an unbounded system at an atomistic level *J. of Phys.: Conf. Series,* **237** 012022

[23] Blum K 1996 *Density Matrix Theory and Applications* (New York: Plenum)

[24] Weiss U 1999 *Quantum Dissipative Systems* (Singapore: World Scientific)

[25] Lindblad G 1976 On the generators of quantum dynamical semigroups *Communications in Mathematical Physics* **48** 119

[26] Kossakowski A 1972 On necessary and sufficient conditions for a generator of a quantum dynamical semi-group *Bulletin of Academy Science and Mathematics* **20** 1021

[27] Kossakowski A 1972 On quantum statistical mechanics of non-Hamiltonian systems *Rep. on Math. Phys.* **3** 247

[28] Zhu W and Rabitz H 2005 Perturbative and nonpertubative master equations for open quantum systems *J. of Math. Phys.* **46** 022105

[29] Loschmidt J 1876 *Sitzungsber. Kais. Akad. Wiss. Wien, Math. Naturwiss. Classe* **73** 128–142

[30] Hatsopoulos G N and Keenan J H 1965 *Principles of General Thermodynamics* (New York: Wiley)

[31] Gyftopoulos E P and Beretta G P 2005 *Thermodynamics – Foundations and Applications* (New York: Dover)

[32] Kukolich S G 1968 Time dependence of quantum-state amplitudes demonstrated by free precession of spin *Am. J. of Phys.* **36** 420

[33] Beretta G P 1985 *International J. of Theoretical Physics* **24** 119

[34] Turchette Q A, Myatt C J, King BE, Sackett CA, Kielpinski D, Itano W M, Monroe C and Wineland D J 2000 Decoherence and decay of motional quantum states of a trapped atom coupled to engineered reservoirs *Phys Rev A* **62** 053807

[35] Smith C E 2012 Intrinsic Quantum Thermodynamics: Application to hydrogen storage on a carbon nanotube and theoretical consideration of non-work interactions *Ph.D. dissertation (advisor: M. R. von Spakovsky, Virginia Tech, Blacksburg)*

[36] Feynman R P, Leighton R B and Sands M 1965 *Feynman Lectures on Physics vol. 3.* (Reading, MA: Addison-Wesley)

[37] Nagel M and Haworth F E 1966 Advanced laboratory experiments on optical pumping of rubidium atoms—part II: Free precession *Am. J. of Phys.* **34** 559

[38] Franzen W 1959 Spin relaxation of optically aligned rubidium vapor *Phys. Rev.***115** 850

[39] Balabas M V, Karaulanov T, Ledbetter M P and Budker D 2010 Polarized alkali vapor with minute-long transverse spin-relaxation time *Phys. Rev. Lett.* **105** 070801

[40] Happer W, Jau, Y-Y and Walker T 2010 *Optically Pumped Atoms* (Weinhem, Germany: Wiley-VCH)

[41] Messiah A 1962 *Quantum Mechanics* (North-Holland Publishing Co: Amsterdam), Vol. 2, XIII, 19.

[42] Messiah A 1999 *Quantum Mechanics* (Dover Publications, Inc.: New York), Vols. 1&2, XIII, 19.

Exceptional orthogonal polynomials and new exactly solvable potentials in quantum mechanics

C Quesne

Physique Nucléaire Théorique et Physique Mathématique, Université Libre de Bruxelles, Campus de la Plaine CP229, Boulevard du Triomphe, B-1050 Brussels, Belgium

E-mail: cquesne@ulb.ac.be

Abstract. In recent years, one of the most interesting developments in quantum mechanics has been the construction of new exactly solvable potentials connected with the appearance of families of exceptional orthogonal polynomials (EOP) in mathematical physics. In contrast with families of (Jacobi, Laguerre and Hermite) classical orthogonal polynomials, which start with a constant, the EOP families begin with some polynomial of degree greater than or equal to one, but still form complete, orthogonal sets with respect to some positive-definite measure. We show how they may appear in the bound-state wavefunctions of some rational extensions of well-known exactly solvable quantum potentials. Such rational extensions are most easily constructed in the framework of supersymmetric quantum mechanics (SUSYQM), where they give rise to a new class of translationally shape invariant potentials. We review the most recent results in this field, which use higher-order SUSYQM. We also comment on some recent re-examinations of the shape invariance condition, which are independent of the EOP construction problem.

1. Introduction

As well known, the classical orthogonal polynomials (COP) of Jacobi, Laguerre and Hermite [1] have a lot of applications in applied mathematics and in physics. Since the early days of quantum mechanics, in particular, they have made their appearance in the bound-state wavefunctions of exactly solvable (ES) potentials.

In this respect, the factorization method, initiated by Schrödinger [2] and developed by Infeld and Hull [3], has played an important role. This technique has been reformulated in the framework of supersymmetric quantum mechanics (SUSYQM) [4], also connected with the Darboux algorithm [5].

The concept of shape invariance (SI), introduced by Gendenshtein [6], has enriched the set of tools available, as its combination with SUSYQM has provided an integration method for the Schrödinger equation. In such a context, the class of translationally SI potentials is especially interesting because it includes all well-known ES potentials found in most textbooks on quantum mechanics. With the addition of some potentials depending upon more than two parameters [7], the classification given in [8, 9] was until recently thought to be complete [4].

In 2008, the introduction of the first families of exceptional orthogonal polynomials (EOP) [10, 11] and the realization of their usefulness in constructing new ES quantum potentials [12], which were so far unknown translationally SI ones, produced a revival in this field.

The purpose of the present contribution is to review the fast-growing branch of EOP and related ES potentials.

2. New families of orthogonal polynomials and quantum mechanics

The COP are often characterized as the polynomial solutions of a Sturm-Liouville problem in connection with the celebrated theorem of Bochner [13]. According to the latter, if an infinite sequence of polynomials $P_n(z)$, $n = 0, 1, 2, \ldots$, satisfies a second-order eigenvalue equation

$$p(z)P_n''(z) + q(z)P_n'(z) + r(z)P_n(z) = \lambda_n P_n(z), \qquad n = 0, 1, 2, \ldots, \tag{1}$$

then $p(z)$, $q(z)$ and $r(z)$ must be polynomials of degree 2, 1 and 0, respectively. In addition, if the sequence is an orthogonal polynomial system, then it has to be (up to an affine transformation) one of the COP of Jacobi, Laguerre or Hermite.

Many generalizations and extensions of these EOP families have been proposed, among which we may quote the Askey scheme of hypergeometric orthogonal polynomials and its q-analogue (see, e.g., [14]), as well as algebraic deformations of SI potentials [15, 16].

More recently, the concept of an exceptional polynomial subspace and the closely related notion of EOP were introduced [10]. As the COP, the latter are eigenfunctions of a second-order differential operator, but in contrast with Bochner's assumption, the first eigenpolynomial of the sequence is of degree $m \geq 1$, though the system of polynomials $P_n(z)$, $n = m, m+1, m+2, \ldots$, denoted by X_m, still forms an orthogonal and complete set with respect to some positive-definite measure. The X_1 case was studied in detail in [10, 11], where it was shown that there exist two such systems of Jacobi or Laguerre type.

As an illustration, let us quote some properties of the latter [11], denoted by $\hat{L}_n^{(\alpha)}(z)$, $n = 1, 2, 3, \ldots$,

$$\hat{L}_1^{(\alpha)}(z) = -z - \alpha - 1, \qquad \hat{L}_2^{(\alpha)}(z) = z^2 - \alpha(\alpha+2), \ldots, \tag{2}$$

$$\left\{z\frac{d^2}{dz^2} - \frac{z-\alpha}{z+\alpha}\left[(z+\alpha+1)\frac{d}{dz} - 1\right]\right\}\hat{L}_n^{(\alpha)}(z) = -(n-1)\hat{L}_n^{(\alpha)}(z), \tag{3}$$

$$\int_0^\infty \hat{L}_{n'}^{(\alpha)}(z)\hat{L}_n^{(\alpha)}(z)\frac{z^\alpha e^{-z}}{(z+\alpha)^2}dz = \delta_{n',n}\frac{\Gamma(n+\alpha+1)}{(n+\alpha-1)(n-1)!}, \tag{4}$$

and contrast them with corresponding properties [1] of conventional Laguerre polynomials $L_n^{(\alpha)}(z)$, $n = 0, 1, 2, \ldots$,

$$L_0^{(\alpha)}(z) = 1, \qquad L_1^{(\alpha)}(z) = -z + \alpha + 1, \ldots, \tag{5}$$

$$\left[z\frac{d^2}{dz^2} + (\alpha + 1 - z)\frac{d}{dz}\right]L_n^{(\alpha)}(z) = -nL_n^{(\alpha)}(z), \tag{6}$$

$$\int_0^\infty L_{n'}^{(\alpha)}(z)L_n^{(\alpha)}(z)z^\alpha e^{-z}dz = \delta_{n',n}\frac{\Gamma(n+\alpha+1)}{n!}. \tag{7}$$

The former can actually be written as linear combinations of three of the latter,

$$\hat{L}_n^{(\alpha)}(z) = nL_n^{(\alpha)}(z) - 2(n+\alpha)L_{n-1}^{(\alpha)}(z) + (n+\alpha)L_{n-2}^{(\alpha)}(z). \tag{8}$$

Both X_1 families of EOP were then shown to have interesting applications in quantum mechanics [12]. For such a purpose, use was made of the point canonical transformation (PCT) method [17, 18], which consists in mapping a Schrödinger equation containing some physically acceptable potential into a second-order differential equation satisfied by some polynomials by making changes of variable and of function.

In the case of equation (3) for Laguerre-type EOP $\hat{L}_n^{(\alpha)}(z)$, the PCT method led to the Schrödinger equation for a rationally-extended radial oscillator (in units $\hbar = 2m = 1$)

$$\left[-\frac{d^2}{dx^2} + V_{l,\mathrm{ext}}(x)\right] \hat{\psi}_\nu^{(l)}(x) = E_\nu^{(l)} \hat{\psi}_\nu^{(l)}(x), \qquad 0 < x < \infty, \tag{9}$$

where

$$V_{l,\mathrm{ext}}(x) = V_l(x) + V_{l,\mathrm{rat}}(x), \qquad V_l(x) = \frac{1}{4}\omega^2 x^2 + \frac{l(l+1)}{x^2}, \tag{10}$$

$$V_{l,\mathrm{rat}}(x) = \frac{4\omega}{\omega x^2 + 2l + 1} - \frac{8\omega(2l+1)}{(\omega x^2 + 2l + 1)^2}. \tag{11}$$

Here $V_l(x)$ is a conventional radial oscillator potential, where ω and l denote the oscillator frequency and the angular momentum quantum number, respectively. The additional term $V_{l,\mathrm{rat}}(x)$ does not modify the behaviour of $V_l(x)$ for large values of x, while for small values it has only an effect for $l = 0$ as it changes $V_0(0) = 0$ into $V_{0,\mathrm{ext}}(0) = -4\omega$. No singularity being introduced on the half-line, the resulting potential $V_{l,\mathrm{ext}}(x)$ is a well-behaved quantum one.

It turns out that $V_{l,\mathrm{ext}}(x)$ has the same bound-state spectrum as $V_l(x)$, namely

$$E_\nu^{(l)} = \omega\left(2\nu + l + \tfrac{3}{2}\right), \qquad \nu = 0, 1, 2, \ldots. \tag{12}$$

Whereas the bound-state wavefunctions of the latter,

$$\psi_\nu^{(l)}(x) \propto x^{l+1} e^{-\frac{1}{4}\omega x^2} L_\nu^{(l+\frac{1}{2})}\left(\tfrac{1}{2}\omega x^2\right) \propto \eta_l(z) L_\nu^{(\alpha)}(z), \qquad \nu = 0, 1, 2, \ldots, \tag{13}$$

$$z = \tfrac{1}{2}\omega x^2, \qquad \alpha = l + \tfrac{1}{2}, \qquad \eta_l(z) = z^{\frac{1}{4}(2\alpha+1)} e^{-\frac{1}{2}z}, \tag{14}$$

can be expressed in terms of Laguerre polynomials, those of the former,

$$\hat{\psi}_\nu^{(l)}(x) \propto \frac{x^{l+1} e^{-\frac{1}{4}\omega x^2}}{\omega x^2 + 2l + 1} \hat{L}_{\nu+1}^{(l+\frac{1}{2})}\left(\tfrac{1}{2}\omega x^2\right) \propto \frac{\eta_l(z)}{z + \alpha} \hat{L}_{\nu+1}^{(\alpha)}(z), \qquad \nu = 0, 1, 2, \ldots, \tag{15}$$

are given in terms of Laguerre-type X_1 EOP.

The same approach used for Jacobi-type X_1 EOP led to a rational extension $V_{A,B,\mathrm{ext}}(x)$ of the well-known Scarf I potential $V_{A,B}(x) = [A(A-1) + B^2]\sec^2 x - B(2A-1)\sec x \tan x$, $-\frac{\pi}{2} < x < \frac{\pi}{2}$, $0 < B < A - 1$,[1] with similar properties [12].

Furthermore, it was proved [12] that the two new ES potentials $V_{l,\mathrm{ext}}(x)$ and $V_{A,B,\mathrm{ext}}(x)$ were translationally SI or, in other words, that their SUSY partner in a first-order SUSYQM approach was a potential of the same type but with different parameters, namely $V_{l+1,\mathrm{ext}}(x)$ and $V_{A+1,B,\mathrm{ext}}(x)$, respectively (see section 5 for more details on SI).

The discovery of this surprising property showed that SUSYQM techniques might prove convenient to build new examples of ES potentials related to X_1 EOP, in the same spirit as previous works on algebraic deformations of SI potentials [15, 16]. This was confirmed by the construction of a rational extension $V_{A,B,\mathrm{ext}}(x)$ of the generalized Pöschl-Teller potential $V_{A,B}(x) = [B^2 + A(A+1)]\,\mathrm{csch}^2 x - B(2A+1)\,\mathrm{csch}\, x \coth x$, $0 < x < \infty$, $B > A + 1 > 1$

[1] Note that some people prefer to use the so-called trigonometric Pöschl-Teller (or Pöschl-Teller I) potential $V_{\bar{A},\bar{B}}(\bar{x}) = \bar{A}(\bar{A}-1)\sec^2 \bar{x} + \bar{B}(\bar{B}-1)\csc^2 \bar{x}$, $0 < \bar{x} < \frac{\pi}{2}$, related to the Scarf I by the changes of parameters and of variable $\bar{A} = A - B$, $\bar{B} = A + B$, $\bar{x} = \frac{1}{2}\left(x + \frac{\pi}{2}\right)$. By the same transformation, $V_{A,B,\mathrm{ext}}(x)$ becomes an extended trigonometric Pöschl-Teller potential $V_{\bar{A},\bar{B},\mathrm{ext}}(\bar{x})$.

[19].[2] In the same work, it was also pointed out that the extension procedure was not successful for any well-known ES potential owing to the possible appearance of a pole in the rational part. This was explicitly shown on the example of the Scarf II potential $V_{A,B}(x) = [B^2 - A(A+1)]\operatorname{sech}^2 x + B(2A+1)\operatorname{sech} x \tanh x$, $-\infty < x < \infty$, $A > 0$. It was then suggested that $\mathcal{PT}$ symmetry might facilitate reconciling the construction of ES rational potentials to the requirement that these be singularity free.

With a similar approach, the first examples of Laguerre- and Jacobi-type X_2 EOP were built in connection with some new rationally-extended radial oscillator and Scarf I potentials [20]. A striking feature was that in each case there appeared two distinct X_2 EOP in contrast with the case of X_1 where only one did exist.

The question of arbitrary large m was successfully addressed by the construction of two distinct families of Laguerre- and Jacobi-type X_m EOP [21, 22], now labelled as type I and type II, and the thorough study of their properties [23, 24]. It then became clear that in the $m = 1$ special case the polynomials of the two families happened to be proportional, hence leaving only a single independent one.

Later on, the X_m EOP were shown to be obtainable through several equivalent approaches to the SUSYQM procedure, such as the Darboux-Crum transformation [25, 26, 27], the Darboux-Bäcklund one [28] and the prepotential method [29].

Very recently, the X_m EOP (and related potentials) were generalized to multi-indexed families $X_{m_1,m_2,\ldots,m_k}$ by making use of multi-step Darboux algebraic transformations [30], the Crum-Adler mechanism [31], higher-order SUSYQM [32, 33] or multi-step Darboux-Bäcklund transformations [34].

3. Rationally-extended radial oscillator and Laguerre-type X_m EOP in first-order SUSYQM

As a example, let us deal with the case of the radial oscillator potential $V_l(x)$, defined in (10), in the framework of first-order SUSYQM.

In such an approach, one considers a pair of SUSY partners [4]

$$\begin{aligned} &H^{(+)} = A^\dagger A = -\frac{d^2}{dx^2} + V^{(+)}(x) - E, \qquad H^{(-)} = AA^\dagger = -\frac{d^2}{dx^2} + V^{(-)}(x) - E, \\ &A^\dagger = -\frac{d}{dx} + W(x), \qquad A = \frac{d}{dx} + W(x), \qquad V^{(\pm)}(x) = W^2(x) \mp W'(x) + E, \end{aligned} \tag{16}$$

which intertwine with the first-order differential operators A and $A^\dagger$ as $AH^{(+)} = H^{(-)}A$ and $A^\dagger H^{(-)} = H^{(+)}A^\dagger$. Here $W(x)$ is the superpotential, which can be expressed as $W(x) = -\left(\log\phi(x)\right)'$ in terms of a factorization function $\phi(x)$, E is the factorization energy and a prime denotes a derivative with respect to x.

The factorization function is a (nodeless) seed solution of the initial Schrödinger equation

$$\left(-\frac{d^2}{dx^2} + V^{(+)}(x)\right)\phi(x) = E\phi(x) \tag{17}$$

with energy E smaller than or equal to the ground-state energy $E_0^{(+)}$ of $V^{(+)}$. One may distinguish three cases. In case i, $E = E_0^{(+)}$, $\phi(x) = \psi_0^{(+)}(x)$ and $V^{(-)}$ has the same spectrum as $V^{(+)}$ except for $E_0^{(+)}$, which is removed. In case ii, $E < E_0^{(+)}$, $\phi(x)$ and $\phi^{-1}(x)$ are both

[2] Note that the hyperbolic Pöschl-Teller (or Pöschl-Teller II) potential $V_{\bar{A},\bar{B}}(\bar{x}) = -\bar{A}(\bar{A}+1)\operatorname{sech}^2\bar{x} + \bar{B}(\bar{B}-1)\operatorname{csch}^2\bar{x}$, $0 < \bar{x} < \infty$, is related to the generalized Pöschl-Teller one by the changes of parameters and of variable $\bar{A} = A + B$, $\bar{B} = B - A$, $\bar{x} = \frac{1}{2}x$. As a consequence, $V_{A,B,\text{ext}}(x)$ is connected with an extended hyperbolic Pöschl-Teller potential $V_{\bar{A},\bar{B},\text{ext}}(\bar{x})$, which can be derived by the same transformation.

nonnormalizable and $V^{(-)}$ has exactly the same spectrum as $V^{(+)}$ (isospectral case). Finally, in case iii, $E < E_0^{(+)}$ and $\phi^{-1}(x)$ is normalizable (while $\phi(x)$ is of course nonnormalizable), so that $V^{(-)}$ has an extra bound state at an energy E below the spectrum of $V^{(+)}$.

Furthermore, from the intertwining relations, it results that the wavefunctions of $V^{(-)}$ can be obtained from those of $V^{(+)}$ with the same energy by applying the operator A.

We shall be interested here in case ii (isospectral case). It turns out that for the radial oscillator potential $V_l(x)$, there are two types of (nodeless) seed functions, whose energy E is less than $E_0^{(+)} = \omega(l + \frac{3}{2})$ and whose inverse is nonnormalizable, namely

$$\phi^{\rm I}_{lm}(x) = \chi^{\rm I}_l(z) L^{(\alpha)}_m(-z) \propto x^{l+1} e^{\frac{1}{4}\omega x^2} L_m^{(l+\frac{1}{2})}(-\tfrac{1}{2}\omega x^2), \tag{18}$$

$$\phi^{\rm II}_{lm}(x) = \chi^{\rm II}_l(z) L^{(-\alpha)}_m(z) \propto x^{-l} e^{-\frac{1}{4}\omega x^2} L_m^{(-l-\frac{1}{2})}(\tfrac{1}{2}\omega x^2), \tag{19}$$

with

$$\chi^{\rm I}_l(z) = z^{\frac{1}{4}(2\alpha+1)} e^{\frac{1}{2}z}, \qquad \chi^{\rm II}_l(z) = z^{-\frac{1}{4}(2\alpha-1)} e^{-\frac{1}{2}z}, \tag{20}$$

and z, α defined in equation (14). The corresponding energies are $E^{\rm I}_{lm} = -\omega(\alpha + 2m + 1)$, $m = 1, 2, 3, \ldots$, and $E^{\rm II}_{lm} = -\omega(\alpha - 2m - 1)$, $m = 1, 2, \ldots, l$ $(< \alpha)$, respectively.

To get as partner potential a rationally-extended potential with a given l, $V^{(-)}(x) = V_{l,{\rm ext}}(x) + C$, where C is some additive constant, we have to start from a conventional radial oscillator potential $V_{l'}(x)$ with a different l'. The rational part of $V_{l,{\rm ext}}(x)$ (see equation (10)) can then be expressed in terms of some mth-degree polynomial in z, $g^{(\alpha)}_m(z)$, as

$$V_{l,{\rm rat}}(x) = -2\omega \left\{ \frac{\dot{g}^{(\alpha)}_m}{g^{(\alpha)}_m} + 2z \left[\frac{\ddot{g}^{(\alpha)}_m}{g^{(\alpha)}_m} - \left(\frac{\dot{g}^{(\alpha)}_m}{g^{(\alpha)}_m} \right)^2 \right] \right\}, \tag{21}$$

where a dot denotes a derivative with respect to z. According to the choice made for $\phi(x)$, we may distinguish the two cases

$$(i) \quad l' = l - 1, \quad \phi = \phi^{\rm I}_{l-1,m}, \quad g^{(\alpha)}_m(z) = L^{(\alpha-1)}_m(-z), \quad C = -\omega, \quad m = 1, 2, 3, \ldots; \tag{22}$$

$$(ii) \quad l' = l + 1, \quad \phi = \phi^{\rm II}_{l+1,m}, \quad g^{(\alpha)}_m(z) = L^{(-\alpha-1)}_m(z), \quad C = \omega, \quad m = 1, 2, \ldots, l+1 (< \alpha + 1). \tag{23}$$

From the properties of Laguerre polynomials [1], it follows that $V_{l,{\rm ext}}(x)$ has no pole on $(0, \infty)$.

For $m = 1$, for instance, we get $g^{(\alpha)}_1(z) = z + \alpha$ for type I and $g^{(\alpha)}_1(z) = -z - \alpha$ for type II. Hence there is a single rationally-extended potential, as obtained in [12, 20] (see equations (10) and (11)). In contrast, for $m = 2$, $g^{(\alpha)}_2(z) = \frac{1}{2}[z^2 + 2(\alpha + 1)z + \alpha(\alpha + 1)]$ or $g^{(\alpha)}_2(z) = \frac{1}{2}[z^2 + 2(\alpha - 1)z + \alpha(\alpha - 1)]$ according to the choice made. Hence there are two distinct extended potentials, as first observed in [20] and confirmed for all $m \geq 2$ in [21, 22].

From the wavefunctions $\psi^{(+)}_\nu(x) \propto \eta_{l'}(z) L^{(\alpha')}_\nu(z)$, $\alpha' = l' + \frac{1}{2}$, $\nu = 0, 1, 2, \ldots$, of $V^{(+)}$, those of $V^{(-)}$ with the same energy are obtained as

$$\psi^{(-)}_\nu(x) \propto A\psi^{(+)}_\nu(x) \propto \frac{\eta_l(z)}{g^{(\alpha)}_m(z)} y^{(\alpha)}_n(z), \qquad n = m + \nu, \qquad \nu = 0, 1, 2, \ldots, \tag{24}$$

where $y^{(\alpha)}_n(z)$ is some nth-degree polynomial in z. The latter results from the application of a first-order differential operator $\mathcal{O}^{(\alpha)}_m$ on the conventional Laguerre polynomial $L^{(\alpha')}_\nu(z)$,

$$y^{(\alpha)}_n(z) \propto \mathcal{O}^{(\alpha)}_m L^{(\alpha')}_\nu(z). \tag{25}$$

Here $\mathcal{O}_m^{(\alpha)} = g_m^{(\alpha)}\left(\frac{d}{dz} - 1\right) - \dot{g}_m^{(\alpha)}$ for type I and $\mathcal{O}_m^{(\alpha)} = g_m^{(\alpha)}\left(z\frac{d}{dz} + \alpha + 1\right) - z\dot{g}_m^{(\alpha)}$ for type II.

On the other hand, by directly inserting equation (24) in the Schrödinger equation for $V^{(-)}(x)$, one arrives at the following second-order differential equation satisfied by $y_n^{(\alpha)}(z)$,

$$\left[z\frac{d^2}{dz^2} + \left(\alpha + 1 - z - 2z\frac{\dot{g}_m^{(\alpha)}}{g_m^{(\alpha)}}\right)\frac{d}{dz} + (z-\alpha)\frac{\dot{g}_m^{(\alpha)}}{g_m^{(\alpha)}} + z\frac{\ddot{g}_m^{(\alpha)}}{g_m^{(\alpha)}}\right] y_n^{(\alpha)}(z) = (m-n)y_n^{(\alpha)}(z). \tag{26}$$

The orthonormality and completeness of $\psi_\nu^{(-)}(x)$, $\nu = 0, 1, 2, \ldots$, on the half-line imply that the polynomials $y_n^{(\alpha)}(z)$, $n = m + \nu$, $\nu = 0, 1, 2, \ldots$, form an orthogonal and complete set with respect to the positive-definite measure $z^\alpha e^{-z}\big(g_m^{(\alpha)}\big)^{-2}dz$. According to the choice made for ϕ, such polynomials belong to the $L1$ or $L2$ family of X_m EOP and are denoted by $L^{\mathrm{I}}_{\alpha,m,n}(z)$ or $L^{\mathrm{II}}_{\alpha,m,n}(z)$. In [21, 22, 24], they were normalized in such a way that their highest-degree term is given by $(-z)^n/[(n-m)!m!]$ multiplied by $(-1)^m$ or 1, respectively. In conformity with the choices made in [10, 11, 33], we will drop the additional factor $(-1)^m$ in case I. Then $L^{\mathrm{I}}_{\alpha,1,n}(z)$ and $L^{\mathrm{II}}_{\alpha,1,n}(z)$ both exactly reduce to $\hat{L}_n^{(\alpha)}(z)$, considered in equations (2), (3) and (4). With this convention, it is easy to see that

$$L^{\mathrm{I}}_{\alpha,m,n}(z) = (-1)^{m-1}\left[g_m^{(\alpha)}\left(\frac{d}{dz} - 1\right) - \dot{g}_m^{(\alpha)}\right] L_\nu^{(\alpha-1)}(z), \tag{27}$$

$$L^{\mathrm{II}}_{\alpha,m,n}(z) = (\alpha + \nu + 1 - m)^{-1}\left[g_m^{(\alpha)}\left(z\frac{d}{dz} + \alpha + 1\right) - z\dot{g}_m^{(\alpha)}\right] L_\nu^{(\alpha+1)}(z), \tag{28}$$

by comparing the highest-degree terms on both sides of the equations.

In the present approach, to get $V_{l,\mathrm{ext}}(x)$ we have had to reparametrize the conventional potential by considering $V_{l'}(x)$ with $l' \neq l$. It is worth mentioning that an isospectral transformation from $V_l(x)$ to $V_{l,\mathrm{ext}}(x)$ with the same parameter can be obtained in the framework of second-order SUSYQM [19, 20].

4. Rationally-extended radial oscillator and Laguerre-type X_{m_1,m_2} EOP in second-order SUSYQM

To illustrate the construction of multi-indexed families of EOP and corresponding rationally-extended potentials, let us consider the case of the radial oscillator potential in second-order SUSYQM (SSUSY) [32].

In such a setting, one starts from a pair of Hamiltonians [35, 36, 37, 38, 39, 40]

$$H^{(1)} = -\frac{d^2}{dx^2} + V^{(1)}(x), \qquad H^{(2)} = -\frac{d^2}{dx^2} + V^{(2)}(x), \tag{29}$$

intertwining with two second-order differential operators

$$\mathcal{A}^\dagger = \frac{d^2}{dx^2} - 2p(x)\frac{d}{dx} + q(x), \qquad \mathcal{A} = \frac{d^2}{dx^2} + 2p(x)\frac{d}{dx} + 2p'(x) + q(x), \tag{30}$$

as $\mathcal{A}H^{(1)} = H^{(2)}\mathcal{A}$ and $\mathcal{A}^\dagger H^{(2)} = H^{(1)}\mathcal{A}^\dagger$. These intertwining relations imply that the functions $p(x)$, $q(x)$, $V^{(1)}(x)$ and $V^{(2)}(x)$ are constrained by the relations

$$\begin{aligned} q(x) &= -p' + p^2 - \frac{p''}{2p} + \left(\frac{p'}{2p}\right)^2 - \frac{c^2}{16p^2}, \\ V^{(1,2)}(x) &= \mp 2p' + p^2 + \frac{p''}{2p} - \left(\frac{p'}{2p}\right)^2 + \frac{c^2}{16p^2}, \end{aligned} \tag{31}$$

where c is some integration constant. Hence the knowledge of $p(x)$ and c determines everything else. SSUSY offers more possibilities of manipulating the initial spectrum than standard SUSYQM, such as deleting or creating one or two levels, moving one level or leaving the spectrum unchanged. Here we shall only be concerned with the last eventuality (isospectral transformation).

In the reducible case to be considered here, $\mathcal{A}^\dagger$ and $\mathcal{A}$ can be factorized into a product of two first-order differential operators, $\mathcal{A}^\dagger = A^\dagger \tilde{A}^\dagger$ and $\mathcal{A} = \tilde{A} A$, where $A^\dagger$ and A are expressed as in (16), while

$$\tilde{A}^\dagger = -\frac{d}{dx} + \tilde{W}(x), \qquad \tilde{A} = \frac{d}{dx} + \tilde{W}(x). \tag{32}$$

Then $A^\dagger$ and A may be used as ladder operators in first-order SUSYQM with $H^{(+)} = A^\dagger A$ and $H^{(-)} = AA^\dagger$ as in (16), while $W(x) = -\big(\log \phi(x)\big)'$ and $H^{(+)}\phi(x) = 0$. Similarly, from $\tilde{A}^\dagger$ and $\tilde{A}$ one may build $\tilde{H}^{(+)} = \tilde{A}^\dagger \tilde{A}$ and $\tilde{H}^{(-)} = \tilde{A}\tilde{A}^\dagger$, corresponding to some potentials $\tilde{V}^{(\pm)}(x)$, factorization function $\tilde{\phi}(x)$ and factorization energy $\tilde{E}$, such that $\tilde{V}^{(\pm)}(x) = \tilde{W}^2(x) \mp \tilde{W}'(x) + \tilde{E}$, $\tilde{W}(x) = -\big(\log \tilde{\phi}(x)\big)'$ and $\tilde{H}^{(+)}\tilde{\phi}(x) = 0$. Whenever $\tilde{V}^{(+)}(x) = V^{(-)}(x)$, so that both pairs $\big(H^{(+)}, H^{(-)}\big)$ and $\big(\tilde{H}^{(+)}, \tilde{H}^{(-)}\big)$ can be glued together, we get a reducible SSUSY system or, equivalently, a second-order parasupersymmetric (PSUSY) one [41, 42].

In the isospectral case, we have to assume $E < E_0^{(+)}$, $\tilde{E} < \tilde{E}_0^{(+)} = E_0^{(+)}$ and to demand that both $\phi^{-1}(x)$ and $\tilde{\phi}^{-1}(x)$ be nonnormalizable (together with $\phi(x)$ and $\tilde{\phi}(x)$).

The connection between the SSUSY and PSUSY approaches is given by the relations

$$\begin{array}{lll} H^{(1)} = H^{(+)} + \frac{c}{2}, & H^{(2)} = \tilde{H}^{(-)} - \frac{c}{2}, & H = H^{(-)} + \frac{c}{2} = \tilde{H}^{(+)} - \frac{c}{2}, \\ p(x) = \frac{1}{2}\big(W + \tilde{W}\big), & c = E - \tilde{E}, & \end{array} \tag{33}$$

showing that the intermediate Hamiltonian H is both partner to $H^{(1)}$ and $H^{(2)}$.

Instead of $\phi(x)$ and $\tilde{\phi}(x)$, we may start from two eigenfunctions $\phi_1(x)$ and $\phi_2(x)$ of the same $V^{(+)}(x)$ with respective energies E_1 and E_2 (less than $E_0^{(+)}$) and such that ϕ_1^{-1} and ϕ_2^{-1} are nonnormalizable. Then, $\phi = \phi_1$, $\tilde{\phi} = A\phi_2 = \mathcal{W}(\phi_1, \phi_2)/\phi_1$, $E = E_1$ and $\tilde{E} = E_2$, where $\mathcal{W}(\phi_1, \phi_2)$ denotes the Wronskian of $\phi_1(x)$ and $\phi_2(x)$. As a consequence,

$$p(x) = -\frac{\mathcal{W}'(\phi_1, \phi_2)}{2\mathcal{W}(\phi_1, \phi_2)} = -\frac{(E_1 - E_2)\phi_1\phi_2}{2\mathcal{W}(\phi_1, \phi_2)}, \tag{34}$$

which entirely determines the SSUSY partner potential through the equation

$$V^{(2)}(x) = V^{(1)}(x) + 4p'(x). \tag{35}$$

In the case of the radial oscillator potential, we know that there are two types of seed functions (18) and (19), which may be used. As the order of ϕ_1 and ϕ_2 is irrelevant as far as the final potential $V^{(2)}(x)$ is concerned, there are three types of possibilities for the pair (ϕ_1, ϕ_2). In all three cases, we have to start from a potential $V^{(+)}(x) = V_{l'}(x)$ with some different l'. The Wronskian $\mathcal{W}(\phi_1(x), \phi_2(x))$ can be written in terms of some μth-degree polynomial in z, $g_\mu^{(\alpha)}(z)$, itself expressible in terms of a Wronskian $\tilde{\mathcal{W}}(f(z), g(z))$ of two appropriate functions of z. The results read

$$\begin{aligned}
(i)\quad & V^{(+)} = V_{l-2}, \quad \phi_1 = \phi^{\rm I}_{l-2,m_1}, \quad \phi_2 = \phi^{\rm I}_{l-2,m_2}, \quad 0 < m_1 < m_2, \\
& \mathcal{W}(\phi_1, \phi_2) = \omega x (\chi^{\rm I}_{l-2})^2 g^{(\alpha)}_\mu(z), \\
& g^{(\alpha)}_\mu(z) = \tilde{\mathcal{W}}\big(L^{(\alpha-2)}_{m_1}(-z), L^{(\alpha-2)}_{m_2}(-z)\big), \quad \mu = m_1 + m_2 - 1; \qquad (36) \\
(ii)\quad & V^{(+)} = V_{l+2}, \quad \phi_1 = \phi^{\rm II}_{l+2,m_1}, \quad \phi_2 = \phi^{\rm II}_{l+2,m_2}, \quad 0 < m_1 < m_2 < \alpha + 2, \\
& \mathcal{W}(\phi_1, \phi_2) = \omega x (\chi^{\rm II}_{l+2})^2 g^{(\alpha)}_\mu(z), \\
& g^{(\alpha)}_\mu(z) = \tilde{\mathcal{W}}\big(L^{(-\alpha-2)}_{m_1}(z), L^{(-\alpha-2)}_{m_2}(z)\big), \quad \mu = m_1 + m_2 - 1; \qquad (37) \\
(iii)\quad & V^{(+)} = V_l, \quad \phi_1 = \phi^{\rm I}_{l,m_1}, \quad \phi_2 = \phi^{\rm II}_{l,m_2}, \quad 0 < m_1, \quad 0 < m_2 < \alpha, \\
& \mathcal{W}(\phi_1, \phi_2) = \frac{2}{x} \chi^{\rm I}_l \chi^{\rm II}_l g^{(\alpha)}_\mu(z), \\
& g^{(\alpha)}_\mu(z) = z\tilde{\mathcal{W}}\big(L^{(\alpha)}_{m_1}(-z), L^{(-\alpha)}_{m_2}(z)\big) - (z+\alpha) L^{(\alpha)}_{m_1}(-z) L^{(-\alpha)}_{m_2}(z), \\
& \mu = m_1 + m_2 + 1. \qquad (38)
\end{aligned}$$

In all three cases, provided $g^{(\alpha)}_\mu(z)$ does not vanish on the half-line, the SSUSY partner potentials can be written as

$$\begin{aligned}
V^{(1)} &= V_{l'} - \frac{1}{2}(E_1 + E_2), \\
V^{(2)} &= V_l - 2\omega \left\{ \frac{\dot{g}^{(\alpha)}_\mu}{g^{(\alpha)}_\mu} + 2z \left[\frac{\ddot{g}^{(\alpha)}_\mu}{g^{(\alpha)}_\mu} - \left(\frac{\dot{g}^{(\alpha)}_\mu}{g^{(\alpha)}_\mu} \right)^2 \right] \right\} - \frac{1}{2}(E_1 + E_2) + C,
\end{aligned} \qquad (39)$$

where $C = -2\omega$, 2ω or 0 in case (i), (ii) or (iii), respectively. Hence, up to some additive constant, they assume essentially the same form as $V^{(+)}$ and $V^{(-)}$ in first-order SUSYQM (see equation (21)) with the only exception that $g^{(\alpha)}_m(z)$ is now replaced by $g^{(\alpha)}_\mu(z)$.

In the PSUSY setting, the potential in the intermediate H is some $V_{l-1,\rm ext}$, $V_{l+1,\rm ext}$ or $V_{l+1,\rm ext}$, as obtained in first-order SUSYQM. If we change the order of ϕ_1 and ϕ_2, then the intermediate potential is different although the final one remains the same. In particular, in case (iii), it becomes a $V_{l-1,\rm ext}$ potential.

It is worth noting that extending the range of m_1, m_2 values to include $m_1 = 0$ in case (i) or (ii) and $m_1 = 0$ or $m_2 = 0$ in case (iii) would not lead to any new result, because the corresponding $g^{(\alpha)}_\mu(z)$ would then reduce to some $g^{(\alpha)}_m(z)$ already found in first order. Even for the m_1 and m_2 values listed in (36), (37) and (38), there are some possibilities of coincidence. As a matter of fact, the lowest-degree new $g^{(\alpha)}_\mu(z)$ polynomial corresponds to $\mu = 3$ and is given by $g^{(\alpha)}_3(z) = [z^3 + 3\alpha z^2 + 3(\alpha-1)(\alpha+1)z + (\alpha-1)\alpha(\alpha+1)]/3$. It may be obtained in case (i) or (ii) for $m_1 = 1$, $m_2 = 2$, or else in case (iii) for $m_1 = m_2 = 1$. Up to second order, there are therefore three distinct cubic-type extended potentials, the first two being those found in first order and associated with either $L^{(\alpha-1)}_3(-z)$ or $L^{(-\alpha-1)}_3(z)$, the third one being that corresponding to the above-mentioned $g^{(\alpha)}_3(z)$. Since this conclusion remains true when considering kth-order SUSYQM with $k > 2$, it has been conjectured that there are exactly μ distinct extended radial oscillator potentials (and corresponding EOP families) of μth type [32]. This assertion has actually been proved for $\mu = 1$, 2 and 3.

Turning now ourselves to the wavefunctions $\psi^{(2)}_\nu(x)$ of $H^{(2)}$, we note that we may obtain them from those of $H^{(1)}$, $\psi^{(1)}_\nu(x) \propto \eta_{l'}(z) L^{(\alpha')}_\nu(z)$, $\nu = 0, 1, 2, \ldots$, by applying the second-order differential operator $\mathcal{A}$. They can be written as

$$\psi_\nu^{(2)}(x) \propto \frac{\eta_l(z)}{g_\mu^{(\alpha)}(z)} y_n^{(\alpha)}(z), \qquad n = \mu + \nu, \qquad \nu = 0, 1, 2, \ldots, \tag{40}$$

where $y_n^{(\alpha)}(z)$ is some nth-degree polynomial in z, satisfying a second-order differential equation similar to equation (26), but with μ and $g_\mu^{(\alpha)}(z)$ substituted for m and $g_m^{(\alpha)}(z)$, respectively. As the polynomials considered in section 3, those obtained here form orthogonal and complete sets with respect to a positive-definite measure of the type $z^\alpha e^{-z}\left(g_\mu^{(\alpha)}\right)^{-2} dz$. We have therefore got (in general) X_{m_1,m_2} EOP belonging to three families, which may be denoted by $L^{\mathrm{I,I}}_{\alpha,m_1,m_2,n}(z)$, $L^{\mathrm{II,II}}_{\alpha,m_1,m_2,n}(z)$ and $L^{\mathrm{I,II}}_{\alpha,m_1,m_2,n}(z)$, respectively. We choose to normalize them in such a way that their highest-degree term is equal to $(-z)^n/[(n-\mu)!m_1!m_2!]$.

From the factorized form $\mathcal{A} = \tilde{A}A$, it can be shown that the X_{m_1,m_2} EOP can be obtained from the X_{m_1} EOP by applying some first-order differential operator, thereby generalizing equations (27) and (28) [33]. The results read

$$L^{\mathrm{I,I}}_{\alpha,m_1,m_2,n}(z) = (-1)^{m_2}(m_2 - m_1)^{-1} \mathcal{O}^{(\alpha)}_{m_1,\mu} L^{\mathrm{I}}_{\alpha,m_1,n_1}(z), \tag{41}$$

$$L^{\mathrm{II,II}}_{\alpha,m_1,m_2,n}(z) = -[(\alpha + \nu + 2 - m_2)(m_2 - m_1)]^{-1} \mathcal{O}^{(\alpha)}_{m_1,\mu} L^{\mathrm{II}}_{\alpha,m_1,n_1}(z), \tag{42}$$

$$L^{\mathrm{I,II}}_{\alpha,m_1,m_2,n}(z) = (\alpha + \nu - m_2)^{-1} \mathcal{O}^{(\alpha)}_{m_1,\mu} L^{\mathrm{I}}_{\alpha,m_1,n_1}(z), \tag{43}$$

where $n = \mu + \nu$, $n_1 = m_1 + \nu$, $\nu = 0, 1, 2, \ldots$, and

$$\mathcal{O}^{(\alpha)}_{m_1,\mu} = \frac{1}{g^{(\alpha-1)}_{m_1}} \left[g^{(\alpha)}_\mu \left(\frac{d}{dz} - 1 \right) - \dot{g}^{(\alpha)}_\mu \right] \tag{44}$$

in (41), while

$$\mathcal{O}^{(\alpha)}_{m_1,\mu} = \frac{1}{g^{(\alpha+1)}_{m_1}} \left[g^{(\alpha)}_\mu \left(z\frac{d}{dz} + \alpha + 1 \right) - z\dot{g}^{(\alpha)}_\mu \right] \tag{45}$$

in (42) or (43).

The approach used in this section has been extended to kth-order SUSYQM [33], in which case one may distinguish $k+1$ families of $X_{m_1,m_2,\ldots,m_k}$ EOP, denoted by $L^{(q,k-q)}_{\alpha,m_1,m_2,\ldots,m_k,n}(z)$ and associated with $k+1$ possible choices $\mathrm{I}^q\mathrm{II}^{k-q}$, $0 \le q \le k$, for the set of k seed functions of type I or II. Here it is assumed that $0 < m_1 < m_2 < \cdots < m_q$ correspond to type I, whereas $0 < m_{q+1} < m_{q+2} < \cdots < m_k$ belong to type II. The EOP degree n has been shown to be given by $n = \mu + \nu$, $\nu = 0, 1, 2, \ldots$, with

$$\mu = \sum_{i=1}^{k} m_i - \frac{1}{2}q(q-1) - \frac{1}{2}(k-q)(k-q-1) + q(k-q). \tag{46}$$

5. Shape invariance of rationally-extended potentials

In first-order SUSYQM, if one knows the spectrum and wavefunctions of $V^{(+)}(x)$ (resp. $V^{(-)}(x)$), then one is able to obtain those of its partner $V^{(-)}(x)$ (resp. $V^{(+)}(x)$) by application of A (resp. $A^\dagger$). For some special class of potentials, however, no previous knowledge is needed to be able to derive the spectrum and wavefunctions of both partners. This is the case of so-called SI potentials [6], i.e., those potentials for which the two partners are similar in shape and differ only in the parameters. More precisely, $V^{(\pm)}(x; a_1)$ are SI if

$$V^{(-)}(x; a_1) = V^{(+)}(x; a_2) + R(a_1), \tag{47}$$

where a_1 is a set of parameters, a_2 is a function of a_1 (i.e., $a_2 = f(a_1)$) and the remainder $R(a_1)$ is independent of x [4]. When combined with unbroken SUSYQM (case i or iii below equation (17)), equation (47) indeed provides an integrability condition so that the two potentials are solvable by purely algebraic methods. Such a condition is only a sufficient one, since many ES potentials that are not SI are known.

In case i, for instance, equation (47) can be rewritten as

$$W^2(x;a_1) + W'(x;a_1) = W^2(x:a_2) - W'(x;a_2) + R(a_1), \tag{48}$$

where $W(x;a_1) = -\left(\log \psi_0^{(+)}(x;a_1)\right)'$ in terms of the ground-state wavefunction $\psi_0^{(+)}(x;a_1)$ of $V^{(+)}(x;a_1)$. Since equation (48) is a difference-differential equation relating the square of the superpotential W and its spatial derivative computed at two sets a_1, a_2 of parameter values, the classification of its solutions is not an easy task.

The simplest case corresponds to translationally SI potentials for which the parameters a_1 and a_2 are related to each other by translation ($a_2 = a_1 + \lambda$) [7, 8, 9]. Other cases have also been studied, such as SI potentials under scaling [43, 44] and cyclic SI potentials [45]. In the last cases, the potentials are only known formally as Taylor series, whereas in the first one, they are explicitly known in a closed form in terms of simple functions.

For the conventional radial oscillator $V_l(x)$, whose ground-state wavefunction is given by $\psi_0^{(l)}(x) \propto \eta_l(z)$ with $\eta_l(z)$ defined in (14), the role of a_1 is played by l and the superpotential $W(x;l) = W_l(x)$ can be written as

$$W_l(x) = \frac{1}{2}\omega x - \frac{l+1}{x} = -\omega x \frac{\dot{\eta}_l}{\eta_l}. \tag{49}$$

As expected, $V_l(x) - E_0^{(l)} = W_l^2(x) - W_l'(x)$, where $E_0^{(l)} = \omega\left(l + \frac{3}{2}\right) = \omega(\alpha+1)$. In such a case, equation (48) translates into

$$W_l^2(x) + W_l'(x) = W_{l+1}^2(x) - W_{l+1}'(x) + 2\omega = \left[V_{l+1}(x) - E_0^{(l+1)}\right] + 2\omega, \tag{50}$$

which illustrates the well-known SI of $V_l(x)$, its partner being $V_{l+1}(x)$ (hence $a_2 = l+1$) and the remainder 2ω being independent of x [4].

It turns out that the corresponding rationally-extended potentials $V_{l,\text{ext}}(x) = V_l(x) + V_{l,\text{rat}}(x)$, where $V_{l,\text{rat}}(x)$ is expressed in terms of some (nonvanishing) μth-degree polynomial $g_\mu^{(\alpha)}(z)$ as in equation (39), have very similar properties [12, 20, 21, 22, 23, 24, 25, 28, 31, 33, 34]. This is actually a consequence of the fact that their ground-state wavefunction can be written as $\bar{\psi}_0^{(l)}(x) \propto \eta_l(z) g_\mu^{(\alpha+1)}(z)/g_\mu^{(\alpha)}(z)$ or, in other words, that the lowest-degree EOP associated with $g_\mu^{(\alpha)}(z)$ is $y_\mu^{(\alpha)}(z) \propto g_\mu^{(\alpha+1)}(z)$. From $\bar{\psi}_0^{(l)}(x)$, we indeed get the superpotential

$$\begin{aligned} &\bar{W}_l(x) = \bar{W}_{l,1}(x) + \bar{W}_{l,2}(x), \\ &\bar{W}_{l,1}(x) = W_l(x), \qquad \bar{W}_{l,2}(x) = -\omega x \left(\frac{\dot{g}_\mu^{(\alpha+1)}}{g_\mu^{(\alpha+1)}} - \frac{\dot{g}_\mu^{(\alpha)}}{g_\mu^{(\alpha)}}\right), \end{aligned} \tag{51}$$

which satisfies the equation $V_{l,\text{ext}}(x) - E_0^{(l)} = \bar{W}_l^2(x) - \bar{W}_l'(x)$, as it should be. Then, with $a_1 = l$, $a_2 = l+1$ and $R(a_1) = 2\omega$, equation (48) becomes

$$\bar{W}_l^2(x) + \bar{W}_l'(x) = \bar{W}_{l+1}^2(x) - \bar{W}_{l+1}'(x) + 2\omega = \left[V_{l+1,\text{ext}}(x) - E_0^{(l+1)}\right] + 2\omega, \tag{52}$$

which is a counterpart of equation (50).

Since similar results have been obtained for extended potentials associated with Jacobi EOP, we conclude that the extensions considered here do not spoil the nice SI property of the conventional potentials, from which they are built. A whole new class of so far unknown translationally SI potentials has therefore been created.

The appearance of such a novel family has motivated two recent re-examinations of the SI condition (48), independently of the EOP construction problem, in the case where the parameters are linked by translation.

In the first one [46, 47], the non-local differential equation for W has been replaced by two local partial differential equations. The first of them is similar to the Euler equation expressing momentum conservation for inviscid fluid flow in one spatial dimension, while the second provides a constraint helping one to determine unique solutions. In units wherein $\hbar$ is not set equal to one (in contrast with what is done here and in most other works), all the solutions of these equations, which do not depend explicitly on $\hbar$, have been derived and proved to include all conventional translationally SI potentials and nothing else. The generalization of the algorithm to superpotentials that contain $\hbar$ explicitly has given rise to some of the extended potentials previously built in connection with EOP [12, 19].

In the second study [48], as in equation (51), the superpotential is separated into two parts, the first one being the superpotential associated with some conventional potential. The second part is then divided into two pieces again and the latter are shown to satisfy some compatibility condition, for which a few solutions are provided. These give rise to already known extended potentials [12, 19] or slight generalizations thereof.

Although the two last works are interesting because they have shed some light on the resolution of the SI condition, they have not led up to now to any really novel example of translationally SI rationally-extended potential (although the authors of [47, 48] claim the opposite). From this viewpoint, the technique based on EOP construction seems to be more powerful. It has the additional advantage of simultaneously providing the solutions of the Schrödinger equation at a low price.

6. Final comments

The purpose of this concluding section is to briefly review some works connected somehow or other with the topics of EOP and of related ES quantum potentials.

Some applications of EOP have been considered to other problems than the resolution of the conventional Schrödinger equation. The inclusion of a position-dependent effective mass in the latter has been dealt with [49], as well as the replacement of the real potential by a complex one, either $\mathcal{PT}$-symmetric or not [19, 50]. The EOP have been studied in the framework of the quantum Hamilton-Jacobi formalism in connection with the supersymmetric WKB quantization condition [51] and in that of type A $\mathcal{N}$-fold supersymmetry and quasi-solvability [52, 53, 54]. They have also been shown to be useful in the context of the Dirac equation coupled minimally or non-minimally with some external field and in that of the Fokker-Planck equation [55].

Some generalizations of the X_m EOP for noninteger m, which are not polynomials but are expressible in terms of the (confluent) hypergeometric function, have been proved to lead to SI potentials [56]. Some of the latter had been studied before and shown to belong to the class of conditionally ES potentials, although their SI had not been recognized [57, 58, 59].

In the course of the construction of the first examples of type I and type II Laguerre and Jacobi X_2 EOP [20], there occurred so-called type III polynomials, which were generalized later on to higher m values, hence giving rise to what is sometimes referred to as the L3 and J3 families [28, 29, 60, 61, 62]. These polynomials, however, do not qualify to be called EOP, since some degrees are missing from the set of them, and the corresponding extended potentials are not SI either. They belong instead to classes of polynomials and potentials often encountered (and periodically rediscovered) in applications (see, e.g., [15, 57, 63, 64]).

Finally, it is worth mentioning that families of EOP have also been constructed in connection with Wilson, Askey-Wilson, Racah and q-Racah polynomials used in the framework of discrete quantum mechanics [60, 65, 66, 67].

Acknowledgments
The author would like to thank Y Grandati for several useful discussions.

References

[1] Erdélyi A, Magnus W, Oberhettinger F and Tricomi F G 1953 *Higher Transcendental Functions* vols 1, 2 (New York: McGraw-Hill)
[2] Schrödinger E 1941 *Proc. R. Ir. Acad.* A **46** 183
[3] Infeld L and Hull T E 1951 *Rev. Mod. Phys.* **23** 21
[4] Cooper F, Khare A and Sukhatme U 2000 *Supersymmetry in Quantum Mechanics* (Singapore: World Scientific)
[5] Darboux G 1888 *Théorie Générale des Surfaces* vol 2 (Paris: Gauthier-Villars)
[6] Gendenshtein L E 1983 *JETP Lett.* **38** 356
[7] Cariñena J F and Ramos A 2000 *J. Phys. A: Math. Gen.* **33** 3467 (*Preprint* hep-th/0003266)
[8] Cooper F, Ginocchio J N and Khare A 1987 *Phys. Rev.* D **36** 2458
[9] Dabrowska J, Khare A and Sukhatme U 1988 *J. Phys. A: Math. Gen.* **21** L195
[10] Gómez-Ullate D, Kamran N and Milson R 2010 *J. Approx. Theor.* **162** 987 (*Preprint* arXiv:0805.3376)
[11] Gómez-Ullate D, Kamran N and Milson R 2009 *J. Math. Anal. Appl.* **359** 352 (*Preprint* arXiv:0807.3939)
[12] Quesne C 2008 *J. Phys. A: Math. Theor.* **41** 392001 (*Preprint* arXiv:0807.4087)
[13] Bochner S 1929 *Math. Z.* **29** 730
[14] Koekoek R, Lesky P A and Swarttouw R F 2010 *Hypergeometric Orthogonal Polynomials and Their q-Analogues* (Berlin Heidelberg: Springer-Verlag)
[15] Gómez-Ullate D, Kamran N and Milson R 2004 *J. Phys. A: Math. Gen.* **37** 1789 (*Preprint* quant-ph/0308062)
[16] Gómez-Ullate D, Kamran N and Milson R 2004 *J. Phys. A: Math. Gen.* **37** 10065
[17] Bhattacharjie A and Sudarshan E C G 1962 *Nuovo Cimento* **25** 864
[18] Lévai G 1989 *J. Phys. A: Math. Gen.* **22** 689
[19] Bagchi B, Quesne C and Roychoudhury R 2009 *Pramana J. Phys.* **73** 337 (*Preprint* arXiv:0812.1488)
[20] Quesne C 2009 *SIGMA* **5** 084 (*Preprint* arXiv:0906.2331)
[21] Odake S and Sasaki R 2009 *Phys. Lett.* B **679** 414 (*Preprint* arXiv:0906.0142)
[22] Odake S and Sasaki R 2010 *Phys. Lett.* B **684** 173 (*Preprint* arXiv:0911.3442)
[23] Odake S and Sasaki R 2010 *J. Math. Phys.* **51** 053513 (*Preprint* arXiv:0911.1585)
[24] Ho C-L, Odake S and Sasaki R 2011 *SIGMA* **7** 107 (*Preprint* arXiv:0912.5447)
[25] Gómez-Ullate D, Kamran N and Milson R 2010 *J. Phys. A: Math. Theor.* **43** 434016 (*Preprint* arXiv:1002.2666)
[26] Gómez-Ullate D, Kamran N and Milson R 2011 *Preprint* arXiv:1101.5584
[27] Sasaki R, Tsujimoto S and Zhedanov A 2010 *J. Phys. A: Math. Theor.* **43** 315204 (*Preprint* arXiv:1004.4711)
[28] Grandati Y 2011 *Ann. Phys. (NY)* **326** 2074 (*Preprint* arXiv:1101.0055)
[29] Ho C-L 2011 *Prog. Theor. Phys.* **126** 185 (*Preprint* arXiv:1104.3511)
[30] Gómez-Ullate D, Kamran N and Milson R 2011 *Preprint* arXiv:1103.5724
[31] Odake S and Sasaki R 2011 *Phys. Lett.* B **702** 164 (*Preprint* arXiv:1105.0508)
[32] Quesne C 2011 *Mod. Phys. Lett.* A **26** 1843 (*Preprint* arXiv:1106.1990)
[33] Quesne C 2011 *Preprint* arXiv:1110.3958
[34] Grandati Y 2011 *Preprint* arXiv:1108.4503
[35] Andrianov A A, Ioffe M V and Nishnianidze D N 1995 *Phys. Lett.* A **201** 103 (*Preprint* hep-th/9404120)
[36] Bagrov V G and Samsonov B F 1995 *Theor. Math. Phys.* **104** 1051
[37] Samsonov B F 1996 *Mod. Phys. Lett.* A **11** 1563 (*Preprint* quant-ph/9611012)
[38] Bagchi B, Ganguly A, Bhaumik D and Mitra A 1999 *Mod. Phys. Lett.* A **14** 27
[39] Aoyama H, Sato M and Tanaka T 2001 *Nucl. Phys.* B **619** 105 (*Preprint* quant-ph/0106037)
[40] Fernández C D J and Fernández-García N 2005 *AIP Conf. Proc.* vol 744 (Melville, NY: Amer. Inst. Phys.) p 236 (*Preprint* quant-ph/0502098)
[41] Rubakov V A and Spiridonov V P 1988 *Mod. Phys. Lett.* A **3** 1337
[42] Khare A 1993 *J. Math. Phys.* **34** 1277
[43] Khare A and Sukhatme U 1993 *J. Phys. A: Math. Gen.* **26** L901 (*Preprint* hep-th/9212147)
[44] Barclay D, Dutt A, Gangopadhyaya A, Khare A, Pagnamenta A and Sukhatme U 1993 *Phys. Rev.* A **48** 2786 (*Preprint* hep-ph/9304313)

[45] Sukhatme U P, Rasinariu C and Khare A 1997 *Phys. Lett.* A **234** 401 (*Preprint* hep-ph/9706282)
[46] Bougie J, Gangopadhyaya A and Mallow J V 2010 *Phys. Rev. Lett.* **105** 210402 (*Preprint* arXiv:1008.2035)
[47] Bougie J, Gangopadhyaya A and Mallow J V 2011 *J. Phys. A: Math. Theor.* **44** 275307 (*Preprint* arXiv:1103.1169)
[48] Ramos A 2011 *J. Phys. A: Math. Theor.* **44** 342001 (*Preprint* arXiv:1106.3732)
[49] Midya B and Roy B 2009 *Phys. Lett.* A **373** 4117 (*Preprint* arXiv:0910.1209)
[50] Bagchi B and Quesne C 2010 *J. Phys. A: Math. Theor.* **43** 305301 (*Preprint* arXiv:1002.4309)
[51] Sree Ranjani S, Panigrahi P K, Khare A, Kapoor A K and Gangopadhyaya A 2010 *Preprint* arXiv:1009.1944
[52] Bagchi B and Tanaka T 2009 *Ann. Phys. (NY)* **324** 2438 (*Preprint* arXiv:0905.3330)
[53] Tanaka T 2010 *J. Math. Phys.* **51** 032101 (*Preprint* arXiv:0910.0328)
[54] Bagchi B and Tanaka T 2010 *Ann. Phys. (NY)* **325** 1679 (*Preprint* arXiv:1002.1766)
[55] Ho C-L 2011 *Ann. Phys. (NY)* **326** 797 (*Preprint* arXiv:1008.0744)
[56] Odake S and Sasaki R 2011 *J. Phys. A: Math. Theor.* **44** 195203 (*Preprint* arXiv:1007.3800)
[57] Junker G and Roy P 1997 *Phys. Lett.* A **232** 155
[58] Junker G and Roy P 1998 *Ann. Phys. (NY)* **270** 155 (*Preprint* quant-ph/9803024)
[59] Dutta D and Roy P 2010 *J. Math. Phys.* **51** 042101
[60] García-Gutiérrez L, Odake S and Sasaki R 2010 *Prog. Theor. Phys.* **124** 1 (*Preprint* arXiv:1004.0289)
[61] Grandati Y 2011 *J. Math. Phys.* **52** 103505 (*Preprint* arXiv:1103.5023)
[62] Ho C-L 2011 *Preprint* arXiv:1105.3670
[63] Cariñena J F, Perelomov A M, Rañada M F and Santander M 2008 *J. Phys. A: Math. Theor.* **41** 085301 (*Preprint* arXiv:0711.4899)
[64] Fellows J M and Smith R A 2009 *J. Phys. A: Math. Theor.* **42** 335303
[65] Odake S and Sasaki R 2009 *Phys. Lett.* B **682** 130 (*Preprint* arXiv:0909.3668)
[66] Odake S and Sasaki R 2010 *J. Phys. A: Math. Theor.* **43** 335201 (*Preprint* arXiv:1004.0544)
[67] Odake S and Sasaki R 2011 *Prog. Theor. Phys.* **125** 851 (*Preprint* arXiv:1102.0812)

Symmetries in Science XV IOP Publishing
Journal of Physics: Conference Series **380** (2012) 012017 doi:10.1088/1742-6596/380/1/012017

Dynamics of Schrödinger cat states

O Castaños and J A López-Saldívar

Instituto de Ciencias Nucleares, Universidad Nacional Autónoma de México, Ciudad Universitaria, Apartado Postal 70-543
04510 México, Distrito Federal

E-mail: ocasta@nucleares.unam.mx

E-mail: julio.lopez@nucleares.unam.mx

Abstract. A review of the definition of Crystallized Schrödinger cat states and their statistical properties is presented. The associated Wigner and Husimi quasi-probability distribution functions are also calculated together with the second moment of the Husimi function. The Hamiltonian formulation of the Nöther's theorem is used to get the linear time dependent invariants of until general quadratic Hamiltonians in the quadratures of the electromagnetic field. By means of these invariants, the evolution of Crystallized Schrödinger cats is studied, thus we have explicit time dependent expressions for the means and dispersions of the quadratures of the electromagnetic field together with the corresponding Wigner and Husimi functions. By means of these results, we establish some conditions on the parameters of the Crystallized cat states to have the description of non-classical light.

1. Introduction

The discovery of the laser in the early sixties gave birth to the research areas of quantum optics and quantum electronics. The last one implies investigations of device oriented problems, and very soon there was an explosion of works that made laser a common word for the people. The development of quantum optics was slower, and it was until the middle of the seventies and the beginning of eighties that the quantum manifestations of the radiation field become amenable to experimental verification [1, 2]. The coherent states of light introduced by Glauber are usually called classical states of light, they minimize the Heisenberg uncertainties relations of the quadratures of the electromagnetic field and have a Poisson distribution function of the number of photons [3, 4, 5].

Non-classical states of light such as Schrödinger cat states, even and odd coherent states, squeezed states, or correlated state have different statistical properties than the mentioned coherent states [6, 7]. There are several proposals to generate and detect non-classical states of light. For example, one can mention the evolution of a coherent state through an amplitude-dispersive medium [8], others use the evolution in a nonlinear birefringent optical model [9, 10], and dinamically through the evolution of an initial state formed by a two-level state direct product a standard coherent state in a optical cavity [11, 12, 13].

The motivation of this work is to study macroscopic superpositions of coherent states that carry an irreducible representation (irrep) of a finite group. These linear combinations of coherent states are written in terms of the simple characters of the elements of a given irreducible representation, and one of the purposes is to establish the conditions that they must satisfy to have non-classical light, that is for example, when the superposition of states have a subpoissonian distribution function of the number of photons or they present the squeezing phenomena [14].

In general, if we have a group with n elements $G : \{g_1, \cdots, g_n\}$, and denote by $\{D(g_1), \cdots, D(g_n)\}$ a representation of the group. Then the trace of the representation $\chi(g_s) = \sum_{\mu=1}^{l} D_{\mu\mu}(g_s)$ is called the characteristic of g_s in the representation D, where l indicates the dimension of the representation [15]. Given two elements (g_i, g_j) of the group G, they are conjugate one to another if there is an element g_k of the group such that $g_k g_i g_k^{-1} = g_j$, then it is immediate that two elements that belong to the same class have the same characteristic. Therefore, to the set of numbers $\{\chi_1, \chi_2, \cdots \chi_k\}$ with the label k denoting the number of classes of the group we are going to call it the character of the representation $\{D\}$.

It is well known that a finite group of n elements can be divided into classes and that the number of classes is equal to the number of irreducible representations, which are denoted by k. The character associated to an irreducible representation λ is called simple, and thus it can also be denoted by $\{\chi_\rho^\lambda\}$ with $\rho = 1, 2, \cdots, k$. The simple characters satisfy the orthogonality relation

$$\frac{1}{n}\sum_{g_s} \chi^{(\lambda)}(g_s)\chi^{*(\lambda')}(g_s) = \delta_{\lambda\lambda'} \,,$$

where the sum is over the elements of the group G. The sum is equivalent to the sum over the classes, then if we denote by C_ρ the class and assume it has n_ρ elements then the orthogonality relation can be rewritten in the form

$$\frac{1}{n}\sum_{\rho=1}^{k} n_\rho\, \chi_\rho^{(\lambda)} \chi_\rho^{*(\lambda')} = \delta_{\lambda\lambda'} \,.$$

The crystallized states are defined using the simple characters of the groups [14], i.e.,

$$\begin{aligned} \left|\psi^{(\lambda)}\right\rangle &= N_\lambda \sum_{s=1}^{n} \chi^{(\lambda)}(g_s) \left|\alpha_s\right\rangle \,, \\ &= N_\lambda \sum_{\rho=1}^{k} \sum_{s=1}^{g_\rho} \chi_\rho^{(\lambda)} \left|\alpha_s\right\rangle \,, \end{aligned} \tag{1}$$

where in the first row $\chi^{(\lambda)}(g_s)$ represents the simple character for the irreducible representation λ corresponding to the g_s element of the group, $\alpha_s = g_s\alpha$ defines the action of the element g_s over the complex parameter α of a coherent state, and N_λ is the normalization constant. In the second row, we replace the sum of the elements by the sum of the classes of the group and the sum of elements in each class, and modify properly the notation of the simple characters of the group.

Notice that states that belong to different irreducible representations are, in general, not orthogonal due to the non-orthogonality property of the coherent states. As it can be seen immediately by considering the scalar product

$$\left\langle \psi^{(\lambda')} | \psi^{(\lambda)} \right\rangle = N_\lambda N_{\lambda'} \sum_{s,s'} \chi^{(\lambda)}(g_s)\chi^{*(\lambda')}(g_{s'}) E_{s's} \,, \tag{2}$$

where the quantity $E_{s's} \equiv \exp\left(-\frac{|\alpha_{s'}|^2}{2} - \frac{|\alpha_s|^2}{2} + \alpha_{s'}^*\,\alpha_s\right)$.

If we replace $E_{s's} = \delta_{s,s'}$ in the expression (2), one can use the orthogonality properties for the simple characters and conclude that states $\left|\psi^{(\lambda)}\right\rangle$ belonging to different irreducible

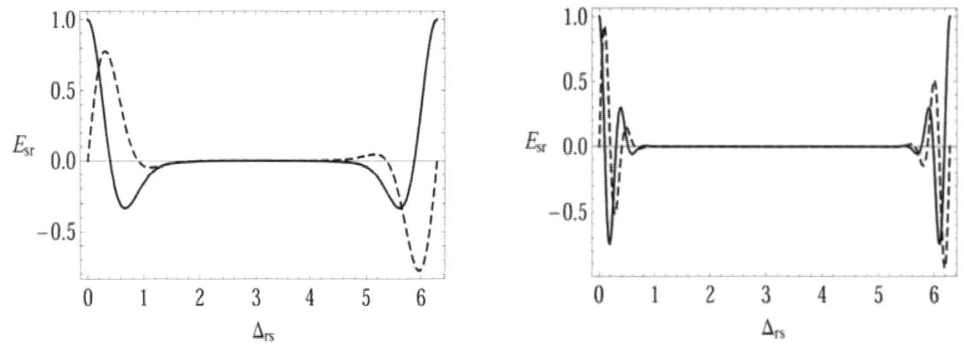

Figure 1. The real (continuous lines) and imaginary (dashed lines) parts of the overlap $E_{sr} = \langle\alpha_s|\alpha_r\rangle$ between two coherent states are shown as a function of their phase diference $\Delta_{rs} = \varphi_r - \varphi_s$ of the complex parameters α_r and α_s. At the left, we take $|\alpha_r| = |\alpha_s| = 2$ and to the right, $|\alpha_r| = |\alpha_s| = 4$.

representations are orthogonal. Then it is important to establish the conditions in the complex parameters of the coherent states to have an orthogonal basis. For the Cyclic and Dihedral groups which are the only ones considered in this work, the parameters of the coherent states $\alpha_s = \rho \exp(i\phi_s)$ have the same magnitude. Thus one has

$$E_{s's} = e^{-\rho^2\left(1-\cos\Delta_{s,s'} - i\sin\Delta_{s,s'}\right)}, \tag{3}$$

where we define the phase difference $\Delta_{s,s'} = \phi_s - \phi_{s'}$. Then it is straightforward that the scalar product is zero when $\Delta_{s,s'} \approx \pi$ and $\rho >> 1$, this behaviour is shown in Fig. (1) for the parameters $|\alpha_{s'}| = |\alpha_s| = 2$ and 4. It can be seen that in a vicinity of $\Delta_{s,s'} = \pi$, the real and the imaginary parts of the overlap are very close to zero.

Cyclic groups

These groups are uniaxial because they have only a rotation axis as element of symmetry. For this reason, the cyclic groups are abelian, all the elements have the same period given by the order of the group, and they have only one dimensional irreducible representations. A physical realization of the cyclic groups appears in the symmetry transformations, rotations, of regular polygons. If we consider the cyclic group of n elements denoted by C_n, a representation of the elements is obtained through the rotation matrices $R_k(\theta_k)$ with $\theta_k = \frac{2\pi k}{n}$, for $k = 0$ we have the identity matrix I. So the n elements of the cyclic group can be given in the following form $C_n : \{I, R_1, R_2, \cdots, R_{n-1}\}$, this two dimensional representation of the cyclic groups is not irreducible, it contains two times the irreps of the cyclic groups and so each element form a class and it has $k = n$ irreducible representations. All the elements can be generated by one element of the group, and so one can select the rotation: $R_1(\frac{2\pi}{n})$.

Then, it is immediate to determine the action on the complex parameter α of the coherent state, that is $\alpha_s = \mu_n^s\alpha$ with $\mu_n = \exp(2\pi i/n)$ and the simple characters of the C_n group is given in Appendix A. Therefore the simple characters can be given by the expression $\chi^{(\lambda)}(g_s) = \mu_n^{(\lambda-1)s}$, with $\lambda = 1, 2, \cdots, n$. So finally the states that carry the irreducible representation λ of C_n can be written in the form

$$\left|\psi^{(\lambda)}\right\rangle = N_\lambda \sum_{s=0}^{n-1} \mu_n^{(\lambda-1)s} \left|\mu_n^s\alpha\right\rangle, \tag{4}$$

where the normalization constant is given by

$$N_\lambda^{-2} = \sum_{s,r=0}^{n-1} \mu_n^{(\lambda-1)(s-r)} \exp\left\{-|\alpha|^2 \left(1-\mu_n^{(s-r)}\right)\right\} .$$

In spite of the non-orthogonality of the coherent states, it can be shown directly that states defined above for different irreducible representations are orthogonal, and thus its scalar product is given by $\langle\psi^{(\lambda')}|\psi^{(\lambda)}\rangle = \delta_{\lambda',\lambda}$.

Dihedral groups

The Dihedral groups are denoted by the symbol D_n, they are generated by an axis of order n and an orthogonal axis C_2. The product and power of their elements give rise to n rotations corresponding to powers of C_n and n second order rotations. This can be illustrated looking for the linear symmerty transformations of a regular polygon into itself. Thus, we establish the coordinates of the n vertices of a regular polygon in the XY plane by the expressions

$$x_k = a\cos\theta_k\,, \quad y_k = a\sin\theta_k\,,$$

where $\theta_k = \frac{2\pi k}{n}$ and $k = 0, 1, 2, \cdots n-1$. If symmetry transformations of regular polygons are considered, one finds: the rotations by an angle θ_k plus rotations of the same type followed by a reflection on the X axis (also called second order rotations).

For $n \geq 2$ the dihedral group can be generated by the elements R_1 and U_0, which define the rotation with an angle θ_1 and the reflection along the X axis. The $2n$ elements can be written in the following form: $\{I, R_1, R_1^2, \cdots, R_1^{n-1}, U_0, U_0R_1, U_0R_1^2, \cdots, U_0R_1^{n-1}\}$, thus the notation $D_n : \{I, R_1, R_2, \cdots, R_{n-1}, U_0, U_1, \cdots, U_{n-1}\}$ is adopted. We can write the two dimensional irreducible representation for both the rotations and rotations-reflections in one equation

$$\begin{pmatrix} \cos\theta_k & \sin\theta_k \\ -\sigma\sin\theta_k & \sigma\cos\theta_k \end{pmatrix} = \begin{cases} R_k & \text{if}\,\sigma = 1, \\ U_k & \text{if}\,\sigma = -1. \end{cases}$$

The action on the complex parameter of the coherent state implies that $\alpha_s = \mu_n^s\alpha$ with $s = 0, 1, 2, 3, \cdots, n-1$, for the rotations (included the identity) while $\alpha_{s+n} = \mu_n^{*s}\alpha^*$ with the same values of s, for the reflections.

In this contribution, we present general expressions for the statistical properties of superpositions of coherent states together with their corresponding Wigner, Husimi function and the area of phase space associated to the Husimi functions which of course can be applied to any point groups, in particular to D_3, D_4, D_5 and D_6, which describe the symmetries of the equilateral triangle, the square, the pentagon and the hexagon, respectively. For this reason we include in the Appendix A their corresponding character tables.

2. Hamiltonian formulation for the Nöther theorem

By means of the Nöther theorem, we are going to construct the linear time dependent invariants of at most quadratic Hamiltonian, in the quadratures of the electromagnetic field,

$$H = \frac{a}{2}p^2 + \frac{b}{2}(q\,p + p\,q) + \frac{c}{2}q^2 + f\,p + g\,q\,, \tag{5}$$

where a, b, c, f, and g can be time dependent functions. Examples of its applications can be found in quantum optics [16], in molecular spectroscopy [17], and to visualize states of the motion of an ion in an electromagnetic trap. In this last case, one has to consider the following choice for the parameters $b = f = g = 0$, $a = 1$ and $c = w^2(t)$ [18]. For this system, one can also use directly the time dependent Scrödinger equation because their solution can be written in terms of an arbitrary gaussian wavepacket written in terms of two functions than satisfy an Ermakov system [25].

In the hamiltonian formulation of the Nöther theorem [20], the variations in the position and momentum variables

$$q \to q + \delta q \,, \qquad p \to p + \delta p \,,$$

constitute a symmetry transformation, if the corresponding variation in the functional of the action, $A(q,p,t) \equiv \dot{q}\,p - H(q,p,t)$, i.e.,

$$\delta A = A(q+\delta q, p+\delta p, t) - A(q,p,t) = \frac{d}{dt}\Omega \,,$$

can be written as a total derivative with respect to time of a function $\Omega(q,p,t)$. Developing the last expression by making a Taylor series expansion, one gets

$$\delta A = \dot{q}\,\delta p + p\,(\dot{\delta q}) - \frac{\partial H}{\partial q}\delta q - \frac{\partial H}{\partial p}\delta p = \frac{d}{dt}\Omega \,. \tag{6}$$

Substituting the time derivatives of the function Ω and $\delta q(q,p,t)$ into the last expression, and equating the coefficients of $\dot{p}$, $\dot{q}$ and the independent terms on both sides of the expression, one gets the following set of differential equations for the function Ω

$$\begin{aligned}
\frac{\partial \Omega}{\partial p} &= \frac{\partial \delta q}{\partial p} p \,, \\
\frac{\partial \Omega}{\partial q} &= \delta p + \frac{\partial \delta q}{\partial q} p \,, \\
\frac{\partial \Omega}{\partial t} &= -\frac{\partial H}{\partial p}\delta p - \frac{\partial H}{\partial q}\delta q + \frac{\partial \delta q}{\partial t} p \,.
\end{aligned} \tag{7}$$

Establishing the integrability condition for Ω, that is the equality of all the cross derivatives, for example $\frac{\partial^2 \Omega}{\partial q \partial p} = \frac{\partial^2 \Omega}{\partial p \partial q}$ and so on, which implies the existence of the generators of the symmetry transformation, the following expressions are obtained

$$\begin{aligned}
\left(\mathcal{O} + \frac{\partial}{\partial t}\right)\delta q - \frac{\partial^2 H}{\partial p^2}\delta p - \frac{\partial^2 H}{\partial p \partial q}\delta q &= 0 \,, \\
\left(\mathcal{O} + \frac{\partial}{\partial t}\right)\delta p + \frac{\partial^2 H}{\partial q \partial p}\delta p - \frac{\partial^2 H}{\partial q^2}\delta q &= 0 \,.
\end{aligned} \tag{8}$$

with the definition $\mathcal{O} = \frac{\partial H}{\partial p}\frac{\partial}{\partial q} - \frac{\partial H}{\partial q}\frac{\partial}{\partial p}$.

Then the conserved quantity associated to the symmetry transformations is given by

$$K = p\delta q - \Omega(q,p,t) \,. \tag{9}$$

We use the expressions (8) and (9) to get the linear time dependent invariants of the Hamiltonian (5), by considering the time dependent variations $\delta q = v(t)$ and $\delta p = u(t)$. To be symmetry transformations they must satisfy the differential equations

$$\begin{aligned}
v' - a\,u - b\,v &= 0 \,, \\
u' + b\,u + c\,v &= 0 \,.
\end{aligned} \tag{10}$$

These expressions remind us the Hamilton equations for the quadratic part of (5) by means of the identifications $u \to p$ and $v \to q$. This implies that to get the linear time dependent invariants one needs to do variation along the classical trajectories of motion. To determine the conserved quantity, one needs to determine Ω through the expressions (7) and the result is

$$K = p\,v - q\,u + \int_0^t (f(\tau)\,u(\tau) + g(\tau)v(\tau))d\tau\,. \tag{11}$$

Since we have a second order differential equation, there are two conserved quantities, named $(Q(t),P(t))$ expressed in matrix form

$$\begin{pmatrix} P \\ Q \end{pmatrix} = \begin{pmatrix} v^{(1)} & -u^{(1)} \\ v^{(2)} & -u^{(2)} \end{pmatrix} \begin{pmatrix} p \\ q \end{pmatrix} + \begin{pmatrix} \delta^{(1)} \\ \delta^{(2)} \end{pmatrix} \tag{12}$$

where $\delta^{(i)} = \int_0^t (f u^{(i)} + g v^{(i)}) d\tau$ with $i = 1, 2$, these terms appear due to the linear parts in the Hamiltonian. By establishing the initial conditions that the constants of the motion are equal to the quadratures of the electromagnetic field $P(0) = p$ and $Q(0) = q$, one gets the initial conditions for the differential equations of the variations, (10), i.e.,

$$\begin{aligned} v^{(1)}(0) &= 1, \qquad u^{(1)}(0) = 0, \qquad \delta^{(1)}(0) = 0\,, \\ v^{(2)}(0) &= 0, \qquad u^{(2)}(0) = -1, \qquad \delta^{(2)}(0) = 0\,. \end{aligned} \tag{13}$$

The P, Q are called linear time dependent invariants of quadratic Hamiltonians. The expressions (12) can be extended to describe multimode correlated light [7, 20]. Also they constitute a canonical transformation in a two dimensional phase space, and their extension to the multidimensional case implies to consider P, Q, p and q as n dimensional column vectors while $v^{(1)}$, $v^{(2)}$, $u^{(1)}$ and $u^{(2)}$ like $n \times n$ matrices [21].

As an example, we consider the linear time dependent invariants of a free particle under the action of a time dependent force, that is $V(q) = f(t)\,q$,

$$\begin{pmatrix} P(t) \\ Q(t) \end{pmatrix} = \begin{pmatrix} 1 & 0 \\ -t & 1 \end{pmatrix} \begin{pmatrix} p \\ q \end{pmatrix} + \begin{pmatrix} \int_0^t f(\tau)\,d\tau \\ -\int_0^t f(\tau)\,\tau\,d\tau \end{pmatrix}. \tag{14}$$

In this contribution, we will use these time dependent constants of the motion to shown the evolution of the statistical properties of crystallized Schrödinger cats. As an explicit example, we will consider the evolution of the cristallized states under the Hamiltonian of a free particle under a constant force, that is when $f(t) = F_0$.

3. Evolution of crystallized cats

From Eq. (12), one can get the quadratures of the electromagnetic field p, q in terms of the time dependent constants of the motion P, Q or equivalently the corresponding creation and annihilation operators $A \equiv \frac{1}{\sqrt{2}}(Q + iP)$, $A^\dagger \equiv \frac{1}{\sqrt{2}}(Q - iP)$, this is

$$\begin{pmatrix} p \\ q \end{pmatrix} = i \begin{pmatrix} -\lambda_q^* & \lambda_q \\ \lambda_p^* & -\lambda_p \end{pmatrix} \begin{pmatrix} A - \delta \\ A^\dagger - \delta^* \end{pmatrix}, \tag{15}$$

where the parameters carry the dynamic associated to the Hamiltonian and are given in terms of the classical solutions of the Hamilton equations, i.e.,

$$\lambda_p = \frac{1}{\sqrt{2}}(v^{(2)} + iv^{(1)})\,, \quad \lambda_q = \frac{1}{\sqrt{2}}(-u^{(2)} - iu^{(1)})\,, \quad \delta = \frac{1}{\sqrt{2}}(\delta^{(2)} + i\delta^{(1)})\,.$$

Therefore to calculate the evolution of the statistical properties, it is convenient to construct the eigenstates of the annihilation operator, that is the correlated coherent states that satisfy [7]

$$A(t)|\alpha, t\rangle = \alpha|\alpha, t\rangle \,.$$

Then it is straightforward to prove that the fluctuations of the quadratures of the electromagnetic field associated to the generalized correlated states of the general quadratic Hamiltonian (5) are given by [22]

$$\begin{aligned} \sigma_p \equiv \langle\alpha_k, t|\hat{p}^2|\alpha_k, t\rangle - \langle\alpha_k, t|\hat{p}|\alpha_k, t\rangle^2 &= |\lambda_q|^2 \,, \\ \sigma_q \equiv \langle\alpha_k, t|\hat{q}^2|\alpha_k, t\rangle - \langle\alpha_k, t|\hat{q}|\alpha_k, t\rangle^2 &= |\lambda_p|^2 \,, \\ \sigma_{pq} \equiv \langle\alpha_k, t|\tfrac{1}{2}\left\{\hat{p}, \hat{q}\right\}|\alpha_k, t\rangle - \langle\alpha_k, t|\hat{p}|\alpha_k, t\rangle\langle\alpha_k, t|\hat{q}|\alpha_k, t\rangle &= -\frac{1}{2}\left(\lambda_p \lambda_q^* + \lambda_p^* \lambda_q\right) \,. \end{aligned} \tag{16}$$

It is important to notice that the results mentioned above are independent of the value of the complex parameter α_k.

For the crystallized cats defined in (1), it is straightforward to evaluate the expectation values of the quadratures of the electromagnetic field,

$$\langle q\rangle_t = N_\lambda^2 \sum_{r,s=1}^{n} \chi^{(\lambda)}(g_r)\,\chi^{(\lambda)*}(g_s)\, q_{sr}(t)\, E_{sr}, \quad \langle p\rangle_t = N_\lambda^2 \sum_{r,s=1}^{n} \chi^{(\lambda)}(g_r)\,\chi^{(\lambda)*}(g_s)\, p_{sr}(t)\, E_{sr} \,. \tag{17}$$

From here on, we are going to defined the matrix element of an arbitrary observable with respect to the standard coherent states associated to the complex parameters α_s y α_r in the form

$$\langle\alpha_s, t|\hat{\mathcal{O}}|\alpha_r, t\rangle \equiv \mathcal{O}_{sr}(t)\, E_{sr} \,.$$

Thus in the previous expression, we use that

$$p_{sr}(t) \;\equiv\; -i\left(\lambda_q^*(\alpha_r - \delta) - \lambda_q(\alpha_s^* - \delta^*)\right) \,, \quad q_{sr}(t) \equiv i\left(\lambda_p^*(\alpha_r - \delta) - \lambda_p(\alpha_s^* - \delta^*)\right) \,. \tag{18}$$

From these expressions for $t = 0$, one gets the behavior of the expectation values for the static properties of the crystallized cats. In this case, it is immediate from expressions (13) that $\lambda_p(0) = i/\sqrt{2}$, $\lambda_q(0) = 1/\sqrt{2}$, and $\delta(0) = 0$. These imply that

$$p_{sr}(0) = -\frac{i}{\sqrt{2}}\left(\alpha_r - \alpha_s^*\right) \,, \quad q_{sr}(0) = \frac{1}{\sqrt{2}}\left(\alpha_r + \alpha_s^*\right) \,, \tag{19}$$

determining in this form the static properties of the crystallized cats.

The calculations of the corresponding dispersions $(\Delta q)^2$ and $(\Delta p)^2$ of the quadratures associated to the the dynamics of the crystallized cats under the Hamiltonian (5)

$$\begin{aligned} (p^2)_{sr}(t) &= -\lambda_q^{*2}\left(\alpha_r - \delta\right)^2 - \lambda_q^2\left(\alpha_s^* - \delta^*\right)^2 + \sigma_p\left[2\left(\alpha_s^* - \delta^*\right)\left(\alpha_r - \delta\right) + 1\right] \,, \\ (q^2)_{sr}(t) &= -\lambda_p^{*2}\left(\alpha_r - \delta\right)^2 - \lambda_p^2\left(\alpha_s^* - \delta^*\right)^2 + \sigma_q\left[2\left(\alpha_s^* - \delta^*\right)\left(\alpha_r - \delta\right) + 1\right] \,, \end{aligned} \tag{20}$$

with these and the expectation values given above, one easily determines in analytic form the fluctuations in the variables p and q for the crystallized cats.

Thus the squeezing coefficients for the quadratures of the electromagnetic field are determined by the ratios of the fluctuations to the uncertainties of the ground state of the electromagnetic field,

$$S_q = \frac{(\Delta q)^2}{1/2} \,, \qquad S_p = \frac{(\Delta p)^2}{1/2} \,,$$

which let us determine if the crystallized states present the squeezing phenomena. When S_q or S_p is less than the unity there is squeezing and the crystallized states describe non-classical light.

At the left side of Fig. (2), we present for the three irreducible representations of the point group C_3 the behaviour of the squeezing parameter S_p as a function of the magnitude of the complex parameter $|\alpha| = \rho$ associated to the coherent state. Notice that, the squeezing parameter is independent of time and of the strength of the repulsive force F_0 considered in the Hamiltonian, besides for large values of ρ the squeezing parameter is independent of the irreducible representation of the group. A similar behaviour is shown at the left hand side of Fig. (3), where the squeezing parameter S_p as a function of ρ is displayed for all the irreps of C_6.

The correlation between q and p, denoted by (Δpq), can be written in terms of the expectation values of p and q together with the time dependent anticommutator function

$$\{p,q\}_{sr}(t) = \lambda_q^* \lambda_p^* (\alpha_r - \delta)^2 + \lambda_q \lambda_p (\alpha_s^* - \delta^*)^2 + 2\,\sigma_{pq} \left[(\alpha_s^* - \delta^*)(\alpha_r - \delta) + \frac{1}{2} \right] . \tag{21}$$

The correlation factor is defined by the expression

$$K = \frac{(\Delta pq)^2}{(\Delta p)(\Delta q)} ,$$

when it is different from zero one gets a correlated state with statistical dependence in the two quadratures of the electromagnetic field. Correlated states for the one mode oscillator have been introduced in [23].

For values of $|\alpha| \geq 2$, the scalar products $E_{sr} \approx \delta_{s,r}$, in that case, we can write $\langle \alpha_s | \alpha_r \rangle = \delta_{s,r}$. Therefore, the normalization constant $N_\lambda = \frac{1}{\sqrt{n}}$ and one can get expressions for all the expectation values mentioned before. In particular, for the dispersions of the quadratures of the electromagnetic field one gets

$$\begin{aligned}
(\Delta q)^2 &= \frac{1}{n}\sum_{r=1}^{n} \left|\chi^{(\lambda)}(g_r)\right|^2 (q^2)_{rr}(t) - \left(\frac{1}{n}\sum_{r=1}^{n} |\chi^{(\lambda)}(g_r)|^2\, q_{rr}(t)\right)^2 ,\\
(\Delta p)^2 &= \frac{1}{n}\sum_{r=1}^{n} \left|\chi^{(\lambda)}(g_r)\right|^2 (p^2)_{rr}(t) - \left(\frac{1}{n}\sum_{r=1}^{n} |\chi^{(\lambda)}(g_r)|^2\, p_{rr}(t)\right)^2 ,\\
(\Delta\, qp) &= \frac{1}{n}\sum_{r=1}^{n} \left|\chi^{(\lambda)}(g_r)\right|^2 \{p,q\}_{rr}(t) - \frac{1}{n^2}\sum_{r',r=1}^{n} |\chi^{(\lambda)}(g_r)|^2\, |\chi^{(\lambda)}(g_{r'})|^2\, q_{rr}(t)\, p_{r'r'}(t) ,
\end{aligned} \tag{22}$$

where the expressions of $(q^2)_{rr}(t)$, $(p^2)_{rr}(t)$, $q_{rr}(t)$, $p_{r'r'}(t)$, and $\{p,q\}_{rr}(t)$ can be evaluated by means of the expressions (18), (19), and (20). Indeed for the one dimensional representations the absolute values of the characters are equal to the unity and so the previous expressions are even simpler.

Now we consider the photon statistics of the crystallized cats, to calculate the expectation value of the photon number and its corresponding fluctuation is convenient to write the constants of the motion in terms of the creation and annihilation operators. Thus we have that

$$\begin{pmatrix} a \\ a^\dagger \end{pmatrix} = \begin{pmatrix} \mu_1^* & -\mu_2 \\ -\mu_2^* & \mu_1 \end{pmatrix} \begin{pmatrix} A - \delta \\ A^\dagger - \delta^* \end{pmatrix} , \tag{23}$$

with the definitions of the parameters $\mu_1 = \frac{1}{\sqrt{2}}(\lambda_q - i\lambda_p)$ and $\mu_2 = \frac{1}{\sqrt{2}}(\lambda_q + i\,\lambda_p)$. The 2×2 matrix in the right side of the expression has determinant equal to the unity. Thus the average

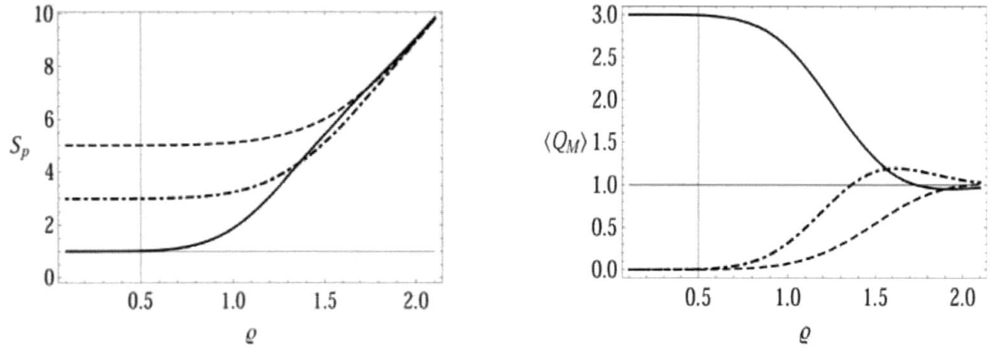

Figure 2. The squeezing parameter for the point group C_3, associated to variable p, is shown at the left side as a function of ϱ while the Mandel parameter is displayed at the right side. In both cases, the cristallized states are subject to a repulsive constant force $F_0 = 10$ in arbitrary units, notice that S_p is independent of time and Q_M is calculated for $t = 0$. The value $Q_M = 1$ is indicating a Poissonian distribution. In both cases, continuous, dashed and dash-dotted lines correspond to the irreducible representations 1, 2, and 3 of C_3, respectively.

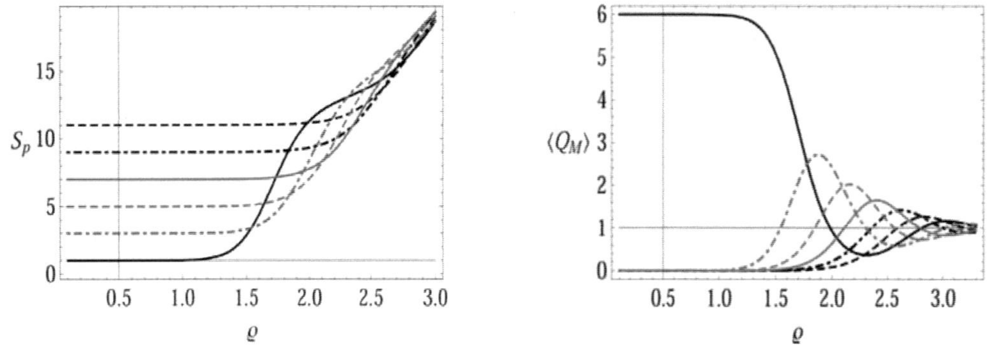

Figure 3. Squeezing and Mandel parameters are shown as functions of ϱ for the cyclic group C_6, when the system is evolving by means of the Hamiltonian of a repulsive constant force. We use $F_0 = 10$ in arbitrary units and the Mandel parameter was evaluated at $t = 0$. Continuous, dashed and dash-dotted black lines correspond to the representations $\lambda = 1$, 2, and 3, respectively while the gray lines are associated to the representations 4, 5 and 6, respectively.

of the photon number operator is given by

$$\langle n \rangle_t = N_\lambda^2 \sum_{r,s=1}^{n} \chi^{(\lambda)}(g_r)\chi^{(\lambda)*}(g_s)\, n_{sr}(t)\, E_{sr},$$

where we define

$$n_{sr}(t) = -\mu_1\,\mu_2\,(\alpha_s^* - \delta^*)^2 - \mu_1^*\,\mu_2^*\,(\alpha_r - \delta)^2 + \left(|\mu_1|^2 + |\mu_2|^2\right)(\alpha_s^* - \delta^*)\,(\alpha_r - \delta) + |\mu_2|^2\,. \quad (24)$$

The dispersion of the photon number operator $(\Delta\hat{n})^2$ of the crystalized states is determined by the standard expression in terms of the expectation values of $\hat{n}$ and $\hat{n}^2$, thus we give the result for the second one

$$\langle \hat{n}^2 \rangle_t = N_\lambda^2 \sum_{r,s=1}^{n} \chi^{(\lambda)}(g_r)\chi^{(\lambda)*}(g_s)\,(n^2)_{sr}(t)\, E_{sr},$$

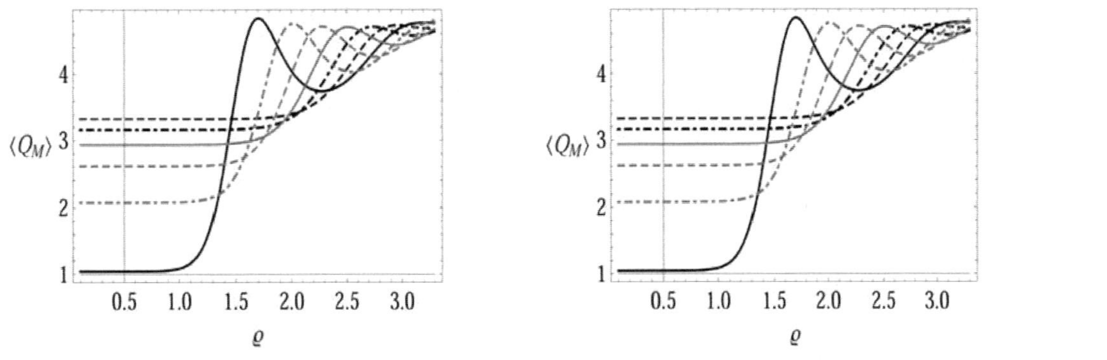

Figure 4. Mandel parameters are shown as functions of ϱ for the point groups C_3 and C_6, when the cristallized states are subject to a repulsive constant force. We use $F_0 = 10$ in arbitrary units, take $t = 0.2$, and the same code lines than in Figs. (2) and (3).

where the time dependent function takes the form

$$\begin{aligned}(n^2)_{sr}(t) &= 2\,|\mu_1|^2\,|\mu_2|^2 - 2\,(|\mu_1|^2 + |\mu_2|^2)\,\left[\mu_1^*\,\mu_2^*\,(\alpha_r - \delta)^2 + \mu_1\,\mu_2\,(\alpha_s^* - \delta^*)^2\right] \\ &+ \left(|\mu_1|^4 + |\mu_2|^4 + 6|\mu_1|^2\,|\mu_2|^2\right)(\alpha_r - \delta)\,(\alpha_s^* - \delta^*) + n_{sr}^2(t)\,. \qquad (25)\end{aligned}$$

The Mandel parameter, $Q_M \equiv R_M - 1$, is the ratio of the dispersion to the photon number mean value. It determines the type of statistics of the considered state. If this ratio is less than the unity, the state satisfies a sub-Poisson statistics while for larger values correspond to a super-Poisson statistics. Of course if the result is equal to the unity, the state has a Poisson statistics. The Mandel parameter can be written in the form

$$R_M = \frac{(\Delta n)^2}{\langle n \rangle} = \frac{\sum_{r,s=1}^{n} \chi^{(\lambda)}(g_r)\chi^{(\lambda)*}(g_s)(n^2)_{sr}(t)\,E_{sr}}{\sum_{r,s=1}^{n} \chi^{(\lambda)}(g_r)\chi^{(\lambda)*}(g_s)n_{sr}(t)\,E_{sr}} - N_\lambda^2 \sum_{r,s=1}^{n} \chi^{(\lambda)}(g_r)\chi^{(\lambda)*}(g_s)n_{sr}(t)\,E_{sr}\,.$$

The corresponding result for the Mandel parameter in the limit of large $|\alpha|^2 \geq 2$, one has the result

$$R_M = \frac{\sum_{r=1}^{n} |\chi^{(\lambda)}(g_r)|^2\,(n^2)_{rr}(t)}{\sum_{r=1}^{n} |\chi^{(\lambda)}(g_r)|^2\,n_{rr}(t)} - \frac{1}{n}\sum_{r=1}^{n} |\chi^{(\lambda)}(g_r)|^2\,n_{rr}(t)\,.$$

In the right side of Fig. (2), the plot of Q_M for the irreps of C_3 as a function of ϱ are shown for $t = 0$. One can see a sub-Poissonian behaviour for the representations $\lambda = 2$ and 3 for small values of the parameter while is super-Poissonian for $\lambda = 1$ For large values of ϱ all the states have a Poissonian photon distribution function. A similar result is found for C_6, as it can be seen in Fig. (3) for all the irreps of the group. A sub-Poissonian distrinution is happening for five of them for small values of ϱ and a super-Poissonian for $\lambda = 1$, newly one gets that all the states have a Poissonian distribution for large values of the parameter. In Fig. (4) for the same point groups C_3 and C_6 the Mandel parameter is plotted as a function of ϱ afterwards the action of a constant force $F_0 = 10$ arbitrary units during a time equal to $t = 0.2$ units, One can see that all the states have now a super-Poissonian photon distributin function. Thus, one has that the constant force Hamiltonian destroys the non-classical properties of the cristallized cats. Similar results are obtained when the cristallized states are under free propagation, one gets sub-Poissonian, and super-Poissonian photon number distributions for small values of ϱ while a Poissonian distribution for large values of the parameter.

4. Quasi-Probability distribution functions

The phase space representation of a quantum system through the Wigner and Husimi distribution functions has been very useful to establish the behavior of the crystallized Schrödinger cat states. As examples, we consider to determine the regions in the phase space where the Wigner function is negative with the well known implications or to find the occupied area in phase space by the Husimi function by means of the evaluation of its second moment.

To obtain the evolution of a Wigner function under the Hamiltonian (5), one can proceed in two forms: By means of the unitary representation of the canonical transformation from the quadratures of the electromagnetic field (q, p) to the linear time dependent constants of motion (Q, P), it has been proved in [24] that

$$W(q, p, t) = W(Q, P, 0)\,.$$

An extension of the work to get the representation of canonical transformations in time-dependent quantum mechanics has been done recently [?].

The second method considers operator equations that satisfy the generalized correlated states [26].

For the crystallized cats, we establish

$$\begin{aligned} A(t)|\alpha_r, t\rangle\langle\alpha_s, t| &= \alpha_r\, |\alpha_r, t\rangle\langle\alpha_s, t|\,, \\ |\alpha_r, t\rangle\langle\alpha_s, t|\, A^\dagger(t) &= \alpha_s^*\, |\alpha_r, t\rangle\langle\alpha_s, t|\,. \end{aligned} \tag{26}$$

Then the constants of the motion $A(t) = (\lambda_p\,\hat{p} + \lambda_q\,\hat{q} + \delta)$ and $A^\dagger(t) = \left(\lambda_p^*\,\hat{p} + \lambda_q^*\,\hat{q} + \delta^*\right)$ are substituted into the operator expressions. We apply the Weyl transform to the resulting expressions, that is multiply by the bra $\langle q - y|$ from the left, by the ket $|q + y\rangle$ from the right, by the exponential $e^{2\,i\,p\,y}$ and integrate the variable y. In this manner, if one defines the operator

$$\hat{\rho}_{rs}(t) = |\alpha_r, t\rangle\langle\alpha_s, t|\,,$$

the following Weyl transforms are obtained

$$\begin{aligned} \hat{p}\,\hat{\rho}_{rs}(t) &\rightarrow \left(p - \frac{i}{2}\frac{\partial}{\partial q}\right) W_{rs}\,, \qquad \hat{\rho}_{rs}(t)\,\hat{p} \rightarrow \left(p + \frac{i}{2}\frac{\partial}{\partial q}\right) W_{rs}\,, \\ \hat{q}\,\hat{\rho}_{rs}(t) &\rightarrow \left(q + \frac{i}{2}\frac{\partial}{\partial p}\right) W_{rs}\,, \qquad \hat{\rho}_{rs}(t)\,\hat{q} \rightarrow \left(q - \frac{i}{2}\frac{\partial}{\partial p}\right) W_{rs}\,. \end{aligned} \tag{27}$$

Using the expressions (26) and (27), the following system of differential equations for the Wigner function is obtained

$$\begin{aligned} (\lambda_p\, p + \lambda_q\, q + \delta)\, W_{rs}(t) + \frac{i}{2}\left(\lambda_q \frac{\partial}{\partial p} - \lambda_p \frac{\partial}{\partial q}\right) W_{rs}(t) &= \alpha_r\, W_{rs}(t)\,, \\ \left(\lambda_p^*\, p + \lambda_q^*\, q + \delta^*\right) W_{rs}(t) - \frac{i}{2}\left(\lambda_q^* \frac{\partial}{\partial p} - \lambda_p^* \frac{\partial}{\partial q}\right) W_{rs}(t) &= \alpha_s^*\, W_{rs}(t)\,. \end{aligned}$$

Making the change of variables $z = \lambda_p\, p + \lambda_q\, q + \delta$ and its complex conjugate, it is straightforward to get

$$\left(z + \frac{1}{2}\frac{\partial}{\partial z^*}\right) W_{rs}(t) = \alpha_r\, W_{rs}(t)\,, \qquad \left(z^* + \frac{1}{2}\frac{\partial}{\partial z}\right) W_{rs}(t) = \alpha_s^*\, W_{rs}(t)\,.$$

This system of differential equations has the following solution:

$$W_{rs}(t) = W_0(\alpha_r, \alpha_s)\, e^{2\alpha_r z^* + 2\alpha_s^* z - 2|z|^2}\,,$$

where $W_0(\alpha_r, \alpha_s)$ is an arbitrary function of the variables α_r, α_s, and its complex conjugate expressions. The time dependence is due to the evolution of the state in the considered Hamiltonian, whose action is reflected in the parameters λ_p, λ_q, and δ.

Then quasi-probabilistic Wigner distribution function for the crystallized Schrödinger cat states can be written as follows

$$W(q,p,t) = N_\lambda^2 \sum_{r,s=1}^{n} \chi^{(\lambda)}(g_r)\chi^{(\lambda)*}(g_s) W_0(\alpha_r, \alpha_s)\, e^{2\alpha_r z^* + 2\alpha_s^* z - 2|z|^2} . \tag{28}$$

Substituting z, z^*, and integrating the expression with respect to the phase space variables q and p, we get

$$1 = N_\lambda^2 \pi \sum_{r,s=1}^{n} \chi^{(\lambda)}(g_r)\, \chi^{(\lambda)*}(g_s)\, W_0(\alpha_r, \alpha_s^*)\, e^{2\alpha_s^* \alpha_r} ,$$

comparing the result with (2) when the irreducible representations of the point group are equal one has the result that

$$W_0(\alpha_r, \alpha_s) = \frac{1}{\pi} E_{sr}\, e^{-2\,\alpha_s^*\,\alpha_r} .$$

Therefore, we find that the time dependent Wigner function associated to the crystallized cats can be written in the following form

$$W(q,p,t) = N_\lambda^2 \sum_{r,s=1}^{n} \chi^{(\lambda)}(g_r)\chi^{(\lambda)*}(g_s) E_{sr}\, e^{\{-2\,\sigma_q\, \tilde{p}_{sr}^2(t) - 2\,\sigma_p\, \tilde{q}_{sr}^2(t) + 4\sigma_{pq}\, \tilde{p}_{sr}(t)\, \tilde{q}_{sr}(t)\}} , \tag{29}$$

where we define $\tilde{p}_{sr}(t) = p - p_{sr}(t)$ and $\tilde{q}_{sr}(t) = q - q_{sr}(t)$, the expressions for $q_{sr}(t)$ and $p_{sr}(t)$ are given in the expression (18). Thus, one can see that each term in the Wigner distribution function is changing in time according to the matrix elements of the quadratures of the electromagnetic field with respect to the generalized correlated states.

The expression of the Wigner distribution function can be simplified one step further by noticing that the argument in the exponential is a quadratic form and thus it can be diagonalized through a rotation in the phase space by an angle

$$\theta(t) = \frac{1}{2} \arctan\left(\frac{2\,\sigma_{pq}}{\sigma_q - \sigma_p}\right) + \theta_0 , \tag{30}$$

where the initial value is used to guarantee that $\theta(t=0) = 0$. Then the Wigner function, in terms of the rotated variables, is given by

$$W(q,p,t) = N_\lambda^2 \sum_{r,s=1}^{n} \chi^{(\lambda)}(g_r)\chi^{(\lambda)*}(g_s) E_{sr}\, e^{\left\{-\frac{q'^2}{(\Delta q')^2} - \frac{p'^2}{(\Delta p')^2}\right\}} ,$$

where the fluctuations in the new variables are independent of the complex parameters α_r and α_s because they are given as follows:

$$(\Delta q')^2 = \left(\sigma_p + \sigma_q + \sqrt{(\sigma_p + \sigma_q)^2 - 1}\right)^{-\frac{1}{2}} ,$$

$$(\Delta p')^2 = \left(\sigma_p + \sigma_q - \sqrt{(\sigma_p + \sigma_q)^2 - 1}\right)^{-\frac{1}{2}} .$$

In this variables one can see that the phase space area of (q', p') stays constant because the product $(\Delta q')^2\,(\Delta p')^2 = 1$.

In conclusion, each term of the Wigner distribution function of a crystallized Schrödinger cat state is moving according the matrix elements of the quadratures of the electromagnetic field and rotating with an angle $\theta(t)$. Besides in the rotating phase space the fluctuations are changing in time but the phase space area is a constant.

Although the Wigner function is very useful sometimes it is convenient to have a non-negative distribution function and for this reason the Husimi function was introduced [27]. The Husimi quasiprobability distribution distribution can be obtained directly by means of the expression

$$\begin{aligned} Q_H(q,p) &= \langle\beta|\psi^{(\lambda)}\rangle\langle\psi^{(\lambda)}|\beta\rangle\,, \\ &= N_\lambda^2 \sum_{s,r=1}^{n} \chi^{*(\lambda)}(g_s)\,\chi^{(\lambda)}(g_r)\,\langle\beta|\alpha_r\rangle\,\langle\alpha_s|\beta\rangle\,, \end{aligned}$$

where $\beta = \frac{1}{\sqrt{2}}(q + ip)$ and in the second row, we substitute the expression for the crystallized cat state, (1). Another form of establishing the Husimi function is smoothing out the Wigner function by averaging over a coarse graining function [28]

$$Q_H(\beta,\beta^*) = \frac{1}{2\pi}\iiint \phi_B(x,x',p) W_{\psi^\lambda}\left(\frac{x+x'}{2}, p\right)\,dx\,dx'\,dp\,,$$

with

$$\Phi(\beta,\beta^*) = \frac{1}{\sqrt{\pi}} e^{ip(x-x') - \frac{1}{2}(\beta^2+\beta^{*2}+2|\beta|^2) - \frac{1}{2}(x^2+x'^2) + \sqrt{2}(\beta^* x + \beta x')}\,.$$

Thus if we substitute the expression for the Wigner function of the crystallized Schrödinger cat states given in (29) and make the integrations indicated in the last expression, one gets

$$Q_H(q,p,t) = N_\lambda^2 \sum_{r,s=1}^{n} \chi^{(\lambda)}(g_r)\,\chi^{(\lambda)*}(g_s)\,E_{sr}\,Q_{sr}(t)\,, \tag{31}$$

where we have defined

$$Q_{sr}(t) = \frac{1}{2\pi}\sqrt{\frac{2}{\sigma_q+\sigma_p+1}}\; e^{\left\{\frac{-(2\sigma_p+1)(q-q_{sr}(t))^2-(2\sigma_q+1)(p-p_{sr}(t))^2+4\sigma_{pq}(q-q_{sr}(t))(p-p_{sr}(t))}{2(\sigma_q+\sigma_p+1)}\right\}}\,. \tag{32}$$

Finally we calculate the second moment of the Husimi distribution function, because it has been proposed as a measure of the complexity of quantum pure states and has similar properties to the Wehrl entropy [29]. Its inverse represents the effective volume in phase space occupied by the Husimi quasiprobability distribution function.

When a quantum state is expanded in a basis for example

$$|\varphi\rangle = \sum_k c_k|\phi_k\rangle\,,$$

the information obtained when one measures the probability $p_k = |c_k|^2$ is obtained through the calculation of the Shannon entropy

$$S = -\sum_k p_k \ln p_k\,.$$

Journal of Physics: Conference Series **380** (2012) 012017 doi:10.1088/1742-6596/380/1/012017

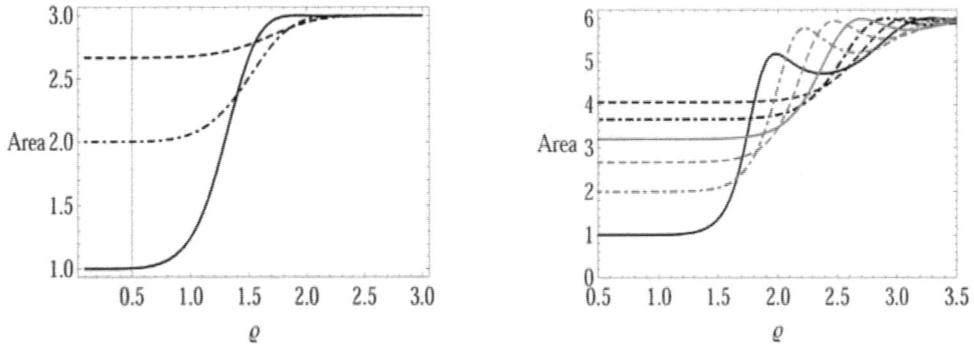

Figure 5. Area of the Husimi function as a function of ϱ for the point groups C_3 and C_6. We use the same code lines than in Figs. (2) and (3).

In addition, the moments of the distribution $M_j = \sum_k p_k^j$ give a measure of the localization with respect to the used basis, in particular the inverse of the second moment determines the number of principal components. As the Husimi function can be considered as a probability distribution function, we are going to calculate its second moment and show the results for the inverse expression, the area, associated to the crystallized cat states. Thus, one has for the second moment of the Husimi function

$$M_H = A \iint_{\infty}^{-\infty} Q^2(q,p)\, dq\, dp \tag{33}$$

$$= N_\lambda^2 \sum_{r,s} \sum_{r',s'} \chi^{(\lambda)}(g_r)\chi^{(\lambda)}(g_{r'})\chi^{(\lambda)*}(g_s)\chi^{(\lambda)*}(g_{s'}) M_{sr}(t) E_{sr} E_{s'r'} \tag{34}$$

with $A = 4\pi\sqrt{2}$ being a constant that we adjust to make the area of a coherent state equal to the unity. The function $M_{sr}(t)$ can be written as follows

$$M_{sr}(t) = \frac{1}{2\pi\sqrt{2\,(1+\sigma_p+\sigma_q)}} \quad e^{-\frac{1}{4(1+\sigma_p+\sigma_q)}\left\{(q_{sr}(t)-q_{s'r'}(t))^2(1+2\sigma_p)+(p_{sr}(t)-p_{s'r'}(t))^2(1+2\sigma_q)\right\}}$$
$$\times \quad e^{-\frac{1}{4(1+\sigma_p+\sigma_q)}\left\{-4\sigma_{pq}(q_{sr}(t)-q_{s'r'}(t))(p_{sr}(t)-p_{s'r'}(t))\right\}}.$$

In Fig. (5) we plot the area in phase space associated to crystallized states that carry the irreducible representations of the point groups C_3 and C_6, for $t = 0$. One can see for small values of ϱ that they have the same area as a coherent state, that $A = 1$ while for larger values the area increases one have a three times the area of a coherent state according to the number of components of the crystallized state for C_3 while one gets $A = 6$ for C_6. This is due to the behavior of the scalar product of two coherent states that they become orthogonal when $|\alpha| >> 1$ as it was shown in Fig. (1).

5. Summary and Conclusions

First of all, we have calculated the expectation values of the quadratures of the electromagnetic field with respect to the crystallized states of cyclic groups under the action of a constant force, and the results are

$$\langle p \rangle = -F_0\, t\,, \quad \langle q \rangle = -F_0\, \frac{t^2}{2}\,, \tag{35}$$

Journal of Physics: Conference Series **380** (2012) 012017 doi:10.1088/1742-6596/380/1/012017

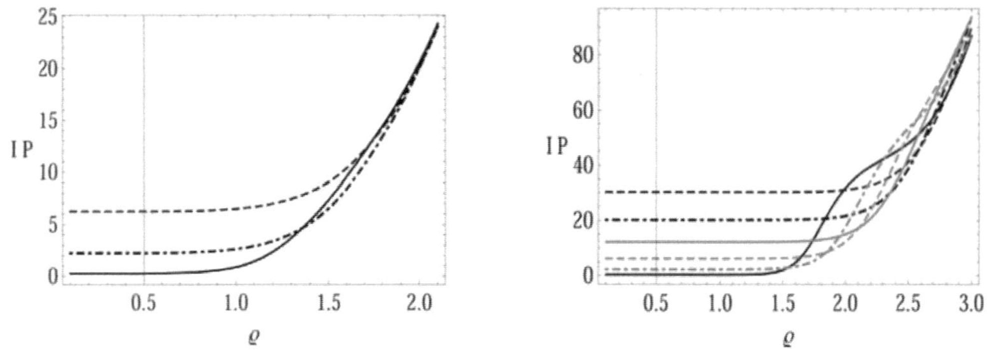

Figure 6. Poincare invariants are shown as functions of ϱ for the point groups C_3 and C_6, when the systems are subject a repulsive constant force $F_0 = 10$ in arbitrary units. They are independent of time, and we use the same code lines than in Figs. (2) and (3).

which have the same value for all the states carrying the irreducible representation of a cyclic group. Their values correspond to the classical evolution of the quadratures. Another interesting result is that the quotient of the fluctuations

$$\frac{\Delta q}{\Delta p} = \sqrt{1+t^2}\,,$$

have a universal value for all the irreducible representation basis states of the cyclic groups.

Another quantity of interest, the universal Poincare-Cartan invariant which is constructed in terms of the quadrature fluctuations of the electromagnetic field, that is

$$I_{PC} = (\Delta q)^2(\Delta p)^2 - (\Delta pq)^2\,. \tag{36}$$

For correlated states that are solutions of the time dependent Schrödinger equation of the Hamiltonian (5) the calculation of this quantity can be easily done. The result is given by $I_{PC} = 1/4$ as can be proved by means of the expressions (16) and using that $\lambda_p\lambda_q^* - \lambda_p^*\lambda_q = i$. Then the correlated states minimize the Schrödinger-Robertson uncertainty relations [7]. In Fig. (6), the Poincare-Cartan invariants for C_3 and C_6 are displayed as function of $\varrho = |\alpha|$ when the corresponding cristallized states are evolving under a constant force Hamiltonian. Of course they are independent of time, always larger than $1/4$, and for large values of ϱ the states of the different irreducible representations have the same bahavior.

The linear time dependent constants of the motion can be obtained by means of the Nöther's theorem. The $(P(t), Q(t))$ constitute a set of inhomogeneous linear canonical transformations and its unitary representation can be used to find the evolution of any wave packet in phase space. For the case of a Wigner distribution function, the unitary representation has the form of a product of Dirac Delta functions. The crystallized Schrödinger cat states can present the squeezing phenomena, correlation between the quadratures, and a subpoissonian photon distribution function.

The evolution of a Wigner function under the action of a general at most quadratic Hamiltonian in the quadratures of the electromagnetic field was established. It moves in phase space according to the matrix elements of the position and momentum operators with respect to coherent states of the parameters α_r associated to the action of the considered point group, it is also rotating by an angle determined by the fluctuations associated to the corresponding generalized correlated states, and in the rotating frame of reference the dispersions are changing but keeping the same area in phase space. This result can be extended to the Wigner function of the crystallized Schrödinger cat states, in this case for each term of the Wigner function one has a similar behavior by replacing the expectation values for the matrix elements, that is the functions $q_{sr}(t)$ and so on. We have the purpose to extend this work to the multimodal case, to be able to calculate quantities of interest in quantum information theory as for example the Von Neumann entropy to measure the entanglement properties of the two modes of the electromagnetic field. calculating the entanglement of cat states, manly to see how changes under evolution of time dependent Hamiltonians.

Acknowledgments
We thank Dieter Schuch for the invitation to participate in the meeting Symmetries in Science XV held in Bregenz, Austria. This work was partially supported by CONACyT-Mexico, project-101541.

Appendix A. Character tables

The character tables for the cyclic group C_n, in this case the irreducible representations are one dimensional and then the characters are given by

	I	R_1	R_2	$\cdots$	R_{n-1}
$\chi^{(0)}$	1	1	1	$\cdots$	1
$\chi^{(1)}$	1	μ_n	μ_n^2	$\cdots$	μ_n^{n-1}
$\chi^{(2)}$	1	μ_n^2	μ_n^4	$\cdots$	$\mu_n^{2(n-1)}$
$\vdots$	$\vdots$	$\vdots$	$\vdots$	$\vdots$	$\vdots$
$\chi^{(n-1)}$	1	μ_n^{n-1}	$\mu_n^{2(n-1)}$	$\cdots$	$\mu_n^{(n-1)(n-1)}$

For the dihedral group D_n we have two cases that behave different: n even and n odd. In the n even case the inversions are divided into two different classes and the rotations of π radians always form its own class. For the n odd case, all the inversions belong to the same class while the rotations form two or more classes. For completeness of the contribution, we give the character tables for the groups: D_3, D_4, D_5, and D_6 taken from the reference [30].

For the dihedral group D_3:

	I	$\left\{R_{\frac{2\pi}{3}}, R_{\frac{4\pi}{3}}\right\}$	$\{U_1, U_2, U_3\}$
$\chi^{(0)}$	1	1	1
$\chi^{(1)}$	1	1	-1
$\chi^{(2)}$	2	-1	0

For the dihedral group D_4:

	I	$\left\{R_{\frac{\pi}{2}}, R_{\frac{3\pi}{2}}\right\}$	R_π	$\{U_1, U_3\}$	$\{U_2, U_4\}$
$\chi^{(0)}$	1	1	1	1	1
$\chi^{(1)}$	1	1	1	-1	-1
$\chi^{(2)}$	1	-1	1	1	-1
$\chi^{(3)}$	1	-1	1	-1	1
$\chi^{(4)}$	2	0	-2	0	0

For the dihedral group D_5:

	I	$\left\{R_{\frac{2\pi}{5}}, R_{\frac{8\pi}{5}}\right\}$	$\left\{R_{\frac{4\pi}{5}}, R_{\frac{6\pi}{5}}\right\}$	$\{U_1, U_2, U_3, U_4, U_5\}$
$\chi^{(0)}$	1	1	1	1
$\chi^{(1)}$	1	1	1	-1
$\chi^{(2)}$	2	γ	γ_2	0
$\chi^{(3)}$	2	γ_2	γ	0

with $\gamma = 2\cos(2\pi/5) = \left(\sqrt{5}-1\right)/2$ the golden ratio and $\gamma_2 = 2\cos(4\pi/5) = -\left(\sqrt{5}+1\right)/2$.

Finally we consider the dihedral group D_6:

	I	$\left\{R_{\frac{\pi}{3}}, R_{\frac{5\pi}{3}}\right\}$	$\left\{R_{\frac{2\pi}{3}}, R_{\frac{4\pi}{3}}\right\}$	R_π	$\{U_1, U_3, U_5\}$	$\{U_2, U_4, U_6\}$
$\chi^{(0)}$	1	1	1	1	1	1
$\chi^{(1)}$	1	1	1	1	-1	-1
$\chi^{(2)}$	1	1	-1	-1	1	-1
$\chi^{(3)}$	1	1	-1	-1	-1	1
$\chi^{(4)}$	2	-1	1	-2	0	0
$\chi^{(5)}$	2	-1	-1	2	0	0

References

[1] Sargent M, Scully M O, and Lamb W E Jr 1975 *Laser Physics* (Reading: Addisson-Wesley)
[2] Meystre P and Walls D F 1991 *Nonclassical effects in Quantum Optics* **4** (New York: American Institute of Physics)
[3] Glauber R J 1963 *Phys. Rev.* **130** 2529
[4] Sudarshan E C G 1963 *Phys. Rev. Lett.* **10** 227
[5] Klauder J R 1963 *J. Math. Phys.* **4** 1055
[6] Dodonov V V, Malkin I A, and Man'ko V I 1974 *Physica* **72** 597
[7] Dodonov V V and Man'ko V I 1989 *Proceedings of the Lebedev Physical Institute* **183** (Commack, NY: Nova Science)
[8] Yurke B and Stoler D 1986 *Phys. Rev. Lett.* **57** 13
[9] Mecozzi A and Tombesi P 1987 *Phys. Rev. Lett.* **58** 1055
[10] Tombesi P and Mecozzi A 1987 *J. Opt. Soc. Am.* B **4** 1700
[11] Gea-Banacloche J 1990 *Phys. Rev. Lett.* **65** 3385
[12] Buzek V and Hladky B 1993 *J. Mod. Phys.* **40** 1309
[13] Gea-Banacloche J 1991 *J. Phys.* A **44** 5913
[14] Castaños O, López-Peña R and Man'ko V I 1995 *J. Russ. Laser Research* **16** 477
[15] Littlewood D E 1962 *University Algebra* (Reading, MA: Addisson-Wesley)
[16] Delgado F, Mielnik C B and Reyes M A 1998 *Phys. Lett.* A **237** 359
[17] Doktorov E V, Malkin I A and Man'ko V I 1977 *J. Mol. Spectrosc.* **64** 302
[18] Schrade G, Man'ko V I, Schleich W P and Glauber R 2010 *Quantum Semiclass. Opt.* **7** 291
[19] Schuch D 2010 *AIP Conf. Proc.* **1334** 307
[20] Castaños O and López-Peña R 1995 in: *Quantum-like Models and Coherent Effects* ed R Fedela and P K Shukla (Singapore: World Scientific) p 3
[21] Moshinsky M and Quesne C 1972 *J. Math. Phys.* **12** 1772.
[22] Padilla E 1996 *Generalized coherent states of quantum systems* (México: Bachelor thesis of the UNAM)
[23] Dodonov V V, Kurmyshev E V and Man'ko V I 1980 *J. Phys.* A **79** 150
[24] García-Calderón G and Moshinsky M 1980 *J. Phys.* A **13** L85
[25] Schuch D and Moshinsky M 2008 *SIGMA* **4** 054
[26] Malkin I A, Man'ko V I and Trifonov D A 1970 *Phys. Rev.* D **2** 1371
[27] Husimi K 1940 *Proc. Phys. Math. Soc. Jpn.* **22** 264
[28] Man'ko V I 1996 *AIP Conf. Proc.* **365** 337
[29] Sugita A and Aiba H 2002 *Phys. Rev.* E **65** 036205
[30] Dresselhaus M S, Dresselhaus G and Jorio A 2008 *Group theory: Application to the physics of Condensed Matter* (Berlin, Heidelberg: Springer-Verlag)

Symmetries in Science XV IOP Publishing
Journal of Physics: Conference Series **380** (2012) 012018 doi:10.1088/1742-6596/380/1/012018

Time delay in the reflection of particles by semi-harmonic wells

Oscar Rosas-Ortiz
Physics Department, Cinvestav, A.P. 14-740, México D.F., Mexico
E-mail: orosas@fis.cinvestav.mx

Sara Cruz y Cruz, Nicolás Fernández-García
Sección de Estudios de Posgrado e Investigación, UPIITA-IPN, Av. IPN 2580, C.P. 07340 México D.F., Mexico
E-mail: sgcruzc@ipn, jfernandezg@ipn.mx

Abstract. The phase time of wave reflection from a one-dimensional semi-harmonic well includes a delay term which can be negative according to the incoming energy and the well parameters in a quantum scattering process. The model is a rectangular well which is embedded in an environment composed by a zero potential energy to its right and a harmonic-like potential to its left.

1. Introduction

The time spent by a particle in traversing an spatial region is one of the most striking features of quantum theory [1, 2]. For instance, the time taken by a particle in the tunneling of a potential barrier is independent of the barrier thickness [3]. This result has stimulated the designing of high-speed devices based on the tunneling properties of semiconductors [1]. Similar effects have been predicted for potential wells [4], where negative phase times $\tau_W = \hbar \frac{d}{dE}\delta$ should be observable for electromagnetic propagation [5].

We recently proposed a one-dimensional well which reduces the scattering process to the case of purely wave reflection [6]. This is a rectangular well with a parabolic potential to its left and zero potential energy to its right. In dimensionless form it reads:

$$V(x;a) = \begin{cases} x^2 & x \leq -a \\ -V_0 & |x| < a \\ 0 & a \leq x \end{cases} \tag{1}$$

The purpose of the semi-harmonic environment is to ensure that all particles coming from $+\infty$ are reflected by the well. As a byproduct, the number N of bound states $(E_0, E_1, \ldots, E_{N-1})$ which are admitted by a given square well is preserved when the semi-harmonic background is added. Yet, these two systems are distinguishable because the corresponding energies are displaced towards the positive threshold in the semi-harmonic case. The effect is also present in the resonances $\epsilon = \varepsilon - i\Gamma/2$ by displacing the semi-harmonic resonance positions ε to the right with respect to their partners of the rectangular well. In our previous work [6], we have shown that the phase time (group delay) associated to the reflection of particles from a semi-harmonic delta well leads to negative time delays (see also [7]). Similar results can be found for reflectionless potentials (see e.g. [8] for the broadening of wave packets which are scattered by a Pöschl-Teller potential). In the present contribution we make a classification of the time delays associated to the semi-harmonic potentials $V(x;a)$.

The organization of the paper is as follows. In Section 2 we present the generalities of the quantum problem associated to the semi-harmonic well. Section 3 is devoted to the analysis of the scattering states. The Jost function is used to construct a pair of physical solutions which lead to the phase time definition. In Section 4 the differences between the motion of a classical particle and the center of a wave packet are analyzed when they are affected by the semi-harmonic well. The contribution of the semi-harmonic environment to the time delay is identified, and some cases of interest are discussed. Final remarks are given in Section 5. The details of the solutions to the Schrödinger equation for the bound states and resonances of this problem have been included in a short appendix at the very end of the paper.

2. General solution

Let us consider the stationary dimensionless Schrödinger equation

$$(H-E)\psi(x;E)=0, \qquad H:=-\frac{d^2}{dx^2}+V(x;a), \tag{2}$$

with $V(x;a)$ the potential defined in (1). The solutions of this equation satisfy the boundary conditions

$$\lim_{x\to-\infty}\psi(x)=0, \tag{3}$$

and

$$\begin{aligned}\psi(-a-0;E)=\psi(-a+0;E); \quad \psi(a-0;E)=\psi(a+0;E),\\ \psi'(-a-0;E)=\psi(-a+0;E); \quad \psi'(a-0;E)=\psi(a+0;E).\end{aligned} \tag{4}$$

Given the energy $E>-V_0$, we take $k=\sqrt{E}$ and $q=\sqrt{V_0+k^2}$, to write the solutions $\psi(x;E)$ as follows [6]:

$$\psi(x;E)=\begin{cases}\frac{\varphi(x,k)}{\varphi(-a,k)}\,e^{(a^2-x^2)/2} & x<-a\\ C(k,a)\sin(qx)+D(k,a)\cos(qx) & |x|\le a\\ \frac{i}{2k}[\mathcal{J}(k,a)e^{-ikx}-\mathcal{J}^*(k,a)e^{ikx}] & a<x\end{cases} \tag{5}$$

where z^* is the complex conjugate of $z\in\mathbb{C}$, and a global constant has been made equal to one for simplicity (see Appendix A). The coefficients $C(k,a)$ and $D(k,a)$ are given in (A.3). In turn, $\varphi(x,k)$ and the Jost Function $\mathcal{J}(k,a)$ are respectively given in equations (A.2) and (A.4). Notice that $\mathcal{J}(k,a)$ is assumed to be analytically continued to the complex k-plane in order to face the spectral problem (2). The analysis for bound states ($-V_0<E<0$) and resonances ($\epsilon=E-i\Gamma/2$) can be consulted in Appendix A.

3. The scattering problem

For scattering states ($E>0$), the Jost function $\mathcal{J}$ is useful to transform the ψ functions (5) into the so called *physical wave solutions*

$$\psi^+(x;E)=\sqrt{\frac{2}{\pi}}\frac{k\psi(x;E)}{\mathcal{J}(k,a)}, \quad \psi^-(x;E)=\sqrt{\frac{2}{\pi}}\frac{k\psi(x;E)}{\mathcal{J}^*(k,a)}. \tag{6}$$

These functions are proportional to each other $\psi^+(x;E)=S(k,a)\psi^-(x;E)$, with

$$S(k,a)=\frac{\mathcal{J}^*(k,a)}{\mathcal{J}(k,a)} \tag{7}$$

the S-matrix (reflection amplitude) of the problem [9, 10]. Thus, the physical wave solutions (6) are eigenfunctions of H belonging to the eigenvalue $E = k^2$ and fulfilling the boundary conditions

$$\lim_{x\to-\infty} \psi^{\pm}(x; E) = 0, \tag{8}$$

$$\psi^{+}(x; E) = \frac{i}{\sqrt{2\pi}}\left[e^{-ikx} - S(k, a)e^{ikx}\right], \quad x > a, \tag{9}$$

$$\psi^{-}(x; E) = -\frac{i}{\sqrt{2\pi}}\left[e^{ikx} - S^{*}(k, a)e^{-ikx}\right], \quad x > a. \tag{10}$$

From (9) and (10) one gets $(\psi^{+})^{*} = \psi^{-}$, so that $|S(k, a)|^2 = 1$. That is, the reflection amplitude $S(k, a)$ can only represent a phase shift since the process is elastic. Taking the polar form of the Jost function $\mathcal{J}(k, a) = |\mathcal{J}(k, a)|e^{-i\delta(k,a)}$, the S-matrix becomes $S(k, a) = e^{2i\delta(k,a)}$, and

$$\psi^{+}(x; E) = \frac{i}{\sqrt{2\pi}}\left[e^{-ikx} - e^{i(kx+2\delta)}\right], \quad x > a. \tag{11}$$

The time-displacement of $\psi^{+}(x; E)$ in $(a, +\infty)$ is given by

$$\psi^{+}(x; E, t) = \frac{i}{\sqrt{2\pi}}\left[e^{-i(kx+\omega\Delta t)} - e^{i(kx+2\delta-\omega\Delta t)}\right], \quad x > a. \tag{12}$$

Here we have taken $E = k^2 = \omega$ and $\Delta t = t - t_0$, with $t_0 \geq 0$ an arbitrary value of the time-parameter t. The function $\psi^{+}_{inc}(x; k, t) = \frac{i}{\sqrt{2\pi}}e^{-i(kx+\omega\Delta t)}$ represents a plane wave propagating towards the left with *phase velocity* $v_{\phi}(k) = \frac{\omega}{k} = k$. In turn, the function $\psi^{+}_{out}(x; k, t) = -\frac{i}{\sqrt{2\pi}}e^{i(kx+2\delta-\omega\Delta t)}$ includes a phase shift 2δ and it corresponds to a plane wave propagating towards the right with phase velocity k. At a given time t, we have $\psi^{+}_{inc} \to \phi_{-}$ and $\psi^{+}_{out} \to \phi_{+}$ as $x \to +\infty$, where

$$\phi_{\pm}(x; k) = \mp\frac{i}{\sqrt{2\pi}}e^{\pm ikx} \tag{13}$$

are the *irregular* eigenfunctions which belong to the continuous eigenvalue $E = k^2 > 0$ of the free Hamiltonian $H_0 = -\frac{d^2}{dx^2}$. To get regular wave solutions let us construct the *incoming* and *outgoing* wave packets (by simplicity we take $t_0 = 0$):

$$\psi_{inc}(x, t) = \int_{-\infty}^{+\infty} dk\Lambda_{inc}(k)\psi^{+}_{inc}(x; k, t) = \frac{i}{\sqrt{2\pi}}\int_{-\infty}^{+\infty} dk\Lambda_{inc}(k)e^{-i(kx+\omega t)}, \quad x > a \tag{14}$$

$$\psi_{out}(x, t) = \int_{-\infty}^{+\infty} dk\Lambda_{out}(k)\psi^{+}_{out}(x; k, t) = \frac{i}{\sqrt{2\pi}}\int_{-\infty}^{+\infty} dk\Lambda_{out}(k)e^{i(kx+2\delta+\pi-\omega t)}, \quad x > a \tag{15}$$

The expressions $\Lambda_{in}(k)$ and $\Lambda_{out}(k)$ stand for functions of k which are narrowly peaked around k_0. For instance, the distribution

$$\Lambda_{inc}(k) = (2\pi\lambda^2)^{1/4}e^{-\frac{\lambda^2}{4}(k-k_0)^2}, \quad \lambda \geq 0 \tag{16}$$

has the width $\lambda\Delta k = 1$, and it gives rise to a Gaussian incoming packet of width $2\Delta x = \lambda$. In the limit $\lambda \to +\infty$, we have $\Delta k = 0$. This means $\Lambda_{inc}(k) \to \delta(k - k_0)$, so that the wave packet $\psi_{inc}(x, t)$ is very broad ($\Delta x \to +\infty$). Finally, the peak value of the wave packets (14) and (15) propagates with a *group velocity* which is twice the phase velocity $v_{\phi}(k_0)$:

$$v_g(k_0) = \left.\frac{d\omega}{dk}\right|_{k=k_0} = 2k_0. \tag{17}$$

Hereafter $\psi_{inc}(x,t)$ will represent the quantum state of a particle with energy $E_0 = k_0^2 > 0$, arriving from $+\infty$ towards the effective zone of the potential $(-\infty, a)$. To represent the quantum state of the reflected particle we shall use $\psi_{out}(x,t)$. Remember that the wave packets ψ_{inc} and ψ_{out} are assumed to have very narrow widths so that they are very broad in coordinate representation. Henceforth, before the center of $\psi_{inc}(x,t)$ arrives at $x = a$, the action of the potential makes the packet wavefront of $\psi_{inc}(x,t)$ turn back. This effect produces the wavefront of $\psi_{out}(x,t)$ to appear in $x \geq a$. Hence, close to the cutoff $x = a$, important interference effects between the incident and reflected wave packets are expected.

4. Phase time and delay time

Applying the stationary phase condition[1] on $\psi_{inc}(x,t)$, one gets the classical kinematical rule for particles moving with constant velocity: $x = -v_g t$. This expression defines the time-dependence of the incoming wave packet center x_{inc}. In a similar form, $x = v_g(t + \tau_W)$ defines the time-dependence of the outgoing wave packet center x_{out} with

$$\tau_W = -2\frac{d}{dE}\delta(E,a) \tag{18}$$

the *Wigner phase time* [11], and $\delta(E,a)$ the *reflection phase shift* defined in (A.7).

Though the time-dependence of the wave packet centers x_{inc} and x_{out} seems to be classical, their motion is clearly different from that of the classical particles. The predictions of classical mechanics are as follows: when the particle has the energy $E \geq 0$ and arrives from the right, it undergoes an abrupt acceleration at $x = a$, then an abrupt deceleration at $x = -a$, and then continues towards the left by decelerating its movement until it reaches the turning point x_0 defined by $E = x_0^2$. The momentum of the particle is instantly reversed at x_0 and the process is repeated in reverse order. In contrast, the motion of x_{inc} does not obey the classical mechanical laws. The center of the incoming wave packet retraces its steps before reaching the classical turning point x_0. As it was explained above, the action of the potential on the ψ_{inc} wavefront is sufficient to make this wave packet turn back before its center x_{inc} arrives at the classical turning point x_0. To verify our statement let us apply the time-dependence rules derived above and (A.7) to write the explicit form of the phase time

$$\tau_W = \tau_f - \tau \equiv \frac{2a}{v_g} - 2\frac{d}{dE}\arctan(\alpha). \tag{19}$$

Here $\alpha(k,a) = \frac{J_i(k,a)}{J_r(k,a)}$, with $J_r(k,a)$ and $J_i(k,a)$ defined in (A.6). Equation (19) shows that τ_W consists on the time $\tau_f = 2a/v_g$ spent by x_{inc} to traverse the distance $2a$ with constant speed $v_g = 2k$, minus the *reflection time delay*

$$\tau = 2\frac{d}{dE}\arctan(\alpha) = \left[\frac{1}{1+\alpha^2(k,a)}\right]\frac{d}{dk}\alpha(k,a). \tag{20}$$

The distance $2a$ corresponds to the trajectory described by a classical free particle in its transit from $x = a$ to the origin of the straight-line $x = 0$, and from $x = 0$ to a. Since $0 \neq x_0$ for any scattering energy $E > 0$, there is a clear difference between the motion of x_{inc} and the motion of a classical particle, both of them subjected to the same initial conditions.

According to these last results, the delay contribution τ to the phase time (18) must be due to both the rectangular well and the semi-harmonic environment. This affirmation is clear by writing the ratio $\alpha = J_i/J_r$ explicitly. From (A.6) one gets the following product

$$\alpha(k,a) = \Theta(k,a)\left[\left(\frac{k^2+q^2}{2kq}\right)\tan(2qa)\right]. \tag{21}$$

[1] That is, the derivative with respect to k of the phase is zero for $k = k_0$.

The term in square brackets of this last equation is the kernel to calculate the phase time of a rectangular well [4, 5, 12], so that the factor $\Theta(k, a)$ is due exclusively to the semi-harmonic environment. This is defined as follows

$$\Theta(k,a) = \left(\frac{2k^2}{k^2+q^2}\right)\left[\frac{\beta_\varphi(-a,k) - a - q\cot(2qa)}{\beta_\varphi(-a,k) - a + q\tan(2qa)}\right]. \tag{22}$$

Let us emphasize that the zeros of $\Theta(k, a)$ coincide with the singularities of $\tan(2qa)$ and viceversa, so that the ratio (21) is a regular function of k and a. The function $\beta_\varphi(k, a)$, on the other hand, plays a relevant role in the defining of $\Theta(k, a)$. That can be zero or even be singular for specific values of a and k which are different from the zeros of $\sin(2qa)$ and $\cos(2qa)$, as it is shown in Figure 1.

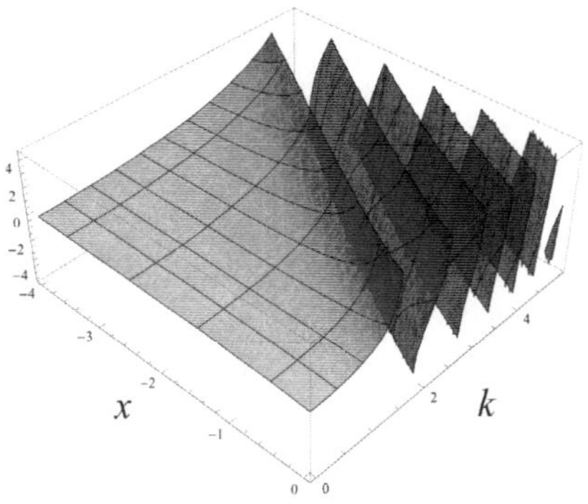

Figure 1. The function $\beta_\varphi(x, k)$ used in the definition of $\Theta(k, a)$ exhibits a diversity of changes of sign according to its zeros and singularities.

In order to analyze the behavior of τ_W and τ notice that the reflection phase shift $\delta = -ka + \arctan(\alpha)$ consists of a negative monotonically decreasing function $-ak = -a\sqrt{E}$, plus the monotonically increasing function $\arctan(\alpha)$. Since this last function is bounded between $-\pi/2$ and $\pi/2$, the phase shift is a negative decreasing function of $E \geq 0$. The periodicity of $\alpha(k, a)$ induces oscillations on the second term, so that a rapid increasing of the phase is expected in the vicinities of the maxima of $\arctan(\alpha)$. These increases should be balanced by the appropriate decreases of $\delta(k, a)$ [11]. Under these considerations, the phase time τ_W defined in (19) is such that the delay contribution (20) can be even negative. The situation is illustrated in Figure 2, where we have depicted the time delay τ for two different semi-harmonic wells of unit area $2aV_0 = 1$. For $a = 4.6257$ the time delay τ is negative in the energy interval $(0, 7.06956 \times 10^{-4})$. However, if $a = 4.8257$ the time delay is positive for any energy $E \geq 0$. In general, for unit area semi-harmonic wells, it can be shown that negative time delays appear for a definite set of parameters $0 \leq a \leq a_0 \approx 4.7257$ [12]. Of particular interest, in the limit $a \to 0$ we obtain the time delay for a semi-harmonic delta well [6] by the substitution of

$$\lim_{a\to 0} \alpha(k,a) = \frac{\Gamma\left(\frac{1-k^2}{4}\right)k}{2\Gamma\left(\frac{3-k^2}{4}\right) - \Gamma\left(\frac{1-k^2}{4}\right)} \tag{23}$$

into the equation (20). It should be pointed out that the semi-harmonic delta well induces negative delays in the energy interval $(0, 0.45727096)$. This is the largest interval in which τ is negative for any of the unit area semi-harmonic wells [7].

The phenomenon of negative delays in the semi-harmonic wells is not exclusive of the rectangular wells of unit area. Indeed, the identification of intervals in which the parameters a and V_0 give rise to negative time delays can be done for arbitrary amounts of bound states. Namely, the unit area potentials (1) admit with certainty a single bound state. However, the set

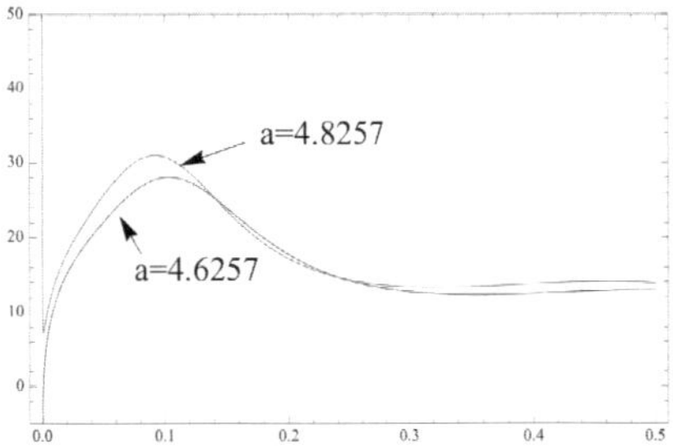

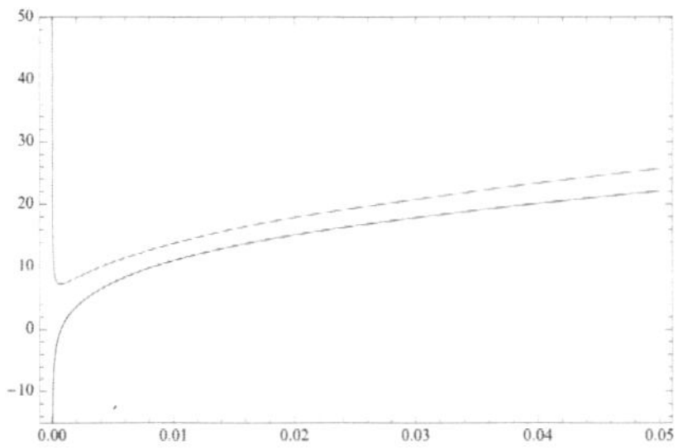

Figure 2. The time delay τ for a semi-harmonic well of unit area with $a = 4.6257$ (blue curve) and $a = 4.8257$ (red curve). These values of the parameter a are close to the critical value $a_0 \approx 4.7257$ which defines the change of sign in τ [12]. A detail of the behavior of τ for low energies is shown at the right.

of semi-harmonic wells which admit a single bound state is wider than the set of unit area wells. The parameters V_0 and a which give rise to negative delays as well as the number of bound states are defined by the area of the well [12]. For example, the Figure 3 shows the case of two different semi-harmonic wells which admit only two bound states. They are distinguishable from each other because only one of them gives rise to negative time delays τ.

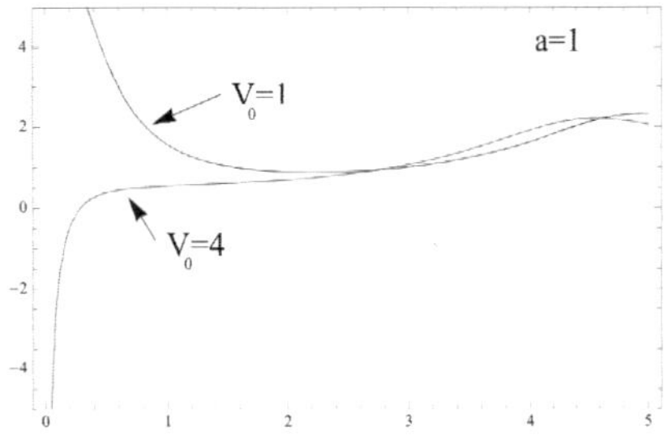

Figure 3. The time delay τ for a semi-harmonic well which admits only two bound states for $a = 1, V_0 = 1$ (blue curve), and $a = 1, V_0 = 4$ (red curve). In this last case τ becomes negative in the energy interval $(0, 0.2813)$.

5. Concluding remarks

It is well known that the quantum mechanical problem of a piecewise constant potential can be always associated with an optical problem [13]. Consider a medium M whose index n has discontinuities of the same type as those of the quantum potential V. The propagation of an electromagnetic wave of angular frequency Ω in M is related to the quantum problem in terms of the identification ([13], pp 36):

$$n(\Omega) = \sqrt{\frac{2mc^2(E-V)}{\hbar^2\Omega^2}}.$$

A transparent medium M corresponds to the region where $E > V$, and the light wave is of the form e^{ipx}. If $E < V$ the index n is pure imaginary and the light wave is evanescent $e^{-p'x}$. This optical analog has been used in the study of particles passing through a rectangular well for which negative phase times have been confirmed [4, 5]. The time delay of the semi-harmonic wells can be studied in a similar form in order to look for observable effects which are attainable to the environment. The detailed study of this problem can be found elsewhere [12].

Appendix A

The solutions of the Schrödinger equation (2) which satisfy the boundary conditions (4) are given by

$$\psi(x;E) = A_1\varphi(-a,k)e^{-a^2/2}\begin{cases} \frac{\varphi(x,k)}{\varphi(-a,k)}e^{(a^2-x^2)/2} & x<-a \\ C(k,a)\sin(qx)+D(k,a)\cos(qx) & |x|\leq a \\ \frac{i}{2k}[\mathcal{J}(k,a)e^{-ikx}-\mathcal{J}^*(k,a)e^{ikx}] & a<x \end{cases} \tag{A.1}$$

where the superscript "*" stands for complex conjugation, and

$$\varphi(x,k) = {}_1F_1\left(\frac{1-k^2}{4},\frac{1}{2};x^2\right)+2x\frac{\Gamma\left(\frac{3-k^2}{4}\right)}{\Gamma\left(\frac{1-k^2}{4}\right)}{}_1F_1\left(\frac{3-k^2}{4},\frac{3}{2};x^2\right), \tag{A.2}$$

with ${}_1F_1(a,c;z)$ and $\Gamma(z)$ respectively the hypergeometric and Euler gamma functions of $z\in\mathbb{C}$. Given the energy E, the wavenumber $k=\sqrt{E}$ and the interaction parameter $q=\sqrt{V_0+k^2}$ take definite values so that the coefficients in (A.1) become

$$\begin{aligned} C(k,a) &= -\frac{[q\sin(qa)+(\beta_\varphi(-a,k)-a)\cos(qa)]}{q}, \\ D(k,a) &= \frac{[q\cos(qa)-(\beta_\varphi(-a,k)-a)\sin(qa)]}{q}. \end{aligned} \tag{A.3}$$

Here, $-\beta_\varphi(x,k)$ stands for the logarithmic derivative of $\varphi(x,k)$ with respect to x. The arbitrary integration constant A_1 has been taken as $A_1=e^{a^2/2}/\varphi(-a,k)$ in the paper. On the other hand, the Jost function $\mathcal{J}(k,a)$ of the problem we are dealing with is defined by

$$e^{-ika}\mathcal{J}(k,a) = \sin(qa)[-ik+q\cot(qa)]C(k,a)-\cos(qa)[ik+q\tan(qa)]D(k,a). \tag{A.4}$$

This last expression can be rewritten in the form

$$e^{-ika}\mathcal{J}(k,a) = J_r(k,a)-iJ_i(k,a), \tag{A.5}$$

with

$$\begin{aligned} J_r(k,a) &= q\left[C(k,a)\cos(qa)-D(k,a)\sin(qa)\right] \\ &= -[q\sin(2qa)+(\beta_\varphi(-a,k)-a)\cos(2qa)]; \\ J_i(k,a) &= k\left[C(k,a)\sin(qa)+D(k,a)\cos(qa)\right] \\ &= \frac{k}{q}\left[q\cos(2qa)-(\beta_\varphi(-a,k)-a)\sin(2qa)\right]. \end{aligned} \tag{A.6}$$

Using $\mathcal{J}(k,a)=|\mathcal{J}(k,a)|e^{-i\delta(k,a)}$, we finally get

$$\delta(k,a) = -ka+\arctan\left(\frac{J_i(k,a)}{J_r(k,a)}\right)\equiv -ka+\arctan[\alpha(k,a)], \tag{A.7}$$

where we have made $\alpha(k,a)=\frac{J_i(k,a)}{J_r(k,a)}$.

The condition for the trapping of particles corresponds to make $\mathcal{J}(k,a)=0$ [6]. That is, the wave functions of either bound states or resonances are written as

$$\psi(x;E) = \begin{cases} \frac{\varphi(x,k)}{\varphi(-a,k)}e^{(a^2-x^2)/2} & x<-a \\ C(k,a)\sin(qx)+D(k,a)\cos(qx) & |x|\leq a \\ \frac{1}{2ik}\mathcal{J}^*(k,a)e^{ikx} & a<x \end{cases} \tag{A.8}$$

The zeros of the Jost function are defined by the roots of the transcendental equation

$$\beta_\varphi(-a,k) = a - ik + \left[\frac{ik\beta_\varphi(-a,k) - ika - q^2}{q}\right]\tan(2qa). \tag{A.9}$$

In particular, if $k = i\kappa$ is a root of (A.9) with $\kappa > 0$, then $E = k^2 = -\kappa^2$ is the energy eigenvalue of a bound state. The number N of these roots is determined by the area $A = 2aV_0$ of the rectangular well. Thereby, the bound states are represented by the square-integrable functions

$$\psi_n(x;E_n) = \begin{cases} \frac{\varphi_n(x,\kappa_n)}{\varphi_n(-a,k_n)}\, e^{(a^2-x^2)/2} & x < -a \\ C(\kappa_n,a)\sin(q_n x) + D(\kappa_n,a)\cos(q_n x) & |x| \leq a \\ \frac{1}{2i\kappa_n}\mathcal{J}^*(\kappa_n,a)e^{i\kappa_n x} & a < x \end{cases} \tag{A.10}$$

where $n = 0, 1, \ldots, N-1$. It is important to remark that the number N of bound energies is the same as that of a rectangular well of the same area. The difference lies on the fact that the energies $E_0, E_1, \ldots, E_{N-1}$ of the semi-harmonic well are displaced towards the positive threshold with respect to their partners of the rectangular well. This property does not depend on the geometry of the rectangle: the wells having the same area admit the same number of bound states. For instance, the single bound energy of a delta well is $E_0 = -0.25$. This becomes less negative $E_0 = -7.97104 \times 10^{-2}$ in the presence of the semi-harmonic background [6].

Acknowledgments

The support of CONACyT grant 152574 and the IPN grants SIP20113705 and SIP20111061 are acknowledged. The authors are indebted to G. García-Calderón and V. Maipo for enlightening discussions. ORO wishes to thank the Organizers of the Symmetries in Science 2011 Symposium for the kind invitation and the warm hospitality at Bregenz, Austria.

References

[1] Muga JG, Sala R and Egusquiza IL (Eds.) 2008 *Time in Quantum Mechanics* 2nd Ed. (Springer)
[2] Muga JG, Ruschhaupt A, and del Campo A (Eds.) 2010 *Time in Quantum Mechanics Vol. 2* (Springer)
[3] Hartman TE 1962 Tunneling of a Wave Packet *J. Appl. Phys.* **33** 3427
[4] Li CF and Wang Q 2000 Negative phase time for particles passing through a potential well *Phys. Lett. A* **275** 287
[5] Vetter RM, Haibel A, and Nimtz G 2001 Negative phase time for scattering at quantum wells: A microwave analogy experiment *Phys. Rev. E* **63** 046701
[6] Fernández-García N and Rosas-Ortiz O 2011 Rectangular Potentials in a Semi-Harmonic Background: Spectrum, Resonances and Dwell Time *SIGMA* **7** 044
[7] Rosas-Ortiz O, Cruz y Cruz S, and Fernández-García N 2012 Negative time delay for wave reflection from a one-dimensional semi-harmonic well, to appear in P Kielanowski et. al. (Eds.) *Proc. of the XXX Workshop on Geom. Meth. Phys.* (Poland)
[8] Nussbaum IG and Kleber M 1988 Delayed broadening in reflectionless scattering *J. Phys. A: Math. Gen* **21** 2953
[9] García-Calderón G 2011 Transient effects in quantum decay *AIP Conf. Proc.* **1334** 84-122
[10] Muga JG 2008 Characteristic Times in One-Dimensional Scattering *Lect. Notes Phys.* **734** 31-72
[11] Wigner EP 1955 Lower limit for the energy derivative of the scattering phase shift *Phys. Rev.* **98** 145
[12] Alonso L, Cruz y Cruz S, Fernández-García N, and Rosas-Ortiz O 2011 *preprint Cinvestav*
[13] Cohen-Tannoudji C, Diu B, and Laloë F 1977 *Quantum Mechanics Vol. 1* (John Wiley)

Journal of Physics: Conference Series **380** (2012) 012019 doi:10.1088/1742-6596/380/1/012019

Coherent state-based generating function approach for Franck–Condon transitions and beyond[1]

J Huh[1,2,3*] **and R Berger**[1,2*]

[1] Clemens-Schöpf Institute, TU Darmstadt, Petersenstr. 22, 64287 Darmstadt, Germany

[2] Frankfurt Institute for Advanced Studies, Goethe University, Ruth-Moufang-Str. 1, 60438 Frankfurt am Main, Germany

[3] Current address: Department of Chemistry and Chemical Biology, Harvard University, Cambridge, Massachusetts 02138, USA

E-mail: *huh@fas.harvard.edu and *robert.berger@tu-darmstadt.de

Abstract. One-photon and multi-photon absorption, spontaneous and stimulated photon emission, resonance Raman scattering and electron transfer are important molecular processes that commonly involve combined vibrational-electronic (vibronic) transitions. The corresponding vibronic transition profiles in the energy domain are usually determined by Franck–Condon factors (FCFs), the squared norm of overlap integrals between vibrational wavefunctions of different electronic states. FC profiles are typically highly congested for large molecular systems and the spectra usually become not well-resolvable at elevated temperatures. The (theoretical) analyses of such spectra are even more difficult when vibrational mode mixing (Duschinsky) effects are significant, because contributions from different modes are in general not separable, even within the harmonic approximation. A few decades ago Doktorov, Malkin and Man'ko [1979 *J. Mol. Spectrosc.* **77**, 178] developed a coherent state-based generating function approach and exploited the dynamical symmetry of vibrational Hamiltonians for the Duschinsky relation to describe FC transitions at zero Kelvin. Recently, the present authors extended the method to incorporate thermal, single vibronic level, non-Condon and multi-photon effects in energy, time and probability density domains for the efficient calculation and interpretation of vibronic spectra. Herein, recent developments and corresponding generating functions are presented for single vibronic levels related to fluorescence, resonance Raman scattering and anharmonic transition.

1. Introduction

One of the fundamental goals of molecular science is to design properties on the molecular level and to obtain perfect control of molecular processes with the help of optical techniques. For this purpose a detailed understanding of the dynamics of vibrational and electronic (vibronic) degrees of freedom (DOF) is necessary, because the interplay of vibronic DOF plays a major role in many molecular processes, which could be either radiative (*e.g.* one-photon absorption (OPA), one-photon emission (OPE) and resonance Raman (rR) scattering) or non-radiative (*e.g.* electron transfer (ET), internal conversion (IC), inter-system crossing (ISC) and conduction of molecular junction).

Usually, polyatomic systems of interest are relatively large (more than 100 atoms, especially for biomolecular applications) such that the corresponding description and analysis of the

[1] This contribution to the proceedings contains (slightly modified) parts from the dissertation of J. Huh in 2010.

dynamics are challenging. Molecular spectra, which carry important information about the underlying microscopic processes, are typically highly congested, such that the dynamics of the individual vibrational DOF can not directly be disentangled. Theoretical analysis tools for complex molecular systems are required to study the individual mode contributions to a given molecular process. Computational difficulties arise for complex systems in evaluating a tremendously large number of multi-dimensional Franck–Condon (FC) integrals. The density of states (DOS) or the number of FC integrals at a transition energy grows steeply with increasing number of vibrational DOF, vibrational excitation energy and temperature [1].

Even in a harmonic approximation for the Born-Oppenheimer (BO) potential energy surface (PES) the computation of FC integrals is still a challenging problem especially for large molecular systems, due to the (partial) inseparability of multi-dimensional integrals caused by Duschinsky mode mixing effects (Ref. [2] and Sec. 2.2). There have been attempts (see *e.g.* Refs. [3–9] for some of the more recent work) to improve the computational efficiency of the FC integral evaluation schemes including the Duschinsky mode mixing effects. Fast evaluation of the FC integrals is essential for tackling large systems. We have suggested an iterative FC integral evaluation scheme exploiting the Magnus expansion for multi-variate Hermite polynomials (MHPs) [10]. The integral evaluation scheme in terms of the MHPs with the Magnus expansion can be expressed as summations of one-dimensional Hermite polynomials, which appears in a simpler form (smaller number of summations) than the existing summation scheme for the multiple products of one-dimensional Hermite polynomials (see *e.g.* Refs. [8, 9]).

It was pointed out, however, that only a small portion of the Franck–Condon factors (FCFs) contributes to the total FC profiles significantly (see *e.g.* Refs. [11–13]). The idea can be used in an *ad hoc* way by limiting for instance the vibrational excitation in each mode by some predefined numbers. Unfortunately, in this brute-force way the computational complexity still increases drastically and becomes infeasible. Instead, if we can construct the most important part of the FC intensity profiles, that accounts say for 99 % of the total integrated FC profile, with a small fraction of FCFs, the computation of FC intensity profiles becomes feasible even for complex systems. For this purpose we must tackle the problem of identifying those relevant subsets of FCFs that contribute to the total FC intensity for about 99 % without having all FCFs at hand. Jankowiak *et al.* [11] exploited this idea by modifying Doktorov et al.'s initial idea of coherent state (CS)-based generating function (GF) [14]. Exploiting the modified GF Jankowiak *et al.* developed an efficient coarse-grained integral prescreening strategy for FC profiles at zero Kelvin.

Vibronic transition. Intramolecular (optical or radiationless) processes are often considered as electronic transitions between two distinct BO PESs via Fermi's golden rule (FGR) [15, 16] resulting from time-dependent (TD) perturbation theory within the BO picture. This vibronic transition model has been used to explain molecular transition processes for OPA (see *e.g.* Refs. [11–14, 17–47]), OPE (see *e.g.* Refs. [48–52]), rR scattering (see *e.g.* Refs. [53–70]), ET (see *e.g.* Refs. [71–83]), IC (see *e.g.* Refs. [84–89]) and ISC (see *e.g.* Ref. [90]). Also other vibronic transitions like electronic circular dichroism (ECD) (see *e.g.* Ref. [91]) and two photon absorption and emission (see *e.g.* Ref. [92]) are usually described via FGR [15, 16].

We may approximate the spectral density function (SDF) within the harmonic oscillator model, which is convoluted with the Lorentzian line shape function ($L(\omega)$) with full width at half maximum (FWHM) of Γ,

$$\rho_{\mathrm{L}}(\omega;T) = \sum_{\underline{v},\underline{v}'} p_{\underline{v}}(T)|\langle\underline{v}'|\hat{\underline{\mu}}(\underline{Q})|\underline{v}\rangle|^2 L(\omega,\omega_0+\omega_{\underline{v}',\underline{v}};\Gamma),$$

$$L(\omega,\overline{\omega};\Gamma) = \frac{1}{\pi}\cdot\frac{(\frac{\hbar}{2}\Gamma)}{[\hbar(\omega-\overline{\omega})]^2+(\frac{\hbar}{2}\Gamma)^2}, \tag{1}$$

where $|\underline{v}\rangle$ and $|\underline{v}'\rangle$ are N-dimensional harmonic oscillator eigenstates of the initial and final electronic states, respectively, in the occupation number vector (ONV) representation. $p_{\underline{v}}(T)$ is the population of the initial vibronic state at temperature T and, here explicitly for electric dipole transitions, $\hat{\underline{\mu}}(\underline{Q})$ is the electronic transition dipole moment (TDM) which depends on the normal coordinates $\underline{Q}$ of the initial state. ω_0 is the adiabatic vibronic transition frequency and $\omega_{\underline{v}',\underline{v}}$ corresponds to the vibrational frequency difference of two harmonic eigenstates excluding the harmonic zero-point frequency difference which has already been included in ω_0.

Within the validity of the FC approximation ($\hat{\underline{\mu}}(\underline{Q}) \simeq \underline{\mu}_0$) and the harmonic oscillator approximation there still remains a challenging problem in evaluating Eq. (1) in the time-independent (TI) manner for two multi-dimensional harmonic oscillators being not only displaced and distorted but also rotated. When the molecule (multi-dimensional harmonic oscillators) undergoes an electronic transition, it experiences in general an equilibrium structural change (displacement), potential energy curvature change (distortion) and normal coordinate variation (rotation). The two sets of normal coordinates are related (approximately) by a linear transformation, the so-called Duschinsky transformation (Ref. [2] and see also Sec. 2.2 for details). The Duschinsky effect is one of the main reasons for asymmetry between absorption and emission spectra. It is responsible for the broadening of vibronic spectra with the effect being typically enhanced (for specific vibrational modes) at elevated temperatures. Small [93] pointed out that the Duschinsky effect is as important as vibronic coupling effects because the mode mixing introduces a quadratic coupling between initial and final vibrational wavefunctions, *i.e.* the Duschinsky effects must be considered for vibronic coupling problems. The importance of Duschinsky mode mixing has been emphasised by many authors in various fields, such as absorption processes (see *e.g.* Refs. [11–13, 40, 42, 94]), resonance Raman scattering (see *e.g.* Refs. [49,54,57,59,60,65–67,69,95]), electron transfer processes (see *e.g.* Refs. [52,76,80,81,96]), radiationless transitions (see *e.g.* Ref. [85,88]), in photoexcited state cooling processes (see *e.g.* Ref. [36]), vibronic coupling of electronic transitions (see *e.g.* Refs. [74, 93]) and molecular junction tunnelling (see *e.g.* Ref. [97]).

When the zeroth order term of the TDM ($\underline{\mu}_0$) is not dominant, one has to consider the coordinate dependence of the electronic TDM ($\hat{\underline{\mu}}(\underline{Q})$), *i.e.* non-Condon effects. The computational problem becomes even harder. FC-forbidden ($|\underline{\mu}_0| = 0$) or weakly allowed FC ($|\underline{\mu}_0| \simeq 0$) transitions are usually described by a vibronic coupling intensity borrowing mechanism of Herzberg and Teller (see *e.g.* Refs. [98–100]). The SDF that is necessary for describing non-Condon effects can be formulated as follows (in a general form)

$$\rho_{\mathrm{L}}(\omega;T)^{(\hat{f},\hat{g})} = \sum_{\underline{v},\underline{v}'} p_{\underline{v}}(T)\langle\underline{v}'|\hat{f}|\underline{v}\rangle\langle\underline{v}'|\hat{g}|\underline{v}\rangle^{*}L(\omega,\omega_0+\omega_{\underline{v}',\underline{v}};\Gamma), \tag{2}$$

which includes general operators $\hat{f}(\hat{\underline{P}},\hat{\underline{Q}})$ and $\hat{g}(\hat{\underline{P}},\hat{\underline{Q}})$, which are functions of momentum ($\hat{\underline{P}}$) and position ($\hat{\underline{Q}}$) operators. When $\hat{f} = \hat{g} = \underline{\mu}_0$ in Eq. (2) the FC SDF is recovered. Momentum operators, coupling terms between momentum and position operators and nonlinear operators could appear in the IC, ISC, anharmonic, rR and vibronic coupling problems. Therefore the ability to access the non-Condon SDF (2) is essential in describing those kinds of molecular transitions beyond the Condon approximation. The term "non-Condon" is, in this article, restricted to any transition problem involving polynomial expression of transition operators. Here, transition operators could be momentum and position operators.

Herein, we extend and modify the CS-based GF idea [11, 14] for the various vibronic transitions involving thermal excitation and non-Condon effects within the Duschinsky approximation. The GF approaches for rR scattering and single vibronic level (SVL) transitions are developed to include the thermal and non-Condon effects in the Duschinsky rotated harmonic oscillator basis. The methods outlined herein are general and can be applied with slight modifications to various kinds of transition processes via FGR in frequency or time domains.

This article is organised as follows: The theoretical background related to the developments of this article is briefly presented in section 2. The OPA and rR scattering are described from the TI and TD perspectives. The basic properties of CSs and the Duschinsky relation are explained therein for the GF development. Franck–Condon–Herzberg–Teller (FCHT) GF developed in Refs. [101, 102] is introduced in this section. In section 3 we devise the coherent-Fock (cF), which is the mixture of harmonic oscillator eigenstates and coherent states (Eq. (58)), GF via the MHP technique for the GFs for rR, SVL and anharmonic transitions in the TI and the TD pictures including non-Condon effects. Finally the conclusion follows in section 4.

2. Theoretical background

To describe the various vibronic transitions theoretically, we employ a computational framework for the Franck–Condon (FC) and non-FC vibronic transitions within the (adiabatic) Born-Oppenheimer (BO) approximation [98, 103]. The inclusion of Duschinsky vibrational mode mixing [2] together with temperature and non-Condon effects renders the computational effort for vibronic processes high. Herein we describe the molecular system in different electronic states as N-dimensional harmonic oscillators. Accordingly, the (radiative) vibronic transition is considered as a transition between multi-dimensional harmonic oscillators. But even in the harmonic oscillator approximation, the computational description of the vibronic transitions is still challenging.

The various representations of Fermi's golden rule (FGR) in frequency, time and phase spaces were exploited in Ref. [45, 101, 104] via the coherent state (CS)-based generating function (GF) approach [11] to reduce the computational effort. Each representation has its own benefits and shortcomings for the evaluation, such that they are typically complementary to each other (see *e.g.* Ref. [105]). One of the main achievements of our developments is that the different aspects of FGR can be combined in one GF based on CSs within the displaced-distorted-rotated harmonic oscillator (Duschinsky) approximation.

In order to present the CS-based GF in some detail, we need to bear in mind the various representations of FGR and the relations among them (Sec. 2.1). As specific examples, the transformation between the time-independent (TI) and time-dependent expressions of the one-photon absorption (OPA) and resonance Raman (rR) scattering cross sections are discussed. With slight modifications also other vibronic transitions, such as one-photon emission (OPE), electron transfer (ET), electronic circular dichroism (ECD), internal conversion (IC) and inter-system crossing (ISC) can be described, so that they can be treated similarly to OPA and rR. The Duschinsky linear approximation and the corresponding unitary transformation (Sec. 2.2) as well as basic properties of CSs (Sec. 2.3), which are necessary for the non-Condon CS-based GF method [11] (Sec. 2.4), are detailed in the corresponding sections.

2.1. One-photon absorption and resonance Raman scattering

In this section we transform the TI spectral density function (SDF) of the OPA and the rR scattering cross section from frequency domain to time domain via the density matrix trace formalism [106]. The SDF of OPA (Eq. (1)) in frequency domain reads (again) as,

$$\rho_{\mathrm{L}}(\omega;T) = \sum_{\underline{v},\underline{v}'=\underline{0}}^{\infty} p_{\underline{v}}(T)|\langle\underline{v}'|\hat{\underline{\mu}}(\underline{Q})|\underline{v}\rangle|^2 L(\omega,\omega_0+\omega_{\underline{v}',\underline{v}};\Gamma), \tag{3}$$

where the vibrational transition (angular) frequency is defined by the corresponding energy levels ($\underline{\epsilon}$ and $\underline{\epsilon}'$) of an N-dimensional harmonic oscillator, *i.e.* ($\omega_{\underline{v}',\underline{v}} = (\underline{\epsilon}'\cdot\underline{v}' - \underline{\epsilon}\cdot\underline{v})/\hbar$). In practice, we evaluate the intensity profiles according to the Dirac delta distribution (stick representation) and convolute the profiles in stick representation with the Lorentzian line shape function. Prime "′"

is used, conventionally, for specifying variables belonging to the final electronic state. Raman scattering is a two-photon process involving the incident (ω) and scattered light (ω_S). From second order time-dependent (TD) perturbation theory for the two-photon process, we can obtain the matrix elements of the polarisability tensor as the (vibronic) Raman scattering [53] amplitude in the frequency domain (see *e.g.* Refs. [66, 92, 107]).

The matrix element of the polarisability tensor $\boldsymbol{\alpha}$ in the molecular vibronic wavefunction basis with polarisation vectors $\underline{e}^{\mathrm{L}}$ and $\underline{e}^{\mathrm{S}}$ of incident and scattered photons, respectively, is the vibrational Raman scattering amplitude and corresponds to the vibrational Raman excitation profile (see *e.g.* [66]). At resonance ($\omega \simeq (\omega_{\underline{v}',\underline{v}_i} + \omega_0)$) the vibrational rR scattering amplitude ($\alpha_{\mathrm{rR}}^{\underline{v}_i \to \underline{v}_f}$) is expressed as follows

$$\alpha_{\mathrm{rR}}^{\underline{v}_i \to \underline{v}_f}(\omega) = \sum_{\underline{v}'=\underline{0}}^{\infty} \frac{\langle \underline{v}_f | \hat{\mu}^{\mathrm{S}}(\underline{Q})^{\dagger} | \underline{v}' \rangle \langle \underline{v}' | \hat{\mu}^{\mathrm{L}}(\underline{Q}) | \underline{v}_i \rangle}{\hbar(\omega - (\omega_{\underline{v}',\underline{v}_i} + \omega_0) + \frac{\mathrm{i}}{2}\Gamma)}, \tag{4}$$

where we have used $\hat{\mu}^{\mathrm{L}}(\underline{Q}) = \hat{\underline{\mu}}(\underline{Q}) \cdot \underline{e}^{\mathrm{L}}$ and $\hat{\mu}^{\mathrm{S}}(\underline{Q}) = \hat{\underline{\mu}}(\underline{Q}) \cdot \underline{e}^{\mathrm{S}}$. In Eq. (4) $|\underline{v}_i\rangle$ and $|\underline{v}_f\rangle$ are the initial and the final vibrational states in the electronic ground state ($|g\rangle$), $|\underline{v}'\rangle$ is the virtual (or intermediate) vibrational state in the excited electronic state ($|e\rangle$), and $\Gamma_{\underline{v}_i,\underline{v}'}$ is the line width of the transition between initial and virtual state. The vibrational transition frequency from the initial state to the virtual vibrational state is given as $\omega_{\underline{v}',\underline{v}_i} = (\underline{\epsilon}' \cdot \underline{v}' - \underline{\epsilon} \cdot \underline{v}_i)/\hbar$. In Eq. (4) we have assumed a common homogeneous line broadening factor $\Gamma_{\underline{v}_i,\underline{v}'} = \Gamma$ for molecules in condensed phases (see *e.g.* [58, 61, 63, 65, 66, 108]).

The SDF of rR scattering is given with the Lorentzian line shape function as in Eq. (3), *i.e.*

$$\rho_{\mathrm{rR,L}}(\omega, \omega_{\mathrm{S}}) = \sum_{\underline{v}_i,\underline{v}_f=\underline{0}}^{\infty} p_{\underline{v}_i}(T) |\alpha_{\mathrm{rR}}^{\underline{v}_i \to \underline{v}_f}(\omega)|^2 L(\omega - \omega_{\mathrm{S}}, \omega_{\underline{v}_f,\underline{v}_i}; \Gamma_{\mathrm{rR}}), \tag{5}$$

where $\omega_{\underline{v}_f,\underline{v}_i} = (\underline{\epsilon} \cdot \underline{v}_f - \underline{\epsilon} \cdot \underline{v}_i)/\hbar$ is the rR Stokes scattering frequency.

Evaluation of equations (3) and (4) in the frequency domain (TI picture) is typically limited to small molecular systems. For larger systems one would need efficient strategies that exploit sum rules (see *e.g.* [11,109]). The complementary TD approach for theoretical molecular spectroscopy has been used for several decades, in electron transfer theory [73, 75, 78, 80, 81], in Raman scattering theory [49, 54, 57, 59, 60, 65–67, 69, 95], in FC absorption processes [28, 29, 48, 110, 111], in non-Condon processes [33, 35, 87–89] and in laser cooling theory [32, 35, 36, 38, 41]. The energy eigenstate-free TD approach has long been appreciated for its computational efficiency and its wavepacket interpretation of the transition process. It describes the transition process via the time-propagation of initial wavepacket on the excited potential energy surface [57]. To express the absorption cross section (Eq. (3)) and the rR scattering amplitude (Eq. (4)) in time domain, one invokes the Fourier transformed representation of the Lorentzian line shape function in Eq. (3) and the half-Fourier transformed (a special case of Laplace transform) representation of the frequency dependent weight function, the denominator in Eq. (4). That is for the absorption process [57, 66]

$$\rho_{\mathrm{L}}(\omega; T) = \hbar^{-1} \int_{-\infty}^{\infty} dt \sum_{\underline{v},\underline{v}'=\underline{0}}^{\infty} p_{\underline{v}}(T) |\langle \underline{v}' | \hat{\underline{\mu}}(\underline{Q}) | \underline{v} \rangle|^2 \exp[\mathrm{i}(\omega - (\omega_{\underline{v}',\underline{v}} + \omega_0))t - \tfrac{\Gamma}{2}|t|], \tag{6}$$

and for the scattering process [57, 66],

$$\alpha_{\mathrm{rR}}^{\underline{v}_i \to \underline{v}_f}(\omega) = -\frac{\mathrm{i}}{\hbar} \int_0^{\infty} \mathrm{d}\tau \sum_{\underline{v}'=\underline{0}}^{\infty} \langle \underline{v}_f | \hat{\mu}^{\mathrm{S}}(\underline{Q})^{\dagger} | \underline{v}' \rangle \langle \underline{v}' | \hat{\mu}^{\mathrm{L}}(\underline{Q}) | \underline{v}_i \rangle \exp[\mathrm{i}(\omega - (\omega_{\underline{v}',\underline{v}_i} + \omega_0))\tau - \tfrac{\Gamma}{2}\tau]. \tag{7}$$

Now, one can recover the vibrational Hamiltonians corresponding to the expressions (6) and (7) from the vibrational frequencies, *e.g.* $e^{\mathrm{i}\underline{\epsilon}\cdot\underline{v}t/\hbar}|\underline{v}\rangle = e^{\mathrm{i}\hat{H}t/\hbar}|\underline{v}\rangle$. The SDF of OPA is rewritten as follows (see *e.g.* Refs. [57,66])

$$\rho_{\mathrm{L}}(\omega;T) = \hbar^{-1}\int_{-\infty}^{\infty} \mathrm{d}t\ \chi(t;\beta)\exp[\mathrm{i}(\omega-\omega_0)t - \tfrac{\Gamma}{2}|t|]\,. \tag{8}$$

$\chi(t;T)$ is the thermal time-correlation function (TCF) in a trace form as, *i.e.*

$$\chi(t;T) = Z_{\mathrm{I}}^{-1}\mathrm{Tr}\Big(\underline{\hat{\mu}}(\underline{Q})^{\dagger}\exp(-\mathrm{i}\hat{H}'t/\hbar)\underline{\hat{\mu}}(\underline{Q})\exp(\mathrm{i}\hat{H}t/\hbar)\exp(-\beta\hat{H})\Big), \tag{9}$$

where the vibrational partition function is $Z_{\mathrm{I}} = \mathrm{Tr}\left(\exp(-\beta\hat{H})\right)$ and the Boltzmann factor $p_{\underline{v}}(T)|\underline{v}\rangle = Z_{\mathrm{I}}^{-1}\exp(-\beta\hbar\omega_{\underline{v}})|\underline{v}\rangle = Z_{\mathrm{I}}^{-1}\exp(-\beta\hat{H})|\underline{v}\rangle$.

Similarly, the rR amplitude is expressed in TD language as

$$\alpha_{\mathrm{rR}}^{\underline{v}_i\to\underline{v}_f}(\omega) = -\frac{\mathrm{i}}{\hbar}\int_0^{\infty} \mathrm{d}\tau\ \chi_\alpha(\tau;\underline{v}_i,\underline{v}_f)\exp[\mathrm{i}(\omega-\omega_0)\tau - \tfrac{\Gamma}{2}\tau], \tag{10}$$

where the rR amplitude TCF (χ_α) is defined with the vibrational transition operator $|\underline{v}_i\rangle\langle\underline{v}_f|$,

$$\chi_\alpha(\tau;\underline{v}_i,\underline{v}_f) = \mathrm{Tr}\left(\hat{\mu}^{\mathrm{S}}(\underline{Q})^{\dagger}\exp(-\mathrm{i}\hat{H}'\tau/\hbar)\hat{\mu}^{\mathrm{L}}(\underline{Q})\exp(\mathrm{i}\hat{H}\tau/\hbar)|\underline{v}_i\rangle\langle\underline{v}_f|\right). \tag{11}$$

where we have used the resolution of identity $\sum_{\underline{v}'=\underline{0}}^{\infty}|\underline{v}'\rangle\langle\underline{v}'| = \hat{1}$. The initial wavepacket propagating on its ground potential energy surface (PES) is excited to the upper PES given by the electronic transition dipole moment (TDM), and the resulting wavepacket propagates subsequently on this excited PES. Then the wavepacket returns to the vibronic state on the ground electronic state by the electronic TDM. The Fourier transform (FT) of the time correlations corresponds to a spectrum in the frequency domain.

The rR SDF can be expressed from Eqs. (5) and (10) in incident and scattered photon frequency domain with one Fourier $[-\infty,\infty]$ transformation for the function L in Eq. (5) and two half-Fourier $[0,\infty]$ transformations from Eq. (10), *i.e.*

$$\begin{aligned}\rho_{\mathrm{rR,L}}(\omega,\omega_{\mathrm{S}}) =& Z_{\mathrm{I}}^{-1}\hbar^{-3}\sum_{\underline{v}_i,\underline{v}_f=\underline{0}}^{\infty}\int_{-\infty}^{\infty}\mathrm{d}t\int_0^{\infty}\mathrm{d}\tau\int_0^{\infty}\mathrm{d}\tau'\chi_\alpha(\tau';\underline{v}_i,\underline{v}_f)^{*}\chi_\alpha(\tau;\underline{v}_i,\underline{v}_f)\\ &\exp[\mathrm{i}(\omega-\omega_0)(\tau-\tau') - \tfrac{\Gamma}{2}(\tau+\tau')]\exp[\mathrm{i}(\omega-\omega_{\mathrm{S}})t - \tfrac{\Gamma_{\mathrm{rR}}}{2}|t|] \qquad (12)\\ =&\hbar^{-3}\int_{-\infty}^{\infty}\mathrm{d}t\int_0^{\infty}\mathrm{d}\tau\int_0^{\infty}\mathrm{d}\tau'\chi_{\mathrm{rR}}(t,\tau,\tau';T)\\ &\exp[\mathrm{i}(\omega-\omega_0)(\tau-\tau') - \tfrac{\Gamma}{2}(\tau+\tau')]\exp[\mathrm{i}(\omega-\omega_{\mathrm{S}})t - \tfrac{\Gamma_{\mathrm{rR}}}{2}|t|], \qquad (13)\end{aligned}$$

where we define a 3-point (t,τ,τ') thermal TCF in a trace form for rR intensity,

$$\begin{aligned}\chi_{\mathrm{rR}}(t,\tau,\tau';T) =& Z_{\mathrm{I}}^{-1}\mathrm{Tr}\Big(\exp(-\mathrm{i}\hat{H}\tau'/\hbar)\hat{\mu}^{\mathrm{L}}(\underline{Q})^{\dagger}\exp(\mathrm{i}\hat{H}'\tau'/\hbar)\hat{\mu}^{\mathrm{S}}(\underline{Q})\exp(-\mathrm{i}\hat{H}t/\hbar)\\ &\hat{\mu}^{\mathrm{S}}(\underline{Q})^{\dagger}\exp(-\mathrm{i}\hat{H}'\tau/\hbar)\hat{\mu}^{\mathrm{L}}(\underline{Q})\exp(\mathrm{i}\hat{H}\tau/\hbar)\exp(\mathrm{i}\hat{H}t/\hbar)\exp(-\beta\hat{H})\Big). \qquad (14)\end{aligned}$$

For the treatment of FC-forbidden or weakly FC-allowed transitions ($|\underline{\mu}_0| = 0$ or $|\underline{\mu}_0| \simeq 0$, respectively), one must go beyond the FC approximation (*i.e.* beyond the assumption of $\underline{\hat{\mu}}(\underline{Q}) = \underline{\mu}_0$) and incorporate the dependence of $\underline{\hat{\mu}}$ on the vibrational degrees of freedom (DOF).

Conventionally this expansion is provided in terms of the initial state coordinates $\underline{Q}$. In this case the electronic TDM is expanded at least to the linear order, *i.e.* the Herzberg–Teller (HT) expansion.

$$\hat{\underline{\mu}}(\underline{Q}) = \underline{\mu}_0 + \sum_k \underline{\mu}'_k \hat{Q}_k + \cdots , \tag{15}$$

where the zero-th order ($\underline{\mu}_0$) and the first order ($\underline{\mu}'$) expansion vectors are determined by the first order perturbation of electron and nuclear Coulombic interaction relative to the nuclear coordinate change. The first order derivative of the electronic TDM with respect to the k-th normal mode Q_k at the equilibrium structure ($\underline{Q} = \underline{Q}_0$) of the electronic ground state, *i.e.* $\underline{\mu}'_k$ can be calculated analytically either via the traditional perturbation theory formulation (see *e.g.* Refs. [17, 38, 98, 112]) or the linear response approach (see *e.g.* Ref. [45]). Usually, the gradient of the electronic TDM was evaluated numerically by shifting the molecular equilibrium structure along the normal modes (see *e.g.* Refs. [31, 98]), which leads, however, to difficulties due to the phase of the electronic TDM.

The time-independent representation for the absorption spectrum of Eq. (3), the SDF, can then be decomposed into the FC and higher order non-Condon contributions, respectively. This term-wise expansion yields

$$\rho_{\mathrm{FC,L}}(\omega;T) = |\underline{\mu}_0|^2 \sum_{\underline{v},\underline{v}'=\underline{0}}^{\infty} p_{\underline{v}}(T) |\langle \underline{v}'|\underline{v}\rangle|^2 L(\omega, \omega_0 + \omega_{\underline{v}',\underline{v}}; \Gamma) , \tag{16}$$

for the FC contribution and

$$\rho_{\mathrm{FC/HT,L}}(\omega;T) = 2\sum_{i=1}^{N} \underline{\mu}_0 \cdot \underline{\mu}'_i \Big[\sum_{\underline{v},\underline{v}'=\underline{0}}^{\infty} p_{\underline{v}}(T) \langle \underline{v}'|\hat{Q}_i|\underline{v}\rangle \langle \underline{v}|\underline{v}'\rangle L(\omega, \omega_0 + \omega_{\underline{v}',\underline{v}}; \Gamma) \Big] , \tag{17}$$

$$\rho_{\mathrm{HT,L}}(\omega;T) = \sum_{i,j=1}^{N} \underline{\mu}'_i \cdot \underline{\mu}'_j \Big[\sum_{\underline{v},\underline{v}'=\underline{0}}^{\infty} p_{\underline{v}}(T) \langle \underline{v}'|\hat{Q}_i|\underline{v}\rangle \langle \underline{v}|\hat{Q}_j|\underline{v}'\rangle L(\omega, \omega_0 + \omega_{\underline{v}',\underline{v}}; \Gamma) \Big] , \tag{18}$$

for the Franck–Condon/Herzberg–Teller interference (FC/HT) and HT contributions. The FCHT term weighted density of states (FCHTW) is then defined as,

$$\begin{aligned} \rho_{\mathrm{L}}(\omega;T) &\simeq \rho_{\mathrm{FCHTW,L}}(\omega;T) \\ &= \rho_{\mathrm{FC,L}}(\omega;T) + \rho_{\mathrm{FC/HT,L}}(\omega;T) + \rho_{\mathrm{HT,L}}(\omega;T) . \end{aligned} \tag{19}$$

Those term-wise expressions and the corresponding TCFs are exploited for the HT GF developments of the TI and TD approaches in Ref. [101]. The rR scattering SDF can also be expressed in the term-wise fashion with the HT expansion. But the explicit expansions will not be shown in this article. The HT expansion expression for rR can be found in many books and articles (see *e.g.* Refs. [53, 107]).

For the evaluation of the SDFs and TCFs in two sets of harmonic oscillator basis, we need to consider, in which coordinate system the two sets of harmonic oscillators are defined. The matrix elements in the TI and TD approach are typically evaluated by integration in position space. When the vibronic wavefunctions in the initial and final electronic states are expressed in the corresponding normal coordinate systems, the relation between the two coordinate systems has to be defined for the integral evaluation. This will be discussed in the following section.

2.2. Duschinsky rotation

In evaluating overlap integrals and matrix elements in the vibrational wavefunction basis of two electronic states within the BO approximation, the choice of the coordinate system (in which the vibrational wavefunctions of two different BO surfaces are defined) is crucial.

If the two BO surfaces can be approximated as two harmonic potential surfaces, it appears beneficial to approximate the vibrational wavefunctions with harmonic oscillators centred at the corresponding equilibrium molecular structures instead of using one common set. One has to evaluate, therefore in the most general case, inseparable multi-dimensional overlap integrals for Franck–Condon factors (FCFs). When the vibrational wavefunctions of the two electronic states were expressed in one centre basis set (*e.g.* harmonic oscillator basis set), it would be trivial to evaluate the overlap integrals. However, at this point one would typically need a larger basis set to describe the vibrational wavefunction properly for both electronic states (where the local potential minima are shifted relative to each other) which increases the computational complexity.

Herein the two reference point approach is adopted within the harmonic and Duschinsky approximation. The corresponding matrix element of an operator $\hat{f}$ in the Duschinsky rotated harmonic oscillator basis set reads

$$\langle \underline{v}' | \hat{f} | \underline{v} \rangle. \tag{20}$$

When $\hat{f} = \hat{1}$, the matrix element simply becomes a FC integral, and when $\hat{f} = \hat{Q}_i$, the matrix element becomes a linear (first order) HT integral of the i-th position operator of the initial electronic state. The initial and final vibrational eigenstates are described in molecule fixed axis systems attached to each equilibrium molecular structure. We choose particular molecule fixed axis systems which minimise the coupling between rotational and vibrational DOF so that we can use a separation approximation ansatz for rotational and vibrational wavefunctions. This assumption could be supported by the Eckart conditions [113–115], which minimise, when fulfilled, the rovibrational coupling in the molecular Hamiltonian (see *e.g.* Ref. [116]), for the molecular system.

The coordinate space representations of the vibrational states can be obtained by projecting onto the position operator eigenstates $|\underline{Q}\rangle$ and $|\underline{Q}'\rangle$, where $\underline{Q}$ and $\underline{Q}'$ are the mass-weighted normal coordinates, *i.e.* $\langle \underline{Q} | \underline{v} \rangle$ and $\langle \underline{Q}' | \underline{v}' \rangle$.

Even if we could successfully separate the vibrational motions from the other DOF by the Eckart transformations [113–115] (which is, however, not possible), we would need to consider the alignment between the two coordinate systems carefully, because sudden axis-switching [117] can cause artifacts in the vibronic spectrum calculations. When the equilibrium structures of the two electronic states are different, the axis systems ($\underline{r}$ and $\underline{r}'$) from the Eckart conditions related to the corresponding equilibrium structures are typically differently oriented. We determine the alignment of the two axis systems by employing the Eckart condition again but with respect to the initial equilibrium structure as reference structure. Then the two axis systems ($\underline{r}$ and $\underline{r}'$) are adjusted to the initial state molecule fixed axes. The Cartesian coordinates are related by a rotation matrix depending on the normal coordinate ($\underline{Q}$), *i.e.*

$$\underline{r}' = \mathbf{T}(\underline{Q})\underline{r}. \tag{21}$$

The rotation matrix is called Eckart transformation matrix (axis-switching matrix) [34, 118] and it is a unitary transformation matrix, *i.e.*

$$\mathbf{T}(\underline{Q})^{\mathrm{t}}\mathbf{T}(\underline{Q}) = \mathbf{I}. \tag{22}$$

Then we have a linear transformation between the two sets of normal coordinates by introducing the coordinate expansion to the Eckart transformation matrices in Eq. (21) and neglecting nonlinear terms (see *e.g.* Refs. [34, 102, 118]). The Duschinsky linear transformation relation is given as

$$\underline{Q}' = \underline{d} + \mathbf{S}\underline{Q} + \mathcal{O}(\underline{Q}^2)\,, \tag{23}$$

which is already a good approximation for vibronic spectra [119]. Here $\mathbf{S}$ is the Duschinsky mode mixing matrix, which rotates the initial normal coordinates $\underline{Q}$, and the displacement vector $\underline{d}$ is associated to the molecular structural changes shifting the origin of the final harmonic oscillators to that of the initial ones.

Ideally, for the N-dimensional harmonic oscillator (complete separation of the rotational and translational motions from the vibrational modes), the Duschinsky rotation matrix $\mathbf{S}$ is an orthogonal matrix and its determinant is unity. Then the Duschinsky relation (23) becomes exactly a linear unitary transformation and the higher order nonlinear expansion terms vanish. However, in polyatomic molecular systems this normal coordinate transformation is generally nonlinear (see *e.g.* Refs. [34, 96, 114, 117, 120–123]). The Eckart transition matrix $\mathbf{T}(\underline{Q})$ depends nonlinearly on the instantaneous displacements from the equilibrium structures.

To evaluate overlap integrals of CSs, FC integrals and non-Condon integrals exploiting the coordinate space representations of CSs and harmonic oscillator eigenstates, we need to transform one coordinate representation to the other according to the Duschinsky relation (23). The Duschinsky linear equation (23) is translated into a unitary transformation operator for the coordinate systems appearing in the overlap integration.

The N-dimensional harmonic oscillator Hamiltonian is defined as for the initial electronic state

$$\begin{aligned}\hat{H} &= \frac{1}{2}\hat{\underline{P}}\cdot\hat{\underline{P}} + \frac{1}{2}\hat{\underline{Q}}^{\mathrm{t}}\mathbf{\Omega}^4\hat{\underline{Q}} - E_{\mathrm{zp}} \\ &= \frac{\hbar}{2}\hat{\underline{a}}^{\dagger\mathrm{t}}\mathbf{\Omega}^2\hat{\underline{a}},\end{aligned} \tag{24}$$

in terms of the annihilation ($\{\hat{a}_i\}$) and the creation ($\{\hat{a}_i^\dagger\}$) operators corresponding to the harmonic oscillators with harmonic energies $\{\epsilon_i\}$. The diagonal matrix of the (square root) harmonic angular frequencies is defined and used,

$$\mathbf{\Omega} = \hbar^{-\frac{1}{2}}\mathrm{diag}(\underline{\epsilon})^{\frac{1}{2}}\,, \tag{25}$$

where "diag" stands for diagonal and transforms a vector to a square matrix with the diagonal elements being identical to the vector and the off-diagonal elements being zero. In addition, the zero-point vibrational energy ($E_{\mathrm{zp}} = \frac{1}{2}\mathrm{Tr}(\mathrm{diag}(\underline{\epsilon}))$) is subtracted from the vibrational Hamiltonian for convenience. The operators satisfy the following relations,

$$\hat{\underline{Q}} = \sqrt{\tfrac{\hbar}{2}}\mathbf{\Omega}^{-1}(\hat{\underline{a}}^\dagger + \hat{\underline{a}}), \tag{26}$$

$$\hat{\underline{P}} = \mathrm{i}\sqrt{\tfrac{\hbar}{2}}\mathbf{\Omega}(\hat{\underline{a}}^\dagger - \hat{\underline{a}}). \tag{27}$$

The commutation relations of the annihilation and creation operators are

$$[\hat{a}_i, \hat{a}_j] = 0, \quad [\hat{a}_i^\dagger, \hat{a}_j^\dagger] = 0, \quad [\hat{a}_i, \hat{a}_j^\dagger] = \delta_{ij}. \tag{28}$$

The harmonic oscillator eigenfunctions in coordinate space contain products of one-dimensional Hermite polynomials ($\mathcal{H}_v(x) = (-1)^v \mathrm{e}^{x^2}\frac{\mathrm{d}^v}{\mathrm{d}x^v}\mathrm{e}^{-x^2}$),

$$\langle\underline{Q}|\underline{v}\rangle = \prod_{i=1}^{N}\frac{1}{\sqrt{2^{v_i}v_i!}}\left(\tfrac{\epsilon_i}{\pi\hbar^2}\right)^{\frac{1}{4}}\mathrm{e}^{-\frac{\epsilon_i Q_i^2}{2\hbar^2}}\mathcal{H}_{v_i}\left(\sqrt{\tfrac{\epsilon_i}{\hbar^2}}Q_i\right), \tag{29}$$

where $|\underline{v}\rangle$ is an N-dimensional occupation number vector (ONV) representation of the N-dimensional harmonic oscillator eigenstates with corresponding vibrational energy $E_{\underline{v}} = \underline{v} \cdot \underline{\epsilon}$. The action of the annihilation and creation operators on the ONV follows

$$\hat{a}_i|\underline{v}\rangle = \sqrt{v_i}|v_1, \ldots, v_i - 1, \ldots, v_N\rangle, \quad \hat{a}_i^\dagger|\underline{v}\rangle = \sqrt{v_i + 1}|v_1, \ldots, v_i + 1, \ldots, v_N\rangle. \tag{30}$$

Similarly the N-dimensional harmonic oscillator Hamiltonian of the final electronic state is given by

$$\begin{aligned} \hat{H}' &= \frac{1}{2}\underline{\hat{P}}' \cdot \underline{\hat{P}}' + \frac{1}{2}\underline{\hat{Q}}'^{\mathrm{t}}\mathbf{\Omega}'^4\underline{\hat{Q}}' - E'_{\mathrm{zp}} \\ &= \frac{\hbar}{2}\underline{\hat{a}}'^{\dagger\mathrm{t}}\mathbf{\Omega}'^2\underline{\hat{a}}', \end{aligned} \tag{31}$$

in terms of the annihilation ($\{\hat{a}'_i\}$) and the creation ($\{\hat{a}_i'^\dagger\}$) operators corresponding to harmonic oscillators of harmonic energies $\{\epsilon'_i\}$. The diagonal matrix of (square root) harmonic angular frequencies is

$$\mathbf{\Omega}' = \hbar^{-\frac{1}{2}}\mathrm{diag}(\underline{\epsilon}')^{\frac{1}{2}}. \tag{32}$$

Again, the zero-point vibrational energy ($E'_{\mathrm{zp}} = \frac{1}{2}\mathrm{Tr}(\mathrm{diag}(\underline{\epsilon}'))$) is subtracted from the vibrational Hamiltonian for convenience.

Doktorov *et al.* [20, 22] defined a unitary operator ($\hat{U}_{\mathrm{Doktorov}}$) which performs the Duschinsky transformation. It is composed of a translation operator ($\hat{U}_{\mathrm{translation}}$), two distortion operators ($\hat{U}_{\mathrm{distortion}}$ and $\hat{U}'_{\mathrm{distortion}}$) and a rotation operator ($\hat{U}_{\mathrm{rotation}}$). The unitary operator of Doktorov *et al.* [20, 22] is given as

$$\hat{U}_{\mathrm{Doktorov}} = \hat{U}_{\mathrm{translation}}\hat{U}'^{\dagger}_{\mathrm{distortion}}\hat{U}_{\mathrm{distortion}}\hat{U}_{\mathrm{rotation}}, \tag{33}$$

where

$$\hat{U}_{\mathrm{translation}} = \mathrm{e}^{\frac{1}{\sqrt{2\hbar}}\underline{d}^{\mathrm{t}}\mathbf{\Omega}'(\underline{\hat{a}}^\dagger - \underline{\hat{a}})} = \mathrm{e}^{-\frac{\mathrm{i}}{\hbar}\underline{d}^{\mathrm{t}}\mathbf{\Omega}'\mathbf{\Omega}^{-1}\underline{\hat{P}}}, \tag{34}$$

$$\hat{U}'_{\mathrm{distortion}} = \mathrm{e}^{-\frac{1}{2}(\underline{\hat{a}}^\dagger + \underline{\hat{a}})^{\mathrm{t}}\ln\{\mathbf{\Omega}'\}(\underline{\hat{a}}^\dagger - \underline{\hat{a}}) + \frac{1}{2}\mathrm{Tr}(\ln\mathbf{\Omega}')} = \mathrm{e}^{\frac{\mathrm{i}}{\hbar}\underline{\hat{Q}}^{\mathrm{t}}\mathbf{\Omega}\ln\{\mathbf{\Omega}'\}\mathbf{\Omega}^{-1}\underline{\hat{P}} + \frac{1}{2}\mathrm{Tr}(\ln\mathbf{\Omega}')}, \tag{35}$$

$$\hat{U}_{\mathrm{distortion}} = \mathrm{e}^{-\frac{1}{2}(\underline{\hat{a}}^\dagger + \underline{\hat{a}})^{\mathrm{t}}\ln\{\mathbf{\Omega}\}(\underline{\hat{a}}^\dagger - \underline{\hat{a}}) + \frac{1}{2}\mathrm{Tr}(\ln\mathbf{\Omega})} = \mathrm{e}^{\frac{\mathrm{i}}{\hbar}\underline{\hat{Q}}^{\mathrm{t}}\ln\{\mathbf{\Omega}\}\underline{\hat{P}} + \frac{1}{2}\mathrm{Tr}(\ln\mathbf{\Omega})}, \tag{36}$$

$$\hat{U}_{\mathrm{rotation}} = \mathrm{e}^{\frac{1}{2}(\underline{\hat{a}}^{\dagger\mathrm{t}}\ln\{\mathbf{S}\}\underline{\hat{a}} - \underline{\hat{a}}^{\mathrm{t}}\ln\{\mathbf{S}\}\underline{\hat{a}}^\dagger)} = \mathrm{e}^{\frac{\mathrm{i}}{2\hbar}[\mathbf{\Omega}\underline{\hat{Q}}, \ln\{\mathbf{S}\}\mathbf{\Omega}^{-1}\underline{\hat{P}}]}. \tag{37}$$

In this article, a unitary operator $\hat{U}$ which also performs the Duschinsky transformation is instead defined [102] in position space similarly to the squeezing operator (see *e.g.* Ref. [124]), *i.e.*

$$\begin{aligned} \hat{U} &= \int \mathrm{d}\underline{Q}|\underline{Q}'\rangle\langle\underline{Q}| \\ &= |\det(\mathbf{S})|^{\frac{1}{2}}\int \mathrm{d}\underline{Q}|\mathbf{S}\underline{Q} + \underline{d}\rangle\langle\underline{Q}|, \end{aligned} \tag{38}$$

$$\hat{U}^\dagger = |\det(\mathbf{S})|^{\frac{1}{2}}\int \mathrm{d}\underline{Q}|\underline{Q}\rangle\langle\mathbf{S}\underline{Q} + \underline{d}|, \tag{39}$$

or equivalently in momentum space as

$$\hat{U} = \left(\int \mathrm{d}\underline{\tilde{P}}|\underline{\tilde{P}}\rangle\langle\underline{\tilde{P}}|\right)\hat{U}\left(\int \mathrm{d}\underline{P}|\underline{P}\rangle\langle\underline{P}|\right)$$

$$= |\det(\mathbf{S})|^{\frac{1}{2}} \int \mathrm{d}\underline{P}|\underline{P}\rangle\langle\mathbf{S}^{\mathrm{t}}\underline{P}|\exp(-\mathrm{i}\underline{P}\cdot\underline{d}/\hbar), \tag{40}$$

$$\hat{U}^{\dagger} = |\det(\mathbf{S})|^{\frac{1}{2}} \int \mathrm{d}\underline{P}|\mathbf{S}^{\mathrm{t}}\underline{P}\rangle\langle\underline{P}|\exp(\mathrm{i}\underline{P}\cdot\underline{d}/\hbar), \tag{41}$$

where the prefactor $|\det(\mathbf{S})|^{\frac{1}{2}}$ is introduced to restore unitarity $\hat{U}^{\dagger}\hat{U} = \hat{1}$. For the ideal N-dimensional harmonic oscillators $|\det(\mathbf{S})| = 1$ holds, but for polyatomic systems the quantity typically slightly deviates from one as discussed. The unitary operator transforms the initial state operators to the final ones and *vice versa*, *i.e.*

$$\underline{\hat{Q}}' = \hat{U}^{\dagger}\underline{\hat{Q}}\hat{U} = \mathbf{S}\underline{\hat{Q}} + \underline{d}, \tag{42}$$

$$\underline{\hat{P}}' = \hat{U}^{\dagger}\underline{\hat{P}}\hat{U} = (\mathbf{S}^{-1})^{\mathrm{t}}\underline{\hat{P}} = \mathbf{S}\underline{\hat{P}}, \tag{43}$$

in which $\mathbf{S}\mathbf{S}^{\mathrm{t}} = \mathbf{I}$ is assumed for the momentum operator transformation, which strictly holds only for ideal N-dimensional harmonic oscillators ($|\det(\mathbf{S})| = 1$) and approximately holds for polyatomic molecules which have a quasi-unitary rotation matrix ($|\det(\mathbf{S})| \simeq 1$). Accordingly we can see the following relations for the position operator eigenstates,

$$\underline{\hat{Q}}|\underline{Q}\rangle = \underline{Q}|\underline{Q}\rangle, \quad \underline{\hat{Q}}'|\underline{Q}'\rangle = \underline{Q}'|\underline{Q}'\rangle,$$

$$\underline{\hat{Q}}'|\underline{Q}\rangle = (\mathbf{S}\underline{Q} + \underline{d})|\underline{Q}\rangle, \quad \underline{\hat{Q}}|\underline{Q}'\rangle = \mathbf{S}^{-1}(\underline{Q}' - \underline{d})|\underline{Q}'\rangle, \tag{44}$$

and for the momentum operator eigenstates,

$$\underline{\hat{P}}|\underline{P}\rangle = \underline{P}|\underline{P}\rangle, \quad \underline{\hat{P}}'|\underline{P}'\rangle = \underline{P}'|\underline{P}'\rangle,$$

$$\underline{\hat{P}}'|\underline{P}\rangle = (\mathbf{S}^{-1})^{\mathrm{t}}\underline{P}|\underline{P}\rangle, \quad \underline{\hat{P}}|\underline{P}'\rangle = \mathbf{S}^{\mathrm{t}}\underline{P}'|\underline{P}'\rangle. \tag{45}$$

The annihilation and creation operators in the initial and the final states are related by a similar equation to the linear Duschinsky expression (23), *i.e.*

$$\begin{pmatrix}\underline{\hat{a}}'\\ \underline{\hat{a}}'^{\dagger}\end{pmatrix} = \frac{1}{2}\begin{pmatrix}\mathbf{J} + (\mathbf{J}^{-1})^{\mathrm{t}} & \mathbf{J} - (\mathbf{J}^{-1})^{\mathrm{t}}\\ \mathbf{J} - (\mathbf{J}^{-1})^{\mathrm{t}} & \mathbf{J} + (\mathbf{J}^{-1})^{\mathrm{t}}\end{pmatrix}\begin{pmatrix}\underline{\hat{a}}\\ \underline{\hat{a}}^{\dagger}\end{pmatrix} + \sqrt{2}\begin{pmatrix}\underline{\delta}\\ \underline{\delta}\end{pmatrix}, \tag{46}$$

where the N-dimensional square matrix $\mathbf{J}$ and the N-dimensional vector $\underline{\delta}$ are defined as

$$\mathbf{J} = \mathbf{\Omega}'\mathbf{S}\mathbf{\Omega}^{-1}, \quad \underline{\delta} = \hbar^{-\frac{1}{2}}\mathbf{\Omega}'\underline{d}. \tag{47}$$

Accordingly the vibrational Hamiltonians are mutually convertible with the Duschinsky unitary transformation, *i.e.*

$$\hat{H}' = \hat{U}^{\dagger}\hat{H}\hat{U}. \tag{48}$$

The primed and unprimed ONVs states ($|\underline{v}'\rangle$ and $|\underline{v}\rangle$, respectively) are the eigenstates of primed and unprimed Hamiltonians ($\hat{H}'$ and $\hat{H}$, respectively). Otherwise, the states will be indicated by their corresponding Hamiltonians as subscripts, for example $|\underline{v}\rangle_{\hat{H}'}$ is an eigenstate of the primed Hamiltonian but with the unprimed ONV and $|\underline{v}'\rangle_{\hat{H}}$ is the opposite case. As a result of the Duschinsky unitary operator $\hat{U}$ the harmonic eigenstates (ONV states) of each Hamiltonian described in the respective coordinate systems are transformed into each other,

$$\hat{H}\left(\hat{U}|\underline{v}'\rangle\right) = \hat{H}\left(|\underline{v}'\rangle_{\hat{H}}\right) = E_{\underline{v}'}\left(\hat{U}|\underline{v}'\rangle\right), \tag{49}$$

$$\hat{H}'\left(\hat{U}^\dagger|\underline{v}\rangle\right) = \hat{H}'\left(|\underline{v}\rangle_{\hat{H}'}\right) E'_{\underline{v}}\left(\hat{U}^\dagger|\underline{v}\rangle\right), \tag{50}$$

where the eigenvalues are given by ONVs belong to different electronic states, *i.e.* $E_{\underline{v}'} = \underline{v}' \cdot \underline{\epsilon}$ and $E'_{\underline{v}} = \underline{v} \cdot \underline{\epsilon}'$. Precisely the harmonic eigenstates of initial and final states in each phase space are transformed to the other phase space by the unitary transformation, *i.e.*

$$\hat{U}|\underline{v}'\rangle = \hat{U}|\underline{v}'\rangle_{\hat{H}'} = |\underline{v}'\rangle_{\hat{H}}, \tag{51}$$

$$\hat{U}^\dagger|\underline{v}\rangle = \hat{U}^\dagger|\underline{v}\rangle_{\hat{H}} = |\underline{v}\rangle_{\hat{H}'}. \tag{52}$$

Then the FC integrals in different coordinate systems can be described in one coordinate system via the Duschinsky unitary transformation operator,

$$\langle\underline{v}'|\underline{v}\rangle = {}_{\hat{H}'}\langle\underline{v}'|\underline{v}\rangle_{\hat{H}} = {}_{\hat{H}}\langle\underline{v}'|\hat{U}|\underline{v}\rangle_{\hat{H}} = {}_{\hat{H}'}\langle\underline{v}'|\hat{U}|\underline{v}\rangle_{\hat{H}}. \tag{53}$$

The CS overlap integral is exploited for the FC or non-Condon integrals because the CS is the GF of harmonic eigenstates.

2.3. Basic properties of coherent states

Glauber's coherent state [125] is a special type of quantum harmonic oscillator state which fulfils the minimal uncertainty relation of position and momentum operators. The mean values of position and momentum of a quantum mechanical Gaussian wavepacket [126], which is evolving in time without spreading in the corresponding harmonic potential, follow the motion of a classical harmonic oscillator in a given harmonic potential. The basic properties of CSs, which are exploited throughout this article, are briefly reviewed in this section. The Duschinsky unitary operator (39) is used for computing the CS overlap integral for Duschinsky related (Eq. (23)) CS. Most of the following relations of CSs can be found in Refs. [125, 127], and in many other articles and books (see *e.g.* Ref. [92]).

The N-dimensional CS $|\underline{\alpha}\rangle$ is defined as an eigenstate of the annihilation operator $\underline{\hat{a}}$,

$$\underline{\hat{a}}|\underline{\alpha}\rangle = \underline{\alpha}|\underline{\alpha}\rangle, \tag{54}$$

where $\underline{\alpha} \in \mathbb{C}^N$ is a complex-valued N-dimensional vector. The CSs can be expanded in the basis of harmonic oscillator eigenstates,

$$|\underline{\alpha}\rangle \doteq \exp(-\tfrac{1}{2}\underline{\alpha}^\dagger\underline{\alpha})\exp(\underline{\alpha}^{\mathrm{t}}\underline{\hat{a}}^\dagger)|\underline{0}\rangle \tag{55}$$

$$= \exp(-\tfrac{1}{2}\underline{\alpha}^\dagger\underline{\alpha})\sum_{\underline{v}=\underline{0}}^{\infty}\widetilde{\prod}_{\underline{v}}^{-\frac{1}{2}}\overline{\prod}_{\underline{\alpha}}^{\underline{v}}|\underline{v}\rangle, \tag{56}$$

where we have defined and used the product notations for factorials and powers of vectors, *i.e.* $\widetilde{\prod}_{\underline{x}_1,\cdots,\underline{x}_N}^{\underline{n}_1,\cdots,\underline{n}_N} = (\prod_k (x_{1,k}!)^{n_{1,k}})\cdots(\prod_k (x_{N,k}!)^{n_{N,k}})$ and $\overline{\prod}_{\underline{x}_1,\cdots,\underline{x}_N}^{\underline{n}_1,\cdots,\underline{n}_N} = \left(\prod_k x_{1,k}^{n_{1,k}}\right)\cdots\left(\prod_k x_{N,k}^{n_{N,k}}\right)$ respectively. The N-dimensional CSs can be partitioned into subspaces X and Y of dimension N_X and $N - N_X$ respectively, *i.e.* $|\underline{\alpha}\rangle = |\underline{\alpha}_X;\underline{\alpha}_Y\rangle$,

$$|\underline{\alpha}_X;\underline{\alpha}_Y\rangle = \exp(-\tfrac{1}{2}|\underline{\alpha}_X|^2 - \tfrac{1}{2}|\underline{\alpha}_Y|^2)\exp(\underline{\alpha}_X^{\mathrm{t}}\underline{a}_X^\dagger + \underline{\alpha}_Y^{\mathrm{t}}\underline{a}_Y^\dagger)|\underline{0}_X;\underline{0}_Y\rangle. \tag{57}$$

Then the coherent-Fock state [124] $|\underline{v}_X;\underline{\alpha}_Y\rangle$ can be constructed by partial derivatives with respect to $\{\alpha_{X;i}\}$, *i.e.*

$$|\underline{v}_X;\underline{\alpha}_Y\rangle = \widetilde{\prod}_{\underline{v}_X}^{-\frac{1}{2}} \hat{\partial}_{\underline{\alpha}_X}^{\underline{v}_X}\Big(\exp(\tfrac{1}{2}|\underline{\alpha}_X|^2)|\underline{\alpha}_X;\underline{\alpha}_Y\rangle\Big)\Big|_{\underline{\alpha}_X=\underline{0}}, \tag{58}$$

where we have defined and used the multi-dimensional partial derivative notation, *i.e.* $\hat{\partial}_{\underline{x}_1,\cdots,\underline{x}_N}^{\underline{n}_1,\cdots,\underline{n}_N} = \left(\frac{\partial^{\sum_k n_{1,k}}}{\prod_k \partial x_{1,k}^{n_{1,k}}}\right)\cdots\left(\frac{\partial^{\sum_k n_{N,k}}}{\prod_k \partial x_{N,k}^{n_{N,k}}}\right)$.

The spatial representation of CSs is given in normalised form

$$\langle \underline{Q}|\underline{\alpha}\rangle = \overline{\prod}_{(\pi\hbar^2)^{-1}\underline{\epsilon}}^{\frac{1}{4}} \exp\left[-\frac{1}{2}\underline{Q}^{\mathrm{t}}\mathbf{\Omega}^2\underline{Q} + \sqrt{2}\underline{\alpha}^{\mathrm{t}}\mathbf{\Omega}\underline{Q} - \frac{1}{2}\underline{\alpha}^{\mathrm{t}}\underline{\alpha} - \frac{1}{2}|\underline{\alpha}|^2\right]. \tag{59}$$

The conjugate momentum space representation reads

$$\langle \underline{P}|\underline{\alpha}\rangle = \overline{\prod}_{\pi\underline{\epsilon}}^{-\frac{1}{4}} \exp\left[-\frac{1}{2}\underline{P}^{\mathrm{t}}\mathbf{\Omega}^{-2}\underline{P} - \mathrm{i}\sqrt{2}\underline{\alpha}^{\mathrm{t}}\mathbf{\Omega}^{-1}\underline{P} + \frac{1}{2}\underline{\alpha}^{\mathrm{t}}\underline{\alpha} - \frac{1}{2}|\underline{\alpha}|^2\right], \tag{60}$$

which is in normalised form as well.

CSs can be defined with the help of a unitary operator, the CS displacement operator, which is defined by

$$\begin{aligned}\hat{D}(\underline{\alpha}) &= \exp(\underline{\alpha}^{\mathrm{t}}\hat{\underline{a}}^\dagger - \underline{\alpha}^\dagger\hat{\underline{a}})\\ &= \exp(-\tfrac{1}{2}\sum_k[\hat{a}_k^\dagger, -\hat{a}_k]|\alpha_k|^2)\exp(\underline{\alpha}^{\mathrm{t}}\hat{\underline{a}}^\dagger)\exp(-\underline{\alpha}^\dagger\hat{\underline{a}})\\ &= \exp(-\tfrac{1}{2}|\underline{\alpha}|^2)\exp(\underline{\alpha}^{\mathrm{t}}\hat{\underline{a}}^\dagger)\exp(-\underline{\alpha}^\dagger\hat{\underline{a}}).\end{aligned} \tag{61}$$

Under the action of $\hat{D}(\underline{\alpha})$ CSs are created from the vibrational ground state $|\underline{0}\rangle$ via

$$\hat{D}(\underline{\alpha})|\underline{0}\rangle = |\underline{\alpha}\rangle. \tag{62}$$

The CS displacement operator satisfies the following phase composition rules:

$$\hat{D}^\dagger(\underline{\alpha}) = \hat{D}^{-1}(\underline{\alpha}) = \hat{D}(-\underline{\alpha}), \tag{63}$$

$$\hat{D}(\underline{\alpha})\hat{D}(\underline{\gamma}) = \hat{D}(\underline{\alpha}+\underline{\gamma})\exp[\tfrac{1}{2}(\underline{\gamma}^\dagger\underline{\alpha} - \underline{\gamma}^{\mathrm{t}}\underline{\alpha}^*)], \tag{64}$$

$$[\hat{D}(\underline{\alpha}), \hat{D}(\underline{\gamma})] = \hat{D}(\underline{\alpha}+\underline{\gamma})\,2\mathrm{i}\,\mathrm{Im}\Big(\exp[\tfrac{1}{2}(\underline{\gamma}^\dagger\underline{\alpha} - \underline{\gamma}^{\mathrm{t}}\underline{\alpha}^*)]\Big). \tag{65}$$

The unitary operation on the annihilation and creation operators look like

$$\begin{aligned}\hat{D}(\underline{\gamma})^\dagger\hat{\underline{a}}\hat{D}(\underline{\gamma}) &= \hat{\underline{a}} + \underline{\gamma},\\ \hat{D}(\underline{\gamma})^\dagger\hat{\underline{a}}^\dagger\hat{D}(\underline{\gamma}) &= \hat{\underline{a}}^\dagger + \underline{\gamma}^*.\end{aligned} \tag{66}$$

Using the phase addition rule of Eq. (64), one finds

$$\begin{aligned}\hat{D}(\underline{\alpha})\hat{D}(\underline{\gamma})|\underline{0}\rangle &= \hat{D}(\underline{\alpha})|\underline{\gamma}\rangle\\ &= \exp[\tfrac{1}{2}(\underline{\gamma}^\dagger\underline{\alpha} - \underline{\gamma}^{\mathrm{t}}\underline{\alpha}^*)]\,|\underline{\alpha}+\underline{\gamma}\rangle,\end{aligned} \tag{67}$$

the phase composition state from the vacuum state.

The operation of the harmonic vibrational Hamiltonian $\hat{H}$ (24) on a CS appears relatively trivial. Unitary (Eq. (68)) and non-unitary (Eq. (69)) transformations of CSs can be shown by applying the transforming operators to the occupation representation of the CSs, Eq. (56). Under the exponential such operators simply shift the CS phase, *e.g.*

$$\mathrm{e}^{-\mathrm{i}\hat{H}t/2\hbar}|\underline{\alpha}\rangle = |\mathbf{z}(t)\underline{\alpha}\rangle, \tag{68}$$

where $\mathbf{z}(t) = \mathrm{diag}(\mathrm{e}^{-\mathrm{i}\epsilon_1 t/2\hbar}, \ldots, \mathrm{e}^{-\mathrm{i}\epsilon_N t/2\hbar})$. The unitary operation like above is called the phase-shifting operation. For non-unitary transformations with a diagonal Hamiltonian in the harmonic oscillator basis, such as for the thermal Boltzmann population of states, an additional factor appears, *i.e.*

$$\mathrm{e}^{-\beta\hat{H}/2}|\underline{\alpha}\rangle = \mathrm{e}^{-\frac{1}{2}\underline{\alpha}^\dagger(\mathbf{I}-\mathbf{\Gamma}^\dagger\mathbf{\Gamma})\underline{\alpha}}|\mathbf{\Gamma}\underline{\alpha}\rangle\,, \tag{69}$$

where $\mathbf{\Gamma} = \mathrm{diag}(\mathrm{e}^{-\beta\epsilon_1/2}, \ldots, \mathrm{e}^{-\beta\epsilon_N/2})$ with the reciprocal temperature $\beta = 1/(k_\mathrm{B}T)$ where T is the temperature and k_B is the Boltzmann constant. Comparing to the unitary operation (68) the CS phase factor is rescaled as well with the Boltzmann related factors ($\mathbf{\Gamma}$) and the non-unitary operation leaves a prefactor (the exponential factor in Eq. (69)) which is related to the vibrational partition function.

Coherent states (CSs) are over-complete basis sets satisfying the following resolution of identity [125],

$$\frac{1}{\pi^N}\int \mathrm{d}^2\underline{\alpha}|\underline{\alpha}\rangle\langle\underline{\alpha}| = \hat{1}. \tag{70}$$

Two CSs are not orthogonal [125],

$$\langle\underline{\gamma}|\underline{\alpha}\rangle = \exp(-\tfrac{1}{2}|\underline{\alpha}|^2 - \tfrac{1}{2}|\underline{\gamma}|^2)\exp(\underline{\gamma}^\dagger\underline{\alpha}). \tag{71}$$

The overlap integral of CSs $|\underline{\alpha}\rangle$ and $|\underline{\gamma}'\rangle$, described by the corresponding normal coordinates $\underline{Q}$ and $\underline{Q}'$ respectively in the Duschinsky relation (Eq. (23)), can be given in the occupation representation [23] exploiting the occupation representation of the coherent states (Eq. (56)) such that

$$\langle\underline{\gamma}'|\underline{\alpha}\rangle = \exp(-\tfrac{1}{2}\underline{\alpha}^\dagger\underline{\alpha} - \tfrac{1}{2}\underline{\gamma}'^\dagger\underline{\gamma}')\sum_{\underline{v},\underline{v}'=\underline{0}}^{\infty}\langle\underline{v}'|\underline{v}\rangle\widetilde{\prod}_{\underline{v},\underline{v}'}^{-\frac{1}{2},-\frac{1}{2}}\overline{\prod}_{\underline{\alpha},\underline{\gamma}'^*}^{\underline{v},\underline{v}'}, \tag{72}$$

where $|\underline{v}\rangle$ and $|\underline{v}'\rangle$ are the harmonic oscillator eigenstates corresponding to the CSs $|\underline{\alpha}\rangle$ and $|\underline{\gamma}'\rangle$ respectively. The occupation representation of the overlap integral of the CSs, Eq. (72) is the GF of the FC integrals with the generating function parameters $\{\alpha_k\}$ and $\{\gamma'_k\}$. The CS overlap integral in the Duschinsky relation is then given by

$$\begin{aligned}\langle\underline{\gamma}'|\underline{\alpha}\rangle &= \langle\underline{\gamma}';\underline{Q}',\underline{P}'|\underline{\alpha};\underline{Q},\underline{P}\rangle = \langle\underline{\gamma}';\underline{Q},\underline{P}|\hat{U}|\underline{\alpha};\underline{Q},\underline{P}\rangle \\ &= |\det(\mathbf{S})|^{\frac{1}{2}}\int \mathrm{d}\underline{Q}\langle\underline{\gamma}'|\mathbf{S}\underline{Q}+\underline{d}\rangle\langle\underline{Q}|\underline{\alpha}\rangle \\ &= \langle\underline{0}'|\underline{0}\rangle\exp(-\tfrac{1}{2}\underline{\xi}^\dagger\underline{\xi})\mathcal{J}\left[\mathbf{W},\underline{r};\underline{\xi}\right]\,,\end{aligned} \tag{73}$$

where the collective CS phase vector is used

$$\underline{\xi} = \begin{pmatrix}\underline{\alpha} \\ \underline{\gamma}'^*\end{pmatrix}, \tag{74}$$

and the exponential function $\mathcal{J}$ is defined as $\mathcal{J}[\mathbf{A},\underline{b};\underline{x}] = \exp(-\frac{1}{2}\underline{x}^{\rm t}\mathbf{A}\underline{x} + \underline{b}^{\rm t}\underline{x})$. The overlap integral of CSs that is represented by the Duschinsky rotated harmonic oscillators can be integrated in closed form by exploiting the spatial representation of CSs, with the corresponding spatial representation (or with the corresponding momentum space representation Eq. (60)) Eq. (59) being expressed by exponential functions. The Duschinsky relation, Eq. (23), is taken into account with the help of Doktorov matrices and vectors [22] as parameters,

$$\mathbf{W} = \begin{pmatrix} \mathbf{I}-2\mathbf{Q} & -2\mathbf{R} \\ -2\mathbf{R}^{\rm t} & \mathbf{I}-2\mathbf{P} \end{pmatrix}, \quad \underline{r} = \sqrt{2}\begin{pmatrix} -\mathbf{R}\underline{\delta} \\ (\mathbf{I}-\mathbf{P})\underline{\delta} \end{pmatrix}. \tag{75}$$

$\mathbf{W}$ is a self-inverse $2N \times 2N$ matrix [5] and the $2N$-dimensional vector denoted by $\underline{r}$ should be distinguished from the position vector $\underline{r}$ of section 2.2. The N-dimensional symmetric positive-definite square matrices $\mathbf{Q}$ and $\mathbf{P}$, and the N-dimensional vector $\underline{\delta}$ are given as

$$\mathbf{Q} = (\mathbf{I}+\mathbf{J}^{\rm t}\mathbf{J})^{-1}, \quad \mathbf{P} = \mathbf{J}\mathbf{Q}\mathbf{J}^{\rm t}, \quad \mathbf{R} = \mathbf{Q}\mathbf{J}^{\rm t}. \tag{76}$$

The Doktorov matrices are related to each other as [5]

$$\mathbf{R}\mathbf{R}^{\rm t} = \mathbf{Q}-\mathbf{Q}^2, \quad \mathbf{R}^{\rm t}\mathbf{R} = \mathbf{P}-\mathbf{P}^2, \tag{77}$$

$$\mathbf{R}\mathbf{P}^{-1}\mathbf{R}^{\rm t} = \mathbf{Q}, \quad \mathbf{R}^{\rm t}\mathbf{Q}^{-1}\mathbf{R} = \mathbf{P}, \quad \mathbf{Q}\mathbf{R}+\mathbf{R}\mathbf{Q} = \mathbf{R}. \tag{78}$$

The vibrational ground state overlap integral ($\langle \underline{0}'|\underline{0}\rangle$) is expressed with the Doktorov matrices and vector as well,

$$\begin{aligned} \langle \underline{0}'|\underline{0}\rangle &= 2^{\frac{N}{2}} \left(\frac{\overline{\prod}^{1}_{\underline{\epsilon}'}}{\overline{\prod}^{1}_{\underline{\epsilon}}}\right)^{\frac{1}{4}} \det(\mathbf{Q})^{\frac{1}{2}} \exp\left(-\frac{1}{2}\underline{\delta}^{\rm t}(\mathbf{I}-\mathbf{P})\underline{\delta}\right) |\det(\mathbf{S})|^{\frac{1}{2}} \\ &= 2^{\frac{N}{2}} |\det(\mathbf{R})|^{\frac{1}{2}} \exp\left(-\frac{1}{2}\underline{\delta}^{\rm t}(\mathbf{I}-\mathbf{P})\underline{\delta}\right). \end{aligned} \tag{79}$$

Equating the occupation representation (72) of CS overlap integral and the expanded spatial expression (73) with respect to the CS phase variables, and comparing CS phase variables to collect same orders, we can find the FC integral evaluation scheme (see *e.g.* Refs. [11,22,31]). The CS phase variables take a role as FC integral generating function parameters. Similar integral evaluation scheme for matrix elements of non-Condon operators were developed in Ref. [102] within the multi-variate Hermite polynomial (MHP) frame work.

2.4. Coherent state-based generating function

In the following subsections we present the generating functions (GFs) for the Franck–Condon/Herzberg–Teller interference (FC/HT) and Herzberg–Teller (HT) contributions of Eq. (17) and Eq. (18) respectively, via the incorporation of coherent state (CS) displacement operators, which were developed for zero Kelvin [11], finite temperature [104] and HT transition [101].

2.4.1. Generating function for one-photon process To determine the explicit functional forms for the contributions corresponding to the Franck–Condon (FC), FC/HT and HT expressions of Eqs. (16)–(18), we now introduce a Boltzmann weighted GF G^K with the operators $\hat{f}(\hat{\underline{P}},\hat{\underline{Q}})$ and $\hat{g}(\hat{\underline{P}},\hat{\underline{Q}})$, that is

$$G^K(\mathbf{Z};\mathbf{\Lambda})^{(\hat{f},\hat{g})} = \mathcal{N}|\langle \underline{0}'|\underline{0}\rangle|^{-2} \sum_{\underline{v},\underline{v}'=\underline{0}}^{\infty} \langle \underline{v}'|\hat{f}|\underline{v}\rangle \langle \underline{v}'|(\hat{g})|\underline{v}\rangle^{*} \overline{\prod}^{2\underline{v},2\underline{v}'}_{\underline{z},\underline{z}'} \mathrm{e}^{-(\underline{v}^{\rm t}\mathbf{B}\underline{\epsilon}+\underline{v}'^{\rm t}\mathbf{B}'\underline{\epsilon}')}, \tag{80}$$

in which a collective block diagonal matrix is introduced $\mathbf{Z} = \text{bldiag}(\mathbf{z}, \mathbf{z}')$ where "bldiag" denotes a matrix in block diagonal form, where the GF parameters are on the unit circle in phase space $|z_k| = 1$, $|z'_k| = 1$. The parameters corresponding to temperatures, $\beta_k = 1/(k_B T_k)$ (with Boltzmann constant k_B and temperature T_k), are combined in the matrix $\mathbf{B} = \text{diag}(\beta_1, \ldots, \beta_k)$ with analogous expressions holding for primed quantities. Assigning different temperature parameters rather than a common value to each mode allows selecting specific modes.

After identification of $\hat{f}$ and $\hat{g}$ with products of CS displacement operators, we can obtain an analytic expression for G^K and subsequently specialise to the desired FC, FC/HT and HT GF expressions (corresponding to Eqs. (16), (17) and (18) respectively) via partial derivatives with respect to CS parameters.

We can evaluate the non-Condon GF G^K, which is casted within the occupancy representation in Eq. (80), analytically through an integral formulation arising from a CS representation [11]. If we identify $\hat{f}$ and $\hat{g}$ with the CS displacement operators (Eq. (61)) within the overlap integrals, the integral form of the GF becomes

$$G^K(\mathbf{Z}; \mathbf{\Lambda}; \underline{\eta}, \underline{\eta}') = \pi^{-2N} |\langle \underline{0}' | \underline{0} \rangle|^{-2} \int \mathrm{d}^2\underline{\alpha}\, \mathrm{d}^2\underline{\gamma}' K(\mathbf{\Lambda}; \underline{\xi}) \langle \mathbf{z}'^* \underline{\gamma}' | \hat{D}_{\mathrm{NC}}(\underline{\eta}) | \mathbf{z}\underline{\alpha} \rangle \langle \mathbf{z}' \underline{\gamma}' | \hat{D}_{\mathrm{NC}}(\underline{\eta}') | \mathbf{z}^* \underline{\alpha} \rangle^* . \tag{81}$$

The integration variables are defined as

$$\mathrm{d}^2\underline{\alpha} = \prod_{k=1}^{N} \mathrm{d}^2\alpha_k , \quad \mathrm{d}^2\underline{\gamma}' = \prod_{k=1}^{N} \mathrm{d}^2\gamma'_k , \tag{82}$$

where

$$\mathrm{d}^2\alpha_k = \mathrm{dRe}(\alpha_k)\mathrm{dIm}(\alpha_k) = \tfrac{1}{2\hbar}\mathrm{d}P_k \mathrm{d}Q_k, \tag{83}$$

$$\mathrm{d}^2\gamma'_k = \mathrm{dRe}(\gamma'_k)\mathrm{dIm}(\gamma'_k) = \tfrac{1}{2\hbar}\mathrm{d}P'_k \mathrm{d}Q'_k, \tag{84}$$

and with a thermal integral kernel K [104],

$$K(\mathbf{\Lambda}; \underline{\xi}) = \mathcal{N} \det(\mathbf{I} + \mathbf{\Lambda}) \exp(-\underline{\xi}^\dagger \mathbf{\Lambda} \underline{\xi}) . \tag{85}$$

Here the parameter matrix $\mathbf{\Lambda}$ is assumed to be diagonal with real-valued entries $\boldsymbol{\lambda} = \text{diag}(\underline{\lambda})$ and $\boldsymbol{\lambda}' = \text{diag}(\underline{\lambda}')$, so that $\mathbf{\Lambda} = \text{bldiag}(\boldsymbol{\lambda}, \boldsymbol{\lambda}')$. This matrix will be associated with the Boltzmann factors ($(1 + \lambda_k)^{-1} = \exp(-\beta_k \epsilon_k)$ and $(1 + \lambda'_k)^{-1} = \exp(-\beta'_k \epsilon'_k)$). $\mathcal{N}$ is a normalisation factor which turns out, later, to be a reciprocal vibrational partition function, such that

$$1 = \mathcal{N} \mathrm{Tr}(\exp(-\underline{\beta} \cdot \underline{\hat{h}}) \exp(-\underline{\beta}' \cdot \underline{\hat{h}}')), \tag{86}$$

where the vibrational Hamiltonians are in mode separated form $\hat{H} = \sum_k^N \hat{h}_k$ and $\hat{H}' = \sum_k^N \hat{h}'_k$ for the initial and final state, respectively. The non-Condon displacement operator is defined as follows

$$\hat{D}_{\mathrm{NC}}(\underline{\eta}) = \hat{D}^{-1}((\mathrm{i}\underline{\eta_{Q'}})^*) \hat{D}^{-1}(\underline{\eta_{P'}}) \hat{D}(\underline{\eta_P}) \hat{D}(\mathrm{i}\underline{\eta_Q}), \tag{87}$$

$$\underline{\tilde{\eta}_P} = \begin{pmatrix} \underline{\eta_P} \\ \underline{\eta_{P'}} \end{pmatrix}, \quad \underline{\tilde{\eta}_Q} = \begin{pmatrix} \underline{\eta_Q} \\ \underline{\eta_{Q'}} \end{pmatrix}, \tag{88}$$

$$\underline{\eta} = \underline{\tilde{\eta}_P} + \mathrm{i}\underline{\tilde{\eta}_Q} . \tag{89}$$

This construction does include additional flexibility, such that restrictions can subsequently be imposed on the resulting GF G^K to treat specific cases of interest. Then Eq. (81) reduces to the analytic form with Gaussian integrals

$$\begin{aligned} G^K(\mathbf{Z};\boldsymbol{\Lambda};\underline{\eta},\underline{\eta}') =& \mathcal{N}\det(\mathbf{I}+\mathbf{Z}\mathbf{W}_T\mathbf{Z})^{-\frac{1}{2}}\det(\mathbf{I}-\mathbf{Z}\mathbf{W}_T\mathbf{Z})^{-\frac{1}{2}}R(\underline{\eta},\underline{\eta}') \\ & \exp((\underline{b}^+)^{\mathrm{t}}\mathbf{Z}(\mathbf{I}+\mathbf{Z}\mathbf{W}_T\mathbf{Z})^{-1}\mathbf{Z}\underline{b}^+)\exp((\underline{b}^-)^{\mathrm{t}}\mathbf{Z}(\mathbf{I}-\mathbf{Z}\mathbf{W}_T\mathbf{Z})^{-1}\mathbf{Z}\underline{b}^-)\,. \end{aligned} \tag{90}$$

where the quantities $\mathbf{W}_T = \mathbf{W}(T)$ and $\underline{r}_T = \underline{r}(T)$ are now temperature dependent. Specifically,

$$\mathbf{W}_T = (\mathbf{I}+\boldsymbol{\Lambda})^{-\frac{1}{2}}\mathbf{W}(\mathbf{I}+\boldsymbol{\Lambda})^{-\frac{1}{2}}, \quad \underline{r}_T = (\mathbf{I}+\boldsymbol{\Lambda})^{-\frac{1}{2}}\underline{r}\,, \tag{91}$$

and the temperature dependence enters via the association made above between λ_k, λ'_k and the Boltzmann factor. Here the quantities in Eq. (90) are defined as

$$\begin{aligned} R(\underline{\eta},\underline{\eta}') =& \exp(-\tfrac{1}{2}\underline{\eta}^\dagger\underline{\eta} - \tfrac{1}{2}\underline{\eta}'^\dagger\underline{\eta}')\exp(-\tfrac{1}{2}\underline{\eta}^{\mathrm{t}}\mathbf{W}\underline{\eta} - \tfrac{1}{2}\underline{\eta}'^\dagger\mathbf{W}\underline{\eta}'^* + \underline{r}^{\mathrm{t}}(\underline{\eta}+\underline{\eta}'^*)) \\ & \exp(-\mathrm{i}\underline{\tilde{\eta}_P}^{\mathrm{t}}\underline{\tilde{\eta}_Q} + \mathrm{i}\underline{\tilde{\eta}'_P}^{\mathrm{t}}\underline{\tilde{\eta}'_Q})\,, \end{aligned} \tag{92}$$

$$(\underline{b}^+)^{\mathrm{t}}(\boldsymbol{\Lambda};\underline{\eta},\underline{\eta}') = \underline{r}_T^{\mathrm{t}} - \tfrac{1}{2}[\underline{\eta}^\dagger + \underline{\eta}'^{\mathrm{t}} + (\underline{\eta}^{\mathrm{t}} + \underline{\eta}'^\dagger)\mathbf{W}](\mathbf{I}+\boldsymbol{\Lambda})^{-\frac{1}{2}}, \tag{93}$$

$$(\underline{b}^-)^{\mathrm{t}}(\boldsymbol{\Lambda};\underline{\eta},\underline{\eta}') = -\tfrac{\mathrm{i}}{2}[\underline{\eta}^\dagger - \underline{\eta}'^{\mathrm{t}} + (\underline{\eta}^{\mathrm{t}} - \underline{\eta}'^\dagger)\mathbf{W}](\mathbf{I}+\boldsymbol{\Lambda})^{-\frac{1}{2}}. \tag{94}$$

Direct substitution leads to $G^K(\mathbf{Z};\boldsymbol{\Lambda})$, which is simply a thermally-weighted Franck–Condon factor (FCF) GF [11],

$$\begin{aligned} G^K(\mathbf{Z};\boldsymbol{\Lambda}) &= G^K(\mathbf{Z};\boldsymbol{\Lambda})^{(\hat{1},\hat{1})} \\ &= G^K(\mathbf{Z};\boldsymbol{\Lambda};\underline{\eta},\underline{\eta}')\big|_{\underline{\eta}=\underline{0},\underline{\eta}'=\underline{0}} \\ &= \mathcal{N}\det(\mathbf{I}+\mathbf{Z}\mathbf{W}_T\mathbf{Z})^{-\frac{1}{2}}\det(\mathbf{I}-\mathbf{Z}\mathbf{W}_T\mathbf{Z})^{-\frac{1}{2}} \\ &\quad \exp(\underline{r}_T^{\mathrm{t}}\mathbf{Z}(\mathbf{I}+\mathbf{Z}\mathbf{W}_T\mathbf{Z})^{-1}\mathbf{Z}\underline{r}_T). \end{aligned} \tag{95}$$

After rearranging the non-Condon GF expression (90) and by specifying the auxiliary parameters of position and momentum operators belonging to initial and final states, we can introduce operators up to arbitrary orders with the help of multi-variate Hermite polynomials (MHPs) (Eq. (104)). The non-Condon GF expression (90) is rewritten as

$$G^K(\mathbf{Z};\boldsymbol{\Lambda};\underline{\eta},\underline{\eta}') = G^K(\mathbf{Z};\boldsymbol{\Lambda})\mathcal{J}\left[\widetilde{\mathbf{W}}_{\mathrm{NCF}}(\mathbf{Z};\boldsymbol{\Lambda}),\underline{\tilde{r}}_{\mathrm{NCF}}(\mathbf{Z};\boldsymbol{\Lambda});\underline{\tilde{\eta}}_{\mathrm{NCF}}\right], \tag{96}$$

with the $8N$-dimensional collective auxiliary parameter vector (Eqs. (88) and (89)),

$$\underline{\tilde{\eta}}_{\mathrm{NCF}} = \begin{pmatrix} \underline{\tilde{\eta}_P} \\ \underline{\tilde{\eta}_Q} \\ \underline{\tilde{\eta}'_P} \\ \underline{\tilde{\eta}'_Q} \end{pmatrix}. \tag{97}$$

Note that non-Condon GF separates into a FCF GF part ($G^K(\mathbf{Z};\boldsymbol{\Lambda})$) and a non-Condon contribution (the exponential function, $\mathcal{J}$) in Eq. (96).

The $8N$-dimensional square matrix $\widetilde{\mathbf{W}}_{\mathrm{NCF}}$ and vector $\underline{\tilde{r}}_{\mathrm{NCF}}$ are defined as

$$\widetilde{\mathbf{W}}_{\mathrm{NCF}}(\mathbf{Z};\mathbf{\Lambda}) = \tag{98}$$

$$\begin{pmatrix} (\mathbf{I}+\mathbf{W})-\frac{1}{2}(\mathbf{I}+\mathbf{W})\widetilde{\mathbf{A}}_T^-(\mathbf{I}+\mathbf{W}) & \frac{\mathrm{i}}{2}(\mathbf{I}+\mathbf{W})+\frac{\mathrm{i}}{2}(\mathbf{I}+\mathbf{W})\widetilde{\mathbf{A}}_T^-(\mathbf{I}-\mathbf{W}) & -\frac{1}{2}(\mathbf{I}+\mathbf{W})\widetilde{\mathbf{A}}_T^+(\mathbf{I}+\mathbf{W}) & -\frac{\mathrm{i}}{2}(\mathbf{I}+\mathbf{W})\widetilde{\mathbf{A}}_T^+(\mathbf{I}-\mathbf{W}) \\ \frac{\mathrm{i}}{2}(\mathbf{I}+\mathbf{W})+\frac{\mathrm{i}}{2}(\mathbf{I}-\mathbf{W})\widetilde{\mathbf{A}}_T^-(\mathbf{I}+\mathbf{W}) & (\mathbf{I}-\mathbf{W})+\frac{1}{2}(\mathbf{I}-\mathbf{W})\widetilde{\mathbf{A}}_T^-(\mathbf{I}-\mathbf{W}) & \frac{\mathrm{i}}{2}(\mathbf{I}-\mathbf{W})\widetilde{\mathbf{A}}_T^+(\mathbf{I}+\mathbf{W}) & -\frac{1}{2}(\mathbf{I}-\mathbf{W})\widetilde{\mathbf{A}}_T^+(\mathbf{I}-\mathbf{W}) \\ -\frac{1}{2}(\mathbf{I}+\mathbf{W})\widetilde{\mathbf{A}}_T^+(\mathbf{I}+\mathbf{W}) & \frac{\mathrm{i}}{2}(\mathbf{I}+\mathbf{W})\widetilde{\mathbf{A}}_T^+(\mathbf{I}-\mathbf{W}) & (\mathbf{I}+\mathbf{W})-\frac{1}{2}(\mathbf{I}+\mathbf{W})\widetilde{\mathbf{A}}_T^-(\mathbf{I}+\mathbf{W}) & -\frac{\mathrm{i}}{2}(\mathbf{I}+\mathbf{W})-\frac{\mathrm{i}}{2}(\mathbf{I}+\mathbf{W})\widetilde{\mathbf{A}}_T^-(\mathbf{I}-\mathbf{W}) \\ -\frac{\mathrm{i}}{2}(\mathbf{I}-\mathbf{W})\widetilde{\mathbf{A}}_T^+(\mathbf{I}+\mathbf{W}) & -\frac{1}{2}(\mathbf{I}-\mathbf{W})\widetilde{\mathbf{A}}_T^+(\mathbf{I}-\mathbf{W}) & -\frac{\mathrm{i}}{2}(\mathbf{I}+\mathbf{W})-\frac{\mathrm{i}}{2}(\mathbf{I}-\mathbf{W})\widetilde{\mathbf{A}}_T^-(\mathbf{I}+\mathbf{W}) & (\mathbf{I}-\mathbf{W})+\frac{1}{2}(\mathbf{I}-\mathbf{W})\widetilde{\mathbf{A}}_T^-(\mathbf{I}-\mathbf{W}) \end{pmatrix},$$

and

$$\underline{\tilde{r}}_{\mathrm{NCF}}(\mathbf{Z};\mathbf{\Lambda}) = \begin{pmatrix} \underline{r}-(\mathbf{I}+\mathbf{W})(\mathbf{I}+\mathbf{\Lambda})^{-\frac{1}{2}}\mathbf{A}^+\underline{r}_T \\ \mathrm{i}\underline{r}+\mathrm{i}(\mathbf{I}-\mathbf{W})(\mathbf{I}+\mathbf{\Lambda})^{-\frac{1}{2}}\mathbf{A}^+\underline{r}_T \\ \underline{r}-(\mathbf{I}+\mathbf{W})(\mathbf{I}+\mathbf{\Lambda})^{-\frac{1}{2}}\mathbf{A}^+\underline{r}_T \\ -\mathrm{i}\underline{r}-\mathrm{i}(\mathbf{I}-\mathbf{W})(\mathbf{I}+\mathbf{\Lambda})^{-\frac{1}{2}}\mathbf{A}^+\underline{r}_T \end{pmatrix}, \tag{99}$$

where

$$\widetilde{\mathbf{A}}^{\pm}(\mathbf{Z};\mathbf{\Lambda}) = \mathbf{A}^+(\mathbf{Z};\mathbf{\Lambda}) \pm \mathbf{A}^-(\mathbf{Z};\mathbf{\Lambda}), \tag{100}$$

$$\mathbf{A}^{\pm}(\mathbf{Z};\mathbf{\Lambda}) = \mathbf{Z}(\mathbf{I} \pm \mathbf{Z}\mathbf{W}_T\mathbf{Z})^{-1}\mathbf{Z}, \tag{101}$$

$$\widetilde{\mathbf{A}}_T^{\pm}(\mathbf{Z};\mathbf{\Lambda}) = (\mathbf{I}+\mathbf{\Lambda})^{-\frac{1}{2}}\widetilde{\mathbf{A}}^{\pm}(\mathbf{Z};\mathbf{\Lambda})(\mathbf{I}+\mathbf{\Lambda})^{-\frac{1}{2}}. \tag{102}$$

We can express the non-Condon GF in terms of MHPs from Eq. (96),

$$\begin{aligned} G^K(\mathbf{Z};\mathbf{\Lambda})^{(\hat{f},\hat{g})} &= \prod_{k=1}^{2N}\left[\mathrm{i}\sqrt{\tfrac{\bar{\epsilon}_k}{2}}\right]^{\tilde{l}_k}\left[\tfrac{1}{\mathrm{i}}\sqrt{\tfrac{\hbar^2}{2\bar{\epsilon}_k}}\right]^{\tilde{m}_k}\left[-\mathrm{i}\sqrt{\tfrac{\bar{\epsilon}_k}{2}}\right]^{\tilde{n}_k}\left[-\tfrac{1}{\mathrm{i}}\sqrt{\tfrac{\hbar^2}{2\bar{\epsilon}_k}}\right]^{\tilde{o}_k} \\ &\quad \hat{\partial}^{\underline{\tilde{l}},\underline{\tilde{m}},\underline{\tilde{n}},\underline{\tilde{o}}}_{\underline{\tilde{\eta}}_P,\underline{\tilde{\eta}}_Q,\underline{\tilde{\eta}}'_P,\underline{\tilde{\eta}}'_Q} G^K(\mathbf{Z};\mathbf{\Lambda};\underline{\eta},\underline{\eta}')\Big|_{\underline{\tilde{\eta}}_{\mathrm{NCF}}=\underline{0}} \\ &= \prod_{k=1}^{2N}\left[\mathrm{i}\sqrt{\tfrac{\bar{\epsilon}_k}{2}}\right]^{\tilde{l}_k}\left[\tfrac{1}{\mathrm{i}}\sqrt{\tfrac{\hbar^2}{2\bar{\epsilon}_k}}\right]^{\tilde{m}_k}\left[-\mathrm{i}\sqrt{\tfrac{\bar{\epsilon}_k}{2}}\right]^{\tilde{n}_k}\left[-\tfrac{1}{\mathrm{i}}\sqrt{\tfrac{\hbar^2}{2\bar{\epsilon}_k}}\right]^{\tilde{o}_k} \\ &\quad \mathcal{H}e_{\underline{\tilde{l}},\underline{\tilde{m}},\underline{\tilde{n}},\underline{\tilde{o}}}(\widetilde{\mathbf{W}}^{-1}_{\mathrm{NCF}}\underline{\tilde{r}}_{\mathrm{NCF}};\widetilde{\mathbf{W}}^{-1}_{\mathrm{NCF}}), \end{aligned} \tag{103}$$

where the operators ($\hat{f}$ and $\hat{g}$) are identified as $\hat{f} = \overline{\prod}^{\underline{m}',\underline{l}',\underline{l},\underline{m}}_{\underline{\hat{Q}}',\underline{\hat{P}}',\underline{\hat{P}},\underline{\hat{Q}}}$ and $\hat{g} = \overline{\prod}^{\underline{o}',\underline{n}',\underline{n},\underline{o}}_{\underline{\hat{Q}}',\underline{\hat{P}}',\underline{\hat{P}},\underline{\hat{Q}}}$, in which the collective indices are used, *i.e.* $\underline{\tilde{l}}^{\mathrm{t}} = (\underline{l}^{\mathrm{t}},\underline{l}'^{\mathrm{t}})$, $\underline{\tilde{m}}^{\mathrm{t}} = (\underline{m}^{\mathrm{t}},\underline{m}'^{\mathrm{t}})$, $\underline{\tilde{n}}^{\mathrm{t}} = (\underline{n}^{\mathrm{t}},\underline{n}'^{\mathrm{t}})$ and $\underline{\tilde{o}}^{\mathrm{t}} = (\underline{o}^{\mathrm{t}},\underline{o}'^{\mathrm{t}})$. The prefactors are multiplied for the corresponding operators in Eq. (103), which can be identified with the collective indices. The MHP, $\mathcal{H}e_{\underline{\tilde{v}}}$, is for convenience defined (unlike the univariate Hermite polynomials used in Eq. 29, which are employed in the conventional form used in quantum mechanics) here as is customary in statistics applications, namely

$$\mathcal{H}e_{\underline{\tilde{v}}}(\underline{x};\mathbf{\Lambda}_{\mathrm{c}}) = (-1)^{\tilde{v}}\exp(\tfrac{1}{2}\underline{x}^{\mathrm{t}}\mathbf{\Lambda}_{\mathrm{c}}^{-1}\underline{x})\hat{\partial}^{\underline{\tilde{v}}}_{\underline{x}}\exp(-\tfrac{1}{2}\underline{x}^{\mathrm{t}}\mathbf{\Lambda}_{\mathrm{c}}^{-1}\underline{x}), \tag{104}$$

where $\tilde{v} = \sum_k \tilde{v}_k$ and $\mathbf{\Lambda}_{\mathrm{c}}$ is a complex symmetric matrix with a symmetric positive definite real part.

By evaluating Eq. (103) the FC/HT GF is explicitly expressed as,

$$G^K(\mathbf{Z};\mathbf{\Lambda})^{(\hat{Q}_i,\hat{1})} = \sqrt{\tfrac{\hbar^2}{2\epsilon_i}}G^K(\mathbf{Z};\mathbf{\Lambda})[\underline{r}+(\mathbf{I}-\mathbf{W})(\mathbf{I}+\mathbf{\Lambda})^{-\frac{1}{2}}\mathbf{A}^+\underline{r}_T]_i\,, \tag{105}$$

and the HT GF is given explicitly by,

$$G^K(\mathbf{Z};\mathbf{\Lambda})^{(\hat{Q}_i,\hat{Q}_j)} = \frac{\hbar^2}{2}\sqrt{\frac{1}{\epsilon_i\epsilon_j}}G^K(\mathbf{Z};\mathbf{\Lambda})$$
$$\Big[[\underline{r} + (\mathbf{I}-\mathbf{W})(\mathbf{I}+\mathbf{\Lambda})^{-\frac{1}{2}}\mathbf{A}^{+}\underline{r}_T]_i[\underline{r} + (\mathbf{I}-\mathbf{W})(\mathbf{I}+\mathbf{\Lambda})^{-\frac{1}{2}}\mathbf{A}^{+}\underline{r}_T]_j$$
$$+\tfrac{1}{2}[(\mathbf{I}-\mathbf{W})\widetilde{\mathbf{A}}_T^{+}(\mathbf{I}-\mathbf{W})]_{ij}\Big]\,. \qquad (106)$$

One can also determine expressions for position operators of the final electronic state, $\{\hat{Q}'_i\}$, by using the $(N+i)$-th component, instead of i-th component, of vectors or matrices in Eqs. (105) and (106) with the corresponding harmonic energy (ϵ'_i).

2.4.2. Spectral density functions By using Eqs. (105) and (106) we can construct the electronic transition dipole moment induced absorption intensity GF with FC, FC/HT and HT contributions (Eq. (19)), *i.e.*

$$\rho_{\mathrm{FCHTW}}(\mathbf{Z};\mathbf{\Lambda}) =$$
$$|\langle\underline{0}'|\underline{0}\rangle|^2\left(|\underline{\mu}_0|^2G^K(\mathbf{Z};\mathbf{\Lambda}) + 2\sum_i\underline{\mu}_0\cdot\underline{\mu}'_iG^K(\mathbf{Z};\mathbf{\Lambda})^{(\hat{Q}_i,\hat{1})} + \sum_{i,j}\underline{\mu}'_i\cdot\underline{\mu}'_jG^K(\mathbf{Z};\mathbf{\Lambda})^{(\hat{Q}_i,\hat{Q}_j)}\right), \qquad (107)$$

associated with the FCHT spectral profile of Eq. (19). Immediately we can exploit these relations within the thermal time-correlation function (TCF) formalism, where we obtain a simple analytic form for the exact thermal FC TCF. Employing Lorentzian line shapes, $L(t)$, one obtains the absorption profile as a one-dimensional Fourier transform (FT),

$$\rho_{\mathrm{FCHTW,L}}(\omega) = \hbar^{-1}\int_{-\infty}^{\infty}\mathrm{d}t\;\rho_{\mathrm{FCHTW}}(\mathbf{Z}(t);\mathbf{\Lambda})L(t)\mathrm{e}^{\mathrm{i}(\omega-\omega_0)t}\,, \qquad (108)$$

where we choose $\mathbf{Z}(t) = \mathrm{bldiag}(\mathbf{z}(t),\mathbf{z}'(t))$ with

$$\mathbf{z}(t) = \mathrm{diag}(\mathrm{e}^{\mathrm{i}\epsilon_1 t/(2\hbar)},\ldots,\mathrm{e}^{\mathrm{i}\epsilon_N t/(2\hbar)}), \qquad (109)$$
$$\mathbf{z}'(t) = \mathrm{diag}(\mathrm{e}^{-\mathrm{i}\epsilon'_1 t/(2\hbar)},\ldots,\mathrm{e}^{-\mathrm{i}\epsilon'_N t/(2\hbar)})\,. \qquad (110)$$

It should be noted here that the current development of non-Condon TCF is not restricted to linear HT expansion. The nonlinear non-Condon TCF can be evaluated by the complex MHPs in Eq. (103). One can compute moments (see *e.g.* Refs. [102, 128, 129]) of the distribution with the generating function

$$\langle E^k_{\underline{\epsilon}',\underline{\epsilon}}\rangle(T) = \Big(-\frac{\hbar}{\mathrm{i}}\Big)^k\frac{\partial^k}{\partial t^k}[\rho_{\mathrm{FCHTW}}(\mathbf{Z}(t);\mathbf{\Lambda})L(t)]\Big|_{t=0}, \qquad (111)$$

where $\langle E^k_{\underline{\epsilon}',\underline{\epsilon}}\rangle(T)$ is the k-th moments at temperature T and the time derivatives can be obtained analytically and numerically [130]. This time-independent cumulant expansion (CE) (cumulants and moments are mutually convertable) method were exploited in Refs. [102, 128, 129] with Edgeworth expanion (see *e.g.* Ref. [131]) for FC transitions to approximate the distribution with the first few cumulants. The method was extended to incorporate non-Condon effects in Ref. [130].

We should mention the work of Islampour and Miralinaghi [87] who devised recently a TCF for internal conversion (IC) rate involving multi-promoting modes (which mediate the intramolecular transition) and vibrational mode mixing effects. These authors exploited second order multi-variate normal moments for the momentum operator matrix elements of the promoting modes to evaluate their trace. However this time-dependent (TD) method is not as generally applicable to various transition problems and the method cannot directly handle nonlinear coupling problems (*cf.* Eq. (103)). Peng *et al.* [88, 89] have reported a similar development to that of Islampour and Miralinaghi [87].

In this section, the methods for inclusion of non-Condon effects were explained. We extended the FC thermal GF to the non-Condon effects via the CS phase displacement operators [101, 102], which allows to use a similar mathematical frameworks as the FC prescreening strategy and FC thermal TCF [104].

3. Single vibronic levels

So far we have assumed the initial states to be in a Boltzmann-distributed ensemble of vibrational states at finite temperature or a vibrationless state at zero Kelvin for the one-photon absorption (OPA) process. Those kinds of initial vibrational states for generating functions (GFs) can be introduced by simple GF parameter assignments (*e.g.* $\underline{z} = \underline{0}$ for the vibrational ground state). The Boltzmann weighted coherent states (CSs) (Eq. (69)) are used in the case of a thermally averaged ensemble of vibrational states. However, for optical processes such as the single vibronic level (SVL) fluorescence (see *e.g.* Refs. [29, 132, 133]) and the resonance Raman (rR) scattering we have to consider specific vibronic levels, which are expressed in the fixed quantum levels during the optical processes.

In SVL fluorescence, a vibronic level which is not necessarily the vibrational ground state in the excited electronic state is populated by tuning the corresponding laser wavelength [132, 133]. The vibronic state on the excited electronic potential energy surface (PES) is called the SVL. The spectral density function (SDF) of SVL fluorescence is identical to the OPA one in Eq. (3) but with the reversed initial and final electronic states and a given fixed initial vibronic state (SVL), which is not included in the summation over vibronic levels. The fluorescence from a vibrationally excited vibronic state shows often vibrationally well-resolved spectra, unlike the normal absorption or emission spectra at finite temperature [132, 133]. The initial and the final vibronic levels belonging to the ground electronic states for the rR amplitude (4) appear in rR scattering process (see Sec. 2.1).

In Ref. [11] we employed the subspaces X and Y given in subsection 2.3 for rigorous prescreening of Franck–Condon (FC) integrals. We considered therein only one-dimensional fixed quantum number spaces X, which corresponds to an SVLs in terms of a fixed quantum number in one vibrational mode, *i.e.* the single vibrational mode excitation (*e.g.* $|v, 0, \cdots, 0\rangle$). We have not used the space X for the SVL time-correlation function (TCF) in the complementary time-dependent (TD) approach, although by virtue of the developments reported in the previous sections such an application is straight forward. For coarse-grained integral prescreening in SVL transition calculations we have to consider at least two-dimensional X spaces. Even for the usual absorption and emission processes, if we want to have the fine-grained integral prescreening strategy, we need to consider multi-dimensional X spaces. When non-Condon effects should be considered, the problem becomes even more difficult. The purpose of this section is thus to modify the GFs developed in the previous section 2.4.1. The possible modifications could be applicable to the rR, SVL, fine-grained integral prescreening and anharmonic FC transitions with the SVLs either in the time-independent (TI) integral prescreening or TD TCF approaches.

There were efforts (see *e.g.* Refs. [29, 58, 65, 67]) to generate SVLs particularly suited for the rR excitation profile computation. But most of the applications are restricted to the single vibrational mode excited states (*e.g.* $|v, 0 \cdots, 0\rangle$), Condon approximation and limited Duschinsky effects. Recently, Santoro *et al.* [109] developed a TI method based on convergence criteria for rR scattering which includes Duschinsky and Herzberg–Teller (HT) effects. The method, however, is limited to TI and linear HT. We develop herein an arbitrary SVL transition GFs with the help of multi-variate Hermite polynomials (MHPs) including the thermal and non-Condon effects. This development is similar to the previous section developments in terms of the non-Condon effects. The TCF for the thermally averaged rR intensity including arbitrary order of non-Condon effects is developed in the form of the overlap integral of two rR amplitude GFs in different time domains. The novelty of our GF approach is that it can provide both TI and TD methods, including arbitrary order of non-Condon effects and arbitrary SVLs. We can evaluate TCFs or sum rules for arbitrary SVLs and non-Condon operators with the help of MHPs.

This section is organised as follows: In the methodology section 3.1, we develop a GF having fixed quantum number occupation number vector (ONV) spaces. The analytic form of the GF is derived via the coherent-Fock (cF) state (58) which is a mixture of harmonic eigenstates and CSs. The development is formulated for the rR profile GFs in section 3.2, SVL transitions in section 3.3 and anharmonic FC/non-Condon transitions in section 3.4.

3.1. Methodology

In this section, we introduce the arbitrary dimensions in X space by exploiting the MHP technique to generate the arbitrary SVLs.

With this partitioning, the GF is decomposed into products of contributions in the orthogonal subspaces X and Y. The subspaces X and Y have dimensions N_X and $N_Y = 2N - N_X$ respectively. The subspace X corresponds to the vibrational modes with fixed quantum numbers. The vibrational modes in subspace Y can be excited infinitely and stay in the CS phase integral space as a result. Integration over one phase variable belonging to one vibrational mode corresponds to complete summation of the individual vibrational mode contributions to the total intensity. The corresponding orthogonal projection operators $\hat{\pi}_X$ and $\hat{\pi}_Y$ satisfy the usual conditions for projection operators. Applying the projection operators to matrices and vectors indicates the corresponding projection labels, *e.g.*

$$\begin{aligned} &\hat{\pi}_X\underline{\xi} = \underline{\xi}_X, \quad \hat{\pi}_Y\underline{\xi} = \underline{\xi}_Y, \\ &\hat{\pi}_X\mathbf{A}\hat{\pi}_X = \mathbf{A}_{XX}, \quad \hat{\pi}_X\mathbf{A}\hat{\pi}_Y = \mathbf{A}_{XY}, \quad \hat{\pi}_Y\mathbf{A}\hat{\pi}_X = \mathbf{A}_{YX}, \quad \hat{\pi}_Y\mathbf{A}\hat{\pi}_Y = \mathbf{A}_{YY}. \end{aligned} \tag{112}$$

The GF that we want to have for an arbitrary number of fixed quantum numbers both in the initial and the final vibronic states would be in the following occupation representation (*cf.* Eq. (80) for empty X space GF) with thermally weighted Y space ($K_{YY} = K(\mathbf{\Lambda}_{YY};\underline{\xi}_Y)$, see Eq. (85)), namely

$$\begin{aligned} &G_{\mathrm{cF}}^{K_{YY}}(\tilde{\mathbf{Z}}_{\mathrm{cF}}, \mathbf{Z}_{YY}; \mathbf{\Lambda}_{YY}; \underline{\tilde{v}}_{\mathrm{KB}}, \underline{\tilde{v}}_{\mathrm{BK}})^{(\hat{f},\hat{g})} \\ &= \mathcal{N}_{YY}|\langle\underline{0}'|\underline{0}\rangle|^{-2} \sum_{\underline{v}_Y,\underline{v}'_Y=\underline{0}}^{\infty} \langle\underline{v_{\mathrm{B}}}_X;\underline{v}_Y|(\hat{g})^*|\underline{v_{\mathrm{K}}}'_X;\underline{v}'_Y\rangle\langle\underline{v_{\mathrm{B}}}'_X;\underline{v}'_Y|\hat{f}|\underline{v_{\mathrm{K}}}_X;\underline{v}_Y\rangle \\ &\overline{\prod}_{\underline{z}_Y,\underline{z}'_Y}^{2\underline{v}_Y,2\underline{v}'_Y} \mathrm{e}^{-(\underline{v}_Y^{\mathrm{t}}\mathbf{B}_{YY}\underline{\epsilon}_Y+\underline{v}_Y'^{\mathrm{t}}\mathbf{B}'_{YY}\underline{\epsilon}'_Y)}\overline{\prod}_{\underline{z_{\mathrm{K}}}_X,\underline{z_{\mathrm{B}}}_X,\underline{z_{\mathrm{B}}}'_X,\underline{z_{\mathrm{K}}}'_X}^{\underline{v_{\mathrm{K}}}_X,\underline{v_{\mathrm{B}}}_X,\underline{v_{\mathrm{B}}}'_X,\underline{v_{\mathrm{K}}}'_X}, \end{aligned} \tag{113}$$

where $\underline{v_{\mathrm{B}}}_X$ and $\underline{v_{\mathrm{K}}}_X$ are N_X-dimensional ONVs and $\underline{v_{\mathrm{B}}}'_X$ and $\underline{v_{\mathrm{K}}}'_X$ are N'_X-dimensional ONVs where B and K stand for "Bra" and "Ket" respectively. $\mathcal{N}_{YY} = 1/\mathrm{Tr}(-\underline{\beta}_Y \cdot \underline{\hat{h}}_Y)$ is a normalising factor corresponding to the thermal integral kernel K_{YY}. The collective ONVs are additionally defined,

$$\underline{\tilde{v}}_{\mathrm{KB}} = \begin{pmatrix} \underline{v_{\mathrm{K}}}_X \\ \underline{v_{\mathrm{B}}}_X \end{pmatrix}, \quad \underline{\tilde{v}}_{\mathrm{BK}} = \begin{pmatrix} \underline{v_{\mathrm{B}}}_X \\ \underline{v_{\mathrm{K}}}_X \end{pmatrix}, \tag{114}$$

and the block-diagonal GF parameter matrix is given by

$$\widetilde{\mathbf{Z}}_{\mathrm{cF}} = \mathrm{bldiag}(\mathbf{z}_{\mathrm{K}}, \mathbf{z}'_{\mathrm{B}}, \mathbf{z}_{\mathrm{B}}, \mathbf{z}'_{\mathrm{K}}), \tag{115}$$

with the matrices $\mathbf{z}_{\mathrm{K}} = \mathrm{diag}(\underline{z_{\mathrm{K}}})$, $\mathbf{z}'_{\mathrm{B}} = \mathrm{diag}(\underline{z_{\mathrm{B}}}')$, $\mathbf{z}_{\mathrm{B}} = \mathrm{diag}(\underline{z_{\mathrm{B}}})$ and $\mathbf{z}'_{\mathrm{K}} = \mathrm{diag}(\underline{z_{\mathrm{K}}}')$ corresponding to the ONVs $\underline{v_{\mathrm{K}}}_X$, $\underline{v_{\mathrm{B}}}'_X$, $\underline{v_{\mathrm{B}}}_X$ and $\underline{v_{\mathrm{K}}}'_X$, respectively. In the OPA SDF (3) we have $\underline{v_{\mathrm{B}}}_X = \underline{v_{\mathrm{K}}}_X$ and $\underline{v_{\mathrm{B}}}'_X = \underline{v_{\mathrm{K}}}'_X$, but for the generality of the development and for the rR GFs we make these vibrational ONVs independent of each other. In section 3.3 the constraints are recovered for the SVL transition to the GF (113).

We express the functional form of the GF (113) by exploiting the cF state (58), which can be generated by taking partial derivatives of the CSs, *i.e.*

$$\begin{aligned}
&G^{K_{YY}}_{\mathrm{cF}}(\widetilde{\mathbf{Z}}_{\mathrm{cF}}, \mathbf{Z}_{YY}; \mathbf{\Lambda}_{YY}; \underline{\tilde{v}}_{\mathrm{KB}}, \underline{\tilde{v}}_{\mathrm{BK}})^{(\hat{f},\hat{g})} \\
&= \prod_{k=1}^{2N} \left[\mathrm{i}\sqrt{\tfrac{\tilde{\epsilon}_k}{2}}\right]^{\tilde{l}_k} \left[\tfrac{1}{\mathrm{i}}\sqrt{\tfrac{\hbar^2}{2\tilde{\epsilon}_k}}\right]^{\tilde{m}_k} \left[-\mathrm{i}\sqrt{\tfrac{\tilde{\epsilon}_k}{2}}\right]^{\tilde{n}_k} \left[-\tfrac{1}{\mathrm{i}}\sqrt{\tfrac{\hbar^2}{2\tilde{\epsilon}_k}}\right]^{\tilde{o}_k} \\
&\left(\hat{\partial}^{\underline{\tilde{l}},\underline{\tilde{m}},\underline{\tilde{n}},\underline{\tilde{o}}}_{\underline{\tilde{\eta}_P},\underline{\tilde{\eta}_Q},\underline{\tilde{\eta}'_P},\underline{\tilde{\eta}'_Q}}\right) \left(\widetilde{\prod}^{-\frac{1}{2},-\frac{1}{2}}_{\underline{v_{\mathrm{B}}}_X,\underline{v_{\mathrm{K}}}_X} \hat{\partial}^{\underline{v_{\mathrm{B}}}_X,\underline{v_{\mathrm{K}}}_X}_{\underline{\alpha_{\mathrm{B}}}^*_X,\underline{\alpha_{\mathrm{K}}}_X}\right) \left(\widetilde{\prod}^{-\frac{1}{2},-\frac{1}{2}}_{\underline{v_{\mathrm{B}}}'_X,\underline{v_{\mathrm{K}}}'_X} \hat{\partial}^{\underline{v_{\mathrm{B}}}'_X,\underline{v_{\mathrm{K}}}'_X}_{\underline{\gamma_{\mathrm{K}}}'_X,\underline{\gamma_{\mathrm{B}}}'^*_X}\right) \\
&G^{K_{YY}}_{\mathrm{cF}}(\widetilde{\mathbf{Z}}_{\mathrm{cF}}, \mathbf{Z}_{YY}; \mathbf{\Lambda}_{YY}; \underline{\tilde{\xi}}_{\mathrm{BK}}, \underline{\tilde{\xi}}_{\mathrm{KB}}; \underline{\eta}, \underline{\eta}')\Big|_{\underline{\tilde{\xi}}_{\mathrm{KB}},\underline{\tilde{\xi}}_{\mathrm{BK}},\underline{\tilde{\eta}}_{\mathrm{NCF}}=\underline{0}},
\end{aligned} \tag{116}$$

where

$$\underline{\tilde{\xi}}_{\mathrm{KB}} = \begin{pmatrix} \underline{\alpha_{\mathrm{K}}}_X \\ \underline{\gamma_{\mathrm{B}}}'^*_X \end{pmatrix}, \quad \underline{\tilde{\xi}}_{\mathrm{BK}} = \begin{pmatrix} \underline{\alpha_{\mathrm{B}}}_X \\ \underline{\gamma_{\mathrm{K}}}'^*_X \end{pmatrix}. \tag{117}$$

In Eq. (116) we have inserted the CS displacement operators ($\hat{D}_{\mathrm{NC}}$, Eq. (87)) to produce non-Condon operators with the corresponding partial derivatives. The collective vectors are used additionally $\underline{\tilde{l}}^{\mathrm{t}} = (\underline{l}^{\mathrm{t}}, \underline{l}'^{\mathrm{t}})$, $\underline{\tilde{m}}^{\mathrm{t}} = (\underline{m}^{\mathrm{t}}, \underline{m}'^{\mathrm{t}})$, $\underline{\tilde{n}}^{\mathrm{t}} = (\underline{n}^{\mathrm{t}}, \underline{n}'^{\mathrm{t}})$ and $\underline{\tilde{o}}^{\mathrm{t}} = (\underline{o}^{\mathrm{t}}, \underline{o}'^{\mathrm{t}})$. The integral form of the CS-based GF, $G^{K_{YY}}_{\mathrm{cF}}(\mathbf{Z}_{XX}, \mathbf{Z}_{YY}; \mathbf{\Lambda}_{YY}; \underline{\tilde{\xi}}_{\mathrm{KB}}, \underline{\tilde{\xi}}_{\mathrm{BK}}; \underline{\eta}, \underline{\eta}')$ is, then, given as

$$\begin{aligned}
&G^{K_{YY}}_{\mathrm{cF}}(\widetilde{\mathbf{Z}}_{\mathrm{cF}}, \mathbf{Z}_{XX}, \mathbf{Z}_{YY}; \mathbf{\Lambda}_{YY}; \underline{\tilde{\xi}}_{\mathrm{KB}}, \underline{\tilde{\xi}}_{\mathrm{BK}}; \underline{\eta}, \underline{\eta}') \\
&= \pi^{-(N_Y+N'_Y)} |\langle \underline{0}'|\underline{0}\rangle|^{-2} \exp(\tfrac{1}{2}|\underline{\tilde{\xi}}_{\mathrm{KB}}|^2 + \tfrac{1}{2}|\underline{\tilde{\xi}}_{\mathrm{BK}}|^2) \\
&\int \mathrm{d}^2\underline{\alpha}_Y \mathrm{d}^2\underline{\gamma}'_Y K(\mathbf{\Lambda}_{YY}; \underline{\xi}_Y) \\
&\langle \mathbf{z}'^*_{\mathrm{B}}\underline{\gamma_{\mathrm{B}}}'_X; \mathbf{z}'^*_{YY}\underline{\gamma}'_Y|\hat{D}_{\mathrm{NC}}(\underline{\eta})|\mathbf{z}_{\mathrm{K}}\underline{\alpha_{\mathrm{K}}}_X; \mathbf{z}_{YY}\underline{\alpha}_Y\rangle \langle \mathbf{z}'_{\mathrm{K}}\underline{\gamma_{\mathrm{K}}}'_X; \mathbf{z}'_{YY}\underline{\gamma}'_Y|\hat{D}_{\mathrm{NC}}(\underline{\eta}')|\mathbf{z}^*_{\mathrm{B}}\underline{\alpha_{\mathrm{B}}}_X; \mathbf{z}^*_{YY}\underline{\alpha}_Y\rangle^*.
\end{aligned} \tag{118}$$

Then the resulting expression after integration is obtained in separated form with the FC contribution, the non-Condon part and the SVL and non-Condon contribution, of the vibrational modes belonging to the space Y, Y and X, respectively,

$$\begin{aligned}
&G^{K_{YY}}_{\mathrm{cF}}(\widetilde{\mathbf{Z}}_{\mathrm{cF}}, \mathbf{Z}_{YY}; \mathbf{\Lambda}_{YY}; \underline{\tilde{\xi}}_{\mathrm{KB}}, \underline{\tilde{\xi}}_{\mathrm{BK}}; \underline{\eta}, \underline{\eta}') = \\
&G^{K_{YY}}(\mathbf{Z}_{YY}; \mathbf{\Lambda}_{YY}) \mathcal{J}[\widetilde{\mathbf{W}}_{\mathrm{NCF};YY}(\mathbf{Z}_{YY}; \mathbf{\Lambda}_{YY}), \underline{\tilde{r}}_{\mathrm{NCF};Y}(\mathbf{Z}_{YY}; \mathbf{\Lambda}_{YY}); \underline{\tilde{\eta}}_{\mathrm{NCF};Y}] \\
&\mathcal{J}[\widetilde{\mathbf{Z}}_{\mathrm{cF}}\widetilde{\mathbf{W}}_{\mathrm{cF}}(\mathbf{Z}_{YY}; \mathbf{\Lambda}_{YY})\widetilde{\mathbf{Z}}_{\mathrm{cF}}, \widetilde{\mathbf{Z}}_{\mathrm{cF}}\underline{\tilde{b}}_{\mathrm{cF}}(\mathbf{Z}_{YY}; \mathbf{\Lambda}_{YY}; \underline{\eta}, \underline{\eta}'); \underline{\tilde{\xi}}_{\mathrm{cF}}],
\end{aligned} \tag{119}$$

with the collective vectors and matrices,

$$\underline{\tilde{\xi}}_{\mathrm{cF}} = \begin{pmatrix} \underline{\xi}_{\mathrm{KB}} \\ \underline{\xi}^*_{\mathrm{BK}} \end{pmatrix}, \tag{120}$$

$$\widetilde{\mathbf{W}}_{\mathrm{cF}}(\mathbf{Z}_{YY};\mathbf{\Lambda}_{YY}) = \frac{1}{2}\begin{pmatrix} (\widetilde{\mathbf{W}}^{+}_{T;XX} + \widetilde{\mathbf{W}}^{-}_{T;XX}) & (\widetilde{\mathbf{W}}^{+}_{T;XX} - \widetilde{\mathbf{W}}^{-}_{T;XX}) \\ (\widetilde{\mathbf{W}}^{+}_{T;XX} - \widetilde{\mathbf{W}}^{-}_{T;XX}) & (\widetilde{\mathbf{W}}^{+}_{T;XX} + \widetilde{\mathbf{W}}^{-}_{T;XX}) \end{pmatrix}, \tag{121}$$

$$\tilde{\underline{b}}_{\mathrm{cF}}(\mathbf{Z}_{YY};\mathbf{\Lambda}_{YY};\underline{\eta},\underline{\eta}') = \begin{pmatrix} \tilde{\underline{b}}^{+}_{X} - \mathrm{i}\tilde{\underline{b}}^{-}_{X} \\ \tilde{\underline{b}}^{+}_{X} + \mathrm{i}\tilde{\underline{b}}^{-}_{X} \end{pmatrix}, \tag{122}$$

where the components of $\widetilde{\mathbf{W}}_{\mathrm{cF}}$ and $\tilde{\underline{b}}_{\mathrm{cF}}$ are defined as follows [11, 102, 104]

$$\widetilde{\mathbf{W}}^{\pm}_{T;XX}(\mathbf{Z}_{YY}) = \mathbf{W}_{T;XX} \mp \mathbf{W}_{T;XY}\mathbf{Z}_{YY}((\mathbf{I} \pm \mathbf{Z}\mathbf{W}_{T}\mathbf{Z})_{YY})^{-1}\mathbf{Z}_{YY}\mathbf{W}_{T;YX}, \tag{123}$$

$$\tilde{\underline{r}}^{+}_{T;X}(\mathbf{Z}_{YY}) = \underline{r}_{T;X} - \mathbf{W}_{T;XY}\mathbf{Z}_{YY}((\mathbf{I} + \mathbf{Z}\mathbf{W}_{T}\mathbf{Z})_{YY})^{-1}\mathbf{Z}_{YY}\underline{r}_{T;Y}\,. \tag{124}$$

We can evaluate the cF GF with Eq. (116). If we can rearrange Eq. (119) to be expressed in terms of MHPs, it will be easy to automatise the calculation. The arrangement is straightforward but lengthy. Herein we present thus only explicitly the FC cF GF, in a simple form, as a special case ignoring the non-Condon operators, *i.e.*

$$\begin{aligned} &G^{K_{YY}}_{\mathrm{cF}}(\widetilde{\mathbf{Z}}_{\mathrm{cF}},\mathbf{Z}_{YY};\mathbf{\Lambda}_{YY};\tilde{\underline{\xi}}_{\mathrm{KB}},\tilde{\underline{\xi}}_{\mathrm{BK}};\underline{\eta},\underline{\eta}')\Big|_{\underline{\eta},\underline{\eta}'=\underline{0}} = G^{K_{YY}}_{\mathrm{cF}}(\widetilde{\mathbf{Z}}_{\mathrm{cF}},\mathbf{Z}_{YY};\mathbf{\Lambda}_{YY};\tilde{\underline{\xi}}_{\mathrm{KB}},\tilde{\underline{\xi}}_{\mathrm{BK}}) \\ &= G^{K_{YY}}(\mathbf{Z}_{YY};\mathbf{\Lambda}_{YY})\mathcal{J}[\widetilde{\mathbf{Z}}_{\mathrm{cF}}\widetilde{\mathbf{W}}_{\mathrm{cF}}(\mathbf{Z}_{YY};\mathbf{\Lambda}_{YY})\widetilde{\mathbf{Z}}_{\mathrm{cF}},\widetilde{\mathbf{Z}}_{\mathrm{cF}}\tilde{\underline{r}}_{\mathrm{cF}}(\mathbf{Z}_{YY};\mathbf{\Lambda}_{YY});\tilde{\underline{\xi}}_{\mathrm{cF}}], \end{aligned} \tag{125}$$

where

$$\tilde{\underline{r}}_{\mathrm{cF}}(\mathbf{Z}_{YY};\mathbf{\Lambda}_{YY}) = \begin{pmatrix} \tilde{\underline{r}}^{+}_{T;X} \\ \tilde{\underline{r}}^{+}_{T;X} \end{pmatrix}, \tag{126}$$

and the vector component is defined in Eq. (124). Eq. (125) can be rearranged to give the FC cF GF in MHPs, *i.e.*

$$\begin{aligned} &G^{K_{YY}}_{\mathrm{cF}}(\widetilde{\mathbf{Z}}_{\mathrm{cF}},\mathbf{Z}_{YY};\mathbf{\Lambda}_{YY};\tilde{\underline{v}}_{\mathrm{KB}},\tilde{\underline{v}}_{\mathrm{BK}}) \\ &= \left(\widetilde{\prod}^{-\frac{1}{2},-\frac{1}{2}}_{\underline{v_{\mathrm{B}}}_X,\underline{v_{\mathrm{K}}}_X}\partial^{\underline{v_{\mathrm{B}}}_X,\underline{v_{\mathrm{K}}}_X}_{\underline{\alpha_{\mathrm{B}}}^{*}_X,\underline{\alpha_{\mathrm{K}}}_X}\right)\left(\widetilde{\prod}^{-\frac{1}{2},-\frac{1}{2}}_{\underline{v_{\mathrm{B}}}'_X,\underline{v_{\mathrm{K}}}'_X}\partial^{\underline{v_{\mathrm{B}}}'_X,\underline{v_{\mathrm{K}}}'_X}_{\underline{\gamma_{\mathrm{K}}}'_X,\underline{\gamma_{\mathrm{B}}}'^{*}_X}\right) \\ &\quad G^{K_{YY}}_{\mathrm{cF}}(\widetilde{\mathbf{Z}}_{\mathrm{cF}},\mathbf{Z}_{YY};\mathbf{\Lambda}_{YY};\tilde{\underline{\xi}}_{\mathrm{BK}},\tilde{\underline{\xi}}_{\mathrm{KB}})\Big|_{\tilde{\underline{\xi}}_{\mathrm{KB}},\tilde{\underline{\xi}}_{\mathrm{BK}}=\underline{0}} \\ &= \left(\widetilde{\prod}^{-\frac{1}{2},-\frac{1}{2}}_{\underline{v_{\mathrm{B}}}_X,\underline{v_{\mathrm{K}}}_X}\right)\left(\widetilde{\prod}^{-\frac{1}{2},-\frac{1}{2}}_{\underline{v_{\mathrm{B}}}'_X,\underline{v_{\mathrm{K}}}'_X}\right) \\ &\quad G^{K_{YY}}(\mathbf{Z}_{YY};\mathbf{\Lambda}_{YY})\mathcal{H}e_{\tilde{\underline{v}}_{\mathrm{KB}},\tilde{\underline{v}}_{\mathrm{BK}}}\left(\left(\widetilde{\mathbf{Z}}_{\mathrm{cF}}\widetilde{\mathbf{W}}_{\mathrm{cF}}\widetilde{\mathbf{Z}}_{\mathrm{cF}}\right)^{-1}\tilde{\underline{r}}_{\mathrm{cF}};\left(\widetilde{\mathbf{Z}}_{\mathrm{cF}}\widetilde{\mathbf{W}}_{\mathrm{cF}}\widetilde{\mathbf{Z}}_{\mathrm{cF}}\right)^{-1}\right), \end{aligned} \tag{127}$$

which can be evaluated recursively or iteratively. We can introduce an arbitrary number of fixed quantum numbers in the GFs as in Eq. (113) exploiting the equations (119) and (125) for non-Condon and Condon processes respectively in combination with the partial derivatives Eq. (116). With the cF GF (119) we try to obtain GFs for the rR (Sec. 3.2), the SVL transition (Sec. 3.3) and the anharmonic OPA transition (Sec. 3.4) in the following sections.

3.2. Application to resonance Raman scattering

We can express the rR amplitude TCF (χ_α) of Eq. (11) in harmonic approximation with the cF GF Eq. (119). In the TCF for the rR excitation profile we have no fixed quantum numbers in the excited electronic state and all vibrational quantum numbers in the ground electronic state are fixed, *i.e.* ($N_X = N, N_Y = 0$) and ($N'_X = 0, N'_Y = N$) such that X represents ground electronic states and Y represents excited electronic states in this section. We can express in the TCF the scalar products of transition dipole moment (TDM) and polarisation vectors ($\mu^{\mathrm{S}}(\underline{Q})$ and $\mu^{\mathrm{L}}(\underline{Q})$ in Eq. (4)) similarly to the linear HT approximation reported in Eq. (15). Then we obtain

$$\begin{aligned}\chi_\alpha(\tau;\underline{v}_i,\underline{v}_f) \simeq |\langle\underline{0}'|\underline{0}\rangle|^2\Big(& (\mu_0^{\mathrm{S}})^\dagger\mu_0^{\mathrm{L}} G_{\mathrm{cF}}^{K_{YY}}(\widetilde{\mathbf{I}}_{\mathrm{cF}},\mathbf{z}'(\tau);\mathbf{\Lambda}_{YY};\tilde{\underline{v}}_{\mathrm{KB}},\tilde{\underline{v}}_{\mathrm{BK}}) \\ & + \sum_i (\mu_i'^{\mathrm{S}})^\dagger\mu_0^{\mathrm{L}} G_{\mathrm{cF}}^{K_{YY}}(\widetilde{\mathbf{I}}_{\mathrm{cF}},\mathbf{z}'(\tau);\mathbf{\Lambda}_{YY};\tilde{\underline{v}}_{\mathrm{KB}},\tilde{\underline{v}}_{\mathrm{BK}})^{(\hat{Q}_i,\hat{1})} \\ & + \sum_i (\mu_0^{\mathrm{S}})^\dagger\mu_i'^{\mathrm{L}} G_{\mathrm{cF}}^{K_{YY}}(\widetilde{\mathbf{I}}_{\mathrm{cF}},\mathbf{z}'(\tau);\mathbf{\Lambda}_{YY};\tilde{\underline{v}}_{\mathrm{KB}},\tilde{\underline{v}}_{\mathrm{BK}})^{(\hat{1},\hat{Q}_i)} \\ & + \sum_{i,j} (\mu_i'^{\mathrm{S}})^\dagger\mu_j'^{\mathrm{L}} G_{\mathrm{cF}}^{K_{YY}}(\widetilde{\mathbf{I}}_{\mathrm{cF}},\mathbf{z}'(\tau);\mathbf{\Lambda}_{YY};\tilde{\underline{v}}_{\mathrm{KB}},\tilde{\underline{v}}_{\mathrm{BK}})^{(\hat{Q}_i,\hat{Q}_j)}\Big), \end{aligned} \tag{128}$$

with proper set up for GF parameters $\mathbf{Z}_{YY}(\tau) = \mathbf{z}'(\tau)$ as in Eq. (110) for the TCF of the vibronic absorption profiles and $\widetilde{\mathbf{I}}_{\mathrm{cF}} = \mathrm{bldiag}(\mathbf{I}_{XX},\mathbf{I}_{XX})$. There were other GF approaches (see *e.g.* Refs. [65] and [67]) including the Duschinsky effects proposed that are similar to the current developments, but the approaches have limitations for the dimension of X and for including thermal effects in a closed formula within the Condon approximation. The current approach can handle arbitrary SVLs and non-Condon (linear and nonlinear HT terms in rR scattering cross section) effects. We can obtain a closed form, with the help of the CS formalism, for the rR intensity TCF (14), namely the 3-point-TCF, *i.e.*

$$\begin{aligned} & G_{\mathrm{rR}}^K(\widetilde{\mathbf{Z}}_{\mathrm{cF}}(t),\mathbf{Z}_{YY}(\tau),\mathbf{Z}_{YY}(\tau');\mathbf{\Lambda}_{YY},\widetilde{\mathbf{\Lambda}}_{\mathrm{cF}};\underline{\eta_1},\underline{\eta_1}',\underline{\eta_2},\underline{\eta_2}') \\ & = \pi^{-2N}\int \mathrm{d}^2\underline{\alpha_{\mathrm{K}}}\mathrm{d}^2\underline{\alpha_{\mathrm{B}}} K(\widetilde{\mathbf{\Lambda}}_{\mathrm{cF}};\tilde{\underline{\xi}}_{\mathrm{cF}})\exp(-|\tilde{\underline{\xi}}_{\mathrm{cF}}|^2) \\ & G_{\mathrm{cF}}^{K_{YY}}(\widetilde{\mathbf{Z}}_{\mathrm{cF}}(t),\mathbf{Z}_{YY}(\tau);\mathbf{\Lambda}_{YY};\tilde{\underline{\xi}}_{\mathrm{KB}},\tilde{\underline{\xi}}_{\mathrm{BK}};\underline{\eta_1},\underline{\eta_1}') \\ & G_{\mathrm{cF}}^{K_{YY}}(\widetilde{\mathbf{Z}}^*_{\mathrm{cF}}(t),\mathbf{Z}_{YY}(\tau');\mathbf{\Lambda}_{YY};\tilde{\underline{\xi}}_{\mathrm{KB}},\tilde{\underline{\xi}}_{\mathrm{BK}};\underline{\eta_2},\underline{\eta_2}')^*, \end{aligned} \tag{129}$$

which is in the form of an overlap between two rR amplitude GFs, in different time domains (τ and τ'), where the thermal parameter matrix $\widetilde{\mathbf{\Lambda}}_{\mathrm{cF}} = \mathrm{bldiag}(\mathbf{\Lambda}_{XX},\mathbf{\Lambda}_{XX})^{\frac{1}{2}}$ is used. It is worth to note here in passing that we suggest a closed integral form of the rR cross section in harmonic approximation including Duschinsky and thermal effects in a functional analogy to the Förster-type energy transfer processes (see *e.g.* Ref. [134]) which is in a form of a convolution between the absorption and the emission spectra. We can transform the integration into a $2N$-dimensional Gaussian integral

$$\mathcal{I}_{2N}[\mathbf{A},\underline{b}] = \det(\mathbf{A})^{-\frac{1}{2}}\exp(\underline{b}^{\mathrm{t}}\mathbf{A}^{-1}\underline{b}) \tag{130}$$

and obtain the Gaussian integral

$$\begin{aligned}
&G^{K}_{\mathrm{rR}}(\widetilde{\mathbf{Z}}_{\mathrm{cF}}(t),\mathbf{Z}_{YY}(\tau),\mathbf{Z}_{YY}(\tau');\mathbf{\Lambda}_{YY},\widetilde{\mathbf{\Lambda}}_{\mathrm{cF}};\underline{\eta_1},\underline{\eta_1}',\underline{\eta_2},\underline{\eta_2}')\\
&=G^{K_{YY}}(\mathbf{Z}_{YY}(\tau);\mathbf{\Lambda}_{YY})G^{K_{YY}}(\mathbf{Z}_{YY}(\tau');\mathbf{\Lambda}_{YY})^*\\
&\mathcal{J}[\widetilde{\mathbf{W}}_{\mathrm{NCF};YY}(\mathbf{Z}_{YY}(\tau);\mathbf{\Lambda}_{YY}),\underline{\tilde{r}}_{\mathrm{NCF};Y}(\mathbf{Z}_{YY}(\tau);\mathbf{\Lambda}_{YY});\underline{\tilde{\eta}1}_{\mathrm{NCF};Y}]\\
&\mathcal{J}[\widetilde{\mathbf{W}}_{\mathrm{NCF};YY}(\mathbf{Z}_{YY}(\tau');\mathbf{\Lambda}_{YY}),\underline{\tilde{r}}_{\mathrm{NCF};Y}(\mathbf{Z}_{YY}(\tau');\mathbf{\Lambda}_{YY});\underline{\tilde{\eta}2}_{\mathrm{NCF};Y}]^*\\
&\mathcal{N}_{\mathrm{cF}}\mathcal{I}_{2N}[\mathbf{I}-\widetilde{\mathbf{Z}}_{\mathrm{rR}}(t)\widetilde{\mathbf{W}}_{\mathrm{rR}}(\mathbf{Z}_{YY}(\tau),\mathbf{Z}_{YY}(\tau');\mathbf{\Lambda}_{YY},\widetilde{\mathbf{\Lambda}}_{\mathrm{cF}})\widetilde{\mathbf{Z}}_{\mathrm{rR}}(t)\\
&,\tfrac{1}{2}\widetilde{\mathbf{Z}}_{\mathrm{rR}}(t)\underline{\tilde{b}}_{\mathrm{rR}}(\mathbf{Z}_{YY}(\tau),\mathbf{Z}_{YY}(\tau');\mathbf{\Lambda}_{YY},\mathbf{\Lambda}_{\mathrm{cF}})],
\end{aligned}\tag{131}$$

where the $4N$-dimensional square matrix and vector are defined as

$$\begin{aligned}
&\widetilde{\mathbf{W}}_{\mathrm{rR}}(\mathbf{Z}_{YY}(\tau),\mathbf{Z}_{YY}(\tau');\mathbf{\Lambda}_{YY},\widetilde{\mathbf{\Lambda}}_{\mathrm{cF}})=\\
&\frac{1}{2}\begin{pmatrix}\left(\widetilde{\mathbf{W}}_{\mathrm{cF};T}(\mathbf{Z}_{YY}(\tau);\mathbf{\Lambda}_{YY})+\widetilde{\mathbf{W}}_{\mathrm{cF};T}(\mathbf{Z}_{YY}(\tau');\mathbf{\Lambda}_{YY})^*\right) & \mathrm{i}\left(\widetilde{\mathbf{W}}_{\mathrm{cF};T}(\mathbf{Z}_{YY}(\tau);\mathbf{\Lambda}_{YY})-\widetilde{\mathbf{W}}_{\mathrm{cF};T}(\mathbf{Z}_{YY}(\tau');\mathbf{\Lambda}_{YY})^*\right)\\ \mathrm{i}\left(\widetilde{\mathbf{W}}_{\mathrm{cF};T}(\mathbf{Z}_{YY}(\tau);\mathbf{\Lambda}_{YY})-\widetilde{\mathbf{W}}_{\mathrm{cF};T}(\mathbf{Z}_{YY}(\tau');\mathbf{\Lambda}_{YY})^*\right) & -\left(\widetilde{\mathbf{W}}_{\mathrm{cF};T}(\mathbf{Z}_{YY}(\tau);\mathbf{\Lambda}_{YY})+\widetilde{\mathbf{W}}_{\mathrm{cF};T}(\mathbf{Z}_{YY}(\tau');\mathbf{\Lambda}_{YY})^*\right)\end{pmatrix},
\end{aligned}\tag{132}$$

$$\begin{aligned}
&\underline{\tilde{b}}_{\mathrm{rR}}(\mathbf{Z}_{YY}(\tau),\mathbf{Z}_{YY}(\tau');\mathbf{\Lambda}_{YY},\widetilde{\mathbf{\Lambda}}_{\mathrm{cF}};\underline{\eta_1},\underline{\eta_1}',\underline{\eta_2},\underline{\eta_2}')\\
&=\begin{pmatrix}\underline{\tilde{b}}_{\mathrm{cF};T}(\mathbf{Z}_{YY}(\tau);\mathbf{\Lambda}_{YY};\underline{\eta_1},\underline{\eta_1}')+\underline{\tilde{b}}_{\mathrm{cF};T}(\mathbf{Z}_{YY}(\tau');\mathbf{\Lambda}_{YY};\underline{\eta_2},\underline{\eta_2}')^*\\ \mathrm{i}\left(\underline{\tilde{b}}_{\mathrm{cF};T}(\mathbf{Z}_{YY}(\tau);\mathbf{\Lambda}_{YY};\underline{\eta_1},\underline{\eta_1}')-\underline{\tilde{b}}_{\mathrm{cF};T}(\mathbf{Z}_{YY}(\tau');\mathbf{\Lambda}_{YY};\underline{\eta_2},\underline{\eta_2}')^*\right)\end{pmatrix},
\end{aligned}\tag{133}$$

and

$$\widetilde{\mathbf{Z}}_{\mathrm{rR}}=\mathrm{bldiag}(\widetilde{\mathbf{Z}}_{\mathrm{cF}},\widetilde{\mathbf{Z}}_{\mathrm{cF}})^{\frac{1}{2}}.\tag{134}$$

The thermally weighted quantities used are

$$\widetilde{\mathbf{W}}_{\mathrm{cF};T}=(\mathbf{I}+\mathbf{\Lambda}_{\mathrm{cF}})^{-\frac{1}{2}}\widetilde{\mathbf{W}}_{\mathrm{cF}}(\mathbf{I}+\mathbf{\Lambda}_{\mathrm{cF}})^{-\frac{1}{2}},\tag{135}$$

$$\underline{\tilde{b}}_{\mathrm{cF};T}=(\mathbf{I}+\mathbf{\Lambda}_{\mathrm{cF}})^{-\frac{1}{2}}\underline{\tilde{b}}_{\mathrm{cF}}.\tag{136}$$

$\mathcal{N}_{\mathrm{cF}}=1/\mathrm{Tr}(\exp(-\underline{\beta}\cdot\underline{\hat{h}}))$ is a normalising factor for the thermal integral kernel, with the N-dimensional vibrational Hamiltonian ($\hat{H}=\sum_i^N\hat{h}_i$).

In the Condon approximation, the expression (Eq. (113)) is simplified as

$$\begin{aligned}
&G^{K}_{\mathrm{rR}}(\widetilde{\mathbf{Z}}_{\mathrm{cF}}(t),\mathbf{Z}_{YY}(\tau),\mathbf{Z}_{YY}(\tau');\mathbf{\Lambda}_{YY},\widetilde{\mathbf{\Lambda}}_{\mathrm{cF}})\\
&=G^{K_{YY}}(\mathbf{Z}_{YY}(\tau);\mathbf{\Lambda}_{YY})G^{K_{YY}}(\mathbf{Z}_{YY}(\tau');\mathbf{\Lambda}_{YY})^*\\
&\mathcal{N}_{\mathrm{cF}}\mathcal{I}_{2N}[\mathbf{I}-\widetilde{\mathbf{Z}}_{\mathrm{rR}}(t)\widetilde{\mathbf{W}}_{\mathrm{rR}}(\mathbf{Z}_{YY}(\tau),\mathbf{Z}_{YY}(\tau');\mathbf{\Lambda}_{YY},\widetilde{\mathbf{\Lambda}}_{\mathrm{cF}})\widetilde{\mathbf{Z}}_{\mathrm{rR}}(t)\\
&,\tfrac{1}{2}\widetilde{\mathbf{Z}}_{\mathrm{rR}}(t)\underline{\tilde{r}}_{\mathrm{rR}}(\mathbf{Z}_{YY}(\tau),\mathbf{Z}_{YY}(\tau');\mathbf{\Lambda}_{YY},\mathbf{\Lambda}_{\mathrm{cF}})],
\end{aligned}\tag{137}$$

where

$$\begin{aligned}
&\underline{\tilde{b}}_{\mathrm{rR}}(\mathbf{Z}_{YY}(\tau),\mathbf{Z}_{YY}(\tau');\mathbf{\Lambda}_{YY},\widetilde{\mathbf{\Lambda}}_{\mathrm{cF}};\underline{\eta_1},\underline{\eta_1}',\underline{\eta_2},\underline{\eta_2}')\Big|_{\underline{\eta_1},\underline{\eta_1}',\underline{\eta_2},\underline{\eta_2}'=\underline{0}}\\
&=\underline{\tilde{r}}_{\mathrm{rR}}(\mathbf{Z}_{YY}(\tau),\mathbf{Z}_{YY}(\tau');\mathbf{\Lambda}_{YY},\widetilde{\mathbf{\Lambda}}_{\mathrm{cF}})\\
&=\begin{pmatrix}\underline{\tilde{r}}_{\mathrm{cF};T}(\mathbf{Z}_{YY}(\tau);\mathbf{\Lambda}_{YY})+\underline{\tilde{r}}_{\mathrm{cF};T}(\mathbf{Z}_{YY}(\tau');\mathbf{\Lambda}_{YY})\\ \mathrm{i}\left(\underline{\tilde{r}}_{\mathrm{cF};T}(\mathbf{Z}_{YY}(\tau);\mathbf{\Lambda}_{YY})-\underline{\tilde{r}}_{\mathrm{cF};T}(\mathbf{Z}_{YY}(\tau');\mathbf{\Lambda}_{YY})\right)\end{pmatrix}.
\end{aligned}\tag{138}$$

$G^K_{\mathrm{rR}}(t,\tau,\tau')$ is a 3-point-TCF in an analytic closed form for the rR intensity including Duschinsky, thermal and non-Condon effects (Albrecht B and C terms [53]). The evaluation is straightforward with the help of MHP technique for non-Condon operators. G^K_{rR} provides not only the TCFs but also the sum rule for the Raman intensity [62] by setting $\mathbf{Z}_{\mathrm{rR}}(t)$ to the identity matrix which can lead to possible prescreening strategies for the rR intensity profile. In the rR calculation the peaks are identified by computing the individual scattering amplitude from the amplitude TCF in Eq. (128). If we know the individual contributions from the sum rule, we can avoid the efforts to compute many scattering amplitudes that have negligible impact on the profile.

The explicit expressions for the rR amplitude and intensity in Condon approximation are given in the Appendix where only the vibrational modes in initial state are allowed to be thermally excited at finite temperature.

3.3. Application to single vibronic level transitions

It is straightforward to derive the GF of the SVL transition from the cF GF in Eq. (116). Herein we introduce constraints $\underline{v}_{\mathrm{B}_X} = \underline{v}_{\mathrm{K}_X}$ and $\underline{v}'_{\mathrm{B}_X} = \underline{v}'_{\mathrm{K}_X}$ which will reduce the dimension of the corresponding MHPs to half, *i.e.*

$$\begin{aligned} &G^{K_{YY}}_{\mathrm{SVL}}(\mathbf{Z}_{XX}, \mathbf{Z}_{YY}; \mathbf{\Lambda}_{YY}; \underline{\xi}_X; \underline{\eta}, \underline{\eta}') = G^{K_{YY}}(\mathbf{Z}_{YY}; \mathbf{\Lambda}_{YY}) \\ &\mathcal{J}[\widetilde{\mathbf{W}}_{\mathrm{NCF};YY}(\mathbf{Z}_{YY}; \mathbf{\Lambda}_{YY}), \underline{\tilde{r}}_{\mathrm{NCF};Y}; \underline{\tilde{\eta}}_{\mathrm{NCF};Y}] \\ &\mathcal{J}[2\mathbf{Z}_{XX}\widetilde{\mathbf{W}}^+_{T;XX}(\mathbf{Z}_{YY}; \mathbf{\Lambda}_{YY})\mathbf{Z}_{XX}, 2\mathbf{Z}_{XX}(\underline{\tilde{b}}^+_X)^{\mathrm{t}}(\mathbf{Z}_{YY}; \mathbf{\Lambda}_{YY}; \underline{\eta}, \underline{\eta}'); \underline{\xi}_X], \end{aligned} \tag{139}$$

where we have assumed that the generating variables are real numbers, *i.e.*

$$\underline{\alpha}_{\mathrm{K}_X} = \underline{\alpha}^*_{\mathrm{B}_X} = \underline{\alpha}_X \in \mathbb{R}, \tag{140}$$

$$\underline{\gamma}'^*_{\mathrm{B}_X} = \underline{\gamma}'_{\mathrm{K}_X} = \underline{\gamma}'_X \in \mathbb{R}. \tag{141}$$

In the SVL transition the ONV of the initial vibronic state is in the fixed quantum number space X, but also some of the final vibrational modes have fixed quantum numbers for possible prescreening applications, *i.e.* $N_X = N$, $N_Y = 0$, $0 \leq N'_X \leq N$ and $N'_Y = N - N'_X$. In Condon approximation the expression is further simplified as

$$\begin{aligned} &G^{K_{YY}}_{\mathrm{SVL}}(\mathbf{Z}_{XX}, \mathbf{Z}_{YY}; \mathbf{\Lambda}_{YY}; \underline{\xi}_X) = G^{K_{YY}}(\mathbf{Z}_{YY}; \mathbf{\Lambda}_{YY}) \\ &\mathcal{J}[2\mathbf{Z}_{XX}\widetilde{\mathbf{W}}^+_{T;XX}(\mathbf{Z}_{YY}; \mathbf{\Lambda}_{YY})\mathbf{Z}_{XX}, 2\mathbf{Z}_{XX}(\underline{\tilde{r}}^+_{T;X})^{\mathrm{t}}(\mathbf{Z}_{YY}; \mathbf{\Lambda}_{YY}); \underline{\xi}_X]. \end{aligned} \tag{142}$$

We can also have a similar MHP expression in Eq. (127) for the FC SVL GF

$$\begin{aligned} &G^{K_{YY}}_{\mathrm{SVL}}(\mathbf{Z}_{XX}, \mathbf{Z}_{YY}; \mathbf{\Lambda}_{YY}; \underline{\tilde{v}}_X) \\ &= \left(\widetilde{\prod}_{\underline{\tilde{v}}_X}^{-\frac{1}{2}}\right) G^{K_{YY}}(\mathbf{Z}_{YY}; \mathbf{\Lambda}_{YY}) \\ &\mathcal{H}e_{\underline{\tilde{v}}_X}\left((2\mathbf{Z}_{XX}\widetilde{\mathbf{W}}^+_{T;XX}\mathbf{Z}_{XX})^{-1}\underline{\tilde{r}}^+_{T;X}; (2\mathbf{Z}_{XX}\widetilde{\mathbf{W}}^+_{T;XX}\mathbf{Z}_{XX})^{-1}\right). \end{aligned} \tag{143}$$

We can obtain the SVL transition TCF from the GF (143) with the TD GF parameter $\mathbf{Z}_{YY} = \mathbf{z}'(t)$ in Eq. (110). The resulting expression is not given in a closed form, but we can evaluate the TCF with the MHPs carrying complex numbers from the TD GF parameters. Additionally we can also apply the SVL transition GF to fine-grained integral prescreening strategies directly with the integral space partitioned expressions (139) and (142) for multi-dimensional X spaces.

The explicit expression for the SVL transition in Condon approximation is given in Appendix, where only the vibrational modes in initial state are allowed to be thermally excited at finite temperature.

3.4. Application to transition between anharmonic oscillators

FC or non-FC transition moments computed for the approximated harmonic PESs are in general not sufficient to describe transitions between states in which for instance large amplitude motions are relevant (see *e.g.* Ref. [123]). In this case one would have to start from the full rovibrational Hamiltonian such as the Watson [135] or Meyer-Günthard [136] molecular Hamiltonian, which include terms coupling momentum and position operators as well as anharmonic potential energy terms.

Numerous approaches for propagating wavepackets on general potential energy surfaces in time domain have been proposed to compute vibronic spectra, for instance the Gaussian wavepacket approach [110, 137], the multi-configurational time dependent Hartree (MCTDH) method [138, 139] and the coupled CS method [140]. In these approaches, the initial wavepacket evolves on the final PES which provides the time correlation functions. The corresponding Fourier transform (FT) gives the vibronic spectrum in frequency domain. One of the major demands for computational approaches, however, is the peak assignment and identification of (exact) excitation energy levels, which can not be given directly by the TD approaches.

In TI approaches one typically obtains the vibrational anharmonic eigenfunctions from the molecular Hamiltonian by perturbation methods (see *e.g.* Refs. [31, 141–146]) or by diagonalising it in a given finite basis set representation for instance in harmonic oscillator states and (complex) Gaussian functions. The anharmonic transition amplitudes can then be calculated by perturbation methods and basis set expansion approaches, respectively.

In all these approaches the TI method for FC transitions between anharmonic vibrational states is computationally expensive because of the multitude of overlap integrals to be evaluated. The situation is even worse, if we need to consider the Duschinsky mode mixing effects (Sec. 2.2). There were some approaches proposed (see *e.g.* Refs. [18, 147–151]) to avoid the mode mixing problem by expanding the wavefunctions with a one-centre basis set. However, the one-centre basis set expansion of the anharmonic vibrational wavefunctions usually requires a larger number of basis functions than the two-reference point expansion approach inducing the Duschinsky rotation between the two basis sets.

The vibrational self-consistent field (VSCF) and vibrational configuration interaction (VCI), which are related to the Hartree–Fock and configuration interaction (CI) methods, respectively, in electronic structure theory, are widely used to describe the anharmonic vibrational wavefunctions with harmonic oscillator eigenfunctions [152–155]. VSCF and VCI wavefunctions, expanded via the harmonic oscillator eigenfunctions, were used to compute the anharmonic Franck–Condon factors (FCFs) of three atomic systems [46, 151, 155]. In the work of Huh *et al.* [46], we computed the potodetachment-photoelectron spectra of HS_2^- and DS_2^- to their neutral electronic ground and first electronically excited states by calculating the FC profiles based on VCI wavefunctions from the Watson molecular Hamiltonian [135] expanded in terms of displaced Gaussian functions [156]. In this method we only need to compute numerous Gaussian overlap integrals (of the type $\langle \underline{0}' | \underline{0} \rangle$, Eq. (79)) including the Duschinsky rotations. One simple prescreening strategy was to discard overlap integrals according to the displacements of the two displaced Gaussian functions, and then to store all important Gaussian overlap integrals at least for the three atomic systems to speed up the calculations. Luis and coworkers [157–160] have set up linear equations, for anharmonic FC integrals and matrix elements of the potential energy difference operator, which lower the computational cost by introducing the Duschinsky relation in the potential energy operator not in the wavefunctions, but these authors considered only a simple diagonal kinetic energy operator.

Lucas [121] proposed a method for evaluation of FCFs also for a nonlinear Duschinsky relation (Sec. 2.2). Nonlinear Duschinsky effects were treated perturbatively by Lucas for the (momentum-position coupled) non-Condon operators. We can easily handle this perturbation approach with the non-Condon integral evaluation scheme developed in Ref. [102].

Luckhaus [123] developed an approach which treats reaction path problems with large curvature. Therein the author solved the vibrational eigenfunction problem for a two-dimensional model system by diagonalising the Hamiltonian in non-orthogonal harmonic oscillator basis sets along the grid points on the reaction path. In this approach one has to compute the non-Condon integrals for matrix elements of the Hamiltonian as well as the FC integrals for the overlap integrals for the non-orthogonal harmonic basis set in the Duschinsky relation. If the reaction path and non-orthogonal basis set approach should be extended to larger systems, the developments of this article appears highly beneficial for this method.

Perhaps the simplest approach would be to introduce the anharmonicity in the potential energy surfaces only for a few vibrational modes neglecting the kinetic energy rovibrational part. Even for the simplest model Hamiltonians it is a computationally difficult task to compute the FCFs including the mode mixing effects due to the extremely large number of integrals to be evaluated. We present herein the simple anharmonic-harmonic block approach to treat systems that have many harmonic degrees of freedom (DOF) but a small number of anharmonic DOF which are still much more computationally expensive than the harmonic transitions. We further assume that the anharmonic and harmonic DOF are separable but the normal coordinates of two electronic states are related by the Duschinsky equation (Sec. 2.2). The anharmonic partition of the vibrational wavefunctions are expanded by a finite number of the harmonic oscillator eigenstates as a basis set and the harmonic basis sets of two electronic states are related in the Duschinsky relation. The important contribution of our development in this section is that it can provide the possible prescreening strategies and TCFs with the the non-Condon effects for anharmonic transitions. The GF developed in this section is in the form of linear combination of the cF GFs in section 3.1.

The anharmonic DOF are denoted as X and the harmonic ones as Y. The n-th vibrationally excited anharmonic wavefunction of the initial electronic state, with the associated harmonic eigenstate $|\underline{v}_Y\rangle$, $|\Psi_{n;\underline{v}_Y}\rangle$ is defined as,

$$\begin{aligned}|\Psi_{n;\underline{v}_Y}\rangle &= \sum_{\{\underline{v_{\mathrm{K}}}_X\}} c_{\underline{v_{\mathrm{K}}}_X;n}|\underline{v_{\mathrm{K}}}_X\rangle \otimes |\underline{v}_Y\rangle \\ &= \sum_{\{\underline{v_{\mathrm{K}}}_X\}} c_{\underline{v_{\mathrm{K}}}_X;n}|\underline{v_{\mathrm{K}}}_X;\underline{v}_Y\rangle, \end{aligned} \tag{144}$$

where the finite number of N_X^{A} anharmonic expansion coefficients for finite basis set ($\{\underline{v_{\mathrm{K}}}_X\}$) satisfy the normalisation condition,

$$\sum_{\{\underline{v_{\mathrm{K}}}_X\}} |c_{\underline{v_{\mathrm{K}}}_X;n}|^2 = 1, \tag{145}$$

and the final state is expressed in the same manner,

$$|\Psi'_{m;\underline{v}'_Y}\rangle = \sum_{\{\underline{v_{\mathrm{K}}}'_X\}} c'_{\underline{v_{\mathrm{K}}}'_X;m}|\underline{v_{\mathrm{K}}}'_X;\underline{v}'_Y\rangle, \tag{146}$$

$$\sum_{\{\underline{v_{\mathrm{K}}}'_X\}} |c'_{\underline{v_{\mathrm{K}}}'_X;m}|^2 = 1. \tag{147}$$

The FCFs of the vibronic transition from the n-th vibrationally excited wavefunction is summed to unity in any complete basis set expansion but it should be close to unity in a proper finite basis set expansion, *i.e.*

$$
\begin{aligned}
1 &= \langle \Psi_{n;\underline{v}_Y} | \Psi_{n;\underline{v}_Y} \rangle \\
&\simeq \sum_{\{\underline{v_{\mathrm{K}}}_X\},\{\underline{v_{\mathrm{B}}}_X\},\{\underline{v_{\mathrm{K}}}'_X\},\{\underline{v_{\mathrm{B}}}'_X\}} \sum_{m=0}^{(N_X^{\mathrm{A}})'-1} c'_{\underline{v_{\mathrm{K}}}'_X;m} c'^{*}_{\underline{v_{\mathrm{B}}}'_X;m} c^{*}_{\underline{v_{\mathrm{B}}}_X;n} c_{\underline{v_{\mathrm{K}}}_X;n} \\
&\quad \times \sum_{\underline{v}'_Y=\underline{0}}^{\infty} \langle \underline{v_{\mathrm{B}}}_X; \underline{v}_Y | \underline{v_{\mathrm{K}}}'_X; \underline{v}'_Y \rangle \langle \underline{v_{\mathrm{B}}}'_X; \underline{v}'_Y | \underline{v_{\mathrm{K}}}_X; \underline{v}_Y \rangle,
\end{aligned} \tag{148}
$$

where $(N_X^{\mathrm{A}})'$ is the finite number of final vibronic levels, which equals the number of harmonic basis functions.

From Eq. (148) and (113) we have the anharmonic transition GF G_{A} which contains the anharmonic transition information associated with the GF parameters,

$$
\begin{aligned}
&G_{\mathrm{A}}(\widetilde{\mathbf{Z}}_{\mathrm{cF}}, \mathbf{Z}_{YY}; \underline{c_n}, \underline{c_m}'; \underline{\tilde{\xi}}_{\mathrm{KB}}, \underline{\tilde{\xi}}_{\mathrm{BK}}; \underline{\eta}, \underline{\eta}') \\
&= \sum_{\{\underline{v_{\mathrm{K}}}_X\},\{\underline{v_{\mathrm{B}}}_X\},\{\underline{v_{\mathrm{K}}}'_X\},\{\underline{v_{\mathrm{B}}}'_X\}} c'_{\underline{v_{\mathrm{K}}}'_X;m} c'^{*}_{\underline{v_{\mathrm{B}}}'_X;m} c^{*}_{\underline{v_{\mathrm{B}}}_X;n} c_{\underline{v_{\mathrm{K}}}_X;n} G_{\mathrm{cF}}^{K_{YY}}(\widetilde{\mathbf{Z}}_{\mathrm{cF}}, \mathbf{Z}_{YY}; \mathbf{0}_{YY}; \underline{\tilde{\xi}}_{\mathrm{KB}}, \underline{\tilde{\xi}}_{\mathrm{BK}}; \underline{\eta}, \underline{\eta}'),
\end{aligned} \tag{149}
$$

where the cF GF in Eq. (119) is used for the possible non-Condon transition. The anharmonic transition GF is evaluated to have specific SVLs and non-Condon operators with Eq. (116), *i.e.*

$$
\begin{aligned}
&G_{\mathrm{A}}(\widetilde{\mathbf{Z}}_{\mathrm{cF}}, \mathbf{Z}_{YY}; \underline{c_n}, \underline{c_m}'; \underline{\tilde{v}}_{\mathrm{KB}}, \underline{\tilde{v}}_{\mathrm{BK}})^{(\hat{f},\hat{g})} \\
&= \sum_{\{\underline{v_{\mathrm{K}}}_X\},\{\underline{v_{\mathrm{B}}}_X\},\{\underline{v_{\mathrm{K}}}'_X\},\{\underline{v_{\mathrm{B}}}'_X\}} c'_{\underline{v_{\mathrm{K}}}'_X;m} c'^{*}_{\underline{v_{\mathrm{B}}}'_X;m} c^{*}_{\underline{v_{\mathrm{B}}}_X;n} c_{\underline{v_{\mathrm{K}}}_X;n} \\
&\left(\widetilde{\prod}_{\underline{v_{\mathrm{B}}}_X,\underline{v_{\mathrm{K}}}_X}^{-\frac{1}{2},-\frac{1}{2}} \hat{\partial}_{\underline{\alpha_{\mathrm{B}}}^{*}_X,\underline{\alpha_{\mathrm{K}}}_X}^{\underline{v_{\mathrm{B}}}_X,\underline{v_{\mathrm{K}}}_X} \right) \left(\widetilde{\prod}_{\underline{v_{\mathrm{B}}}'_X,\underline{v_{\mathrm{K}}}'_X}^{-\frac{1}{2},-\frac{1}{2}} \hat{\partial}_{\underline{\gamma_{\mathrm{K}}}'_X,\underline{\gamma_{\mathrm{B}}}'^{*}_X}^{\underline{v_{\mathrm{B}}}'_X,\underline{v_{\mathrm{K}}}'_X} \right) \\
&\prod_{k=1}^{2N} \left[\mathrm{i}\sqrt{\frac{\tilde{\epsilon}_k}{2}} \right]^{\tilde{l}_k} \left[\frac{1}{\mathrm{i}}\sqrt{\frac{\hbar^2}{2\tilde{\epsilon}_k}} \right]^{\tilde{m}_k} \left[-\mathrm{i}\sqrt{\frac{\tilde{\epsilon}_k}{2}} \right]^{\tilde{n}_k} \left[-\frac{1}{\mathrm{i}}\sqrt{\frac{\hbar^2}{2\tilde{\epsilon}_k}} \right]^{\tilde{o}_k} \left(\hat{\partial}_{\underline{\tilde{\eta}_P},\underline{\tilde{\eta}_Q},\underline{\tilde{\eta}'_P},\underline{\tilde{\eta}'_Q}}^{\underline{\tilde{l}},\underline{\tilde{m}},\underline{\tilde{n}},\underline{\tilde{o}}} \right) \\
&G_{\mathrm{cF}}^{K_{YY}}(\widetilde{\mathbf{Z}}_{\mathrm{cF}}, \mathbf{Z}_{YY}; \mathbf{0}_{YY}; \underline{\tilde{\xi}}_{\mathrm{BK}}, \underline{\tilde{\xi}}_{\mathrm{KB}}; \underline{\eta}, \underline{\eta}') \Big|_{\underline{\tilde{\xi}}_{\mathrm{KB}},\underline{\tilde{\xi}}_{\mathrm{BK}},\underline{\tilde{\eta}}_{\mathrm{NCF}}=\underline{0}}.
\end{aligned} \tag{150}
$$

We can devise prescreening strategies [11, 101, 102, 104] for an anharmonic transition from n-th vibrationally excited initial state to the m-th vibrationally excited final state with the GF including non-Condon effects. The TCF can also be evaluated by the GF with the TD GF parameters $\mathbf{Z}_{YY}(t)$ in Eqs. (109) and (110). In this development we assume a transition from a single vibronic initial state to a manifold of final vibronic states, but it can easily be generalised to account for thermally averaged initial vibronic states.

The explicit expression for the anharmonic transition in Condon approximation is given in the Appendix where only the vibrational modes in initial state are allowed to be thermally excited at finite temperature.

4. Conclusion and outlook

Herein, we reviewed our coherent state (CS)-based generating function (GF) method [11, 101, 102, 104] and presented the extensions for the resonance Raman (rR) scattering and intensity profile (Sec. 3.2), the single vibronic level (SVL) transition (Sec. 3.3) and the anharmonic Franck–Condon (FC)/non-Condon transition (Sec. 3.4). Those GFs are built upon the coherent-Fock (cF) GF, see section 3.1. These can introduce arbitrary SVLs via the corresponding partial derivatives.

The rR developments in section 3.2 differ from other works [49,54,57,59,60,65–67,69,95]. The difference appears in the time-correlation functions (TCFs) from the GFs. We can introduce non-Condon and thermal effects to our CS GF approach within the Duschinsky linear transform approximation. We emphasise herein that we can relatively easily include the non-Condon effects and the arbitrary SVLs at the same time by the multi-variate Hermite polynomial (MHP) technique. We have derived the analytic expression for the 3-point TCF for the rR intensity profile within the Duschinsky approximation including the non-Condon and thermal effects.

The SVL transition GF is shown to be only a special case of the cF GF in section 3.1. It has constraints on identical vibrational occupation number vectors (ONVs) in the ket and bra vibronic states. As a result, the SVL transition GF is simplified to a reduced dimensional form. From this development, we can obtain the SVL transition TCF, the integral prescreening strategy for the SVL transition, and the fine-grained FC/non-Condon prescreening strategies of multi-dimensional fixed quantum number space X. As a last development, in section 3.4, we have applied the cF GF of section 3.1 to one anharmonic transition problem that can be approximated with a few anharmonic degrees of freedom (DOF), whereas the remaining DOF are treated harmonically. With the developments we can construct the FC/non-Condon TCFs and integral prescreening conditions [11, 101, 102, 104] from the anharmonic GF. We have made the GF with a separation ansatz of anharmonic and harmonic DOF, and the benefit of the GF decreases as the number of anharmonic DOF grows.

In diagram 1 we summarise the developments we have made starting from the work of Doktorov *et al.* [14] and show the relations between the developments described herein. Based on the fundamental developments (the round boxes at the bottom of the diagram) the GF approaches have been extended (following the arrows in the diagram) to account for thermal effects, non-Condon effects and SVL transitions. An invariant quantity for a quantum mechanical problem is a powerful tool to understand the system (see *e.g.* Refs. [14, 20–24, 161]). The invariant constraints the system dynamics. We can extract useful data about the quantum mechanical processes from the invariant functional, *i.e.* herein the GF which is characterised by its GF parameters (see Ref. [11]), with proper modifications or mathematical manipulations to the functional (see the bottom of Fig. 1). The GF appears to be an analogous tool like the partition function ($\mathrm{Tr}(\exp(-\beta\hat{H}))$), which is invariant and from which we can obtain thermodynamic quantities. Analogously the GF can provide the quantity of interest, when (proper) operations are applied on it, *e.g.* partial derivatives are taken with respect to the GF parameters to obtain various intensity sum rules.

The special contributions of our present developments are the combination of a TCF approach and a GF approach [11], extended by non-Condon effects, in a unified framework. The former can simply be adopted to the CS-based GFs by assigning the time-dependent (TD) GF parameters. We have taken the non-Condon effects into account with the CS displacement operator. With the help of the phase displacement operator we have used the same mathematical machinery of the Franck–Condon factor (FCF) GF for the non-Condon GF. The phase displacement operator introduces auxiliary phase parameters for momentum and position operators to be used for non-Condon operators extracted by partial derivatives with respect to the auxiliary parameters.

The theoretical description of most of the molecular vibronic transitions involving the Duschinsky rotated harmonic oscillators boils down to a common mathematical problem, *i.e.* the

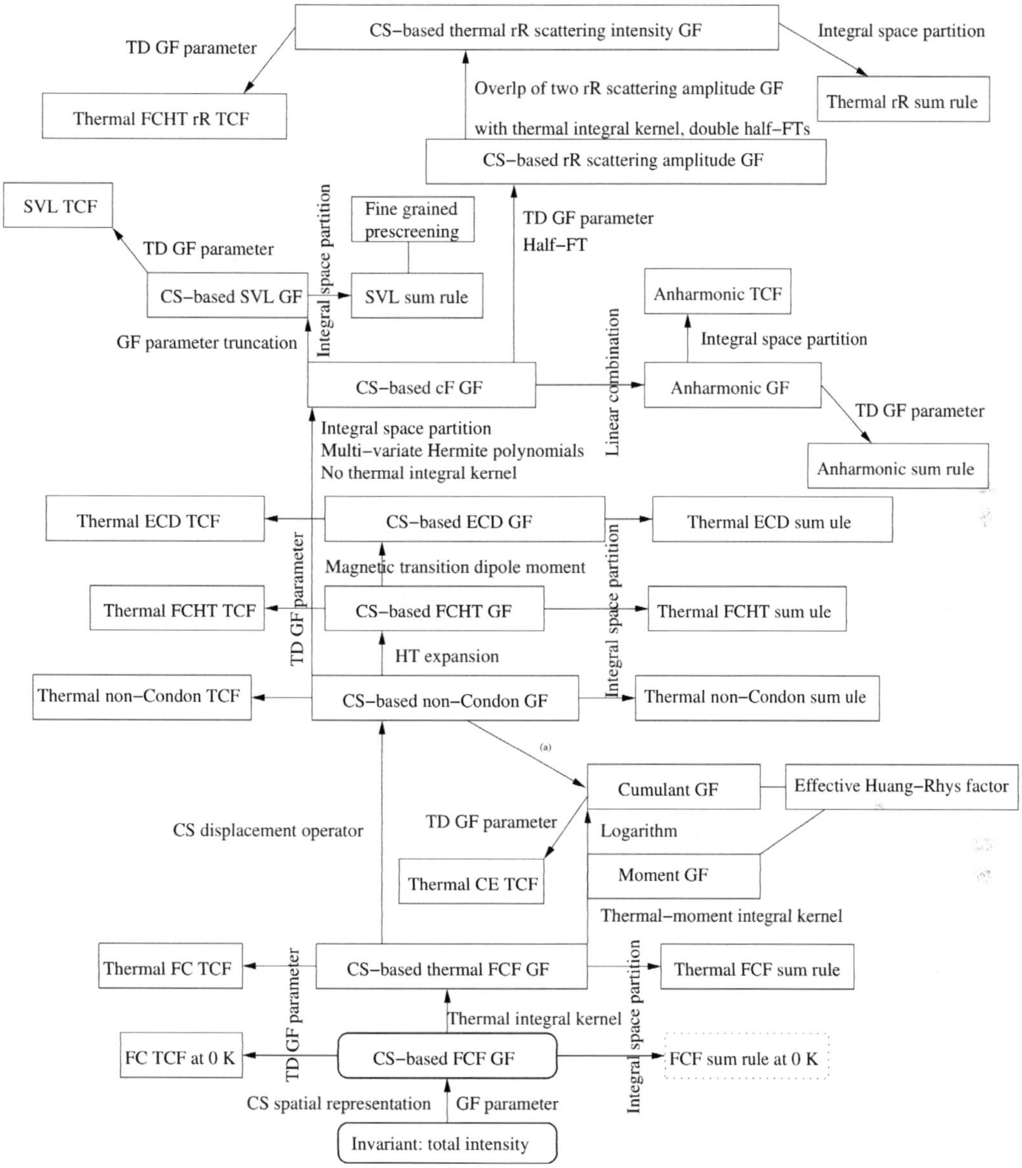

Figure 1. Summary of the CS-based GF developments for various molecular vibronic transitions. The round boxes refer to fundamental quantities. The dotted box represents work that was not part of this article. The solid square boxes indicate the developments within the article. The arrows indicate the functional transformations. The properties of the GF are shown in a box that are inherited from the previous (lower) boxes, otherwise details are mentioned on the arrows. The single solid lines without arrows imply only one of several possible usages for the corresponding functions. (a) It is straightforward to obtain the mean value including the non-Condon effects but for the higher order statistical quantities a numerical algorithm has to be developed.

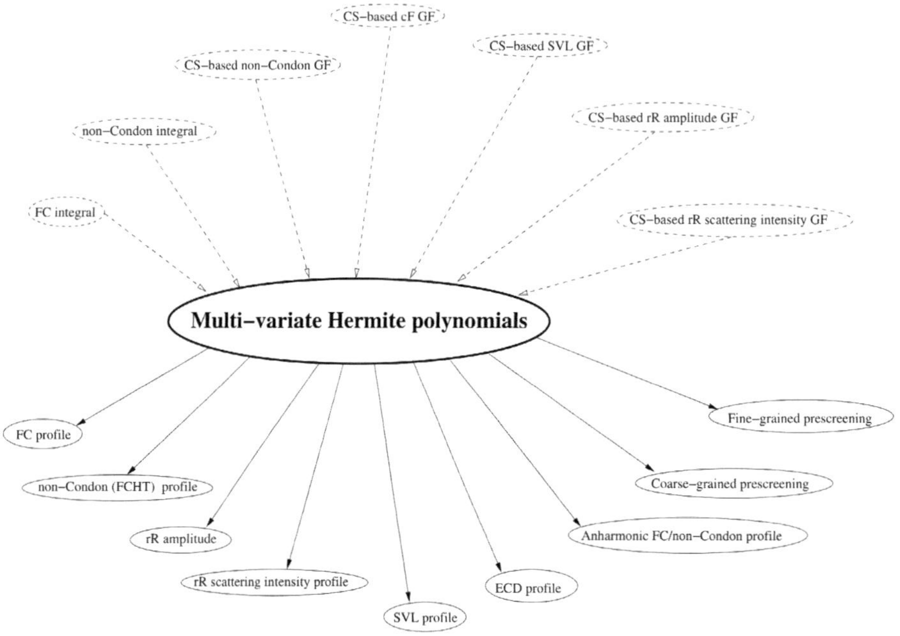

Figure 2. Most of the working equations in this article are translated into the MHP evaluation problem. The functions in dashed-ellipse are cast into MHPs and the corresponding output can be obtained after the MHP evaluation.

MHPs evaluation (see diagram 2). The calculation of the MHPs appears as a computationally hard problem (see *e.g.* Refs. [11, 102]). Nevertheless, translation of the molecular transition problem in harmonic approximation to an equivalent MHP evaluation problem enables us to analyse various transition processes by the same mathematical tool. We have exploited the MHP technique to modify the FCF GF and the non-Condon GF for SVLs. From this development, we can now have GFs including thermal effects, arbitrary non-Condon effects and arbitrary SVLs in the Duschinsky oscillator basis. As special applications to the cF GF, SVL transition, rR scattering and anharmonic transitions are introduced. But the GF approach can possibly contribute to the vibronic coupling, multi-photon transition, general anharmonic problems which could have intrinsic harmonic structures. The application is shown to be not only restricted to molecular transitions but it could be any quantum mechanical process involving harmonic oscillators.

The developments in this article are generally applicable to transition problems in the harmonic approximation (see Fig. 2). With slight modifications the same approach can be exploited for other problems arising in various flavours of mass spectrometry (see *e.g.* Ref. [162]), non-equilibrium FC processes in molecular junction (see *e.g.* Ref. [163]) and the Förster-type energy transfer processes (see *e.g.* Ref. [134]). The internal conversion (IC) process requires the matrix elements of momentum operators which can easily be handled by the non-Condon developments (Sec. 2.4). The non-Condon developments can be applied to vibronic coupling effects (see *e.g.* Refs. [43,74,82,90]) including for instance spin-orbit coupling (see *e.g.* Ref. [90]), Jahn-Teller effects (see *e.g.* Ref [98]), nonadiabatic electron transfer (see *e.g.* Ref. [164]) and dissipation due to environmental effects (see *e.g.* Ref. [134]). The rR developments in this article could be extended for the surface enhancement effects, *i.e.* the surface enhanced Raman scattering (SERS) (see *e.g.* Refs. [165–168]) and the surface enhanced Raman optical activity

(SEROA) (see *e.g.* Ref. [169]) with the non-Condon effects and the finite temperature effects if we can make a proper description on the metal surfaces with for example jellium model (see *e.g.* Ref. [170]). Quantum tomography could be an application of the current method as well because the Duschinsky operation can be generalized to allow the mixing between position and momentum operators. In Ref. [171], for example, the resulting multivariate (four dimensional) Hermite polynomials were exploited for the photon-number tomogram of stimulated Raman scattering (see also Refs. [172, 173]). An extension to general anharmonic problems, however, will be crucial, because the harmonic approximation is too crude in many cases. As mentioned in the section on anharmonic effects (Sec. 3.4), the non-Condon development and the matrix element evaluation procedure are suitable to address anharmonic problems possibly within the reaction path approach [123] of Luckhaus. Our CS-based TCF could be used for propagating wavepackets in time domain on general anharmonic surfaces (see *e.g.* Refs. [110, 137–140]) via the complex MHP evaluation.

To this end, we remark that a unified description of vibronic transitions in frequency and time domains, presented in this article, can facilitate the detailed analysis of experimental spectra and paves the way for a wealth of theoretical applications for large molecular systems.

The CS-based GF approach leaves room for future developments, *e.g.* nonadiabatic coupling, fully anharmonic, dissipative systems and non-equilibrium problems to name but a few, which are not treated herein.

Acknowledgments
We are deeply indebted to Jason Stuber and Dieter Schuch for numerous discussions. Financial support by the Volkswagen Foundation and computer time provided by the Center for Scientific Computing (CSC) Frankfurt are gratefully acknowledged. Before having been presented in detail on the symposium Symmetries in Science XV in Bregenz 2011, the present work and corresponding developments were in part reported on the XIV European Seminar on Computational Methods in Quantum Chemistry, on Elba in 2008.

Appendix

General remarks

In this appendix we present explicit expressions for the developments of section 3, as for a special case of thermally averaged initial state because the expressions in Sec. 3 are general expressions allowing for individual temperatures for each mode. Herein, only the initial state is thermally excited and all vibrational modes are at a finite temperature T and non-Condon effects are ignored. The N-dimensional space X belongs to the initial state and the N-dimensional space Y does to the final state in this appendix.

Resonance Raman scattering

- Amplitude

  The first (Condon) term of resonance Raman (rR) amplitude in Eq. (128) is given as follows,

$$G_{\mathrm{cF}}^{K_{YY}}(\widetilde{\mathbf{I}}_{\mathrm{cF}}, \mathbf{z}'(\tau); \mathbf{\Lambda}_{YY}; \underline{\tilde{v}}_{\mathrm{KB}}, \underline{\tilde{v}}_{\mathrm{BK}}) = G_{\mathrm{cF}}^{K_{YY}}(\mathbf{I}, \mathbf{z}'(\tau); \mathbf{0}; \underline{\tilde{v}}_{\mathrm{KB}}, \underline{\tilde{v}}_{\mathrm{BK}}), \tag{151}$$

  where we do not have thermal excitation ($\mathbf{\Lambda}_{YY} = \mathbf{0}$, an N-dimensional square zero matrix) for single vibronic levels ($\underline{\tilde{v}}_{\mathrm{KB}}, \underline{\tilde{v}}_{\mathrm{BK}}$). $\mathbf{z}'(\tau)$ is given in Eq. (110). The explicit expression reads then as

$$\begin{aligned} &G_{\mathrm{cF}}^{K_{YY}}(\mathbf{I}, \mathbf{z}'(\tau); \mathbf{0}; \underline{\tilde{v}}_{\mathrm{KB}}, \underline{\tilde{v}}_{\mathrm{BK}}) \\ &= \left(\widetilde{\prod}_{\underline{v_{\mathrm{B}}}_X, \underline{v_{\mathrm{K}}}_X}^{-\frac{1}{2}, -\frac{1}{2}} \right) \left(\widetilde{\prod}_{\underline{v_{\mathrm{B}}}'_X, \underline{v_{\mathrm{K}}}'_X}^{-\frac{1}{2}, -\frac{1}{2}} \right) G_Y(\mathbf{z}'(\tau)) \mathcal{H}e_{\underline{\tilde{v}}_{\mathrm{KB}}, \underline{\tilde{v}}_{\mathrm{BK}}}(\widetilde{\mathbf{W}}_{\mathrm{cF}}^{-1} \underline{\tilde{r}}_{\mathrm{cF}}; \widetilde{\mathbf{W}}_{\mathrm{cF}}^{-1}). \end{aligned} \tag{152}$$

The first part is expressed with determinant and exponential functions

$$G_Y(\mathbf{z}'(\tau)) = \det(\mathbf{I}+\mathbf{z}'(\tau)\mathbf{W}_{YY}\mathbf{z}'(\tau))^{-1/2}\det(\mathbf{I}-\mathbf{z}'(\tau)\mathbf{W}_{YY}\mathbf{z}'(\tau))^{-1/2}$$
$$\exp(\underline{r}_Y^{\mathrm{t}}\mathbf{z}'(\tau)(\mathbf{I}+\mathbf{z}'(\tau)\mathbf{W}_{YY}\mathbf{z}'(\tau))^{-1}\mathbf{z}'(\tau)\underline{r}_Y)\,, \tag{153}$$

with the Doktorov matrices and vectors in Eq. (75),

$$\mathbf{W}_{YY} = \mathbf{I}-2\mathbf{P}, \quad \underline{r}_Y = \sqrt{2}(\mathbf{I}-\mathbf{P})\underline{\delta}\,. \tag{154}$$

For the multi-variate Hermite polynomial (MHP) part, the parameters are given as follows,

$$\widetilde{\mathbf{W}}_{\mathrm{cF}}(\mathbf{z}'(\tau);\mathbf{0}) = \frac{1}{2}\begin{pmatrix}(\widetilde{\mathbf{W}}^{+}_{XX}+\widetilde{\mathbf{W}}^{-}_{XX})(\mathbf{z}'(\tau)) & (\widetilde{\mathbf{W}}^{+}_{XX}-\widetilde{\mathbf{W}}^{-}_{XX})(\mathbf{z}'(\tau))\\ (\widetilde{\mathbf{W}}^{+}_{XX}-\widetilde{\mathbf{W}}^{-}_{XX})(\mathbf{z}'(\tau)) & (\widetilde{\mathbf{W}}^{+}_{XX}+\widetilde{\mathbf{W}}^{-}_{XX})(\mathbf{z}'(\tau))\end{pmatrix}, \tag{155}$$

$$\underline{\tilde{r}}_{\mathrm{cF}}(\mathbf{z}'(\tau);\mathbf{0}_{YY}) = \begin{pmatrix}\underline{\tilde{r}}^{+}_X(\mathbf{z}'(\tau))\\ \underline{\tilde{r}}^{+}_X(\mathbf{z}'(\tau))\end{pmatrix}, \tag{156}$$

with the quantities

$$\widetilde{\mathbf{W}}^{\pm}_{XX}(\mathbf{Z}_{YY}) = \mathbf{W}_{XX} \mp \mathbf{W}_{XY}\mathbf{Z}_{YY}((\mathbf{I}\pm\mathbf{ZWZ})_{YY})^{-1}\mathbf{Z}_{YY}\mathbf{W}_{YX}, \tag{157}$$
$$\underline{\tilde{r}}^{+}_X(\mathbf{Z}_{YY}) = \underline{r}_X - \mathbf{W}_{XY}\mathbf{Z}_{YY}((\mathbf{I}+\mathbf{ZWZ})_{YY})^{-1}\mathbf{Z}_{YY}\underline{r}_Y\,, \tag{158}$$

in which

$$\mathbf{W}_{XX} = \mathbf{I}-2\mathbf{Q}, \quad \mathbf{W}_{XY} = -2\mathbf{R}, \quad \mathbf{W}_{YX} = -2\mathbf{R}^{\mathrm{t}}, \tag{159}$$
$$\underline{r}_X = -\sqrt{2}\mathbf{R}\underline{\delta}, \tag{160}$$

are used.

- Intensity
  The explicit rR intensity time-correlation function (TCF) expression of Eq. (137) in Condon approximation is given for the thermally averaged initial state, *i.e.*

$$G^K_{\mathrm{rR}}(\mathrm{bldiag}(\mathbf{z}(t),\mathbf{z}(t)),\mathbf{z}'(\tau),\mathbf{z}'(\tau');\mathbf{0},\mathrm{bldiag}(\boldsymbol{\lambda},\boldsymbol{\lambda})^{\frac{1}{2}})$$
$$= \frac{1}{\mathrm{Tr}(\exp(-\beta\hat{H}))}G_Y(\mathbf{z}'(\tau))G_Y(\mathbf{z}'(\tau))^*$$
$$\mathcal{I}_{2N}\Big[\mathbf{I}-\mathrm{bldiag}\Big(\mathbf{z}(t),\mathbf{z}(t),\mathbf{z}(t),\mathbf{z}(t)\Big)\widetilde{\mathbf{W}}_{\mathrm{rR}}\Big(\mathbf{z}'(\tau),\mathbf{z}'(\tau');\mathbf{0},\mathrm{bldiag}(\boldsymbol{\lambda},\boldsymbol{\lambda})^{\frac{1}{2}}\Big)\,\mathrm{bldiag}\Big(\mathbf{z}(t),\mathbf{z}(t),\mathbf{z}(t),\mathbf{z}(t)\Big)$$
$$,\tfrac{1}{2}\mathrm{bldiag}\Big(\mathbf{z}(t),\mathbf{z}(t),\mathbf{z}(t),\mathbf{z}(t)\Big)\underline{\tilde{r}}_{\mathrm{rR}}\Big(\mathbf{z}'(\tau),\mathbf{z}'(\tau');\mathbf{0},\mathrm{bldiag}(\boldsymbol{\lambda},\boldsymbol{\lambda})^{\frac{1}{2}}\Big)\Big], \tag{161}$$

  where $\mathcal{I}_{2N}$ is the Gaussian integral defined in Eq. (130), and $\widetilde{\mathbf{W}}_{\mathrm{rR}}(\mathbf{z}'(\tau),\mathbf{z}'(\tau');\mathbf{0},\mathrm{bldiag}(\boldsymbol{\lambda},\boldsymbol{\lambda})^{\frac{1}{2}})$ and $\underline{\tilde{r}}_{\mathrm{rR}}(\mathbf{z}'(\tau),\mathbf{z}'(\tau');\mathbf{0},\mathrm{bldiag}(\boldsymbol{\lambda},\boldsymbol{\lambda})^{\frac{1}{2}})$ can be found in Eq. (132) and Eq. (138), respectively.

Single vibronic level transitions
In this section we present the explicit expression for Eq. (143) of non-thermally excited final state, *i.e.*

$$G^{K_{YY}}_{\mathrm{SVL}}(\mathbf{z},\mathbf{z}';\mathbf{0};\underline{\tilde{v}}_X)$$
$$= \left(\widetilde{\prod}_{\underline{\tilde{v}}_X}^{-\frac{1}{2}}\right) G_Y(\mathbf{z}')\mathcal{H}e_{\underline{\tilde{v}}_X}\left((2\mathbf{z}\widetilde{\mathbf{W}}^{+}_{XX}\mathbf{z})^{-1}\underline{\tilde{r}}^{+}_X;(2\mathbf{z}\widetilde{\mathbf{W}}^{+}_{XX}\mathbf{z})^{-1}\right), \tag{162}$$

where the quantities ($\widetilde{\mathbf{W}}^{+}_{XX}$, $\underline{\tilde{r}}^{+}_X$) defined in Eqs. (157) and (158) are used with Eqs. (159) and (160).

Transitions between anharmonic oscillators
Eq. (150) can be evaluated with Eq. (152) in Condon approximation, *i.e.*

$$\begin{aligned} & G_{\mathrm{A}}(\mathbf{I},\mathbf{z}';\underline{c_n},\underline{c_m}';\underline{\tilde{v}}_{\mathrm{KB}},\underline{\tilde{v}}_{\mathrm{BK}}) \\ & = \sum_{\{\underline{v_{\mathrm{K}}}_X\},\{\underline{v_{\mathrm{B}}}_X\},\{\underline{v_{\mathrm{K}}'}_X\},\{\underline{v_{\mathrm{B}}'}_X\}} c'_{\underline{v_{\mathrm{K}}'}_X;m} c'^{*}_{\underline{v_{\mathrm{B}}'}_X;m} c^{*}_{\underline{v_{\mathrm{B}}}_X;n} c_{\underline{v_{\mathrm{K}}}_X;n} \\ & \left(\widetilde{\prod}^{-\frac{1}{2},-\frac{1}{2}}_{\underline{v_{\mathrm{B}}}_X,\underline{v_{\mathrm{K}}}_X}\right)\left(\widetilde{\prod}^{-\frac{1}{2},-\frac{1}{2}}_{\underline{v_{\mathrm{B}}'}_X,\underline{v_{\mathrm{K}}'}_X}\right) G_Y(\mathbf{z}')\mathcal{H}e_{\underline{\tilde{v}}_{\mathrm{KB}},\underline{\tilde{v}}_{\mathrm{BK}}}(\widetilde{\mathbf{W}}^{-1}_{\mathrm{cF}}\underline{\tilde{r}}_{\mathrm{cF}};\widetilde{\mathbf{W}}^{-1}_{\mathrm{cF}}). \end{aligned} \tag{163}$$

References

[1] Berger R and Klessinger M 1997 *J. Comput. Chem.* **18** 1312–1319
[2] Duschinsky F 1937 *Acta Physicochim. URSS* **7** 551–566
[3] Malmqvist P Å and Forsberg N 1998 *Chem. Phys.* **228** 227–240
[4] Islampour R, Dehestani M and Lin S H 1999 *J. Mol. Spectrosc.* **194** 179–184
[5] Liang J and Li H 2005 *Mol. Phys.* **103** 3337–3342
[6] Liang J, Zheng H, Zhang X, Li R and Cui Z 2007 *Mol. Phys.* **105** 1903
[7] Borrelli R and Peluso A 2008 *J. Chem. Phys.* **129** 064116
[8] Chang J L 2008 *J. Chem. Phys.* **128** 174111
[9] Liang J, Liu C, Wang C and Cui Z 2009 *Mol. Phys.* **107** 2601
[10] Kan R 2008 *J. Multivar. Anal.* **99** 542–554
[11] Jankowiak H C, Stuber J L and Berger R 2007 *J. Chem. Phys.* **127** 234101
[12] Santoro F, Improta R, Lami A, Bloino J and Barone V 2007 *J. Chem. Phys.* **126** 084509
[13] Santoro F, Lami A, Improta R and Barone V 2007 *J. Chem. Phys.* **126** 184102
[14] Doktorov E V, Malkin I A and Man'ko V I 1979 *J. Mol. Spectrosc.* **77** 178–194
[15] Dirac P A M 1927 *Proc. Roy. Soc. (London) A* **114** 243–265
[16] Fermi E 1950 *Nuclear Physics* (Chicago: University of Chicago Press.)
[17] Orlandi G and Siebrand W 1973 *J. Chem. Phys.* **58** 4513–4525
[18] Cederbaum L and Domcke W 1976 *J. Chem. Phys.* **64** 603–611
[19] Heller E J 1978 *J. Chem. Phys.* **68** 2066
[20] Doktorov E V, Malkin I A and Man'ko V I 1975 *J. Mol. Spectrosc.* **56** 1–20
[21] Doktorov E V, Malkin I A and Man'ko V I 1976 *J. Phys. B.* **9** 507–514
[22] Doktorov E V, Malkin I A and Man'ko V I 1977 *J. Mol. Spectrosc.* **64** 302–326
[23] Doktorov E V, Malkin I A and Man'ko V I 1977 *Chem. Phys. Lett.* **46** 183–187
[24] Doktorov E V, Malkin I A and Man'ko V I 1979 *Sov. Phys. JETP* **49** 638–645
[25] Heider N and Fischer S F 1984 *Chem. Phys.* **88** 209
[26] Islampour R and Mukamel S 1984 *Chem. Phys. Lett.* **239** 107
[27] Islampour R and Mukamel S 1984 *J. Chem. Phys.* **80** 5487
[28] Mukamel S, Abe S, Yan Y J and Islampour R 1985 *J. Phys. Chem.* **89** 201
[29] Yan Y J and Mukamel S 1986 *J. Chem. Phys.* **85** 5908
[30] Islampour R 1989 *Chem. Phys.* **133** 425
[31] Berger R, Fischer C and Klessinger M 1998 *J. Phys. Chem. A* **102** 7157–7167
[32] Wadi H and Pollak E 1999 *J. Chem. Phys.* **110** 11890
[33] Shiu Y J, Hayashi M, Mebel A M, Chen Y T and Lin S H 2001 *J. Chem. Phys.* **115** 4080
[34] Watson J K G 2001 *J. Mol. Spec.* **207** 276
[35] He Y and Pollak E 2001 *J. Phys. Chem. A* **105** 10961
[36] Ianconescu R and Pollak E 2004 *J. Phys. Chem. A* **108** 7778
[37] Dierksen M and Grimme S 2004 *J. Chem. Phys.* **120** 3544–3554
[38] Banerjee S and Gangopadhyay G 2005 *J. Chem. Phys.* **123** 114304
[39] Dierksen M and Grimme S 2005 *J. Chem. Phys.* **122** 244101
[40] Hizhnyakov V and Tehver I 2008 *J. Lumin.* **128** 1001
[41] Tatchen J and Pollak E 2008 *J. Chem. Phys.* **128** 164303
[42] Santoro F, Lami A, Improta R, Bloino J and Barone V 2008 *J. Chem. Phys.* **128** 224311
[43] Schuurman M S and Yarkony D R 2008 *J. Chem. Phys.* **128** 044119
[44] Barone V, Bloino J, Biczysko M and Santoro F 2009 *J. Chem. Theory Comput.* **5** 540
[45] Coriani S, Kjærgaard T, Jøgensen P, Ruud K, Huh J and Berger R 2010 *J. Chem. Theory Comp.* **6** 1028
[46] Huh J, Neff M, Rauhut G and Berger R 2010 *Mol. Phys.* **108** 409

[47] Bloino J, Biczysko M, Santoro F and Barone V 2010 *J. Chem. Theory Comput.* **6** 1256
[48] Mukamel S 1982 *J. Chem. Phys.* **77** 173
[49] Tannor D J, Blanco M and Heller E J 1984 *J. Phys. Chem.* **88** 6240
[50] Gustav K and Storch M 1990 *Inter. J. Quantum Chem.* **38** 25–39
[51] Patwari G N, Wategaonkar S and Prasad M D 2001 *Chem. Phys. Lett.* **344** 229
[52] Peng Q, Yi Y, Shuai Z and Shao J 2007 *J. Am. Chem. Soc.* **129** 9333
[53] Albrecht A C 1961 *J. Chem. Phys.* **34** 1476
[54] Lee S Y and Heller E J 1979 *J. Chem. Phys.* **71** 4777
[55] Hassing S and Mortensen O S 1981 *J. Mol. Spectros.* **87** 1
[56] Champion P M and Albrecht A C 1982 *Ann. Rev. Phys. Chem.* **33** 353
[57] Tannor D J and Heller E J 1982 *J. Chem. Phys.* **77** 202
[58] Myers A B, Nathies R A, Tannor D J and Heller E J 1982 *J. Chem. Phys.* **77** 3857
[59] Heller E J, Sundberg R and Tannor D 1982 *J. Phys. Chem.* **86** 1822
[60] Chan C K and Page J B 1983 *J. Chem. Phys.* **79** 5234
[61] Stallard B R, Champion P M, Callis P R and Albrecht A C 1983 *J. Chem. Phys.* **78** 712
[62] Harris R A, Mathies R and Myers A 1983 *Chem. Phys. Lett.* **94** 327
[63] Svendsen C, Mortensen O S and Clark R J H 1994 *Chem. Phys.* **187** 349
[64] Svendsen C, Mortensen O S and Henriksen N E 1996 *Chem. Phys. Lett.* **260** 627
[65] Islampour R, Hayashi M and Lin S H 1997 *J. Raman Spectros.* **28** 331
[66] Myers A B 1997 *J. Raman Spectros.* **28** 389
[67] Islampour R, Dehestani M and Lin S H 2000 *Mol. Phys.* **98** 101
[68] Neugebauer J and Hess B A 2004 *J. Chem. Phys.* **120** 11564
[69] Petrenko T and Neese F 2007 *J. Chem. Phys.* **127** 164319
[70] Kiewisch K, Neugebauer J and Reiher M 2008 *J. Chem. Phys.* **129** 204103
[71] Lami A and Santoro F 1997 *J. Chem. Phys.* **106** 94
[72] Islampour R, Alden R G, Wu G Y C and Lin S H 1993 *J. Phys. Chem.* **97** 6793
[73] Tang J 1994 *Chem. Phys.* **188** 143
[74] Peluso A, Santoro F and del Re G 1997 *Int. J. Quant. Chem.* **63** 233–244
[75] Pollak E and Pilmak L 2001 *J. Chem. Phys.* **115** 1867
[76] Sando G M, Spears K G, Hupp J T and Ruhoff P T 2001 *J. Phys. Chem. A* **105** 5317
[77] Segev B 2003 *J. Opt. B: Quantum Semiclass. Opt.* **5** S381
[78] Banerjee S and Gangopadhyay G 2007 *J. Chem. Phys.* **126** 034102
[79] Scherer P O J and Tachiya M 2003 *J. Chem. Phys.* **118** 4149
[80] Tang J, Lee M T and Lin S H 2003 *J. Chem. Phys.* **119** 7188
[81] Liang K K, Mebel A M, Lin S H, Hayashi M, Selzle H L, Schlag E W and Tachiya M 2003 *Phys. Chem. Chem. Phys.* **5** 4656
[82] Scherer P O J and Fischer S F 2006 *Chem. Phys. Lett.* **421** 427
[83] Lin N, Zhao X, Rizzo A and Luo Y 2007 *J. Chem. Phys.* **126** 244509
[84] Lin S H 1966 *J. Chem. Phys.* **44** 3759–3767
[85] Mebel A M, Hayashi M, Liang K K and Lin S H 1999 *J. Phys. Chem. A* **103** 10674
[86] Kallush S, Segev B, Sergeev A V and Heller E J 2002 *J. Phys. Chem. A* **106** 6006
[87] Islampour R and Miralinaghi M 2007 *J. Phys. Chem. A* **111** 9454
[88] Peng Q, Yi Y, Shuai Z and Shao J 2007 *J. Chem. Phys.* **126** 114302
[89] Peng Q, Niu Y, Deng C and Shuai Z 2010 *Chem. Phys.* **370** 215
[90] Toniolo A and Persico M 2001 *J. Comput. Chem.* **22** 968–975
[91] Santoro F and Barone V 2010 *Int. J. Quant. Chem.* **110** 476
[92] Craig D P and Thirunamachandran T 1984 *Molecular quantum electrodynamics: an introduction to radiation – molecule interactions* (New York: Dover Publications, Inc.)
[93] Small G J 1971 *J. Chem. Phys.* **54** 3300–3306
[94] Müller C W, Newby J J, Liu C P, Rodrigo C P and Zwier T S 2010 *Phys. Chem. Chem. Phys.* **12** 2331
[95] Tonks D L and Page J B 1981 *Chem. Phys. Lett.* **79** 247
[96] Sando G M and Spears K G 2001 *J. Chem. Phys. A* **105** 5326–5333
[97] Chang C T, Sethna J P, Pasupathy A N, Park J, Ralph D C and McEuen P L 2007 *Phys. Rev. B* **76** 045435
[98] Fischer G 1984 *Vibronic Coupling* (London: Academic Press Inc.)
[99] Klessinger M and Michl J 1995 *Excited States and Photochemistry of Organic Molecules* (New York: VCH)
[100] Atkins P 2005 *Molecular quantum mechanics* 4th ed (Oxford: Oxford University Press)
[101] Huh J, Stuber J L and Berger R to be published
[102] Huh J 2010 *Unified description of vibronic transitions with coherent states* Ph.D. thesis Johann Wolfgang Goethe-University, Frankfurt am Main

[103] Born M and Oppenheimer R 1927 *Ann. Phys.* **84** 457–463
[104] Huh J, Jankowiak H C, Stuber J L and Berger R to be published
[105] Improta R, Lami A, Barone V and Santoro F 2010 *Int. J. Quant. Chem.* **110** 624
[106] Feynman R P 1972 *Statistical mechanics: a set of lectures* (Redwood City: Addison-Wesley)
[107] Long D A 2002 *The Raman effect* (Chichester: John Wiley & Sons Ltd)
[108] Lu H M and Page B 1986 *Chem. Phys. Lett.* **131** 87
[109] Santoro F, Cappelli C and Barone V 2011 *J. Chem. Theory Comp.* **7** 1824–1839
[110] Heller E J 1978 *J. Chem. Phys.* **68** 3891
[111] Warshel A, Stern P S and Mukamel S 1983 *J. Chem. Phys.* **78** 7498–7500
[112] Orlandi G and Siebrand W 1972 *Chem. Phys. Lett.* **15** 465–468
[113] Eckart C 1935 *Phys. Rev.* **47** 552–558
[114] Pickett H M and Strauss H L 1970 *J. Am. Chem. Soc.* **92** 7281–7290
[115] Wilson E B, Decius J C and Cross P C 1980 *Molecular vibrations* (New York: Dover)
[116] Bunker P R 1998 *Molecular Symmetry and Spectroscopy* 2nd ed (Ottawa: NRC Research Press)
[117] Hougen J T and Watson J K G 1965 *Can. J. Phys.* **43** 298–320
[118] Watson J K G 1993 *Can. J. Chem.* **71** 1556
[119] Özkan I 1990 *J. Mol. Spectrosc.* **139** 147
[120] Warshel A and Karplus M 1972 *Chem. Phys. Lett.* **17** 7–14
[121] Lucas N J D 1973 *J. Phys. B* **6** 155–163
[122] Kohn D W, Robles E S J, Logan C F and Chen P 1993 *J. Phys. Chem.* **97** 4936–4940
[123] Luckhaus D 2008 *Phys. Chem. Chem. Phys.* **10** 6215
[124] Fan H 2003 *J. Opt. B: Quantum Semiclass. Opt.* **5** R147
[125] Glauber R J 1963 *Phys. Rev.* **131** 2766–2788
[126] Heller E J 1975 *J. Chem. Phys.* **62** 1544
[127] Malkin I A, Man'ko V I and Trifonov D A 1970 *Phys. Rev. D.* **2** 1371–1385
[128] Huh J, Stuber J L and Berger R to be published
[129] Huh J and Berger R 2011 *Faraday Discuss.* **150** 363–373
[130] Huh J and Berger R 2011 (*Preprint* `chem-phys/1111.3052`)
[131] Blinnikov S and Moessner R 1998 *Astron. Astrophys. Suppl. Ser.* **130** 193
[132] Hollas J M 1998 *High resolution spectroscopy* 2nd ed (Chichester: John Wiley & Sons, Ltd)
[133] Hollas J M 2004 *Modern spectroscopy* 4th ed (Chichester: John Wiley & Sons, Ltd)
[134] May V and Kühn O 2003 *Charge and energy transfer dynamics in molecular systems* 2nd ed (Weinheim: WILEY-VCH Verlag GmbH & Co. KGaA)
[135] Watson J K G 1968 *Mol. Phys.* **15** 479–490
[136] Meyer R and Günthard H H 1968 *J. Chem. Phys.* **49** 1510
[137] Davis M J and Heller E J 1979 *J. Chem. Phys.* **71** 3383
[138] Worth G A, Meyer H D and Cederbaum L S 1998 *J. Chem. Phys.* **109** 3518–3529
[139] Raab A, Worth G A, Meyer H D and Cederbaum L S 1999 *J. Chem. Phys.* **110** 936–946
[140] Shalashilin D V and Child M S 2004 *J. Chem. Phys.* **121** 3563
[141] Hutchisson E 1930 *Phys. Rev.* **36** 410–420
[142] Hutchisson E 1931 *Phys. Rev.* **37** 45–50
[143] Storch M and Gustav K 1989 *Z. Phys. Chem. L* **270** 1041–1046
[144] Chau F T, Dyke J M, Lee E P F and Wang D C 1998 *J. Electron Spectrosc. Relat. Phenom.* **97** 33–47
[145] Ervin K M, Ramond T M, Davico G E, Schwartz R L, Casey S M and Lineberger W C 2001 *J. Phys. Chem. A* **105** 10822–10831
[146] Eisfeld W 2006 *J. Phys. Chem. A* **110** 3903–3910
[147] Domcke W, Cederbaum L, Köppel H and von Niessen W 1977 *Mol. Phys.* **34** 1759–1770
[148] Cederbaum L S and Domcke W 1977 *Adv. Chem. Phys.* **36** 205–344
[149] Hazra A, Chang H H and Nooijen M 2004 *J. Chem. Phys.* **121** 2125–2136
[150] Hazra A and Nooijen M 2005 *Phys. Chem. Chem. Phys.* **7** 1759–1771
[151] Rodriguez-Garcia V, Yagi K, Hirao K, Iwata S and Hirata S 2006 *J. Chem. Phys.* **125** 014109
[152] Bowman J M 1978 *J. Chem. Phys.* **68** 608
[153] Bowman J 1986 *Acc. Chem. Res* **19** 202–208
[154] Christoffel K M and Bowman J M 1982 *Chem. Phys. Lett.* **85** 220
[155] Bowman J M, Huang X, Harding L B and Carter S 2006 *Mol. Phys.* **104** 33–45
[156] Hamilton I and Light J 1986 *J. Chem. Phys.* **84** 306
[157] Luis J M, Bishop D M and Kirtman B 2004 *J. Chem. Phys.* **120** 813–822
[158] Luis J M, Kirtman B and Christiansen O 2006 *J. Chem. Phys.* **125** 154114
[159] Luis J M, Torrent-Sucarrat M, Sol'a M, Bishop D M and Kirtman B 2005 *J. Chem. Phys.* **122** 184104

[160] Bonness S, Kirtman B, Huix M, Sanchez A J and Luis J M 2006 *J. Chem. Phys.* **125** 014311
[161] Schuch D 1992 *Int. J. Quant. Chem.* **42** 663
[162] Roithová J, Schröder D, Loos J, Schwarz H, Jankowiak H C, Berger R, Thissen R and Dutuit O 2005 *J. Chem. Phys.* **122** 094306
[163] Seldenthuis J S, van der Zant H S J, Ratner M A and Thijssen J M 2008 *ACS NANO* **2** 1445
[164] Fischer S F 1982 *Chem. Phys. Lett.* **91** 367
[165] Lombardi J R, Birke R L, Lu T and Xu J 1986 *J. Chem. Phys.* **84** 4174
[166] Lombardi J R and Birke R L 2007 *J. Chem. Phys.* **126** 244709
[167] Lombardi J R and Birke R L 2008 *J. Phys. Chem. C* **112** 5605–5617
[168] Lombardi J R and Birke R L 2009 *Acc. Chem. Res.* **42** 734–742
[169] Jensen L, Autschbach J, Krykunov M and Schatz G C 2007 *J. Chem. Phys.* **127** 134101
[170] Hughes R I G 2006 *Perspectives on Science* **14** 457
[171] Man'ko O V and Tcherniega N V 2006 *Proc. of SPIE* **6256** 62560W
[172] Kuznetsov S V, Man'ko O V and Tcherniega N V 2003 *J. Opt. B* **5** S503–S512
[173] Kuznetsov S V, Kyusev A V, Man'ko O V and Tcherniega N V 2004 *Acta Phys. Hung. B* **20** 11–14

Scale Invariance, Self Similarity and Critical Behavior in Classical and Quantum Systems

Irving O Morales
Instituto de Ciencias Nucleares, Universidad Nacional Autónoma de México, Apartado Postal 70-543, 04510 México, D.F., Mexico

E Landa
Instituto de Ciencias Nucleares, Universidad Nacional Autónoma de México, Apartado Postal 70-543, 04510 México, D.F., Mexico

R Fossion
Instituto de Ciencias Nucleares, Universidad Nacional Autónoma de México, Apartado Postal 70-543, 04510 México, D.F., Mexico

A Frank
Instituto de Ciencias Nucleares, Universidad Nacional Autónoma de México, Apartado Postal 70-543, 04510 México, D.F., Mexico

Abstract. Symmetry and self-affinity or scale invariance are related concepts. We explore the fractal properties of fluctuations in dynamical systems, using some of the available tools in the context of time series analysis. We carry out a power spectrum study in the Fourier domain, the method of detrended fluctuation analysis and the investigation of autocorrelation function behavior. Our study focuses on two particular examples, the logistic module-1 map, which displays properties of classical dynamical systems, and the excitation spectrum of a schematic shell-model Hamiltonian, which is a simple system exhibiting quantum chaos.

1. Introduction

The concept of symmetry is closely related to self-similarity. In a broad sense symmetry implies that a system under the action of some transformation remains invariant, i.e, the system remains similar (or identical) to itself. The concept of self-similarity in a more mathematical sense is related with a special case of transformation, the re-scaling of the system. A self-similar object is statistically (or exactly) similar to a part of itself, which in turn remains similar to a smaller part of itself, and so on. Fractals are systems which present such self-similar behavior and the examples in nature are many.

Coastlines and natural frontiers have been found to be *fractal*. A coastline is *scale invariant*, if it looks self-similar at any scale and does not have a characteristic length. Specifically, the length of a coastline $\mathcal{L}$ is a function of the length of the measuring rod l and follows a *power law* of the form[1, 2],

$$\mathcal{L} = l^{1-D}, \tag{1}$$

where D is the fractal dimension. A coastline with $D = 1$ would appear straight in the map, whereas a coastline with $D = 2$ would consist of an infinite self-similar sequence of bays within bays without upper or lower limits. Natural coastlines have a fractal dimension $1 < D < 2$.

We are interested in using the self-similarity of a dynamical system in order to study some of its properties. A *dynamical system* [3] can be any real or hypothetical system that evolves over time. Its mathematical description consists of a *state vector* (a set of real numbers) that describes the system's state in phase space, and a *function* or *rule* that governs in a deterministic way the evolution of the state vector in time. Many dynamical systems are composite systems, and consist of locally interacting parts. A *complex system* [4] is a special type of dynamical composite system, where under critical circumstances new collective behavior emerges from the short-range interactions between the constituent parts.

It is not always possible to fully determine all the components of the state vector of the dynamical system. However, an elegant way to study a large variety of dynamical systems is through the use of *time series*, that follow the evolution of a specific observable as a sequence of data points, typically measured at successive times spaced at uniform time intervals. Not only do time series offer valuable information on the internal dynamics of the systems that produce them, but have the additional advantage that the same theoretical methods can be used to analyze time series of very different systems. In this contribution we study time series associated with both classical and quantum systems, although in the latter case we shall actually use energy instead of time, as we explain below. Time series can be fractal, if segments of it remain self-similar at different time scales. An interesting example is provided by the heartbeat interval series, that remains self-similar for time scales ranging from 10 seconds up to several hours [5, 6].

If the time series is fractal, then the power spectrum $S(f)$ must also be scale invariant and behave as a power law,

$$S(f) \sim f^{\beta}, \tag{2}$$

where β is the *spectral density exponent*, with $\beta = 0$ for white noise and $\beta < 0$ for reddened noise. In a log-log representation, the power law translates into a straight line, with β as the slope. Such broadband power spectra, without a characteristic frequency, have indeed been observed for time series of a wide variety of dynamical systems, including the interval fluctuations of a healthy heart [7].

The spectral density exponent β is a measure of the amount of correlations present in the time series. There is no correlation for white noise ($\beta = 0$) and a large one for brownian noise ($\beta = -2$). The power-law behaviour of the power spectrum $P(f) \sim f^{\beta}$ of eq. (2) implies that the power spectrum of f^{β} noises is *scale invariant*. In white noise ($\beta = 0$), if one integrates the power of the power spectrum per frequency decade, then one finds that the contribution increases a factor ten per decade. White noise is convergent at small frequencies but diverges at higher frequencies. This translates into the fact that the mean value of a specific white-noise time-series converges as the value is averaged over longer and longer time intervals, but its instantaneous value is undefined. On the other hand, in brownian noise ($\beta = -2$), the power contribution decreases a factor of ten per decade. Brownian noise is convergent at higher frequencies but diverges at lower frequencies. A specific brownian time series is thus well defined at every single time, but it does not have a well defined mean value for large time intervals, as the random-walk function wanders farther and farther away from the initial value. In the power spectrum of $1/f$ noise, every frequency decade contributes with exactly the same amount of power (proportional to log10) [8, 9], a perfect scaling property that we already recognized in the time scaling of the autocorrelation function, see eq. (6). $1/f$ noise thus constitutes a "golden mean", in between white noise and brownian noise. The spectrum is divergent both towards low and high frequencies, and as a consequence a specific $1/f$ time series does not have a well-defined long-term mean nor a well-defined value at a single point. However, as the divergence is slow (logarithmic), frequency cutoffs can be applied to the power spectrum at low or high frequencies, without the time series changing its appearance [10].

In a sense, $1/f$ noise can be seen as the fingerprint of a system that has *maximized its efficiency* within its own limits. This statement is based on the assumption that a biological system operates so as to maximize its efficiency [9, 11], and on the empirical observation that many (if not all) biological time series that correspond to involuntary or autonomous signals display $1/f$ behavior when they are healthy. In contrast, with age, disease or bad habits a deviation from $1/f$ is observed towards either more random or more regular regimes. Health in biological systems is thought to function in a critical regime, characterized by $1/f$ signals.

It can be shown that the autocorrelation function $C(\tau)$ of these systems must also be scale invariant. This can be understood in the following way. According to the Wiener-Khinchin theorem, the autocorrelation function is the inverse Fourier transform of the power spectrum,

$$C(\tau) = F^{-1}\left(S(f)\right). \tag{3}$$

If the power spectrum obeys the power law of eq. (2), and we apply a scale transformation in the time domain, $\tau \to \tau' = a\tau$, then,

$$C(a\tau) \;=\; F^{-1}\left(\frac{1}{a}\, S(\frac{f}{a})\right) \;=\; a^{-\beta-1}\, F^{-1}\left(f^{\beta}\right) \;=\; a^{-\beta-1}\, C(\tau), \tag{4}$$

and thus the autocorrelation function conserves its behavior after time transformations (up to a scaling factor $a^{-\beta-1}$). The general solution of this equation is that the autocorrelation function $C(\tau)$ itself is also a power law. White noise ($\beta = 0$) is a particular case, where

$$C(a\tau) = \frac{1}{a}\, C(\tau), \tag{5}$$

so that the autocorrelation function for white noise is a delta function, $C(\tau) = \delta(\tau)$, since delta functions obey the scaling law $\delta(a\tau) = |a|^{-1}\delta(\tau)$. Flicker noise or $1/f$ noise ($\beta = -1$) is another particular case, where

$$C(a\tau) = C(\tau), \tag{6}$$

with solutions $C(\tau)$ =constant, or the physically more significant case $C(\tau) = -\log(\tau)$. Rescaling a logarithmic function gives $\log(a\tau) = \log(\tau) + \log(a)$, where $\log(a)$ is a constant that does not contribute to the time evolution of the autocorrelation function, so that the scaling relation of eq. (6) is approximatively satisfied, $C(a\tau) \cong C(\tau)$.

For discrete time series, apart from spectral analysis, there are other tools to study scale transformations, including the Detrended Fluctuations Analysis (DFA) method [12]. The DFA method is used to estimate long-range power-law correlation exponents in diverse kinds of signals [13]. In order to apply the DFA the time series is integrated and then divided into nonoverlaping N/l boxes of equal length containing l data points. Then a local trend is defined for each box, i.e. a linear fit to the integrated time series. Then the variance between the integrated time series and the local fit is computed for each box and averaged over all the boxes of size l. The average variance $F(l)$ depends on the box size. The analysis of this dependence allows to study the scaling properties of the system and the presence of long-range correlations. A linear relationship between $\mathrm{Log}(F(l))$ and $\mathrm{Log}(l)$ will indicate a scaling behavior. The slope α_{DFA} in the $\mathrm{Log}(F(l))$ vs $\mathrm{Log}(l)$ plot characterizes the scaling properties of the time series because it reflects a power law of the form $F(l) \sim l^{\alpha_{\mathrm{DFA}}}$. In the case of a regular uncorrelated system the theoretical value of the slope is $\alpha_{\mathrm{DFA}} = 1/2$. The parameter α_{DFA}, obtained by means of the DFA, which in this case is related to the self-similarity properties, can be related with the exponent β of the spectral analysis through the simple relation

$$\beta = 2\alpha_{\mathrm{DFA}} - 1, \tag{7}$$

valid in the range $0.5 < \alpha_{\mathrm{DFA}} < 1.0$. The relationship between the exponent of the autocorrelation function γ and α_{DFA} is given by $\gamma = 2 - 2\alpha_{\mathrm{DFA}}$ [14]. In particular, for time series with $1/f$-noise ($\beta = 1$), $\alpha_{DFA} = 1$ and $\gamma \to 0$ (see e.g. [15]).

2. Module-1 Logistic Map

The *logistic map*, $x_{n+1} = kx_n(1 - x_n)$, is a map of the unit interval [0,1] onto itself, which produces time series that are periodic or chaotic depending on the value of the control parameter $k \in theinterval[0, 4]$ [16]. Complex behavior, such as $1/f$, can only be expected at the *edge of chaos* at the borderline between two different regimes [17]. The *mod-1 map*, also called the *Manneville iteration map*, $x_{n+1} = x_n + kx_n^2(\mathrm{Mod1})$, on the other hand, produces self-affine time series and has been used as a model for intermittency (see Fig. 1). Moreover, the time series exhibit $1/f$ power spectral density for many values of its control parameter k [18, 19, 20, 21]. Both discrete maps belong to the class of discrete-time dynamical deterministic systems, corresponding to one-dimensional maps of a given interval, which are perhaps the simplest models displaying a sensitive dependence on initial data. We construct a new map, the *mod-1 logistic map*,

$$x_{n+1} = k(x_n + x_n^2)(\mathrm{mod}\ 1), \tag{8}$$

where the dynamic behavior of the resulting time series depends on the value of the control parameter k. We focus on the behavior of the time series close to a particular point, where the system undergoes a sudden order to chaos transition for a minimal change of the control parameter. We study the dependence on initial conditions by repeated iteration of the map, Eq. (8), which yields a sequence where the value x_{n+1} at step $n+1$ depends explicitly only on the value x_n of the previous step n, and maps the $[0, 1]$ interval onto itself. The changes in long-range correlations of the time series with this map are analyzed as the control parameter k is varied with steps of 0.0125. In the following, we consider ensembles for different values of k, with 10 different initial conditions and time series of length 3000. We present below the results for each ensemble. For $k < 1$, the time series x_n is a monotonic decreasing function that te nds to zero after a few hundred iterations. For $k = 1$, the map corresponds to the original mod-1 map and the time series is thus self-affine and intermittent. For $k > 1$, the time series loses its intermittent behavior for increasing k. To quantify regular and chaotic behaviour, we calculate an estimated value for the Lyapunov exponent λ for the $x_n(k)$ time series as a function of k,

$$\lambda(k) = \lim_{m\to\infty} \frac{1}{m} \sum_{n=0}^{m-1} \log\left|x'_{n+1}(k)\right|, \tag{9}$$

where $x'_{n+1} = k(1+2x_n)$ is the derivative of x_{n+1} of Eq. (8) with respect to the previous value x_n.

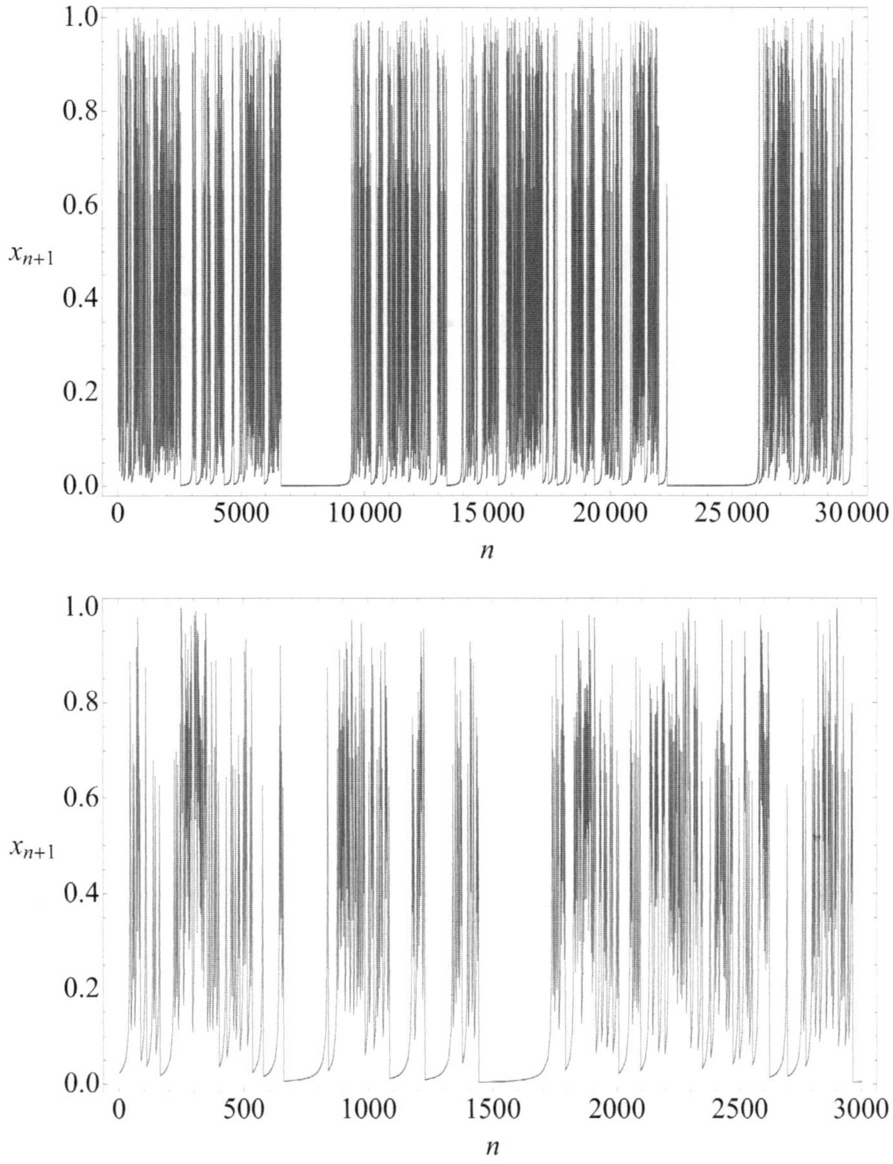

Figure 1: The self-affine time series of the Module-1 Logistic Map of eq. (8) with $k = 1$, consists of bunches of peaks (upper panel) that contain additional bunches on a smaller scale (bottom panel).

We obtain $\lambda < 0$ for $k < 1$, which confirms that the corresponding time series $x_n(k)$ are regular, whereas $\lambda(k) > 0$ for $k > 1$ indicates that the corresponding time series $x_n(k)$ are increasingly chaotic for larger k. The edge of chaos, $\lambda_k = 0$, is obtained for $k = 1$. We now study the evolution of the correlation strength of the time series $x_n(k)$ as a function of the control parameter, using spectral analysis and the DFA method, with the exponent β of Eq. (2) and $\beta = 2\alpha_{\text{DFA}} - 1$ of Eq. (7), respectively. For $k \neq 1$ ($k > 1$), the power spectral density flattens out at the smallest frequencies, and this flat part tends to be larger as k moves further away from 1, indicating that a portion of the large-range correlations are lost. We then study the part of the spectral density where the correlations are conserved. The spectral analysis and the DFA give comparable results, see Fig. 2. We point out that before k acquires the value one, it does not make sense to fit the power spectral density for a periodic regime. On the other hand, for the chaotic time series $k > 1$, the non-flat part of the spectral density at higher frequencies loses gradually its correlation strength, $\beta \to 0$, (white-noise like) for larger k. At the edge of chaos, for $k = 1$, a $1/f$ behavior ($\beta = 1$) is observed that extends through the whole range of the frequencies. In conclusion, the transitional point $k = 1$ in this simple classical system in between regularity and chaos is characterized by long-range correlations of the $1/f$ type, whereas its limiting regimes are characterized by strong and weak correlation strengths, respectively, but with a shorter range.

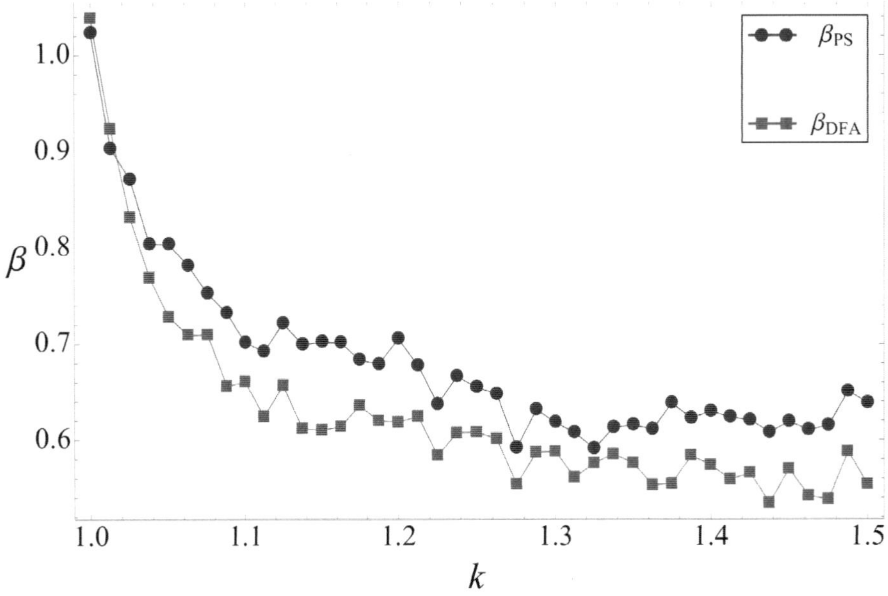

Figure 2: The correlation strength β of time series $x_n(k)$ as a function of k in the map of Eq. (8). The line connecting the blue dots is the spectral density exponent β (in the figure as $\beta_{\rm PS}$) from Eq. (2), whereas the line connecting the red squares is $\beta = 2\alpha_{\rm DFA} - 1$ (in the figure as $\beta_{\rm DFA}$) from Eq. (7).

3. Quantum excitation spectra as a time series

3.1. Unfolding and fluctuations in the quantum spectra

Quantum chaos studies the correlations present in the excitation spectra of quantum systems, which correspond to a monotonic series of levels at increasing energies, $E_1, E_2, E_3, \ldots$. In this case we analyse their *fluctuations* with respect to an equidistant (harmonic oscillator) energy spectrum. The procedure used to extract the information of the fluctuating part of the quantum spectrum is called *unfolding*: the global features of the energy level density are usually modeled by a smooth function, which is subtracted from the total level density to leave only the oscillating part. The unfolding is a delicate procedure [22, 23]. The level density is usually approximated by a smooth function (*e.g.* polynomial function) which can, however, affect the results, especially the sensitive long-range correlations [24]. Instead of following this procedure, in this paper we employ the Empirical Mode Decomposition (EMD) method to perform the unfolding procedure in quantum excitation spectra. This method originated in the theory of signal analysis, where the determination of the trend and the corresponding detrending is a fundamental task. Following [25] we use a simple definition of the global trend for arbitrary nonlinear and non-stationary time series. One of the important properties of the method is its adaptability to a given time series. It can reveal natural time scales of the signal and is able to extract various trends at different time scales [25]. In this sense the EMD method interprets the trend as an integral part of the data, *i.e.* it is driven by the same mechanisms that give rise to the rest of the signal. Our unfolding procedure combines a small degree polynomial fit (fifth-order) to remove the basic trend of the data around their average, followed by t he empirical mode decomposition, which subtracts the rest of the trend in a very accurate way. We have found that these procedure is independent of the degree of the original polynomial. A detailed analysis of the EMD method to determine the fluctuations in the quantum spectra has been recently published [26].

The fluctuations in the quantum spectra can be formally interpreted as a discrete time series [22] by using the quantity

$$\delta_n = \sum_{i=1}^{n}(s_i - \langle s \rangle) = [\epsilon_{n+1} - \epsilon_1] - n\langle s \rangle \,, \tag{10}$$

where $s_i = \epsilon_{i+1} - \epsilon_i$. The stochastic discrete function δ_n measures the deviations of the distance between the $(n+1)$-th unfolded state, with respect to the corresponding state in a uniform (equally spaced) sequence.

3.2. Energy fluctuations in shell model nuclear calculations

Following Relaño et al. [22], we use a spectral analysis to study the long-range correlations. We also apply the DFA method to the quantum spectral fluctuations, in a way similar to that of reference [27]. We performed calculations for large-scale shell-model calculations with realistic interactions (KB3) [28] in the full fp shell for ^{48}Ca in the subspaces $J^\pi = 0^+, 1^+, \ldots, 8^+$ by means of the ANTOINE code [29]. In the present study we repeated the DFA and spectral analysis of the time series, but using the EMD procedure to improve the unfolding. We obtained a nearly $1/f$ power-law behavior, that is confirmed by results with $\alpha_{\mathrm{DFA}} = 1$ for the DFA, the largest deviation being of $\sim 10\%$ (see Table 1). These results are in agreement with the ones of Relaño et al. and our previous results [24]. We also studied the energy fluctuations of the two-body random-ensemble (TBRE) [30, 31] shell-model calculations for ^{48}Ca in the subspace $J^\pi = 0^+$. We found $\beta = 1.01$ and $\alpha_{\mathrm{DFA}} = 1.02$. This value is very similar to the case of realistic calculations that is shown in Table 1. Relaño *et al.* [22] performed the first study of the behavior of the power spectral density of the energy fluctuations of TBRE random shell-model calculations for ^{24}Mg and ^{32}Na and found that they obey a $1/f$ scaling.

3.3. Energy fluctuations in a phase-transitional model

Large-scale shell-model calculations and TBRE calculations carried out to describe real nuclei are found to produce $1/f$ time series, but these calculations are very complicated and involve many degrees of freedom. In this section, we use a schematic Hamiltonian in a simplified version of the shell model, with which we can follow a transition in-between different regimes, in order to study how the power spectral density evolves, with particular emphasis on what happens at the transitional point.

We use the Hamiltonian

$$H = H_0 + \chi \hat{Q} \cdot \hat{Q} \,, \tag{11}$$

Journal of Physics: Conference Series **380** (2012) 012020 doi:10.1088/1742-6596/380/1/012020

Table 1: Self-similarity parameter $\alpha_{\rm DFA}$ obtained using a linear DFA method, and the β exponent in the power spectral density of the energy fluctuations in the shell-model calculations of ^{48}Ca with realistic interactions in different subspaces J^{π}. The dimension N of each subspace is also shown. The unfolding was done using the EMD method.

^{48}Ca			
0^+	1.03	1.09	347
1^+	0.99	1.10	880
2^+	1.02	1.03	1390
3^+	1.04	1.13	1627
4^+	1.01	1.09	1755
5^+	1.10	1.13	1617
6^+	1.06	1.08	1426
7^+	1.03	1.05	1095
8^+	1.05	1.07	808

with an external adjustable parameter χ. This schematic Hamiltonian is analogous to simple boson Hamiltonians used recently to study quantum phase transitions [32]. The unperturbed single-particle (s.p.) Hamiltonian H_0 describes non-interacting fermions in the mean field of an appropriate spherical core which is known to be integrable [33]. The second term, $\chi\hat{Q}\cdot\hat{Q}$, describes a residual quadrupole-quadrupole two-body interaction, which in this case acts on 8 valence neutrons in the fp shell, and this part by itself is also integrable. We carry out calculations in the subspace $J^{\pi} = 3^+$ of ^{48}Ca. The coefficient χ in Eq. (11) modulates the intensity of the quadrupole-quadrupole interaction and gives rise to the level repulsion in the nuclear spectrum. Varying the intensity of this control parameter in the range $0.01 < \chi < 0.40$, we find that the power spectral density of the δ_n time s eries behaves as $P(f) \sim 1/f^{\beta}$ with β=1.92 (for $\chi = 0.01$), $\beta = 1.10$ (for $\chi = 0.21$) and $\beta = 1.43$ (for $\chi = 0.40$), values that are confirmed by an analysis with DFA, see Fig. 3.

The fluctuations in the quantum spectra with the realistic KB3 interaction are in good agreement with the spectral fluctuations obtained with the schematic Hamiltonian when the intensity parameter χ takes values close to 0.21. Thus, we find that the fluctuations of both integrable limits tend to approach a brownian-like $1/f^2$ power spectral density, whereas a density close to $1/f$ is found at the transitional point.

4. Conclusions

Self-similarity is a very important property of dynamical systems which can be analyzed using the usual tools of time-series analysis. For the dynamical systems presenting scale invariance the power spectral density behaves as a power law $P(f) \sim 1/f^{\beta}$. The $1/f$ noise ($\beta = 1$) can be seen as a particular type of self-similar noise. It corresponds to signals that maximize the range of their correlations. We suggest in this paper that time series of both classical and quantum dynamical systems that undergo a transition between two regimes, exhibit $1/f$ behavior near the transitional point. We report in detail on two specific examples: the Module-1 Logistic Map for the classical case, and a schematic nuclear shell-model Hamiltonian for the quantum case. We studied the corresponding time series by spectral analysis, and with the DFA method.

In the case of the module-1 logistic map, for values of the control parameter $k < 0$, we found regular time series (with Lyapunov exponent $\lambda_k < 0$), for a value of $k = 1$, a correlated

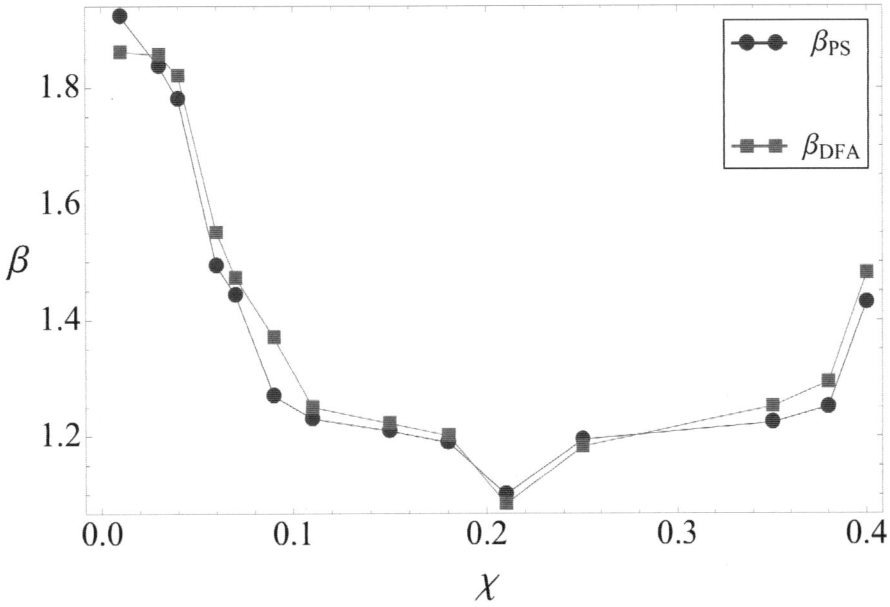

Figure 3: (Color online) The values of β calculated directly from the fit to the slope of the power spectral density (blue dots) to each δ_n in function of χ in the Hamiltonian of Eq. (11). The red squares are the $\beta's$ obtained using the relation with $\alpha_{\rm DFA}$.

non-periodic time series (with Lyapunov exponent $\lambda_k = 0$), and for control parameters $k > 1$ we observe chaotic time series (with Lyapunov exponent $\lambda_k > 0$). We find $1/f$ behavior for the correlated non-periodic time series at the transitional point, whereas the regular and chaotic time series correspond to $1/f^\beta$ ($\beta \neq 1$) power spectral density. We describe a generic nuclear excitation spectrum, using a schematic Hamiltonian. The Hamiltonian has two competing terms: a single-particle term, and a residual quadrupole-quadrupole term. Each term individually is integrable. A control parameter allows for a smooth transition between the two extreme regimes. Both extreme integrable excitation spectra correspond to $1/f^2$ (brownian) power laws, whereas the transitional excitation spectrum corresponds with a $1/f$ power spectral density. For two specific systems, we found a generic $1/f$ behavior of the corresponding time series, where the long-range correlations are maximized, exactly at the point where the transition occurs. We are currently applying these techniques to other systems, such as simple coupled pendula and photon counting rates in different kinds of emitting light sources [34]. We believe that time series analysis can provide relevant information in both physical and biological systems, including early warning signals in diverse phenomena [35].

Acknowledgements
We thank Pavel Stransky for his critical reading of this manuscript and his important comments. This work was supported in part by grants from CONACyT-Mexico, project I0017 - CB-2010-01 - 156663 and by the project FP7-PEOPLE-2009-IRSES-247541-MATSIQEL.

References

[1] Richardson L F 1961 *General Systems Yearbook* **6** 139
[2] Mandelbrot B 1967 *Science* **156** 636
[3] Scheinerman E R 1995 *Invitation to Dynamical Systems* (Upper Saddle River, NJ: Prentice Hall) p 1
[4] Bak P and Paczuski M 1993 *Physics World* (Dec 1993) 39
[5] Goldberger A L, Rigney D R and West B J 1990 *Sci. Am.* **262** 34

[6] Kaplan D T and Talajic M 1991 *Chaos* **1** 251
[7] Goldberger A L Rigney D R, Mietus J, Antman E M and Greenwald M 1988 *Experientia* **44** 983
[8] Halley J M 1996 *TREE* **11** 33
[9] Halley J M and Inchausti P 1996 *Fluct. Noise Lett.* **4** R1
[10] Press W H 1978 *Comments Astrophys.* **7** 103
[11] Deering W and West B J 1992 *IEEE Eng. Med. Biol.* **11** 40
[12] Peng C-K, Havlin S, Stanley H E and Goldberger A L 1995 Chaos **5** 82
[13] Talkner P and Weber R O 2000 *Phys. Rev.* E **62** 150
[14] Beran J 1994 *Statistic for Long-Memory processes* (New York: Chapman and Hall)
[15] Montroll E W and Shlesinger M F 1984 in: *Nonequilibrium Phenomena II. From Stochastics to Hydrodynamics* (Amsterdam: North-Holland) p 1
[16] May R M 1976 *Nature* **261** 459
[17] Bak P 1996 *How Nature works* (New York: Springer-Verlag)
[18] Milotti E 2002 *e-print* physics/0204033
[19] Manneville P 1980 *J. Physique* **41** 1235
[20] Ben-Mizrachi A, Procaccia I, Rosenberg N, Schmidt A and Schuster H G 1985 *Phys. Rev.* A **31** 1830
[21] Procaccia I and Schuster H 1983 *Phys. Rev.* A **28** 1210
[22] Relaño A, Gómez J M G, Molina R A, Retamosa J and Faleiro E 2002 *Phys. Rev. Lett.* **89** 244102
[23] Brody T A, Flores J, French J B, Mello P A, Pandey A and Wong S S M 1981 *Rev. Mod. Phys.* **53** 385
[24] Landa E, Morales I, Hernández C, López Vieyra J C, Frank A and Velázquez V 2008 *Rev. Mex. Fis.* **S-54** 48
[25] Huang N E, Shen Z, Long S R, Wu M C, Shih E H, Zheng Q, Tung C C and Liu H H 1998 *Proc. R. Soc. Lond.* A **454** 903
[26] Morales I O, Landa E, Stránský P and Frank A 2011 *Phys. Rev.* E. **84** 016203
[27] Santhanam M S, Bandyopadhyay J N and Angom D 2006 *Phys. Rev.* E **73** 015201(R)
[28] Kuo T T S and Brown G E 1968 *Nucl. Phys.* A **114** 241
[29] Caurier E, Zuker A P and Poves A 1991 in: *Nuclear Structure of Light Nuclei far from Stability. Experiment ad Theory, Proceedings of the Workshop, Obrnai, France, Nov. 27–29, 1989* ed G Klotz (Strasbourg: Kuster)
[30] French J B and Wong S S M 1970 *Phys. Lett.* B **23** 449
[31] Wong S S M and French J B 1972 *Nucl. Phys.* A **198** 188
[32] Cejnar P, Jolie J and Casten R F 2010 *Rev. Mod. Phys.* **82** 2155
[33] Caurier E, Martínez-Pinedo G, Nowacki F, Poves A and Zuker A P 2005 *Rev. Mod. Phys.* **77** 427
[34] Landa E *Ph.D. Thesis* to be published
[35] Scheffer M, Bascompte J, Brock W A, Brovkin V, Carpenter S R, Dakos V, Held H, van Nes E H, Rietkerk M and Sugihara G 2009 *Nature* **461** 7260

Fractals as macroscopic manifestation of squeezed coherent states and brain dynamics

Giuseppe Vitiello

Facoltà di Scienze MM FF NN and INFN Università di Salerno, I-84100 Salerno, Italy and INFN, Gruppo Collegato di Salerno, Università di Salerno, I-84100 Salerno, Italy
http://www.sa.infn.it/giuseppe.vitiello

E-mail: vitiello@sa.infn.it

Abstract. Recent results on the relation between self-similarity and squeezed coherent states are presented. I consider fractals which are generated iteratively according to a prescribed recipe, the so-called deterministic fractals. Fractal properties are incorporated in the framework of the theory of the entire analytical functions and deformed coherent states. Conversely, fractal properties of squeezed coherent states are recognized. This sheds some light on the understanding of the dynamical origin of fractals and their global nature emerging from local deformation processes. The self-similarity in brain background activity suggested by laboratory observations of power-law distributions of power spectral densities of electrocorticograms is also discussed and accounted in the frame of the dissipative many-body model of brain.

1. Introduction

In this paper I will show that a functional representation of self-similarity (as the one occurring in fractals) is provided by squeezed coherent states. Deterministic fractals, which are generated iteratively according to a prescribed recipe, are studied and their self-similarity property, which is considered to be the *most important property* of fractals (p. 150 in [1]), is framed in the theory of the entire analytical functions and coherent states. Conversely, we also recognize, in some specific sense, fractal properties of coherent states. This provides an insight into the dynamical origin of fractals. It also illustrates the geometrical (fractal) properties of coherent states. The conclusion is that fractals are global systems arising from local deformation processes. Therefore they cannot be purely geometric objects. The so called "random fractals", i.e. those fractals obtained by randomization processes introduced in their iterative generation, will not be considered. Since self-similarity is still a characterizing property in many of such random fractals, my conjecture is that also in such cases there must exist a connection with the algebraic structure of coherent states.

The study of the fractal self-similarity property has been also motivated by the laboratory observation of the brain background activity [2]. It results, indeed, that phase patterns observed in neocortical electroencephalograms (EEG) and electrocorticograms (ECoGs) exhibit power-law distributions with no detectable minima. In other words, the neocortical activity appears to be scale-free [2]-[9]. In my presentation I will frame these observations in the context of the dissipative quantum model of brain [10] and non-homogeneous boson condensation in quantum field theory (QFT) [11].

The conjecture that a relation between fractals and the algebra of coherent states exists was originally presented in [12]. As a matter of facts, in several instances fractal properties have been suspected that might be related in someway to coherent structures of the system under study. However, a mathematical proof of such a relation has never been worked out. The formal proof that this relation between fractals and coherent states actually exists and which ones are its specific features have been presented firstly in [13] and [14] (see also [15, 16]), which I will closely follow in the present paper. Certainly, in view of the ubiquitous presence of fractals in nature, the fact that they appear as *macroscopic quantum systems*, namely as macroscopic manifestation of the underlying quantum coherent dynamics, opens a wide horizon of possible applications in physics as well as in biology and in other disciplines, with related boost to interdisciplinary studies. In particular in biology, quantum theories might then introduce new perspectives in the study of basic dynamic phenomena underlying the rich phenomenology of biochemistry, as indeed it has been and it is pursued already in brain modeling and in living matter physics (for an extended discussion and references see [7, 10, 17]).

As we will see in the following discussion, the fact that fractals appear or emerge as macroscopic manifestation of the underlying coherent dynamics is described in a precise and explicit way, so that the word *emergence* acquires a mathematically well defined sense, in contrast to its somewhat vague and even mysterious way it is used often in the literature.

The paper is organized as follows: the formal relation between self-similarity and the coherent states is obtained in Section 2, where the fractal operator is defined. Its relation with the coherent state squeezing generator is presented in Section 3, where some aspects of noncommutative geometry implicit in the fractal generation are also considered. Final remarks and conclusions are presented in Section 4.

2. Self-similarity and the Fock-Bargmann representation of coherent states

Let me start by considering the familiar cases of lengths, surfaces and volumes. For example, a square S whose side is L_0 scales to $(1/2^2)S$ when $L_0 \to \lambda L_0$ with $\lambda = 1/2$. A cube V of same side with same rescaling of L_0 scales to $(1/2^3)V$. Of course, the length L_0 scales to $(1/2)L_0$. Note that $S(\frac{1}{2}L_0)/S(L_0) \equiv p = 1/4$ and $V(\frac{1}{2}L_0)/V(L_0) \equiv p = 1/8$, respectively, so that in both cases $p = \lambda^d$, with $d = 2$ and $d = 3$ for surfaces and volumes, respectively. For the length L_0, it is $p = 1/2 = 1/2^d = \lambda^d$ and of course $d = 1$. In general, by denoting lengths, surfaces or volumes by $\mathcal{H}(L_0)$ the scaling law

$$\mathcal{H}(\lambda L_0) = \lambda^d \mathcal{H}(L_0) \tag{1}$$

holds when lengths are (homogeneously) scaled as $L_0 \to \lambda L_0$.

Assume now that such a scaling law is valid in the case of any other "hypervolume" $\mathcal{H}$. One considers then the ratio

$$\frac{\mathcal{H}(\lambda L_0)}{\mathcal{H}(L_0)} \equiv p\ , \tag{2}$$

and due to Eq. (1) one obtains

$$p\, \mathcal{H}(L_0) = \lambda^d \mathcal{H}(L_0)\ , \tag{3}$$

i.e. $p = \lambda^d$.

Let me apply this to the *Koch curve* (Fig. 1). The discussion can be extended to other iteratively constructed fractals. In the case of the Koch curve one starts by considering a one-dimensional, $d = 1$, segment u_0 of unit length L_0, called the *initiator* [18]. This is called the step, or stage, of order $n = 0$. The length L_0 is then divided by the reducing factor $s = 3$, and the rescaled unit length $L_1 = (1/3)L_0$ is adopted to construct the new "deformed segment" u_1 made of $\alpha = 4$ units L_1 (step of order $n = 1$). u_1 is called the *generator* [18]. Note that such a "deformation" of the u_0 segment is only possible provided that one "gets out" of the

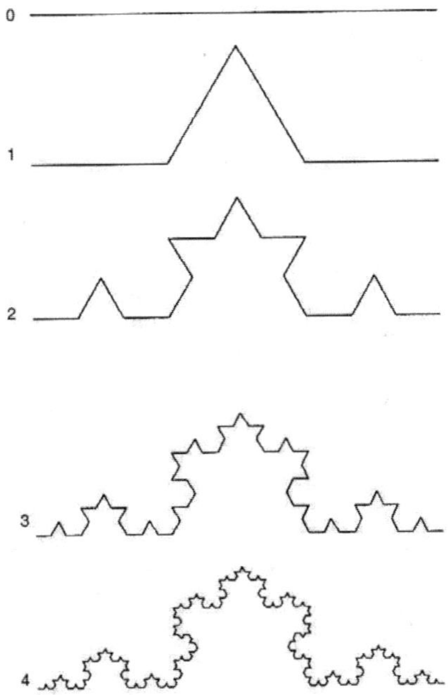

Figure 1. The first five stages of Koch curve.

one-dimensional straight line r to which the u_0 segment belongs: this means that in order to construct the u_1 "shape" the one-dimensional constraint $d = 1$ has to be relaxed: the shape, made of $\alpha = 4$ units L_1, lives in some $d \neq 1$ dimensions. Thus we write

$$u_{1,q}(\alpha) \equiv q\,\alpha\,u_0, \qquad q = \frac{1}{3^d}, \qquad d \neq 1, \tag{4}$$

where d has to be determined. Setting $\alpha \equiv 1/p = 4$ and $q \equiv \lambda^d = 1/3^d$, $p = \lambda^d$ gives

$$q\alpha = 1\ , \quad \text{where} \quad \alpha = 4, \quad q = \frac{1}{3^d}\ . \tag{5}$$

The meaning of Eq. (4) is that in the "deformed space", to which $u_{1,q}$ belongs, the set of four segments of which $u_{1,q}$ is made "equals" (is equivalent to) the three segments of which u_0 is made in the original "undeformed space". In other words, we require that the measure of the deformed segment $u_{1,q}$ with respect to the undeformed segment u_0 be 1: $u_{1,q}/u_0 = 1$, namely $\alpha q = 4/3^d = 1$.

Steps of higher order n, $n = 2, 3, 4, ..\infty$, can be obtained by iteration of the deformation process keeping $q = 1/3^d$ and $\alpha = 4$. For the nth order deformation we have

$$u_{n,q}(\alpha) \equiv (q\,\alpha)\,u_{n-1,q}(\alpha)\ , \quad n = 1, 2, 3, ... \tag{6}$$

i.e., for any $n \in \mathcal{N}_+$

$$u_{n,q}(\alpha) = (q\,\alpha)^n\,u_0\ . \tag{7}$$

By proceeding by iteration, or, equivalently, by requiring that $u_{n,q}(\alpha)/u_0$ be 1 for any n, gives $(q\,\alpha)^n = 1$ (in the following, for brevity I will thus set $u_0 = 1$, whenever no misunderstanding arises). From this and Eq. (5) we obtain, for any n,

$$d = \frac{\ln 4}{\ln 3} \approx 1.2619 \; , \tag{8}$$

which is called the *fractal dimension*, or the *self-similarity dimension* [1]. The (fractal) dimension d is the solution of the relation $1/\alpha = 1/4 = 1/3^d = q$, which for $d = 1$ would be trivially wrong. d is a measure of the "deformation" of the $u_{n,q}$-space *with respect to the* u_0*-space*. It is important to stress that the fractal is mathematically defined in the limit $n \to \infty$ of the deformation process. The fractal dimension is indeed defined by starting from $(q\alpha)^n = 1$ in the $n \to \infty$ limit. Also notice that since $L_n \to 0$ for $n \to \infty$, the Koch fractal is a curve which is everywhere non-differentiable [1].

The *self-similarity* property of a large class of fractals (the Sierpinski gasket and carpet, the Cantor set, etc.) is expressed in analytical form by Eqs. (6) and (7): "cutting a piece of a fractal and magnifying it isotropically to the size of the original, both the original and the magnification look the same" [18]. In this sense fractals are "scale free", namely viewing a picture of part of a fractal one cannot deduce its actual size if the unit of measure is not given in the same picture [19]. Note that self-similarity does not hold when considering only a finite number n of iterations. Self-similarity is defined only in the $n \to \infty$ limit.

I also observe that the deformation of u_0 into $u_{1,q}$ has been done by varying from 3 to 4 the number α of rescaled unit segments. α and its derivative $\frac{d}{d\alpha}$ are thus expected to play a role in the fractal structure. $(\alpha, \frac{d}{d\alpha})$ play indeed the role of conjugate variables, as we will see (cf. Eq. (11). Invariance, in the limit of $n \to \infty$ iterations, only under anisotropic magnification is called self-affinity. The results here presented can be extended to self-affine fractals. The measure of lengths in fractals, the Hausdorff measure, the fractal "mass" and other fractal properties will not be discussed in this paper.

The main observation is now that Eq. (7) expresses the deep formal connection with the theory of coherent states and the related algebraic structure. Let me make explicit such a connection. Consider in full generality the complex α-plane. The functions

$$u_n(\alpha) = \frac{\alpha^n}{\sqrt{n!}} \; , \qquad u_0(\alpha) = 1 \; , \qquad\qquad n \in \mathcal{N}_+ \; , \quad \alpha \in \mathcal{C} \; , \tag{9}$$

form in the vector space $\mathcal{F}$ of the entire analytic functions a basis which is orthonormal under the gaussian measure $d\mu(\alpha) = \dfrac{1}{\pi} e^{-|\alpha|^2} d\alpha d\bar{\alpha}$. The normalization condition with respect to the gaussian measure is ensured by the factor $\frac{1}{\sqrt{n!}}$, $n \in \mathcal{N}_+$, in Eq. (9). Apart from such a normalization factor, the functions $u_{n,q}(\alpha)|_{q \to 1}$ in Eq. (7) are thus immediately recognized to be nothing but the restriction to real α of the functions in Eq. (9). The study of the fractal properties may be thus carried on in the space $\mathcal{F}$ of the entire analytic functions, by restricting, at the end, to real α, $\alpha \to Re(\alpha)$. Since in Eq. (7) it is $q \neq 1$ $(q < 1)$, we also need to consider the "q-deformed" algebraic structure of which the space $\mathcal{F}$ provides a representation.

Actually the space $\mathcal{F}$ provides the Fock-Bargmann representation (FBR) [20, 21] of the Weyl–Heisenberg algebra generated by the set of operators $\{a, a^\dagger, \mathbf{1}\}$:

$$[a, a^\dagger] = \mathbf{1} \; , \qquad [N, a^\dagger] = a^\dagger \; , \qquad [N, a] = -a \; , \tag{10}$$

where $N \equiv a^\dagger a$, with the identification:

$$N \to \alpha \frac{d}{d\alpha} \; , \quad a^\dagger \to \alpha \; , \quad a \to \frac{d}{d\alpha} \; . \tag{11}$$

The functions (vectors) $u_n(\alpha)$ (Eq. (9)) are eigenkets of N with integer (positive and zero) eigenvalues. The FBR is the Hilbert space $\mathcal{K}$ generated by the $u_n(\alpha)$, i.e. the whole space $\mathcal{F}$ of the entire analytic functions. Any vector $|\psi\rangle$ in $\mathcal{K}$ is associated, in a one-to-one correspondence, with a function $\psi(\alpha) \in \mathcal{F}$ and is thus described by the set $\{c_n;\ c_n \in \mathcal{C},\ \sum_{n=0}^{\infty} |c_n|^2 = 1\}$ defined by its expansion in the complete orthonormal set of eigenkets $\{|n\rangle\}$ of N:

$$|\psi\rangle = \sum_{n=0}^{\infty} c_n |n\rangle \quad \rightarrow \quad \psi(\alpha) = \sum_{n=0}^{\infty} c_n u_n(\alpha), \tag{12}$$

$$\langle\psi|\psi\rangle = \sum_{n=0}^{\infty} |c_n|^2 \quad = \quad \int |\psi(\alpha)|^2 d\mu(\alpha) = ||\psi||^2 = 1, \tag{13}$$

$$|n\rangle \quad = \quad \frac{1}{\sqrt{n!}} (a^\dagger)^n |0\rangle \ , \tag{14}$$

where $|0\rangle$ denotes the vacuum vector, $a|0\rangle = 0$, $\langle 0|0\rangle = 1$. The series expressing $\psi(\alpha)$ in Eq. (12) converges uniformly in any compact domain of the α-plane due to the condition $\sum_{n=0}^{\infty} |c_n|^2 = 1$ (cf. Eq. (13)), confirming that $\psi(\alpha)$ is an entire analytic function.

It is well known that the Fock–Bargmann representation provides a simple frame to describe the usual coherent states (CS) [20, 22] $|\alpha\rangle$. One defines the unitary displacement operator $\mathcal{D}(\alpha)$:

$$\mathcal{D}(\alpha) = \exp(\alpha a^\dagger - \bar{\alpha} a) = \exp\Big(-\frac{|\alpha|^2}{2}\Big) \exp(\alpha a^\dagger) \exp(-\bar{\alpha}\ a) \ . \tag{15}$$

so that

$$\mathcal{D}^{-1}(\alpha)\ a\ \mathcal{D}(\alpha) = \ a \ + \ \alpha \ . \tag{16}$$

$\mathcal{D}(\alpha)$ is a bounded operator defined on the whole $\mathcal{K}$. It provides a representation of the Weyl–Heisenberg group usually denoted by W_1 [20]. The coherent state $|\alpha\rangle$ is then given by

$$|\alpha\rangle = \mathcal{D}(\alpha)|0\rangle \ , \quad \alpha \in \mathcal{C} \ , \tag{17}$$

$$|\alpha\rangle = \exp\Big(-\frac{|\alpha|^2}{2}\Big) \sum_{n=0}^{\infty} \frac{\alpha^n}{\sqrt{n!}} |n\rangle = \exp\Big(-\frac{|\alpha|^2}{2}\Big) \sum_{n=0}^{\infty} u_n(\alpha)|n\rangle, \tag{18}$$

with

$$a|\alpha\rangle = \alpha|\alpha\rangle \ , \quad \alpha \in \mathcal{C} \ . \tag{19}$$

Eq. (18) and (14) show that the state $|\alpha\rangle$ "is made" of the superposition of states $|n\rangle$ of any number n of a-particles (a-quanta or excitations). $|\alpha\rangle$ is said to be a "condensate" of a-particles. The transformation (16) induced by $\mathcal{D}(\alpha)$ on a (or equivalently on the vacuum state $|0\rangle$, Eq. (17)) is thus understood to induce a "condensation" of a-particles in the vacuum $|0\rangle$. Eq. (19) shows that the annihilation of an a-particle in $|\alpha\rangle$ produces a rearrangement of the states in the sum in Eq. (18) such that the state $|\alpha\rangle$ is again reobtained (modulus the overall coefficient α). This expresses the "coherence" of the condensate and the "phase" α (cf. Eq. (15)) is a measure of the strength or degree of coherence, which is thus the result of the phase correlation binding (phase locking) the whole set of modes a condensed in the state $|\alpha\rangle$. We have thus a correlation mediated by a phase (not by a force). Since coherent states are defined only in the limit $n \rightarrow \infty$, the summation in Eq. (18) cannot be approximated by any finite number of terms and we thus see that coherent states cannot be obtained by perturbation theory. In quantum field theory one can clearly show that coherence is a non-perturbative phenomenon [11].

The set $\{|\alpha\rangle\}$, for all $\alpha \in \mathcal{C}$, is an overcomplete set of states. In order to extract a complete set of CS from the overcomplete set it is necessary to introduce in the α-complex plane a regular lattice L, called the von Neumann lattice [20]. For a general discussion on this subject we address the reader to Refs. [20] and [23]. In this last one the von Neumann lattice is discussed also in connection with the deformation of the Weyl-Heisenberg algebra introduced below. Finally, the relation between the CS and the entire analytic function basis $\{u_n(\alpha)\}$ is:

$$u_n(\alpha) = e^{\frac{1}{2}|\alpha|^2} \langle n|\alpha\rangle \ . \tag{20}$$

In order to account for the q-parameter (cf. Eq. (7)) I now introduce the finite difference operator $\mathcal{D}_q$ defined by [24]:

$$\mathcal{D}_q f(\alpha) = \frac{f(q\alpha) - f(\alpha)}{(q-1)\alpha} \ , \qquad f(\alpha) \in \mathcal{F} \ , \quad q = e^{\zeta} \ , \quad \zeta \in \mathcal{C}. \tag{21}$$

$\mathcal{D}_q$ reduces to the standard derivative for $q \to 1$ ($\zeta \to 0$). In the space $\mathcal{F}$ the commutation relations hold:

$$[\mathcal{D}_q, \alpha] = q^{\alpha\frac{d}{d\alpha}} \quad , \quad \left[\alpha\frac{d}{d\alpha}, \mathcal{D}_q\right] = -\mathcal{D}_q \quad , \quad \left[\alpha\frac{d}{d\alpha}, \alpha\right] = \alpha \ , \tag{22}$$

which, as for Eq. (11), lead us to the identification

$$N \to \alpha\frac{d}{d\alpha} \quad , \quad \hat{a}_q \to \alpha \quad , \quad a_q \to \mathcal{D}_q \ , \tag{23}$$

with $\hat{a}_q = \hat{a}_{q=1} = a^\dagger$ and $\lim_{q\to 1} a_q = a$ on $\mathcal{F}$. The algebra (22) is the q-deformation of the algebra (10). More details on the properties of $\mathcal{D}_q$ and the q-deformed algebra (22) can be found in Refs. [23]. I only recall that the operator q^N acts on the whole $\mathcal{F}$ as

$$q^N f(\alpha) = f(q\alpha) \ , \quad f(\alpha) \in \mathcal{F} \ , \tag{24}$$

which, when applied to the coherent state functional (18), gives

$$q^N|\alpha\rangle = |q\alpha\rangle = \exp\Big(-\frac{|q\alpha|^2}{2}\Big) \sum_{n=0}^{\infty} \frac{(q\alpha)^n}{\sqrt{n!}} |n\rangle \ . \tag{25}$$

Since $q\alpha \in \mathcal{C}$, from Eq. (17),

$$a|q\alpha\rangle = q\alpha|q\alpha\rangle \ , \quad q\alpha \in \mathcal{C} \ . \tag{26}$$

We thus see that the nth fractal iteration (cf. Eq. (7) where we set $u_0 \equiv 1$) is obtained by projecting out the nth component of $|q\alpha\rangle$ and restricting to real $q\alpha$,

$$u_{n,q}(\alpha) = (q\alpha)^n = \sqrt{n!}\ \exp\Big(\frac{|q\alpha|^2}{2}\Big)\langle n|q\alpha\rangle, \quad q\alpha \to Re(q\alpha) \ , \quad \text{for any } n \in \mathcal{N}_+ \ , \tag{27}$$

which, by using $\langle n| = \langle 0| \ \frac{(a)^n}{\sqrt{n!}}$, gives

$$u_{n,q}(\alpha) = (q\alpha)^n = \exp\Big(\frac{|q\alpha|^2}{2}\Big)\langle 0|(a)^n|q\alpha\rangle, \quad q\alpha \to Re(q\alpha), \quad n \in \mathcal{N}_+, \tag{28}$$

This shows that the operator $(a)^n$ acts as a "magnifying" lens [18]: the nth iteration of the fractal can be "seen" by applying $(a)^n$ to $|q\alpha\rangle$ and restricting to real $q\alpha$:

$$\langle q\alpha|(a)^n|q\alpha\rangle = (q\alpha)^n = u_{n,q}(\alpha), \quad q\alpha \to Re(q\alpha). \tag{29}$$

Let me close this Section by observing that Eq. (26) expresses the invariance of the coherent state (representing the fractal) under the action of the operator $\frac{1}{q\alpha}a$. Such an invariance property allows us to consider the coherent functional $\psi(q\alpha)$ as an "attractor" in $\mathcal{C}$. In this sense, Eq. (26) reminds us of the fixed point equation $W(A) = A$, where W is the Hutchinson operator [18], characterizing the iteration process for the fractal A in the $n \to \infty$ limit.

In summary, the operator q^N applied to $|\alpha\rangle$ "produces" the fractal in the functional form of the coherent state $|q\alpha\rangle$ (cf Eq. (25)). The nth fractal stage of iteration, $n = 0, 1, 2, .., \infty$ is represented, in a one-to-one correspondence, by the nth term in the coherent state series in Eq. (25). I call q^N *the fractal operator.*

Eqs. (27), (28) and (29) formally establish the searched connection between fractal self-similarity and the (q-deformed) algebra of the coherent states.

3. The fractal operator and the squeezing generator

In this Section I consider the relation between the fractal operator q^N and the squeezing transformation for coherent states. In the Hilbert space identified with the space $\mathcal{F}$ of the entire analytic functions $\psi(\alpha)$ the following identity holds

$$2\alpha\frac{d}{d\alpha}\psi(\alpha) = \left\{\frac{1}{2}\left[\left(\alpha + \frac{d}{d\alpha}\right)^2 - \left(\alpha - \frac{d}{d\alpha}\right)^2\right] - 1\right\}\psi(\alpha)\ , \tag{30}$$

and I will put

$$c = \frac{1}{\sqrt{2}}(\alpha + \frac{d}{d\alpha})\quad ,\quad c^\dagger = \frac{1}{\sqrt{2}}(\alpha - \frac{d}{d\alpha})\quad ,\quad [c, c^\dagger] = \mathbf{1}\ . \tag{31}$$

Their relation with the FBR operators a and $a^\dagger$, previously introduced in Eq. (11), is

$$\alpha = \frac{1}{\sqrt{2}}(c + c^\dagger)\ \to a^\dagger\ ,\quad \frac{d}{d\alpha} = \frac{1}{\sqrt{2}}(c - c^\dagger)\ \to a\ . \tag{32}$$

In $\mathcal{F}$, $c^\dagger$ is the conjugate of c [20, 23]. By setting $\alpha \equiv x + iy$, and considering the holomorphy conditions holding for $f(\alpha) \in \mathcal{F}$

$$\frac{d}{d\alpha}f(\alpha) = \frac{d}{dx}f(\alpha) = -i\frac{d}{dy}f(\alpha)\ , \tag{33}$$

in the $y \to 0$ limit we get form (31)

$$c \to \frac{1}{\sqrt{2}}(x + ip_x) \equiv \hat{z}\ ,\quad c^\dagger \to \frac{1}{\sqrt{2}}(x - ip_x) \equiv \hat{z}^\dagger\ ,\quad [\hat{z}, \hat{z}^\dagger] = 1\ , \tag{34}$$

where $p_x = -i\frac{d}{dx}$. $\hat{z}^\dagger \leftrightarrow a^\dagger$ and $\hat{z} \leftrightarrow a$ are the usual creation and annihilation operators in the configuration representation. Thus we see that in the limit $\alpha \to Re(\alpha)$, i.e. $y \to 0$, c and $c^\dagger$ turn into the conventional annihilation and creator operators, respectively, associated with x and p_x in the canonical configuration representation.

I now observe that the fractal operator q^N can be realized in $\mathcal{F}$ as:

$$q^N\psi(\alpha) = \frac{1}{\sqrt{q}}\ \exp\left(\frac{\zeta}{2}(c^2 - c^{\dagger 2})\right)\psi(\alpha) \equiv \frac{1}{\sqrt{q}}\hat{\mathcal{S}}(\zeta)\psi(\alpha) \equiv \frac{1}{\sqrt{q}}\psi_s(\alpha)\ , \tag{35}$$

where $q = e^\zeta$ (for simplicity, assumed to be real) and as usual $N = \alpha\, d/d\alpha$. $\hat{\mathcal{S}}(\zeta)$ is the squeezing generator well known in quantum optics [23, 25]. $\zeta = \ln q$ is called the squeezing parameter and $\psi_s(\alpha)$ denotes the squeezed states in the FBR. From Eq. (35) we see that q^N acts in $\mathcal{F}$, as well as in the configuration representation in the limit $y \to 0$, as the squeezing operator $\hat{\mathcal{S}}(\zeta)$, up to the numerical factor $1/\sqrt{q}$.

Since $q^N\psi(\alpha) = \psi(q\alpha)$ (cf. Eq. (24)), from Eq. (35) we conclude that the q-deformation process, which we have seen is associated to the fractal generation process, is equivalent to the squeezing transformation.

The squeezing generator $\hat{\mathcal{S}}(\zeta)$ is an $SU(1,1)$ group element. In fact, the operators $K_- = (1/2)c^2$, $K_+ = (1/2)c^{\dagger 2}$, $K_\alpha = (1/2)(c^\dagger c + 1/2)$, close the algebra $su(1,1)$. We thus may write the $SU(1,1)$ Bogoliubov (squeezing) transformations for the c's operators:

$$\hat{\mathcal{S}}^{-1}(\zeta)\; c\; \hat{\mathcal{S}}(\zeta) = c\; \cosh\zeta - c^\dagger\; \sinh\zeta\;, \tag{36}$$

$$\hat{\mathcal{S}}^{-1}(\zeta)\; c^\dagger\; \hat{\mathcal{S}}(\zeta) = c^\dagger\; \cosh\zeta - c\; \sinh\zeta\;, \tag{37}$$

and in the $y \to 0$ limit

$$\hat{\mathcal{S}}^{-1}(\zeta)\; \alpha\; \hat{\mathcal{S}}(\zeta) \;=\; \frac{1}{q}\alpha \to \frac{1}{q}x\;, \tag{38}$$

$$\hat{\mathcal{S}}^{-1}(\zeta)\; p_\alpha\; \hat{\mathcal{S}}(\zeta) \;=\; qp_\alpha \to qp_x\;, \tag{39}$$

where $p_\alpha \equiv -i\frac{d}{d\alpha}$, and

$$\int d\mu(\alpha)\bar{\psi}(\alpha)\hat{\mathcal{S}}^{-1}(\zeta)\; \alpha\; \hat{\mathcal{S}}(\zeta)\psi(\alpha) \;\to\; \frac{1}{q} < x > \tag{40}$$

$$\int d\mu(\alpha)\bar{\psi}(\alpha)\hat{\mathcal{S}}^{-1}(\zeta)\; p_\alpha\; \hat{\mathcal{S}}(\zeta)\psi(\alpha) \;\to\; q < p_x > \tag{41}$$

so that the root mean square deviations Δx and Δp_x satisfy

$$\Delta x \Delta p_x = \frac{1}{2} \quad , \quad \Delta x = \frac{1}{q}\sqrt{\frac{1}{2}} \quad , \quad \Delta p_x = q\sqrt{\frac{1}{2}}\;, \tag{42}$$

which confirms that the q-deformation plays the role of squeezing transformation. Note that the action variable $\int p_x\; dx$ is invariant under the squeezing transformation.

Eq. (38) shows that $\alpha \to (1/q)\alpha$ under squeezing transformation, which, in view of the fact that $q^{-1} = \alpha$ (cf. Eq. (5)), means that $\alpha \to \alpha^2$, i.e. under squeezing we proceed further in the fractal iteration process. In conclusion, the fractal iteration process can be described in terms of the coherent state squeezing transformation.

Use of (38) and (39) leads to

$$\hat{z}_q = \frac{1}{q\sqrt{2}}(x + iq^2 p_x)\;, \quad \hat{z}_q^\dagger = \frac{1}{q\sqrt{2}}(x - iq^2 p_x), \quad [\hat{z}_q, \hat{z}_q^\dagger] = 1\;. \tag{43}$$

Notice that the $\hat{z}$ and $\hat{z}^\dagger$ algebra is preserved under the squeezing transformation, which is indeed a canonical transformation. In the "deformed" phase space let us denote the coordinates $(x, q^2 p_x)$ as (x_1, x_2), where $x_1 \equiv x$ and $x_2 \equiv q^2 p_x$. Coordinates do not commute in this (deformed) space:

$$[x_1, x_2] = iq^2\;. \tag{44}$$

We thus recognize that q-deformation introduces non-commutative geometry in the (x_1, x_2)-space. In such a space the noncommutative Pythagora's theorem gives the distance D:

$$D^2 = x_1^2 + x_2^2 = 2q^2(\hat{z}_q^\dagger \hat{z}_q + \frac{1}{2})\;. \tag{45}$$

In $\mathcal{F}$, in the $y \to 0$ limit, from the known properties of creation and annihilation operators we then get

$$D_n^2 = 2q^2(n + \frac{1}{2}) , \quad n = 0, 1, 2, 3... , \tag{46}$$

i.e. in the space (x_1, x_2) associated to the coherent state fractal representation, the (x_1, x_2)-distance is quantized according to the unit scale set by q. Eq. (46) also shows that in the space (x_1, x_2) we have quantized "disks" of squared radius vector D_n^2. It is interesting to observe that the "smallest" of such disks has non-zero radius given by the deformation parameter q (recall that $q = \frac{1}{3^d}$ when Koch fractal is considered). We also realize that Eq. (46) is nothing but the expression of the energy spectrum of the harmonic oscillator, i.e. one could write Eq. (46) as $D_n^2 = 2q^2(n + \frac{1}{2}) \equiv E_n$, $n = 0, 1, 2, 3...$, where E_n might be thought as the "energy" associated to the nth fractal stage.

4. Scale-free laws and brain waves

In laboratory observations the mesoscopic neural activity of neocortex appears to consist of the dynamical formation of spatially extended neuronal domains in which widespread cooperation supports brief epochs of patterned synchronized oscillations [26]. In QFT information is stored in ordered patterns maintained by long range correlations mediated by massless quanta, the so-called Nambu-Goldstone (NG) modes or particles. Memory could be then described as a printed pattern of order supported by long range correlations. In 1967 Umezawa proposed indeed a model of brain dynamics based on the dynamical generation of such long range correlation in the way they are normally treated in many-body theory [27]. Already in the early 1940s Lashley remarked, with reference to the diffuse, non-local nature of brain activity which he observed in laboratory: "...Here is the dilemma. Nerve impulses are transmitted ...from cell to cell through definite intercellular connections. Yet, all behavior seems to be determined by masses of excitation...within general fields of activity, without regard to particular nerve cells... What sort of nervous organization might be capable of responding to a pattern of excitation without limited specialized path of conduction? The problem is almost universal in the activity of the nervous system" (pp. 302–306 of Ref. [28]). Pribram, soon after the discovery of the laser light in the early 1960s [29, 22], proposed his holographic hypothesis [30]. In Umezawa's many-body model for the brain two main ingredients appear: the notion of "field" introduced by Lashley in his puzzling dilemma and the notion of "coherence", intrinsic to the laser theory inspiring Pribram view. Both these notions are basic ones in the QFT dynamics generating ordered patterns.

With the advent of modern advanced technologies, the amplitude modulated (AM) patterns and the phase modulated (PM) patterns in the brain are of common observational access. Laboratory observations show that the observed cortical collective activity cannot be fully accounted for by the electric field of the extracellular dendritic current or the extracellular magnetic field from the high-density electric current inside the dendritic shafts, which are much too weak, or by the chemical diffusion, which is much too slow [7, 8, 31]. Patterns of phase-locked oscillations are intermittently present in resting, awake subjects as well as in the same subject actively engaged in cognitive tasks requiring interaction with environment, so they are best described as properties of the background activity of brains that is modulated upon engagement with the surround. These "packets of waves" extend over spatial domains covering much of the hemisphere in rabbits and cats [3, 32, 33, 34], and over regions of linear size of about 19 cm [35] in human cortex with near zero phase dispersion [6, 36, 37].

Umezawa's many-body model [27] and its extension to dissipative dynamics, the dissipative many-body model of brain [38], are based on the QFT notion of spontaneous breakdown of symmetry (SBS). When the symmetry of the Lagrangian is not the symmetry of the least energy state (the ground state or the vacuum) of the system, the symmetry is said to be spontaneously broken and mathematical consistency requires the existence of NG massless particles [11, 39]. Examples are the phonon modes in the crystals, the magnon modes in ferromagnets, etc. They can be described, as customary in a quantum theory, as the quanta associated to certain waves, such as the elastic wave in crystals, the spin wave in ferromagnets. The role of such waves is the one of establishing long range correlation among the system constituents. For example, in magnets, the elementary magnetic dipoles are forced to oscillate "in phase" under the correlation established by the spin waves, i.e. by the magnon quanta spanning the extended domain which thus gets characterized by its macroscopic magnetization. The mechanism of spontaneous breakdown of symmetry is at the origin of the *change of scale*: from the microscopic scale of the elementary constituent dynamics to the macroscopic scale of the system magnetization, which is therefore a measure of the ordering of the elementary constituents and is for that reason called the "order parameter". The external field (the trigger or stimulus) is responsible of the "phase transition" from the normal (zero magnetization) phase to the magnetic phase. Then the mathematical structure of the theory must be adequate to allow physically distinct phases (technically called unitarily inequivalent representations of the quantum algebra). QFT possesses indeed such a mathematical structure. The ground states corresponding to physically distinct phases are characterized by distinct degrees of ordering which are described by different densities of NG modes *condensed* in them. Such a condensation of NG modes in the ground states is a *coherent* condensation, which physically expresses the "in phase", i.e. synchronized, elementary constituent oscillations. In quantum mechanics (QM), on the contrary, all the state representations are physically equivalent (unitarily equivalent) and therefore QM is not useful to describe phase transitions [40, 41].

I want to stress that in Umezawa's model and in the dissipative many-body model the neuron and the glia cells are treated as classically behaving systems. The quantum variables relevant to the many-body model of brain have been identified [38, 42] to be the electric dipole vibrational modes of the water molecules which constitutes the matrix in which neurons and glia cells and other mesoscopic units are embedded. The spontaneous breakdown of the rotational symmetry of electrical dipoles of water and other molecules implies the existence of NG modes. In such a context, these NG modes have been called the dipole wave quanta (DWQ) [38, 42, 43]. The system ground state thus is obtained in terms of coherent condensation of the DWQ, which are the agents responsible of the neuronal coordination. In particular it is found that the "memory state" is a squeezed coherent state for the basic quantum variables, whose mesoscopic order parameter expresses, at the synaptic level, the amplitude and phase modulation of the carrier signal observed in EEGs and ECoEGs.

The original Umezawa's brain model suffers of a very much limited memory capacity. Any subsequent stimulus cancels the one produced by the preceding stimulus (overprinting). It has been shown [38] that such a problem may find a solution when the model is modified so to include the crucial property of brain system of being an open system. In this paper I will not insist further in the description of the dissipative many-body model, which can be found in the existing literature [7, 8, 38], [44] - [47]. I will observe only that the dissipative many-body model describes the scale-free and self-similarity property of the brain dynamics. As said, laboratory observations show that self-similarity characterizes the brain ground state. Measurements of the durations, recurrence intervals and diameters of neocortical EEG phase patterns have power-law distributions with no detectable minima. The power spectral densities in both time and space of ECoGs from surface arrays conform to power-law distributions [32, 33, 48, 49, 50]. This suggests that the activity patterns generated by neocortical neuropil might be scale-free [2]-[9], thus explaining the self-similarity in ECoGs patterns over distances ranging from hypercolumns to an entire cerebral hemisphere. On the other hand, one of the main actors in the dissipative model is the (brain) ground state, which turns out to be a squeezed coherent state like the ones described in the previous Sections as q-deformed coherent state [38]. Since a functional representation of self-similarity is provided by squeezed coherent states, we see that the dissipative model of brain accounts for the self-similarity in brain background activity.

5. Conclusion

Let me close the paper with a couple of comments. Certainly, connecting fractal self-similarity and coherent states opens the way to a series of possible theoretical developments and practical applications in a number of research sectors, including the field of complex dynamical systems.

From the theoretical point of view, embedding the fractal study in the framework of the entire analytical functions may contribute to the treatment of fractal properties which present non-trivial difficulties due to the lack of a tractable mathematical representation; for example, the Koch curve is known to be everywhere non-differentiable [1], which makes it greatly interesting, but also difficult to deal with. From the standpoint of coherent states, there might be interesting applications in laser physics and quantum optics, where Glauber coherent states play a dominant role. Since coherent states always emerge as a result of boson condensation (see, e.g., Eq. (16) and the subsequent remarks), recognizing their fractal structure may contribute to the understanding of physical properties and behaviors in condensed matter systems endowed with a condensed ground state. Here, of course, the perspective opens towards the large class of phenomena based on the mechanism of spontaneous breakdown of symmetry. There the physical meaning of the process of coherent condensation of Nambu-Goldstone modes is the one of the appearance of long range correlation in the system, namely of the dynamical formation of ordered patterns in the ground state. The common feature underlying fractal formation and ordered pattern formation is thus in the dynamical emergence of long range correlation modes. This is certainly the case in quantum field theory systems characterized by ordered patterns in the ground state. The discussion presented above seems to suggest that this might be also the case for fractals. If so, this would provide an interesting example of interplay between the mesoscopic/macroscopic dimension of fractals and the microscopic dynamics out of which they are formed. It also provides an interesting example of a relation of global features (long range correlation at mesoscopic/macroscopic level) with local deformation processes (i.e. dynamical properties at microscopic level). On such a kind of relation is based our understanding of the formation of topologically non-trivial defects (also called extended objects), such as vortices, domain walls, boundary defects, and other soliton-like defects in condensed matter physics and high energy physics [13, 52]. These extended objects behave as macroscopic classical objects, but are generated by the microscopic (quantum) dynamics. On the basis of our discussion, now one might propose that perhaps the class of these extended objects might be enlarged so to include also fractals.

The relation between fractal self-similarity and squeezing is also interesting since it is known that squeezed coherent states are also related with dissipative systems [23, 53, 54]. Thus a link might be established between dissipation, which plays a relevant role in complex dynamical systems, and fractals. An interesting result is, for example, the one provided by the measurement [55] of the *ph* as a function of conductivity of samples of water in contact with hydrophilic films of substances such as the so called nafion. Considering that the *ph* is the $Log_{10}[H^+]$, where $[H^+]$ denotes the proton concentration in water, the plot *ph*-$Log_{10}\,\chi$, with χ denoting the conductivity, is a log-log plot and it is found [55] to be a straight line (with negative angular coefficient). This signals self-similarity and scale-free dynamics. Thus, according to the discussion here presented, it signals the presence of coherent structures in water, formed as a result of the water molecule interaction with the nafion film.

Of course, a large field of applications might be offered by biological systems. An interesting example in such a direction is provided by brain studies [2]-[9], [14]. According to the result of the present paper, deformed coherent states provide the functional representation of self-similarity observed in neuro-phenomenological data (for a discussion on the relation of fractals with brain/mind, although from a perspective different from the one presented in this paper, see also Ref. [56]).

I recall that the Weyl-Heisenberg representations are labeled by the q-parameter by means of the squeezing transformations [54]. In the infinite volume limit (infinite degrees of freedom) representations labeled by different values of the q-deformation parameter are unitarily inequivalent representations [23]. Changes of the value of the q-parameter induce transitions through unitarily inequivalent representations. The trajectories induced over the space of the representations are shown to be, under quite general conditions, chaotic trajectories [57]. Besides the scale parameter, one might consider also phase parameters and translation parameters (which characterize generalized coherent states, such as $SU(2)$, $SU(1,1)$, etc. coherent states) and relate them to the deformation q-parameter. By changing these parameters in a *deterministic iterated function process*, also called *multiple reproduction copy machine* process [1], the Koch curve may be then transformed into another fractal, e.g. into Barnsley's fern [1]. In the framework presented in this paper, these fractals are then described by corresponding unitarily inequivalent representations in the limit of infinitely many degrees of freedom (infinite volume limit) [13]. In this way one might recover the richness of the variety of "different" fractal shapes obtainable by changing the parameters of the fractal one starts with [1].

In conclusion, by limiting my discussion to deterministic fractals, I have presented the functional realization of fractal self-similarity in terms of the q-deformed algebra of coherent states. Fractal study can be thus incorporated in the theory of the entire analytical functions. From the discussion it appears that the reverse is also true: under convenient choice of the q-deformation parameter and by a suitable restriction to real α, coherent states exhibit fractal properties in the q-deformed space of the entire analytical functions.

References

[1] Peitgen H O, Jürgens H and Saupe D 1986 *Chaos and fractals. New frontiers of Science* (Berlin: Springer-Verlag)
[2] Freeman W J 2005 *Biol. Cybern.* **92/6** 350
[3] Freeman W J 2005 *Clin. Neurophysiol.* **116** (5) 1117
[4] Freeman W J, Rogers L J, Holmes M D and Silbergeld D L 2000 *J. Neurosci. Meth.* **95** 111
[5] Wang X F and Chen G R 2003 *IEEE Circuits Syst.* **31** 6
[6] Bassett D S, Meyer-Lindenberg A, Achard S, Duke T, Bullmore E 2006 *PNAS* **103** 19518
[7] Freeman W J and Vitiello G 2006 *Phys. of Life Reviews* **3** 93
http://www.arxiv.org/find [Freeman]; q-bio.OT/0511037.
[8] Freeman W J and Vitiello G 2008 *J. Phys. A: Math. Theor.* **41** 304042
http://Select.iop.org, q-bio.NC/0701053v1.
[9] Beggs J M and Plenz D 2003 *The Journal of Neuroscience* **23** (35) 11167
Gireesh E D and Plenz D 2008 *PNAS* **105** 7576
Petermanna T, Thiagarajana T C *et al* 2009 *PNAS* **106** 15921
[10] Vitiello G 2001 *My Double Unveiled* (Amsterdam: John Benjamins Pub. Co.).
[11] Blasone M, Jizba P and Vitiello G 2011 *Quantum field theory and its macroscopic manifestations* (London: Imperial College Press)
[12] Vitiello G 2005 *Brazilian J. of Physics* **35** 351
[13] Vitiello G 2008 Topological defects, fractals and the structure of quantum field theory, *E. J. Theor. Phys.* (*in print* arXiv:0807.2164v2) [hep-th].
[14] Vitiello G 2009 *New Mathematics and Natural Computing* **5** 245
[15] Vitiello G 2009 *Quantum Interaction* eds Bruza P, Sofge D *et al*
Lect. Notes in Artificial Intelligence eds Goebel R, Siekmann J, Wahlster W (Berlin: Springer-Verlag) p 6
[16] Vitiello G 2010 *Computing Anticipatory Systems* vol 1303, ed Daniel Dubois (New York: American Institute of Physics) p 168
[17] Montagnier L, Aissa J, Del Giudice E, Lavallee C, Tedeschi A, Vitiello G 2011 *J. Phys.: Conf. Series* **306** 012007
[18] Bunde A and Havlin S (eds.) 1995 *Fractals in Science* (Berlin: Springer -Verlag)
[19] Bak P and Creutz M 1995 *Fractals in Science* eds Bunde A and Havlin S (Berlin: Springer -Verlag)
[20] Perelomov A 1986 *Generalized Coherent States and Their Applications* (Berlin: Springer -Verlag)

[21] Bargmann V 1961 *Comm. Pure and Appl. Math.* **14** 187
[22] Klauder J R and Sudarshan E C 1968 *Fundamentals of Quantum Optics* (New York: Benjamin)
Klauder J R and Skagerstam B 1985 *Coherent States* (Singapore: World Scientific)
[23] Celeghini E, De Martino S, De Siena S, Rasetti M and Vitiello G 1995 *Ann. Phys.* **241** 50
Celeghini E, Rasetti M, Tarlini M and Vitiello G 1989 *Mod. Phys. Lett.* **B3** 1213
Celeghini E, Rasetti M and Vitiello G 1991 *Phys. Rev. Lett.* **66** 2056
[24] Biedenharn L C and Lohe M A 1995 *Quantum Group Symmetry And Q-Tensor Algebras* (Singapore: World Scientific)
Fivel D I 1991 *J. Phys. A: Math. Gen.* **24** 3575
[25] Yuen H P 1976 *Phys. Rev.* **A13** 2226
[26] Freeman W J 1975/2004 *Mass Action in the Nervous System* (New York: Academic Press)
Freeman W J 2000 *Neurodynamics. An Exploration of Mesoscopic Brain Dynamics* (London: Springer-Verlag)
Freeman W J 2001 *How Brains Make Up Their Minds* (New York: Columbia University Press).
[27] Ricciardi L M and Umezawa H 1967 *Kibernetik* **4** 44
Reprint in *Brain and Being* eds Globus G, Pribram K H and Vitiello G 2004 (Amsterdam: John Benjamins) p 255
Stuart C I J, Takahashi Y and Umezawa H 1978 *J.Theor. Biol.* **71** 605
Stuart C I J, Takahashi Y and Umezawa H 1979 *Found. Phys.* **9** 301
[28] Lashley K 1948 *The Mechanism of Vision, XVIII, Effects of Destroying the Visual "Associative Areas" of the Monkey* (Provincetown, MA: Journal Press).
[29] Glauber R J 1963 *Phys. Rev.* **131** 2766
[30] Pribram K H 1971 *Languages of the Brain* (Engelwood Cliffs, NJ: Prentice-Hall)
Pribram K H 1991 *Brain and Perception* (Hillsdale, NJ: Lawrence Erlbaum Associates Publ)
[31] Freeman W J 2005 *J. Integrative Neuroscience* **4** 407
[32] Freeman W J 2004 *Clin. Neurophysiol.* **115** 2077
[33] Freeman W J 2004 *Clin. Neurophysiol.* **115** 2089
[34] Freeman W J 2006 *Clin. Neurophysiol.* **117**(3) 572
[35] Freeman W J, Burke B C, Holmes M D and Vanhatalo S 2003 *Clin. Neurophysiol.* **114** 1055
[36] Freeman W J, Gaál G and Jornten R 2003 *Intern. J. Bifurc. Chaos* **13** 2845
[37] Freeman W J and Rogers L J 2003 *Intern. J. Bifurc. Chaos* **13** 2867
[38] Vitiello G 1995 *Int. J. Mod. Phys.* **B9** 973
[39] Itzykson C and Zuber J 1980 *Quantum field theory* (New York: McGraw-Hill)
[40] Vitiello G 1998 *Toward a science of consciousness II. The second Tucson Discussions and debates* eds Hameroff S R, Kaszniak A W and Scott A C (Cambridge: MIT Press) p 231
[41] Umezawa H and Vitiello G 1985 *Quantum Mechanics* (Naples: Bibliopolis)
[42] Jibu M and Yasue K 1995 *Quantum brain dynamics and consciousness* (Amsterdam: John Benjamins)
Jibu M, Pribram K H and Yasue K 1996 *Int. J. Mod. Phys.* **B 10** 1735
[43] Del Giudice E, Doglia S, Milani M and Vitiello G 1985 *Nucl. Phys. B* **251**(FS 13) 375
Del Giudice E, Doglia S, Milani M and Vitiello G 1986 *Nucl. Phys. B* **275**(FS 17) 185
Del Giudice E, Preparata G and Vitiello G 1988 *Phys. Rev. Lett.* **61** 1085
[44] Alfinito E and Vitiello G 2000 *Int. J. Mod. Phys.* **B14** 853[Erratum-ibid. 2000 **B14** 1613].
[45] Pessa E and Vitiello G 2003 *Mind and Matter* **1** 59
[46] Pessa E and Vitiello G 2004 *Int. J. Mod. Phys.* **B18** 841
[47] Freeman W J and Vitiello G 2010 *Int. J. Mod. Phys.* **B 24** 3269 arXiv:0802.3854v1 [q-bio.NC]
Freeman W J, Livi R, Obinata M and Vitiello G 2010 Cortical phase transitions, non-equilibrium thermodynamics and the time-dependent Ginzburg-Landau equation (*Preprint* arXiv:1110.3677)
[48] Braitenberg V and Schüz A 1991 *Anatomy of the Cortex: Statistics and Geometry* (Berlin: Springer-Verlag)
[49] Linkenkaer-Hansen K, Nikouline V M, Palva J M, Iimoniemi R J 2001 *J Neurosci* **15** 1370
[50] Hwa R C and Ferree T 2002 *Phys. Rev.* **E 66** 021901.
[51] Freeman W J 2005 *Biol. Cybern.* **92/6** 350
[52] Umezawa H 1993 *Advanced Field Theory: Micro, Macro and Thermal Physics* (New York: American Institute of Physics).
[53] Iorio A and Vitiello G 1995 *Annals of Phys.(N.Y.)* **241** 496
[54] Iorio A and Vitiello G 1994 *Mod. Phys. Lett.* **B 8** 269
[55] Elia V and Napoli E *et al. in preparation*
[56] Stamenov M I 1996 *Fractals of Brain, Fractals of Mind* eds Mac Cormac E and Stamenov M I (Amsterdam: John Benjamins) p 273
[57] Vitiello G 2004 *Int. J. Mod. Phys.* **B18** 785

Journal of Physics: Conference Series **380** (2012) 012022 doi:10.1088/1742-6596/380/1/012022

On the emergence of flavor mixing through interaction with an external vector field

Massimo Blasone[†1] **and Petr Jizba**[‡2]

[†] INFN and Università di Salerno, Via Ponte don Melillo, 84084 Fisciano (SA), Italy

[‡] FNSPE, Czech Technical University in Prague, Břehová 7, 115 19 Praha 1, Czech Republic

Abstract. We consider the possibility that two-flavor neutrino mixing can be viewed as resulting from the interaction of two non-mixed flavor neutrinos with an external vector field. Two distinct scenarios of the origin of such a vector field are presented. First we argue that the vector field might be understood as an $su(2)$ gauge field. In the second scenario we put forward the idea that the vector field can be identified with a Nambu–Goldstone vector field which results form a spontaneous symmetry breakdown of the Lorentz symmetry. Phenomenological implications of these pictures are briefly discussed.

1. Introduction

Particle mixing and neutrino oscillations in particular [1] are currently rapidly developing areas in both theoretical and experimental physics. Much of what is known about neutrino oscillations comes from the many experiments [2] about solar, atmospheric or reactor neutrinos. This has, in turn, produced a vast literature trying to work out possible extensions of the original Standard Model which does not accommodate non-zero neutrino masses and mixings. In spite of this, the true origin of the mixing is still rather elusive, although it is generally believed that it is the result of physics occurring at much higher energies with respect to the electroweak scale. An interesting alternative proposal was put forward recently in Ref. [3], considering mixing as arising from the low energy behavior of the interacting fermion fields.

Here we discuss another scenario which originates in the context of the quantum-field theoretical treatment of particle mixing presented in Refs. [4, 5]. There a nontrivial vacuum structure has been discovered in connection with mixing. In this approach, flavor states for mixed particles are consistently defined as eigenstates of the flavor charges [6, 7]. This leads to the derivation of exact oscillation formulas [8] which exhibit corrections with respect to the usual ones. One of the difficulties of this approach resides in the fact that the above flavor states are not representations of the Poincaré group [9]. This in turn leads to problems in formulating a perturbative expansion [10].

Interestingly enough, these difficulties can be bypassed by considering mixing not as an inherent property of the fields (or their associated particles), but as the result of an interaction of non-oscillatory flavor fields with an additional background vector field [5, 11]. In this paper we put forward two possible scenarios in which such a background field can find a natural basis.

[1] Corresponding author: blasone@sa.infn.it

[2] Corresponding author: p.jizba@fjfi.cvut.cz

First we argue that two-flavor neutrino oscillations can be equivalently phrased in terms of a minimally coupled $su(2)$-valued vector field which interacts with two non-oscillatory flavor fields. This reveals a non Abelian gauge-like structure in the mixing of fields. We show that in this way mixed particles can be considered to be on-shell just like ordinary particles. This treatment also removes an arbitrariness present in the formalism of quantum field mixing [12].

The second possibility is to identify the background field with a Nambu–Goldstone vector field which results form a spontaneous symmetry breakdown of the Lorentz symmetry down to its rotation subgroup. This is, in a sense, a natural choice in the light of both the time-like and fixed-norm nature of the involved vector field. Models with such type of vector fields are known as *aether σ-models* [13] and they constitute a typical framework in which one investigates violations of Lorentz invariance at low energies.

The structure of the paper is as follows. To set the stage we recall in the following section some fundamentals of neutrino mixing. Section 3 is devoted to reformulation of usual two-flavor mixing in terms of the $su(2)$-valued vector field which interacts via minimal coupling with on-shell like (i.e., non-oscillatory) flavor neutrinos. This leads to a picture in which flavor mixing is viewed as a non-abelian gauge theory. Peculiarities of the Lorentz and Poincaré symmetries that are related with this gauge theory are relegated to Section 4. Some of the ensuing phenomenological implications are discussed in Section 5. In Section 6 we promote the idea that the vector field is in fact the Nambu–Goldstone vector field. Since Nambu–Goldstone fields naturally contribute into *dark matter*, the neutrinos serve as a probe of such a dark matter. Various remarks and generalizations are proposed in the concluding section. For the reader's convenience the paper is supplemented with one appendix which clarify some finer technical details.

2. Neutrino mixing: essentials

Let us start with the Lagrange density describing two mixed neutrino fields. This can be written in the form

$$\begin{aligned} \mathcal{L} &= \bar{\nu}_e\,(i\,\partial\!\!\!/ - m_e)\,\nu_e + \bar{\nu}_\mu\,(i\,\partial\!\!\!/ - m_\mu)\,\nu_\mu \;-\; m_{e\mu}\,(\bar{\nu}_e\,\nu_\mu \,+\, \bar{\nu}_\mu\,\nu_e) \\ &= \bar{\nu}_f\,(i\,\partial\!\!\!/ - M)\,\nu_f\,. \end{aligned} \tag{1}$$

Here the flavor doublet is $\nu_f = (\nu_e, \nu_\mu)^T$ and the mass matrix is $M = \begin{pmatrix} m_e & m_{e\mu} \\ m_{e\mu} & m_\mu \end{pmatrix}$. The standard treatment of the problem is based on the observation that this Lagrangian, being quadratic, can be diagonalized by means of the *mixing transformation* as

$$\begin{aligned} \nu_e &= \nu_1 \cos\theta + \nu_2 \sin\theta\,, \\ \nu_\mu &= -\nu_1 \sin\theta + \nu_2 \cos\theta\,. \end{aligned} \tag{2}$$

The latter allows to rewrite (1) as the sum of two free Dirac Lagrangians:

$$\mathcal{L} \;=\; \bar{\nu}_m\,(i\,\partial\!\!\!/ - M_d)\,\nu_m\,. \tag{3}$$

Here $\nu_m = (\nu_1, \nu_2)^T$, $M_d = \mathrm{diag}(m_1, m_2)$, θ is the mixing angle and

$$\begin{aligned} m_e &= m_1 \cos^2\theta + m_2 \sin^2\theta\,, \\ m_\mu &= m_1 \sin^2\theta + m_2 \cos^2\theta\,, \\ m_{e\mu} &= (m_2 - m_1)\sin\theta\cos\theta\,. \end{aligned} \tag{4}$$

Solar neutrino experiments combined with KamLAND data allow to evaluated the mixing angle to be approximately $\theta = 33.9°$. From the Lagrangian (1) one can derive the canonical energy-momentum tensor:

$$\begin{aligned} T_{\rho\sigma} &= \frac{\partial \mathcal{L}}{\partial(\partial_\rho \nu_f)} \partial_\sigma \nu_f - \eta_{\rho\sigma}\mathcal{L} \\ &= \bar{\nu}_e i\gamma_\rho \partial_\sigma \nu_e - \eta_{\rho\sigma}\bar{\nu}_e(i\gamma^\lambda\partial_\lambda - m_e)\nu_e + \bar{\nu}_\mu i\gamma_\rho\partial_\sigma\nu_\mu \\ &\quad - \eta_{\rho\sigma}\bar{\nu}_\mu(i\gamma^\lambda\partial_\lambda - m_\mu)\nu_\mu + \eta_{\rho\sigma} m_{e\mu}(\bar{\nu}_e\nu_\mu + \bar{\nu}_\mu\nu_e) \\ &= \bar{\nu}_1 i\gamma_\rho\partial_\sigma\nu_1 - \eta_{\rho\sigma}\bar{\nu}_1(i\gamma^\lambda\partial_\lambda - m_1)\nu_1 + \bar{\nu}_2 i\gamma_\rho\partial_\sigma\nu_2 - \eta_{\rho\sigma}\bar{\nu}_2(i\gamma^\lambda\partial_\lambda - m_2)\nu_2\,, \end{aligned} \tag{5}$$

where $\eta_{\rho\sigma} = \text{diag}(1,-1,-1,-1)$ is the usual Minkowski metric tensor. From Eq. (5) follows that the total Hamiltonian $H = \int d^3\mathbf{x}\, T_{00}$ is just the sum of the two free field Hamiltonians, i.e., $H = H_1 + H_2$. The same situation holds also for the total 3-momentum operator $P_i = \int d^3\mathbf{x}\, T_{0i}$.

The conserved (Noether) charges associated with the global symmetry of the free fields ν_1 and ν_2 are:

$$Q_j = \int d^3\mathbf{x}\, \nu_j^\dagger(x)\, \nu_j(x)\,, \qquad j = 1,2\,, \tag{6}$$

with the total charge $Q = Q_1 + Q_2$. The analysis of symmetries of the Lagrangian in the flavor basis Eq. (1) leads to the identification of the (non conserved) flavor charges [6]:

$$Q_\sigma(x_0) = \int d^3\mathbf{x}\, \nu_\sigma^\dagger(x)\, \nu_\sigma(x)\,, \qquad \sigma = e, \mu\,, \tag{7}$$

with $Q_e(x_0) + Q_\mu(x_0) = Q$. Here it should be stressed that it is only the sum of the charges which is the genuine integral of the motion. The respective flavor charges in the sum are not.

3. Flavor mixing as a non-abelian gauge theory

In this section we show that the mixing interaction can be alternatively understood as the interaction of the *flavor* neutrino fields with a constant minimally coupled external vector field. The case of a dynamical vector field will be discussed in Section 6.

The most direct way to seeing this goes through the Euler–Lagrange equations corresponding to the Lagrangian (1). These can be compactly written as [11]

$$iD_0\nu_f = (-i\boldsymbol{\alpha}\cdot\boldsymbol{\nabla} + \beta\widetilde{M}_d)\nu_f\,, \tag{8}$$

where $\nu_f = (\nu_e, \nu_\mu)^T$ is the flavor doublet and $\widetilde{M}_d = \text{diag}(m_e, m_\mu)$ is a diagonal mass matrix. In Eq. (8) we have defined the (non-abelian) covariant derivative:

$$D_0 \equiv \partial_0 + i\, m_{e\mu}\, \beta\, \sigma_1\,, \tag{9}$$

where $m_{e\mu} = \frac{1}{2}\tan 2\theta\, \delta m$ and $\delta m = m_\mu - m_e$ is the mass differences parameter.

So flavor mixing can be seen as an interaction of the flavor fields with an $su(2)$ valued constant vector field having the following structure:

$$A_\mu \equiv \frac{1}{2}A_\mu^a\sigma_a = n_\mu \delta m\, \frac{\sigma_1}{2}, \qquad n^\mu = (1,0,0,0)^T\,. \tag{10}$$

The covariant derivative is then:

$$D_\mu = \partial_\mu + i\, g\, \beta\, A_\mu\,, \tag{11}$$

where $g = \tan 2\theta$ acts as the coupling constant. Note that in the case of maximal mixing ($\theta = \pi/4$), the coupling constant grows to infinity while δm goes to zero. We thus see some hints of an approximate non–abelian gauge symmetry hidden in neutrino mixing. For this reason we will often refer to the vector field just introduced as a gauge field. Since the vector field is a constant, with just one non-zero component in group space, its field strength vanishes identically:

$$F^a_{\mu\nu} \;=\; \epsilon^{abc} A^b_\mu A^c_\nu \;=\; 0\,, \tag{12}$$

with $a, b, c = 1, 2, 3$. The fact that, despite $F_{\mu\nu}$ vanishes identically, the gauge field has physical effects, suggests an analogy with the Bohm–Aharonov effect [14].

The equations of motion for the mixed fields in a covariant form are:

$$(i\gamma^\mu D_\mu - \widetilde{M}_d)\nu_f \;=\; 0\,, \tag{13}$$

and the Lagrangian density (1) has the form of the one describing a doublet of Dirac fields in interaction with an external gauge field:

$$\mathcal{L} \;=\; \bar{\nu}_f(i\gamma^\mu D_\mu - \widetilde{M}_d)\nu_f\,. \tag{14}$$

The energy momentum tensor associated with the flavor neutrino fields in interaction with the external gauge field is given by [11]:

$$\widetilde{T}_{\rho\sigma} = \bar{\nu}_f i\gamma_\rho D_\sigma \nu_f - \eta_{\rho\sigma}\bar{\nu}_f(i\gamma^\lambda D_\lambda - \widetilde{M}_d)\nu_f\,. \tag{15}$$

By comparing with the canonical energy momentum tensor given in Eq. (5) we see that the difference between the two is just the presence of the interaction terms in the "00" component, i.e., $T_{00} - \widetilde{T}_{00} = m_{e\mu}(\bar{\nu}_e\nu_\mu + \bar{\nu}_\mu\nu_e)$, while we have $T_{0i} = \widetilde{T}_{0i}$, $T_{ij} = \widetilde{T}_{ij}$.

Following the usual procedure, we now define a 4-momentum operator $\widetilde{P}^\mu$ from $\widetilde{T}^{\mu\nu}$. We obtain a conserved 3–momentum operator:

$$\widetilde{P}^i \;=\; \int d^3\mathbf{x}\, \widetilde{T}^{0i} \;=\; i\int d^3\mathbf{x}\, \nu_f^\dagger \partial^i \nu_f \;\equiv\; \widetilde{P}^i_e(x_0) + \widetilde{P}^i_\mu(x_0), \qquad i \;=\; 1,2,3\,, \tag{16}$$

and a non conserved Hamiltonian operator:

$$\begin{aligned}\widetilde{P}^0(x_0) \;\equiv\; \widetilde{H}(x_0) \;&=\; \int d^3\mathbf{x}\, \widetilde{T}^{00} \;=\; \int d^3\mathbf{x}\, \bar{\nu}_f\left(i\gamma_0 D_0 - i\gamma^\mu D_\mu + \widetilde{M}_d\right)\nu_f \\ &\equiv\; \widetilde{H}_e(x_0) + \widetilde{H}_\mu(x_0)\,.\end{aligned} \tag{17}$$

We remark that both the Hamiltonian and the momentum operators split into two contributions, one involving only the electron neutrino field and the other one only the muon neutrino field. We have thus the Hamiltonian and the momentum operators for each flavor field. Notice that the tilde Hamiltonian is *not* the generator of time translations. This role competes to the complete Hamiltonian $H = \int d^3\mathbf{x}\, T^{00}$, which includes the interaction term.

4. Flavor neutrino states and Lorentz invariance

We are now in the position to construct flavor neutrino states which are simultaneous eigenstates of the above 4–momentum operators and of the flavor charges. Of course, this is a highly nontrivial task. We will see, however, that such states can indeed be constructed, but it will involve a nontrivial redefinition of the flavor vacuum which will also erase any reference to the ν_1 and ν_2 fields.

As shown in Appendix, the flavor neutrino field operators can be expanded in the same basis as the free fields with masses m_1 and m_2. With the convention $(\sigma, j) = (\{e,1\},\{\mu,2\})$ we can write

$$\nu_\sigma(x) = \int \frac{d^3\mathbf{k}}{(2\pi)^{3/2}} \sum_r \left[u^r_{\mathbf{k},j}(x_0)\alpha^r_{\mathbf{k},\sigma}(x_0) + v^r_{-\mathbf{k},j}(x_0)\beta^{r\dagger}_{-\mathbf{k},\sigma}(x_0)\right] e^{i\mathbf{k}\cdot\mathbf{x}}, \tag{18}$$

where $\alpha_{\mathbf{k},\sigma}$ and $\beta_{-\mathbf{k},\sigma}$ are the flavor ladder operators [4]. In the same Appendix we show that flavor neutrino states, defined as $|\nu^r_{\mathbf{k},\sigma}\rangle = \alpha^{r\dagger}_{\mathbf{k},\sigma}|0\rangle_{e,\mu}$, are eigenstates of the flavor charge operators Q_σ, at a given time. They turn out also to be eigenstates of the momentum operators $P^i = \int d^3\mathbf{x}\, T^{0i}$. However, since the Hamiltonian operator H does not commute with the charges Q_σ, the above flavor states do not have definite energies.

We will now show that this problem can be solved by noting that the expansion (18) actually relies on a special choice of the bases of spinors, namely those referring to the free field masses m_1, m_2. It is however always possible to perform a Bogoliubov transformation in order to expand the field operators in a different basis of spinors, referring to an arbitrarily chosen couple of mass parameters [12]. Let μ_σ, $\sigma = e, \mu$ be such a couple of arbitrary parameters.

The Bogoliubov transformation to be performed is the following:

$$\begin{pmatrix} \widetilde{\alpha}^r_{\mathbf{k},\sigma}(x_0) \\ \widetilde{\beta}^{r\dagger}_{-\mathbf{k},\sigma}(x_0)\end{pmatrix} = J^{-1}_\mu(x_0)\begin{pmatrix} \alpha^r_{\mathbf{k},\sigma}(x_0) \\ \beta^{r\dagger}_{-\mathbf{k},\sigma}(x_0)\end{pmatrix} J_\mu(x_0), \tag{19}$$

whose generator is [12]:

$$J_\mu(x_0) = \prod_{\mathbf{k},r} \exp\left\{ i \sum_{(\sigma,j)} \xi^{\mathbf{k}}_{\sigma,j}\left[\alpha^{r\dagger}_{\mathbf{k},\sigma}(x_0)\beta^{r\dagger}_{-\mathbf{k},\sigma}(x_0) + \beta^r_{-\mathbf{k},\sigma}(x_0)\alpha^r_{\mathbf{k},\sigma}(x_0)\right]\right\}, \tag{20}$$

and $(\sigma, j) = (\{e,1\},\{\mu,2\})$, $\xi^{\mathbf{k}}_{\sigma,j} = (\chi_\sigma - \chi_j)/2$ and $\chi_\sigma = \arctan(\mu_\sigma/|\mathbf{k}|)$, $\chi_j = \arctan(m_j/|\mathbf{k}|)$. We have thus a whole family of flavor vacua, denoted with a tilde and parameterized by the couples (μ_e, μ_μ):

$$|\widetilde{0}(x_0)\rangle_{e\mu} = J^{-1}_\mu(x_0)|0(x_0)\rangle_{e\mu}. \tag{21}$$

The original flavor vacuum is of course the one associated with the couple (m_1, m_2).

Notice that the flavor charges, as well as the exact oscillation formulae are invariant under the above Bogoliubov transformations [15], i.e., $\widetilde{Q}_\sigma = Q_\sigma$, with:

$$\widetilde{Q}_\sigma(x_0) = \sum_r \int d^3\mathbf{k}\left(\widetilde{\alpha}^{r\dagger}_{\mathbf{k}\sigma}(x_0)\widetilde{\alpha}^r_{\mathbf{k}\sigma}(x_0) - \widetilde{\beta}^{r\dagger}_{-\mathbf{k}\sigma}(x_0)\widetilde{\beta}^r_{-\mathbf{k}\sigma}(x_0)\right). \tag{22}$$

In the context of the above reformulation of mixing as the interaction with an external field, it seems natural to expand the flavor fields in the bases corresponding to the couple of masses (m_e, m_μ). We will see that with this choice the flavor states will be eigenstates of the Hamiltonian operator.

The new spinors are defined as the solutions of the equations:

$$\begin{aligned}(-\alpha\cdot\mathbf{k} + m_\sigma\beta)u^r_{\mathbf{k},\sigma} &= \omega_{\mathbf{k},\sigma}u^r_{\mathbf{k},\sigma}, \\ (-\alpha\cdot\mathbf{k} + m_\sigma\beta)v^r_{-\mathbf{k},\sigma} &= -\omega_{\mathbf{k},\sigma}v^r_{-\mathbf{k},\sigma},\end{aligned} \tag{23}$$

where $\omega_{\mathbf{k},\sigma} = \sqrt{\mathbf{k}^2 + m_\sigma^2}$. These are the momentum-space version of the *free* Dirac equation with mass m_σ. The flavor field operators are then expanded as follows:

$$\nu_\sigma(x) = \int \frac{d^3\mathbf{k}}{(2\pi)^{3/2}} \sum_r \left[u^r_{\mathbf{k},\sigma}(x_0)\tilde{\alpha}^r_{\mathbf{k},\sigma}(x_0) + v^r_{-\mathbf{k},\sigma}(x_0)\tilde{\beta}^{r\dagger}_{-\mathbf{k},\sigma}(x_0)\right] e^{i\mathbf{k}\cdot\mathbf{x}}, \qquad \sigma = e, \mu, \tag{24}$$

with

$$u^r_{\mathbf{k},\sigma}(x_0) = u^r_{\mathbf{k},\sigma} e^{-i\omega_{\mathbf{k},\sigma} x_0} \quad \text{and} \quad v^r_{-\mathbf{k},\sigma}(x_0) = v^r_{-\mathbf{k},\sigma} e^{i\omega_{\mathbf{k},\sigma} x_0}. \tag{25}$$

Here and throughout the tilde operators are those corresponding to the specific couple (m_e, m_μ). With these definitions all the calculations at a fixed instant of time x_0 can be performed in exactly the same way they are done in the free field case. The explicit time dependence of the creation and destruction operators is due to the interaction with the external field and it does not create problems as the states which are acted upon by the operators are evaluated at the same time as the operators themselves and the commutators are all considered at equal times.

In terms of the tilde flavor ladder operators, the Hamiltonian and momentum operators Eqs. (16), (17) read:

$$\widetilde{\mathbf{P}}_\sigma(x_0) = \sum_r \int d^3\mathbf{k}\, \mathbf{k} \left(\tilde{\alpha}^{r\dagger}_{\mathbf{k},\sigma}(x_0)\tilde{\alpha}^r_{\mathbf{k},\sigma}(x_0) + \tilde{\beta}^{r\dagger}_{\mathbf{k},\sigma}(x_0)\tilde{\beta}^r_{\mathbf{k},\sigma}(x_0)\right), \tag{26}$$

$$\widetilde{H}_\sigma(x_0) = \sum_r \int d^3\mathbf{k}\, \omega_{\mathbf{k},\sigma} \left(\tilde{\alpha}^{r\dagger}_{\mathbf{k},\sigma}(x_0)\, \tilde{\alpha}^r_{\mathbf{k},\sigma}(x_0) - \tilde{\beta}^r_{\mathbf{k},\sigma}(x_0)\, \tilde{\beta}^{r\dagger}_{\mathbf{k},\sigma}(x_0)\right). \tag{27}$$

The new flavor states are defined by the action of the tilde creation operator on the tilde flavor vacuum:

$$|\tilde{\nu}^r_{\mathbf{k},\sigma}(x_0)\rangle = \tilde{\alpha}^{r\dagger}_{\mathbf{k},\sigma}(x_0)|\tilde{0}(x_0)\rangle_{e\mu}. \tag{28}$$

We easily find the result that these single particle states are eigenstates of both the Hamiltonian and the momentum operator:

$$\begin{pmatrix} \widetilde{H}_\sigma(x_0) \\ \widetilde{\mathbf{P}}_\sigma(x_0) \end{pmatrix} |\tilde{\nu}^r_{\mathbf{k},\sigma}(x_0)\rangle = \begin{pmatrix} \omega_{\mathbf{k},\sigma} \\ \mathbf{k} \end{pmatrix} |\tilde{\nu}^r_{\mathbf{k},\sigma}(x_0)\rangle, \tag{29}$$

making explicit the 4−vector structure.

The flavor charges commute with the tilde Hamiltonian operator: $[\widetilde{Q}_\sigma(x_0), \widetilde{H}(x_0)] = 0$, as a consequence of $[\widetilde{Q}_\sigma(x_0), \widetilde{H}_{\sigma'}(x_0)] = 0$, $\sigma, \sigma' = e, \mu$, and of the absence of an interaction term in $\widetilde{H}$. This ensures the existence of a common set of eigenstates of these operators. In this set we find the flavor states (28), which are straightforwardly seen to be also eigenstates of the flavor charges:

$$\widetilde{Q}_\sigma(x_0)|\tilde{\nu}^r_{\mathbf{k},\sigma}(x_0)\rangle = |\tilde{\nu}^r_{\mathbf{k},\sigma}(x_0)\rangle. \tag{30}$$

Let us now consider the algebra of the generators descending from the energy-momentum tensor (15). All the generators are defined in the usual way. Besides the translation generators defined by Eqs. (16) and (17), we have the Lorentz generators, defined as:

$$\widetilde{M}^{\lambda\rho}(x_0) = \int d^3\mathbf{x} \left(\widetilde{T}^{0\rho}x^\lambda - \widetilde{T}^{0\lambda}x^\rho\right) + \frac{1}{2}\int d^3\mathbf{x}\, \nu_f^\dagger \sigma^{\lambda\rho}\nu_f = \widetilde{M}_e^{\lambda\rho}(x_0) + \widetilde{M}_\mu^{\lambda\rho}(x_0), \tag{31}$$

where $\sigma^{\mu\nu} = -\frac{i}{2}[\gamma^\mu, \gamma^\nu]$. The algebra of (equal-time) commutators of these generators will be just the direct sum of two Poincaré algebras (we omit the specification of the instant of time):

$$[\widetilde{P}^\mu_\sigma, \widetilde{P}^\nu_{\sigma'}] = 0, \qquad [\widetilde{M}^{\mu\nu}_\sigma, \widetilde{P}^\lambda_{\sigma'}] = i\delta_{\sigma\sigma'}\left(\eta^{\mu\lambda}\widetilde{P}^\nu_\sigma - \eta^{\nu\lambda}\widetilde{P}^\mu_\sigma\right), \qquad \sigma, \sigma' = e, \mu,$$
$$[\widetilde{M}^{\mu\nu}_\sigma, \widetilde{M}^{\lambda\rho}_{\sigma'}] = \imath\delta_{\sigma\sigma'}\left(\eta^{\mu\lambda}\widetilde{M}^{\nu\rho}_\sigma - \eta^{\nu\lambda}\widetilde{M}^{\mu\rho}_\sigma - \eta^{\mu\rho}\widetilde{M}^{\nu\lambda}_\sigma + \eta^{\nu\rho}\widetilde{M}^{\mu\lambda}_\sigma\right). \tag{32}$$

Note that the above construction and the consequent Poincaré invariance, holds *at a given time* x_0. Thus, for each different time, we have a different Poincaré structure. The relation of neutrino mixing and oscillations with Lorentz as well as CPT violation has been the focus of much work in the last decade, see for example [17]. Connection of neutrino mixing with the breakdown of Lorentz symmetry will be further discussed in Section 6.

5. Phenomenological implications

The above results reflect into a number of phenomenological consequences which could as well serve for falsifying the proposed scenario. In Ref. [16] some of them have been discussed: in particular, it has been pointed out that very precise direct measurements of neutrino masses, as those done by looking at tritium decay, could reveal if the mass eigenstates ν_i enter in fundamental interactions, or rather the flavor eigenstates are the basic objects, as here is proposed. In the latter case, the measured masses should be the average masses m_e and m_μ above introduced.

Another test can be done by looking at possible violations of lepton charge conservation at tree level. This effect is by definition absent when considering neutrinos as flavor charge eigenstates as we have done above. However, the usual Pontecorvo states:

$$|\nu^r_{\mathbf{k},e}\rangle_P = \cos\theta\, |\nu^r_{\mathbf{k},1}\rangle + \sin\theta\, |\nu^r_{\mathbf{k},2}\rangle \tag{33}$$
$$|\nu^r_{\mathbf{k},\mu}\rangle_P = -\sin\theta\, |\nu^r_{\mathbf{k},1}\rangle + \cos\theta\, |\nu^r_{\mathbf{k},2}\rangle\,, \tag{34}$$

are *not* eigenstates of the flavor charges, as can be easily checked.

Thus, if we consider the expectation values of the above defined flavor charges onto the Pontecorvo states, we see how much the lepton charge is violated (at tree level) by the usual quantum mechanical states. One finds [10]:

$${}_P\langle\nu^r_{\mathbf{k},e}|\widetilde{Q}_e(x_0)|\nu^r_{\mathbf{k},e}\rangle_P = \cos^4\theta + \sin^4\theta + 2|U_{\mathbf{k}}|\sin^2\theta\cos^2\theta + \sum_r \int d^3\mathbf{k}, \tag{35}$$

and

$${}_{1,2}\langle 0|\widetilde{Q}_e(x_0)|0\rangle_{1,2} = \sum_r \int d^3\mathbf{k}, \tag{36}$$

The diverging terms can be removed by normal ordering the flavor charges with respect to the mass vacuum $|0\rangle_{1,2}$. One has then

$${}_P\langle\nu^r_{\mathbf{k},e}| : \widetilde{Q}_e(x_0) : |\nu^r_{\mathbf{k},e}\rangle_P = \cos^4\theta + \sin^4\theta + 2|U_{\mathbf{k}}|\sin^2\theta\cos^2\theta < 1, \tag{37}$$

$\forall\theta \neq 0$, $m_1 \neq m_2$, $\mathbf{k} \neq 0$, and

$${}_{1,2}\langle 0| : \widetilde{Q}_e(x_0) : |0\rangle_{1,2} = 0, \tag{38}$$

i.e. a violation of lepton charge of the order of $(1-|U_{\mathbf{k}}|)\sin^2(2\theta)$ is present in the Pontecorvo flavor neutrino states.

We also note that the corresponding quantum fluctuations are divergent:

$$_{1,2}\langle 0|(:\widetilde{Q}_e(x_0):)^2|0\rangle_{1,2} = 4\sin^2\theta\cos^2\theta\int d^3\mathbf{k}|V_{\mathbf{k}}|^2, \tag{39}$$

$$_P\langle\nu^r_{\mathbf{k},e}|(:\widetilde{Q}_e(x_0):)^2|\nu^r_{\mathbf{k},e}\rangle_P = \cos^6\theta+\sin^6\theta+\sin^2\theta\cos^2\theta\left[2|U_{\mathbf{k}}|+|U_{\mathbf{k}}|^2+4\int d^3\mathbf{k}|V_{\mathbf{k}}|^2\right]. \tag{40}$$

6. Flavor mixing and Spontaneous Symmetry Breakdown of the Lorentz symmetry

In this section we provide yet another interpretation of the static vector field A^μ. We have seen, the structure of A^μ explicitly violates the Lorentz symmetry. It is well known that in a nongravitational setting, it suffices to specify fixed background fields violating Lorentz symmetry in order to formulate the Lorentz violating (LV) matter dynamics. However, fixed background fields break general covariance. If one wishes to preserve the successes of general relativity in accounting for gravitational phenomena, breaking general covariance is not an option. The obvious alternative is to promote the LV background fields to dynamical fields, governed by a generally covariant action. By allowing the vector field to vary over space-time, the mixing and ensuing flavor oscillations will not be necessarily uniform over space-time. There are several candidates for the covariant kinetic term. At this stage we realize (taking into account Eq. (10)) that in the general frame $n^\mu n_\mu = 1$, $n_0 > 0$. So the dynamics of A^μ takes place on the Lobachevsky 4D target space. This, in fact, corresponds to a quotient $\mathcal{M} = SO(3,1)/SO(3)$ which effectively describes the target space of soft modes (Nambu–Goldstone modes) after the Lorentz symmetry is broken to the rotational symmetry. The latter target-space constraint might be introduced as a constraint in the path-integral measure, such that

$$\begin{aligned}\mathcal{D}n^\mu\delta[n^\mu n_\mu - 1] &= \mathcal{D}\lambda\mathcal{D}n^\mu \exp\left(i\int d^4x\,\lambda(x)(n^\mu(x)n_\mu(x)-1)\right)\\ &= \mathcal{D}\lambda\mathcal{D}n^\mu \exp\left(i\int d^4x\,\mathcal{L}_{\text{constr.}}\right).\end{aligned} \tag{41}$$

Massless field theories where the target space is the group coset space are commonly known as non-linear σ models. Nonlinear σ models typically describe the dynamics of Nambu–Goldstone bosons. The sigma model in question is sometimes called the *space-time σ-model* [19] or *Einstein aether σ-model* [13]. It is well known, that with a suitable choice of the interaction Hamiltonian the generalized coherent state functional integrals describe low-energy effective field theories — Nambu–Goldstone fields, including their mutual interactions [5, 18].

At this stage it should be reminded that in contrast to the spontaneous symmetry breakdown (SSB) of an internal symmetry which produces Nambu–Goldstone scalar fields, the corresponding SSB of Lorentz symmetry yields Nambu–Goldstone vector fields [20]. Thus here we do not regard the above field as a $SU(2)$ gauge field, as done in the previous section, but simply as a massless vector field.

Let us now consider the corresponding $SO(3,1)$ coherent-state functional. This represents a natural playground for discussing the quantum dynamics of Nambu–Goldstones [5, 18]. This can be done in terms of the unit-vector parameters n_μ. The kinetic term for the gauge field (now dynamically epitomized via n^μ) is a Wess–Zumino–Witten term in $3+1$ dimensions. This term can be written in the form (see, e.g., Ref. [13])

$$\mathcal{L}_{\text{WZW}} = \frac{1}{2}(\nabla_\mu n_\nu)(\nabla^\mu n^\nu)\,. \tag{42}$$

Here the gradient ∇_μ should not be mistaken with a covariant derivative. The ensuing total Lagrangian has thus the form

$$\mathcal{L}_{\text{tot}} = \mathcal{L}_{\text{f}} + \mathcal{L}_{\text{WZW}} + \mathcal{L}_{\text{constr.}}, \tag{43}$$

where $\mathcal{L}_{\text{f}}$ is the field Lagrangian (14). It can be shown [13, 21] that this version of the aether σ-theory is stable under small perturbations and the Hamiltonian is globally bounded.

So finally we see that the original Lagrangian (1) for oscillating neutrinos (where mass and flavor states are simultaneously not well defined) can be equally formulated as the Lagrangian for definite mass and definite flavor neutrinos interacting with the Nambu–Goldstone vector field. The natural way ho to proceed with a quantization of this model would be to consider the corresponding $SO(3,1)$ coherent-state functional. Here we shall refrain from doing this. More detailed discussion of this issue will be published elsewhere.

7. Conclusions

In the framework of the quantum field theoretical formalism for flavor mixing, we have discussed a vector-field-like structure associated to two-flavor neutrino mixing. In this connection two natural strategies could be identified. In the first case, flavor neutrino fields are assumed to be primitive entities describing on-shell particles with definite masses m_e and m_μ, which are different from those of the mass eigenstates m_1 and m_2 of the standard approach. Flavor oscillations then arise as a consequence of the interaction with the external $su(2)$ gauge field, which acts as a sort of refractive medium. An interesting consequence of such an analysis is that we recover, locally in time, a Poincaré structure for the flavor neutrino states. Several phenomenological implications can be worked out from such a picture [16].

In the second case, we put forward a proposal in which the vector field is a Nambu–Goldstone vector field which naturally emerges when the Lorentz symmetry $SO(3,1)$ is broken down to the rotation subgroup $SO(3)$. To furnish a dynamics to such a vector field one can take advantage of the fact that Nambu–Goldstone fields are described via non-linear σ-models. The non-linear σ-models have a naturally induced kinetic term — a Wess–Zumino–Witten term. In the present case the σ-model is also known as aether σ-model and the Hamiltonian is globally bounded when only two-derivative terms are included in the action. We have also presented the ensuing Lagrangian which can serve as a starting point for a path-integral quantization. Further work in this direction is in progress.

It would be definitely important to generalize our two approaches also to bosonic systems, such as $K^0 - \overline{K^0}$, $D^0 - \overline{D^0}$ or $B^0 - \overline{B^0}$ flavor oscillations. The extension to 3-flavor mixing is still an open issue, though preliminary results are already available [22].

Finally, we comment on the recent data from the OPERA collaboration [23], exhibiting an apparent superluminal behaviour of neutrinos. In Ref. [24] it has been shown that such a feature can be accomodated in the framework of flavor neutrino states satisfying modified dispersion relations as discussed in Refs. [9], provided at least one of the mass eigenstates being tachyonic. A nice feature of this approach is that one does not encounter the usual runaway pathologies of tachyons. In the present approach, it is not clear yet if superluminal neutrino propagation can arise. Further study in this direction is in progress.

Acknowledgments

We thank H. Kleinert, M. Di Mauro and G. Vitiello for fruitful discussions. M.B. acknowledges partial financial support from MIUR and INFN. P.J. is supported by the Ministry of Education of the Czech Republic under the Grant No. CFRJS 1507001.

Appendix A. Mixing of quantum neutrino fields

In this Appendix we briefly recall some fundamentals of the quantization for mixed Fermi fields, as given in Refs. [4, 6, 8]. We start from the free fields ν_1 and ν_2, whose Fourier expansions are

$$\nu_j(x) = \int \frac{d^3\mathbf{k}}{(2\pi)^{3/2}} \sum_r \left[u^r_{\mathbf{k},j}(x_0)\alpha^r_{\mathbf{k},j} + v^r_{-\mathbf{k},j}(x_0)\beta^{r\dagger}_{-\mathbf{k},j}\right] e^{i\mathbf{k}\cdot\mathbf{x}}, \qquad j = 1,2, \tag{A.1}$$

with

$$u^r_{\mathbf{k},j}(x_0) = u^r_{\mathbf{k},j}e^{-i\omega_{\mathbf{k},j}x_0}, \quad v^r_{-\mathbf{k},j}(x_0) = v^r_{-\mathbf{k},j}e^{i\omega_{\mathbf{k},j}x_0}, \tag{A.2}$$

and $\omega_{\mathbf{k},j} = \sqrt{\mathbf{k}^2 + m_j^2}$. The operators $\alpha^r_{\mathbf{k},j}$ and $\beta^r_{-\mathbf{k},j}$, with $j = 1,2$ and $r = 1,2$ that appear in the mode expansion (A.1) are the annihilation operators for the vacuum state $|0\rangle_{1,2} = |0\rangle_1 \otimes |0\rangle_2$, i.e.,

$$\alpha^r_{\mathbf{k},j}|0\rangle_{1,2} = 0, \qquad \beta^r_{-\mathbf{k},j}|0\rangle_{1,2} = 0. \tag{A.3}$$

The canonical anticommutation rules assume the form

$$\{\nu_i^\alpha(x), \nu_j^{\beta\dagger}(y)\}_{x_0=y_0} = \delta^3(\mathbf{x}-\mathbf{y})\delta_{\alpha\beta}\delta_{ij}, \qquad \alpha,\beta = 1,\ldots,4. \tag{A.4}$$

In terms of the creation and annihilation operators these read as

$$\{\alpha^r_{\mathbf{k},i}, \alpha^{s\dagger}_{\mathbf{q},j}\} = \delta_{\mathbf{kq}}\delta_{rs}\delta_{ij}, \quad \{\beta^r_{\mathbf{k},i}, \beta^{s\dagger}_{\mathbf{q},j}\} = \delta_{\mathbf{kq}}\delta_{rs}\delta_{ij}, \qquad i,j = 1,2, \tag{A.5}$$

with all other anticommutators being zero. The corresponding orthonormality and completeness relations are

$$\begin{aligned} u^{r\dagger}_{\mathbf{k},j}u^s_{\mathbf{k},j} &= v^{r\dagger}_{\mathbf{k},j}v^s_{\mathbf{k},j} = \delta_{rs}, \quad u^{r\dagger}_{\mathbf{k},j}v^s_{-\mathbf{k},j} = v^{r\dagger}_{-\mathbf{k},j}u^s_{\mathbf{k},j} = 0, \\ \sum_r (u^r_{\mathbf{k},j}u^{r\dagger}_{\mathbf{k},j} &+ v^r_{-\mathbf{k},j}v^{r\dagger}_{-\mathbf{k},j}) = 1. \end{aligned} \tag{A.6}$$

We construct the generator for the mixing transformations (2) as:

$$\begin{aligned} \nu^\alpha_\sigma(x) &= G^{-1}_\theta(x_0)\nu^\alpha_j G_\theta(x_0), \qquad (\sigma, j) = (\{e,1\},\{\mu,2\}), \\ G_\theta(x_0) &= \exp\left[\theta \int d^3\mathbf{x} \left(\nu_1^\dagger(x)\nu_2(x) - \nu_2^\dagger(x)\nu_1(x)\right)\right]. \end{aligned} \tag{A.7}$$

At finite volume G_θ is a unitary operator: $G^{-1}_\theta(x_0) = G_{-\theta}(x_0) = G^\dagger_\theta(x_0)$, preserving the canonical anticommutation relations. $G^{-1}_\theta(x_0)$ maps the Hilbert space for the free fields $\mathcal{H}_{1,2}$ to the Hilbert space for the mixed fields $\mathcal{H}_{e,\mu}$: $G^{-1}_\theta(x_0) : \mathcal{H}_{1,2} \to \mathcal{H}_{e,\mu}$. In particular, the vacuum $|0\rangle_{1,2}$ is mapped into a new state:

$$|0(x_0)\rangle_{e,\mu} = G^{-1}_\theta(x_0)\, |0\rangle_{1,2}, \tag{A.8}$$

which is refereed to as the *flavor vacuum*. In the limit $V \to \infty$, the flavor vacuum becomes orthogonal to the vacuum of the free fields, which means that the two Hilbert spaces are *unitarily inequivalent*. From (A.8) we can deduce that the annihilating operators for the flavor vacuum are

$$\alpha^r_{\mathbf{k},\sigma}(x_0) \equiv G^{-1}_\theta(x_0)\, \alpha^r_{\mathbf{k},j}\, G_\theta(x_0), \qquad (\sigma, j) = (\{e,1\},\{\mu,2\}). \tag{A.9}$$

The Hermitian conjugation of the former yields the corresponding relation for antiparticles. Even so, the representation (A.7) (or A.9) of the canonical transformation loses its meaning for an (infinite) field systems, in as much as the exponential operator occurring in it has no domain on the representation space involved. This fact, however, has no direct bearing on our approach which uses the well-defined form (2). It just states that the operator-algebra representation which is used in the particle mixing is unitarily inequivalent to the Fock-space representation. In other words, the flavor vacuum $|0(x_0)\rangle_{e,\mu}$ and the ordinary vacuum $|0\rangle_{1,2}$ do not belong to the same Hilbert space.

The flavor fields can be expanded in the same bases as the fields ν_i:

$$\nu_\sigma(x) = \int \frac{d^3\mathbf{k}}{(2\pi)^{3/2}} \sum_r \left[u^r_{\mathbf{k},j}(x_0)\alpha^r_{\mathbf{k},\sigma}(x_0) + v^r_{-\mathbf{k},j}(x_0)\beta^{r\dagger}_{-\mathbf{k},\sigma}(x_0) \right] e^{i\mathbf{k}\cdot\mathbf{x}}, \qquad (A.10)$$

where $(\sigma, j) = (\{e,1\}, \{\mu, 2\})$. With the help of Eq. (7) it can be seen [6] that the flavor charges have the following expansion

$$Q_\sigma(x_0) \;=\; \sum_r \int d^3\mathbf{k} \left(\alpha^{r\dagger}_{\mathbf{k},\sigma}(x_0)\alpha^r_{\mathbf{k},\sigma}(x_0) - \beta^{r\dagger}_{-\mathbf{k},\sigma}(x_0)\beta^r_{-\mathbf{k},\sigma}(x_0) \right). \qquad (A.11)$$

The flavor neutrino states are then defined as eigenstates of the flavor charges. In particular, we can write that $|\nu^{\mathbf{k}}_\sigma(x_0)\rangle = \alpha^{r\dagger}_{\mathbf{k},\sigma}(x_0)|0(x_0)\rangle_{e,\mu}$, and similar ones for antiparticles.

Note that the flavor charges do not commute with the Hamiltonian of the system

$$[H, Q_\sigma(x_0)] \neq 0, \qquad \sigma = e, \mu \qquad (A.12)$$

with the consequence that they are not conserved by time evolution. This is of course the hallmark of the flavor oscillation phenomenon. The corresponding flavor oscillation formulas are derived by computing the expectation value of the flavor charge operators in the flavor state [8].

References

[1] Bilenky S M and Pontecorvo B 1978 *Phys. Rept.* **41** 225
[2] Smy M B (Super-Kamiokande) 2003 *Nucl. Phys. Proc. Suppl.* **118** 25 (2003)
Hallin A L et al. 2003 *Nucl. Phys. Proc. Suppl.* **118** 3
Eguchi K et al. (KamLAND) 2003 *Phys. Rev. Lett.* **90** 021802
Hampel W et al. (GALLEX) 1999 *Phys. Lett.* B **447** 127
Altmann M et al. (GNO) 2000 *Phys. Lett.* B **490** 16
Abdurashitov J N et al. (SAGE) 2002 *J. Exp. Theor. Phys.* **95** 181
[3] Terranova F 2011 Int. J. Mod. Phys. A **26**, 4739
[4] Blasone M and Vitiello G 1995 *Annals Phys.* **244** 283
[5] Blasone M, Jizba P and Vitiello G 2010 *Quantum Field Theory and its Macroscopic Manifestations* (London: World Scientific & ICP)
[6] Blasone M, Jizba P and Vitiello G 2001 *Phys. Lett.* B **517** 471
[7] Blasone M, Capolupo A, Ji C R and Vitiello G 2010 *Int. J. Mod. Phys.* A **25** 4179
[8] Blasone M, Henning P A and Vitiello G 1999 *Phys. Lett.* B **451** 140
[9] Blasone M, Magueijo J and Pires-Pacheco P 2005 *Europhys. Lett.* **70** 600
Blasone M, Magueijo J and Pires-Pacheco P 2005 *Braz. J. Phys.* **35** 447
[10] Blasone M, Capolupo A, Terranova F and Vitiello G 2005 Phys. Rev. D **72** 013003.
[11] Blasone M, Di Mauro M and Vitiello G 2010 (*Preprint* hep-ph/1003.5812)
[12] Fujii K, Habe C and Yabuki T 2001 *Phys. Rev.* D **59** 113003
Fujii K, Habe C and Yabuki T 2001 *Phys. Rev.* D **59** 013011
[13] Carroll S M, Dulaney T R, Gresham M I, and Tamx H 2009 *Phys. Rev.* D **79** 065012
[14] Aharonov Y and Bohm D 1959 *Phys. Rev.* **115** 48
[15] Blasone M and Vitiello G 1999 *Phys. Rev.* D **60** 111302
[16] Blasone M 2011 J. Phys. Conf. Ser. **306** 012037

[17] Pakvasa S 2001 (*Preprint* hep-ph/0110175)
Lambiase G 2003 *Phys. Lett.* B **560** 1
Kostelecky V A and Mewes M 2004 *Phys. Rev.* D **69** (2004) 016005
Kostelecky V A and Mewes M 2004 (ibid.) *Phys. Rev.* D **70** 031902
Klinkhamer F R 2006 *Int. J. Mod. Phys.* A **21** 161
Di Grezia E, Esposito S and Salesi G 2006 *Mod. Phys. Lett.* A **21** 349
Hooper D, Morgan D and Winstanley E 2005 *Phys. Rev.* D **72** 065009
Ellis J R, Harries N, Meregaglia A, Rubbia A and Sakharov A 2008 *Phys. Rev.* D **78** 033013
Diaz J S, Kostelecky V A and Mewes M 2009 *Phys. Rev.* D **80** 076007
[18] Blasone M and Jizba P 2011 (*Preprint* hep-th/1111.3228)
[19] Froggatt C D and Nielsen H B 1991*Origin of Symmetries* (Singapore: World Scientific)
[20] Chkareuli J L, Froggatt C D and Nielsen H B 2001 *Phys. Rev. Lett.* **87** 091601
Chkareuli J L, Froggatt C D and Nielsen H B 2001 *Nucl. Phys.* B **609**46
[21] Kostelecky V A and Lehnert R 2001 *Phys. Rev.* D **63** 065008
Elliott J W, Moore G D and Stoica H 2005 *J. High Energy Phys.* **08** 066
Dulaney T R, Gresham M I and Wise M B 2008 *Phys. Rev.* D **77** 083510
[22] Blasone M, Terranova F and Vitiello G, work in progress.
[23] Adam T *et al.* [OPERA Collaboration] 2011 (*Preprint* hep-ex/1109.4897)
[24] Magueijo J 2011 (*Preprint* hep-ph/1109.6055)

Soliton surfaces associated with $\mathbb{C}P^{N-1}$ sigma models

A M Grundland[1 2] **and S Post**[1]

[1] Centre de Recherches Mathématiques. Université de Montréal. Montréal CP6128 (QC) H3C 3J7, Canada

[2] Department of Mathematics and Computer Sciences, Université du Quebec, Trois-Rivières. CP500 (QC)G9A 5H7, Canada

E-mail: grundlan@crm.umontreal.ca, post@crm.umontreal.ca

Abstract. Soliton surfaces associated with the $\mathbb{C}P^{N-1}$ sigma model are constructed using the Generalized Weierstrass and the Fokas-Gel'fand formulas for immersion of 2D surfaces in Lie algebras. The considered surfaces are defined using continuous deformations of the zero-curvature representation of the model and its associated linear spectral problem. The theoretical framework is discussed in detail and several new examples of such surfaces are presented.

1. Introduction

Integrable models and their continuous deformations under various types of dynamics have produced much interest and activity in several areas of mathematics, physics and biology. Furthermore soliton surfaces associated with integrable models, and with the $\mathbb{C}P^{N-1}$ sigma model in particular, have been shown to play an essential role in many problems with physical applications (see e.g. [8, 15, 25, 28, 29]). The possibility of using a linear spectral problem (LSP) to represent a moving frame on the surface has yielded many new results concerning the intrinsic geometric properties of such surfaces (see e.g. [1, 2, 3, 4, 21, 22, 24]). In this vein, it has recently proved fruitful to extend such characterization of soliton surfaces via their immersion function in Lie algebras. The construction of such surfaces, related to the completely integrable $\mathbb{C}P^{N-1}$ sigma model, has been accomplished by representing the equations of motion for the model as a conservation law which in turn provides a closed differential for the surface. This is the so called generalized Weierstrass formula for immersion (GWFI) [14, 19, 20, 23]. Another method for the construction of such soliton surfaces makes use of the conformal invariance of the zero-curvature (or ZS-AKNS) representation of the model with respect to the spectral parameter. This immersion function is known as the Sym-Tafel formula for immersion [30, 31] and a remarkable result, to be described later in this paper, is that the Sym-Tafel formula and the GWFI coincide in the case of finite action solutions of the $\mathbb{C}P^{N-1}$ sigma model.

More generally, the LSP for the integrable model and its symmetries has been employed by Fokas and Gel'fand [11] and later with Finkel and Liu [12], to construct families of soliton surfaces. Recently, the authors have reformulated and extended the Fokas-Gel'fand immersion formula using the formalism of generalized vector fields and their action on jet space and have given the necessary and sufficient conditions for the existence and explicit integration of soliton surfaces in terms of the symmetry criterion for vector fields [16, 17, 18].

The paper contains a survey of these recent results for the immersion of soliton surfaces, particularly as applied to the completely integrable $\mathbb{C}P^{N-1}$ sigma model. The basics of the model are reviewed in section 2. Section 3 contains an exposition of the GWFI and its relation to the Sym-Tafel formula. The Fokas-Gelfand formula for immersion, in the formalism of generalized vector fields, and its associated surfaces are presented in section 4. Section 5 gives new examples of the application of such immersion formulas to the case of the $\mathbb{C}P^1$ sigma model.

2. The $\mathbb{C}P^{N-1}$ sigma model: Definition and linear spectral problem

Over the past decades, there has been significant progress in the study of general properties of the $\mathbb{C}P^{N-1}$ sigma model and the techniques for finding associated 2D-soliton surfaces immersed in multidimensional space. The most fruitful approach to this subject has been achieved through the descriptions of the model in terms of rank-one Hermitian projectors. A matrix P is said to be a rank-one Hermitian projector if

$$P^2 = P, \qquad tr(P) = 1, \qquad P^\dagger = P. \tag{1}$$

The target space of the projector P is determined by a complex line in $\mathbb{C}^N$, i.e. by the one-dimensional vector function $f(\xi_+, \xi_-)$ given by

$$P = \frac{f \otimes f^\dagger}{f^\dagger f}, \tag{2}$$

where f is a mapping

$$\mathbb{C} \supseteq \Omega \ni \xi_\pm = x \pm iy \mapsto f = (f_0, f_1, \dots f_{N-1}) \in \mathbb{C}^N \setminus \{0\}.$$

In fact, the formula (2) gives an isomorphism between the equivalence classes of $\mathbb{C}P^{N-1}$ and the set of rank-one, Hermitian projectors. The projector formalism automatically encodes the scaling invariance of the vectors and ensures that the maps will be free from removable singularities which could occur in the unnormalized vector fields f [13]. Furthermore, the equations of motion and other properties of the model take a compact form when written in the projector formalism.

2.1. The $\mathbb{C}P^{N-1}$ sigma model in the projector formalism

In terms of a rank-one Hermitian projector P, the Lagrangian density is given by

$$L(P) = tr(\partial_+ P \partial_- P) \tag{3}$$

and the action functional is

$$S(P) = \int_{\mathbb{C}} L(P) d\xi_+ d\xi_-. \tag{4}$$

Here the holomorphic and anti-holomorphic derivatives are defined as

$$\partial_+ = \frac{\partial}{\partial \xi_+} = \frac{1}{2}\left(\frac{\partial}{\partial x} - \frac{\partial}{\partial y}\right), \qquad \partial_- = \frac{\partial}{\partial \xi_-} = \frac{1}{2}\left(\frac{\partial}{\partial x} + \frac{\partial}{\partial y}\right).$$

The Euler-Lagrange (E-L) equation for the model is given by

$$[\partial_+ \partial_- P, P] = \varnothing \tag{5}$$

which is equivalent to a conservative form

$$\partial_+[\partial_- P, P] + \partial_-[\partial_+ P, P] = \varnothing. \tag{6}$$

Here $\varnothing$ is the zero matrix. Another important physical quantity of the model is the topological charge density

$$q(P) = tr(P\partial_+ P\partial_- P - P\partial_- P\partial_+ P), \tag{7}$$

so called because its integral over the plane, the topological charge, is a total divergence and hence

$$Q(P) = \frac{1}{\pi}\int_{\mathbb{C}} q(P) d\xi_+ d\xi_- \tag{8}$$

is an integer [10].

2.2. Raising and lowering operators

As was first proven by [9, 10], see also [33], any solution of the $\mathbb{C}P^{N-1}$ sigma model, defined on the extended complex plane, with finite action can be written as a raising operator acting on a holomorphic vector (or a lowering operator acting on an anti-holomorphic vector). In [27], the authors presented this proof in the projector formalism. In this context, the raising and lowering operators are given by [13]

$$\Pi_\pm(P) \equiv \begin{cases} \frac{\partial_\pm P P\partial_\mp P}{tr(\partial_\pm P P\partial_\mp P)} & \partial_\pm P P\partial_\mp P \neq \varnothing, \\ \varnothing & \partial_\pm P P\partial_\mp P = \varnothing, \end{cases} \tag{9}$$

and have the property that a rank-one projector P projects onto an equivalence class in $\mathbb{C}P^{N-1}$ with a holomorphic representative if and only if

$$\Pi_- P = \varnothing. \tag{10}$$

Analogously, $\Pi_+ P = \varnothing$ if and only if P maps on to an equivalence class in $\mathbb{C}P^{N-1}$ with an anti-holomorphic representative. The operators $\Pi_\pm$ are called raising and lowering operators because they are contracting operators which map between solutions of the E-L equation [13]. They satisfy the following properties. Suppose that a rank-one Hermitian projector P is a solution of the E-L equation with finite action, then:

(i) $\Pi_\pm P$ is either $\varnothing$ or is itself a rank-one Hermitian projector solution of the E-L equation with finite action.

(ii) The raising and lowering operators are mutual inverses for solutions of the E-L equation, i.e.

$$\Pi_\pm \Pi_\mp P = P, \qquad \text{whenever } \Pi_\mp P \neq \varnothing. \tag{11}$$

(iii) The raising and lowering operators are mutually orthogonal

$$\Pi_+^j P \Pi_+^k P = \delta_{jk} \Pi_+^k P, \qquad j,k \in \mathbb{Z}, \tag{12}$$

where the quantities $\Pi_+^j P$ are defined inductively by

$$\Pi_+^0 P = 0, \qquad \Pi_+^{-1} P = \Pi_-(P),$$
$$\Pi_+^{n+1} P = \Pi_+(\Pi_+^n P), \quad \Pi_+^{n-1} P = \Pi_-(\Pi_+^n P).$$

(iv) The raising and lowering operators are contracting operators on P. That is, there exist natural numbers $j,k \in \mathbb{N}$ with $0 \leq j+k \leq N-1$ such that

$$\Pi_-^{j+1} P = \varnothing, \qquad \Pi_+^{k+1} P = \varnothing.$$

Journal of Physics: Conference Series **380** (2012) 012023 doi:10.1088/1742-6596/380/1/012023

(v) The images of the finite set of rank-one projectors

$$\{\Pi_-^j P, \ldots \Pi_- P, P, \Pi_+ P, \ldots \Pi_+^k P\} \tag{13}$$

form an orthogonal basis for the subspace $\mathbb{C}^{j+k+1} \subset \mathbb{C}^N$.

If it is the case that $j+k+1 < N$, then the $\mathbb{C}P^{N-1}$ model can be embedded in a $\mathbb{C}P^{j+k}$ model. Thus, for the remainder of the paper, it is assumed that $j+k+1 = N$ and so

$$\Pi_-^{j+1} P = \varnothing, \qquad \Pi_+^{N-j} P = \varnothing,$$

which in turn implies that $\Pi_-^j P$ is a holomorphic and $\Pi_+^{N-1-j} P$ and anti-holomorphic projector.

2.3. The linear spectral problem

It is well known that the $\mathbb{C}P^{N-1}$ sigma model admits a linear spectral problem [32, 33]. Defining matrix functions

$$U^1 \equiv \frac{2}{1+\lambda}[\partial_+ P, P], \qquad U^2 \equiv \frac{2}{1-\lambda}[\partial_- P, P], \tag{14}$$

the Euler-Lagrange (E-L) equations are equivalent to

$$\Delta' \equiv [\partial_+ \partial_- P, P] = \partial_- U^1 - \partial_+ U^2 + [U^1, U^2] = 0, \tag{15}$$

which are exactly the compatibility conditions for the LSP of the form

$$\partial_+ \Phi = U^1 \Phi, \qquad \partial_- \Phi = U^2 \Phi, \tag{16}$$

where λ is a complex spectral parameter.

In is worth noting that, for the $\mathbb{C}P^{N-1}$ sigma model defined on Euclidean space, the wave function Φ can be explicitly integrated for an arbitrary solution of the E-L equation with finite action. The wave function Φ is given in terms of the set of rank-one Hermitian projectors (13) by [13, 33]

$$\Phi = \left(\mathbb{I} + \frac{4\lambda}{(1-\lambda)^2} \sum_{j=1}^{k} \Pi_-^j P - \frac{2}{1-\lambda} P \right), \qquad \Pi_-^{k+1} P = \varnothing, \tag{17}$$

where $\mathbb{I}$ is the identity matrix on $\mathbb{C}^N$. If the spectral parameter λ is purely imaginary, Φ is an element of the group $SU(N)$ and the inverse of the wave function is given by

$$\Phi^{-1} = \left(\mathbb{I} - \frac{4\lambda}{(1+\lambda)^2} \sum_{j=1}^{k} \Pi_-^j P - \frac{2}{1+\lambda} P \right). \tag{18}$$

The exact form of the wave function Φ will be important later in the construction of the soliton surfaces.

3. Generalized Weierstrass representation of surfaces associated with $\mathbb{C}P^{N-1}$ sigma models

From the form of the E-L equation (6), it is possible to construct a closed, skew-Hermitian differential

$$dF = -i\left([\partial_+ P, P] d\xi_+ - [\partial_- P, P] d\xi_- \right). \tag{19}$$

Integrating the differential along a curve $\gamma \subset \mathbb{C}$ gives a surface immersed in the $su(N)$ algebra associated with P, a solution of the $\mathbb{C}P^{N-1}$ sigma model. These results are present in the following theorem [19].

Theorem 1 *If P is a rank-1 Hermitian projector solution of the Euler-Lagrange equations then there exists a surface F whose immersion is defined by*

$$F = -i\int_{\gamma}[\partial_+ P, P]d\xi_+ - [\partial_- P, P]d\xi_-, \qquad F^\dagger = -F \in su(N). \tag{20}$$

The immersion function defined as in (20) is called the Generalized Weierstrass formula for immersion (GWFI). Alternatively, the surface F can be defined by its tangent vectors

$$\frac{dF}{d\xi_+} = -i[\partial_+ P, P] \qquad \frac{dF}{d\xi_-} = i[\partial_- P, P], \tag{21}$$

whose compatibility conditions is exactly the E-L matrix equations (6). The Killing form on $su(N)$,

$$(X, Y) = -\frac{1}{2}tr(X \cdot Y), \qquad X, Y \in su(N) \tag{22}$$

gives an inner-product on the tangent vectors to a 2D-surface immersed in $su(N)$.

With this metric, it is possible to show that the surfaces defined as in (20) are conformally parameterized [20]. The following theorem was proven in [27].

Theorem 2 *For a finite action solution, P, of the Euler-Lagrange equations (15), the surface F defined by (20) is conformally parameterized and the conformal factor in the first fundamental form is proportional to the Lagrangian density*

$$I(F) = \frac{1}{2}L(P)d\xi_+ d\xi_-. \tag{23}$$

Thus, the area of the surface is given by the action functional of the model and, in particular, the surface will have finite area.

Other geometric quantities of the surface can be written in terms of the physical properties of the model. In complex coordinates, the Gaussian curvature becomes

$$K(F) = -\frac{2\partial_+\partial_-(ln(L(P)))}{L(P)}. \tag{24}$$

The mean curvature vector, written in matrix form, is given by

$$H(F) = \frac{-4i}{L(P)}[\partial_+ P, \partial_- P] \tag{25}$$

and thus is traceless. The norm of H can be written in terms of the Lagrangian density and the topological charge density,

$$(H(F), H(F)) = \frac{8}{L(P)^2}tr([\partial_+ P, \partial_- P]^2) = \frac{4}{L(P)^2}\left(L(P)^2 + 3q(P)^2\right). \tag{26}$$

It is possible to express the Willmore functional in terms of the action, the Lagrangian density, and the topological charge density,

$$W(F) = \frac{1}{2}S(P) + \frac{3}{2}\int_{\mathbb{R}^2}\frac{q(P)^2}{L(P)}d\xi_+ d\xi_-. \tag{27}$$

The Euler-Poincare character can be written terms of only the Lagrangian density

$$\Delta(F) = -\frac{1}{\pi}\int_{\mathbb{R}^2}\partial_+\partial_- ln(L(P))d\xi_1 d\xi_2. \tag{28}$$

It was shown in [20] that the surfaces, defined as in theorem 1 can be integrated explicitly in terms of the set of orthogonal projectors (13).

Theorem 3 *Under the assumptions of theorem 1, the surface has an associated integer k and holomorphic projector P_0 so that the following holds*

$$F = F_k \equiv -i\left(P_k + 2\sum_{j=0}^{k-1} P_j - \frac{1+2k}{N}\mathbb{I}\right), \qquad P_\ell = \Pi_+^\ell P_0. \tag{29}$$

The inverse formula for the projectors P_k in terms of the surface F_k (29) has been derived in [20]

$$P_k = F_k^2 - 2i\left(\frac{1+2k}{N} - 1\right)F_k - \frac{1+2k}{N}\left(\frac{1+2k}{N} - 2\right)\mathbb{I}.$$

The exact integrated form of the surfaces (29) allows one to obtain the minimal polynomials for the surfaces, dimensionality of the spanned subspaces of $\mathbb{R}^{N^2-1} \equiv su(N)$, and the angle between the immersion functions F_k and F_j of the 2D surfaces [20].

3.1. The Generalized Weierstrass and the Sym-Tafel formulas for immersion

An interesting fact is that the GWFI is equivalent to another formula for the immersion of 2D-surfaces in Lie algebas, namely the Sym-Tafel formula for immersion [30, 31]. This immersion formula of a 2D surface is given by

$$F^{ST} = \alpha(\lambda)\Phi^{-1}\frac{\partial}{\partial\lambda}\Phi \in \mathfrak{g}, \tag{30}$$

whenever the tangent vectors

$$\partial_+ F^{ST} = \Phi^{-1}\frac{\partial U^1}{\partial\lambda}\Phi, \qquad \partial_- F^{ST} = \Phi^{-1}\frac{\partial U^2}{\partial\lambda}\Phi \tag{31}$$

are linearly independent.

From the specific form of Φ (17) and Φ^{-1} (18), it is possible to directly compute the Sym-Tafel immersion formula as [20]

$$F^{ST} = \frac{2\alpha(\lambda)}{1-\lambda^2}\left(P_k + 2\sum_{j=0}^{k-1} P_j - \frac{1+2k}{N}\mathbb{I}\right) \in su(N), \tag{32}$$

which coincides with the GWFI, up to a multiplicative factor.

4. Fokas-Gel'fand immersion formula for surfaces associated with $\mathbb{C}P^{N-1}$ sigma models

In light of the results obtained for the Generalized Weierstrass formula for immersion and its equivalent form, the Sym-Tafel formula for immersion, it seems worthwhile to consider the larger class of immersion functions defined via the Fokas-Gel'fand formula for immersion. Such surfaces were recently considered by the authors in a series of papers [16, 17, 18]. The necessary and sufficient conditions for the existence of such surfaces was formulated in terms of the symmetry criterion for generalized vector fields and their action on jet space [26].

4.1. The Fokas-Gel'fand immersion formula for surfaces

The considered surfaces are defined via infinitesimal deformations of the zero-curvature condition

$$\Delta[u] = u^1_2 - u^2_1 + \left[u^1, u^2\right] = 0, \tag{33}$$

with independent variables ξ_i, $i = 1, 2$ and dependent matrix variables u^α, $\alpha = 1, 2$ which take their values in a Lie algebra $\mathfrak{g}$. In the case of the $\mathbb{C}P^{N-1}$ sigma model defined on Euclidean space, the independent variable take the form $\xi+ = \xi_1$ and $\xi_- = \xi_2$ and the Lie algebra is $\mathfrak{g} = su(N)$. The PDE (33) can be considered as a function on the jet space $M \equiv (\xi_1, \xi_2, u^\alpha, u^\alpha_J)$, where the derivatives of u^α are given by

$$\frac{\partial^n}{\partial\xi_{j_1}\ldots\partial\xi_{j_n}} u^\alpha \equiv u^\alpha_J, \qquad J = (j_1, \ldots j_n), \quad j_i = 1, 2. \tag{34}$$

Define the set of smooth functions on the jet space M to be $\mathcal{A} \equiv C^\infty(M)$. The abbreviated notation $f(\xi_1, \xi_2, u^\alpha, u^\alpha_J) \equiv f[u] \in \mathcal{A}$ will be used. The zero-curvature condition $\Delta[u] = 0$ can be realized as the compatibility conditions for a wave function $\Psi[u]$ defined by its tangent vectors

$$D_\alpha\Psi = u^\alpha\Psi, \qquad \alpha = 1, 2, \tag{35}$$

where the total derivatives D_i are defined as

$$D_i = \frac{\partial}{\partial\xi_i} + u^{\alpha,j}_J \frac{\partial}{\partial u^{\alpha,j}_J}, \qquad \alpha, i = 1, 2. \tag{36}$$

Here $u^{\alpha,j}$ are the components of the u^α in a basis for $\mathfrak{g}$, i.e. $u^\alpha = u^{\alpha,j}e_j$ with $j = 1...n$.

Suppose now that it is possible to parameterize the matrices $u^\alpha = U^\alpha([\theta], \lambda) \in \mathfrak{g}$ in terms of some set of dependent variables $\theta^j(\xi_+, \xi_-)$, which depend on the independent variables ξ^j. The potential matrices $U^\alpha([\theta], \lambda)$ are assumed to depend on a spectral parameter λ in such a way that the zero-curvature condition is equivalent to an integrable PDE in dependent variables θ^j which is independent of the spectral parameter

$$\Delta'[\theta] = D_1 U^2[\theta] - D_2 U^1[\theta] + \left[U^1[\theta], U^2[\theta]\right] = 0. \tag{37}$$

In the case of the $\mathbb{C}P^{N-1}$ sigma model, the scalar functions θ^j represent the components of the projector P.

Define a new jet space N to be the jet space associated with θ^j and its derivatives and $\mathcal{B}$ to be the space of smooth functions on this jet space, also possibly depending on the spectral parameter λ. The elements of $\mathcal{B}$ are denoted $f(\lambda, \xi_+, \xi_-, \theta^j_J) = f([\theta], \lambda) \in \mathcal{B}$. Let τ be the mapping between the function spaces $\mathcal{A}$ and $\mathcal{B}$ defined by taking $\tau(u^\alpha_J) = D_J\left(U^\alpha([\theta], \lambda)\right)$:

$$\begin{array}{ccc} u^\alpha \in \mathfrak{g} & \rightarrow & U^\alpha([\theta]) \in \mathfrak{g} \\ \downarrow & & \downarrow \\ & \tau & \\ \mathcal{A} = C^\infty(M) & \rightarrow & \mathcal{B} = C^\infty(N \times \mathbb{C}). \end{array} \tag{38}$$

Because of the structure of jet space, the mapping τ commutes with total derivatives. Under the mapping τ, the LSP becomes

$$D_\alpha\Phi = U^\alpha\Phi, \qquad \alpha = 1, 2, \tag{39}$$

where the wave function is given by $\Phi([\theta], \lambda) = \tau(\Psi[u])$.

This spectral problem was used by Fokas and Gel'fand in [11] to construct immersion functions for surfaces in Lie algebras whose Gauss-Mainardi-Codazzi (GMC) equations are given by

$$D_1A - D_2B + [A, U^2] + [U^1, B] = 0, \tag{40}$$

an infinitesimal deformation of (37). Here A and B are elements of the Lie algebra $\mathfrak{g}$ with components in the function space $\mathcal{B}$. The immersion functions are defined by their tangent vectors

$$D_1F = \Phi^{-1}A\Phi, \qquad D_2F = \Phi^{-1}B\Phi. \tag{41}$$

It is straightforward to verify that the compatibility conditions of (41) are satisfied whenever the LSP (39) and the GMC equations (40) hold.

The possible forms of such matrices A and B have been studied by several authors, e.g. [7, 12, 16, 17, 18]. In this case, there are three types of admissible symmetries:

- generalized symmetries of the zero-curvature condition itself $\Delta[u] = 0$,
- generalized symmetries of the corresponding integrable PDE $\Delta'[\theta] = 0$,
- invariance of the integrable PDE with respect to a conformal transformation in the spectral parameter. This is the Sym-Tafel formula for immersion.

In order to construct such symmetries, it is expedient to consider generalized fields on jet space. A vector field $\vec{v}_Q$ defined on the jet space M in evolutionary form is given by

$$\vec{v}_Q = Q^{\alpha,j}[u]\frac{\partial}{\partial u^{\alpha,j}}. \tag{42}$$

Note that $Q^\alpha \equiv Q^{\alpha,j}e_j$ is an element of the Lie algebra $\mathfrak{g}$. The prolongation of $\vec{v}_Q$ is defined in the standard way

$$pr\vec{v}_Q = \vec{v}_Q + D_J(Q^{\alpha,j}[u])\frac{\partial}{\partial u_J^{\alpha,j}}. \tag{43}$$

A vector field $\vec{w}_{\mathcal{Q}}$ defined on the jet space N in evolutionary form is denoted by

$$\vec{w}_{\mathcal{Q}} = \mathcal{Q}^j[\theta]\frac{\partial}{\partial \theta^j}. \tag{44}$$

The notation $\mathcal{Q}$ is used to distinguish the characteristics of a vector field $\vec{v}_Q$ on M from a vector field $\vec{w}_{\mathcal{Q}}$ on N. The prolongation of $\vec{w}_{\mathcal{Q}}$ is defined in the standard way by analogy with (43).

A generalized vector field $\vec{v}_Q$ on jet space M is said to be a generalized symmetry of $\Delta[u] = 0$ if and only if

$$pr\vec{v}_Q\,(\Delta[u]) = 0, \qquad \text{whenever } \Delta[u] = 0. \tag{45}$$

From the definition of the prolongation (43), it is immediate to verify that (45) is equivalent to

$$D_2Q^1 - D_1Q^2 + [Q^1, u^2] + [u^1, Q^2] = 0 \tag{46}$$

Thus, the $\mathfrak{g}$-valued functions on jet space N

$$A = \tau(Q^1), \qquad B = \tau(Q^2), \tag{47}$$

satisfy (40) whenever $\vec{v}_Q$ is a symmetry of $\Delta[u] = 0$.

Similarly, a generalized vector field $\vec{w}_{\mathcal{Q}}$ on jet space N is said to be a generalized symmetry of $\Delta'[\theta] = 0$ if and only if

$$pr\vec{w}_{\mathcal{Q}}\,(\Delta'[\theta]) = 0, \qquad \text{whenever } \Delta'[\theta] = 0. \tag{48}$$

From the definition of the prolongation (43), it is immediate to verify that (45) is equivalent to

$$D_2(pr\vec{w}_Q U^1) - D_1(pr\vec{w}_Q U^2) + [pr\vec{w}_Q U^1, U^2] + [U^1, pr\vec{w}_Q U^2] = 0. \tag{49}$$

Thus, the $\mathfrak{g}$-valued functions on jet space N

$$A = pr\vec{w}_Q U^1, \qquad B = pr\vec{w}_Q U^2, \tag{50}$$

satisfy (40) whenever $\vec{w}_Q$ is a symmetry of $\Delta'[\theta] = 0$.

The third possible choice of A and B are associated with the invariance of $\Delta'[\theta] = 0$ with respect to conformal symmetries of the spectral parameter λ. In this case, the choice of matrices A and B are given by

$$A = \frac{\partial}{\partial\lambda} U^1, \qquad B = \frac{\partial}{\partial\lambda} U^2,$$

which satisfy (40) whenever $\Delta'[\theta] = 0$ is independent of the spectral parameter. In this case, the tangent vectors given by (41) coincide with those for the Sym-Tafel formula for immersion (31). Thus, the surfaces take the integrated form given in (30).

4.2. Integrated form of the surfaces immersed in Lie algebras

As in the case of the Sym-Tafel formula for immersion, it is often possible to give explicit integrated forms for the immersion function for the surfaces. As proven in [18], any symmetry of the zero-curvature condition $\Delta[u] = 0$ can be written in terms of a gauge function $S[u]$ and the associated immersion function is given by

$$F = \Phi^{-1} \tau\left(S[u]\right) \Phi \in \mathfrak{g}. \tag{51}$$

For example, the zero-curvature condition $\Delta[u] = 0$ is invariant under a scaling of the independent and dependent variables associated with the vector field $\vec{v}_Q$ with

$$Q^1 = D_1\left(\xi_1 u^1\right) + \xi_2 D_2\left(u^1\right), \qquad Q^2 = \xi_1 D_1\left(u^2\right) + D_2\left(\xi_2 u^2\right). \tag{52}$$

The gauge associated with this symmetry is [18]

$$S[u] = \xi_1 u^1 + \xi_2 u^2$$

and the corresponding surface is

$$F = \Phi^{-1}\left(\xi_1 U^1 + \xi_2 U^2\right) \Phi \in \mathfrak{g}. \tag{53}$$

It is straightforward to verify that the tangent vectors to this surface are given as in (41) with

$$A = \tau(Q^1) = D_1\left(\xi_1 U^1\right) + \xi_2 D_2\left(U^1\right), \qquad B = \tau(Q^2) = \xi_1 D_1\left(U^2\right) + D_2\left(\xi_2 U^2\right).$$

As a second example, consider the following generalized symmetry of $\Delta[u] = 0$ whose generalized vector field $\vec{v}_Q$ has characteristics

$$Q^1 = u^1_{11} + u^1_{22} + [u^1_1, u^1] + [u^1_2, u^1], \qquad Q^2 = u^2_{11} + u^2_{22} + [u^2_1, u^1] + [u^2_2, u^2]. \tag{54}$$

The gauge associated to this symmetry is given by

$$S[u] = u^1_1 + u^2_2$$

and the corresponding surface takes the form

$$F = \Phi^{-1}\left(D_1U^1 + D_2U^2\right)\Phi \in \mathfrak{g}. \tag{55}$$

Similarly, as in the previous case, the tangent vectors to this surface are given as in (41) with

$$\begin{aligned} A &= \tau(Q^1) = D_1^2U^1 + D_2^2U^1 + [D_1U^1, U^1] + [D_2U^1, U^1], \\ B &= \tau(Q^2) = D_1^2U^2 + D_2^2U^2 + [D_2U^2, U^2] + [D_1U^2, U^2]. \end{aligned}$$

There also exist soliton surfaces associated with generalized symmetries of the integrable model, such as in the case of conformal symmetries of the $\mathbb{C}P^{N-1}$ sigma model. As was proven in [16], the surface associated with such symmetries, i.e. with A and B of the form (50), can be integrated explicitly as

$$F = \Phi^{-1} pr\vec{w}_Q \Phi \in \mathfrak{g}, \tag{56}$$

whenever $\vec{w}_Q$ is a generalized symmetry of the LSP in the sense that

$$pr\vec{w}_Q\left(D_\alpha\Phi - U^\alpha\Phi\right) = 0 \text{ whenever } D_\alpha\Phi - U^\alpha\Phi = 0. \tag{57}$$

In general, the requirement (57) is difficult to verify since it requires a knowledge of the integrated form of the wave function Φ. However, in the case of finite action solutions of the $\mathbb{C}P^{N-1}$ sigma model, the wave function is known explicitly (17) and it has been directly shown that conformal symmetries of the independent variables are also symmetries of the LSP [16]. Thus, the surface F as in (56) takes the particular form

$$F = \Phi^{-1}\left(g(\xi_+)U^1 + \overline{g}(\xi_-)U^2\right)\Phi \in su(N), \tag{58}$$

associated with vector field

$$\vec{w} = -g(\xi_+)\partial_+ - \overline{g}(\xi_-)\partial_-,$$

which is a conformal symmetry of the model.

5. Soliton surfaces associated with the $\mathbb{C}P^1$ sigma model

To illustrate these theoretical considerations, examples of soliton surfaces associated with the $\mathbb{C}P^1$ sigma model are presented. In this model, the only solutions with finite action are holomorphic and anti-holomorphic projectors. Nevertheless, as will be demonstrated below, there are various forms of surfaces which can be constructed in this manner.

For these simple examples, the holomorphic projector is chosen as the following matrix, based on the Veronese sequence $f = (1, \xi_+)$,

$$P_0 = \frac{1}{(1+|\xi|^2)^2}\begin{bmatrix} 1 & \xi_- \\ \xi_+ & |\xi|^2 \end{bmatrix}, \qquad |\xi|^2 \equiv \xi_+\xi_-. \tag{59}$$

The wave function in the LSP for this model is simply

$$\Phi_0 = \left(\mathbb{I} - \frac{2}{1-\lambda}P\right). \tag{60}$$

The surface corresponding to the Sym-Tafel formula for immersion is written as

$$F^{ST} = \frac{i}{(1+|\xi|^2)}\begin{bmatrix} 1 & \xi_- \\ \xi_+ & \xi_+\xi_- \end{bmatrix},$$

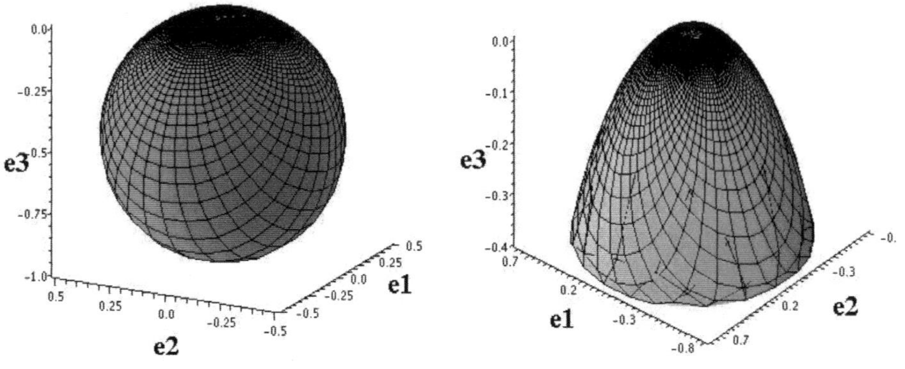

F^{ST} F^{P}

Figure 1. Surfaces F^{ST} and F^{P} with $\lambda = i/2$ and $\xi_\pm = x \pm iy$ with x and $y \in [-20, 20]$. The axes indicate the components of the immersion function in the basis (61)

where the scaling factor is chosen as $i(1-\lambda^2)/2$ so as to coincide with the GFWI. This surface has constant positive Gaussian and mean curvature given by

$$K = H = 4,$$

and so is in fact a sphere, as demonstrated in figure 1. The surfaces are represented in terms of their components in the following basis for $su(2)$

$$e_1 = \begin{bmatrix} 0 & i \\ i & 0 \end{bmatrix}, \quad e_2 = \begin{bmatrix} 0 & -1 \\ 1 & 0 \end{bmatrix}, \quad e_3 = \begin{bmatrix} i & 0 \\ 0 & -i \end{bmatrix}. \tag{61}$$

The surface associated with the (scaling) point symmetry of the zero-curvature condition $\vec{v}_Q$ (42) has the integrated form

$$\begin{aligned} F^P &= \Phi^{-1}\left(\xi_1 U^1 + \xi_2 U^2\right)\Phi \\ &= \frac{1}{(1-\lambda^2)(1+|\xi|^2)^2}\begin{bmatrix} 4\lambda\,\xi_+\xi_- & 2\left((\lambda+1)\,\xi_+\xi_- + 1 - \lambda\right)\xi_- \\ 2\left((\lambda-1)\,\xi_+\xi_- - 1 - \lambda\right)\xi_+ & -4\lambda\,\xi_+\xi_- \end{bmatrix}. \end{aligned}$$

This surface, F^P, also has constant positive Gaussian and mean curvatures

$$K = -4\lambda^2, \qquad H = -4i\lambda, \qquad i\lambda \in \mathbb{R}$$

though the surface is not a sphere since it has a boundary. A graph of the surface is shown in figure 1.

In the case of the surface associated with the generalized symmetry of the zero-curvature condition (55), the immersion function is given by

$$\begin{aligned} F^G &= \Phi^{-1}\left(D_1 U^1 + D_2 U^2\right)\Phi \\ &= \frac{4}{(1-\lambda^2)(1+|\xi|^2)^3}\begin{bmatrix} -\lambda({\xi_+}^2 + {\xi_-}^2) + {\xi_+}^2 - {\xi_-}^2 & -(\lambda+1){\xi_-}^3 + (\lambda-1)\xi_+ \\ (1-\lambda)\,{\xi_+}^3 + (1+\lambda)\xi_- & +\lambda({\xi_+}^2 + {\xi_-}^2) - {\xi_+}^2 + {\xi_-}^2 \end{bmatrix}. \end{aligned}$$

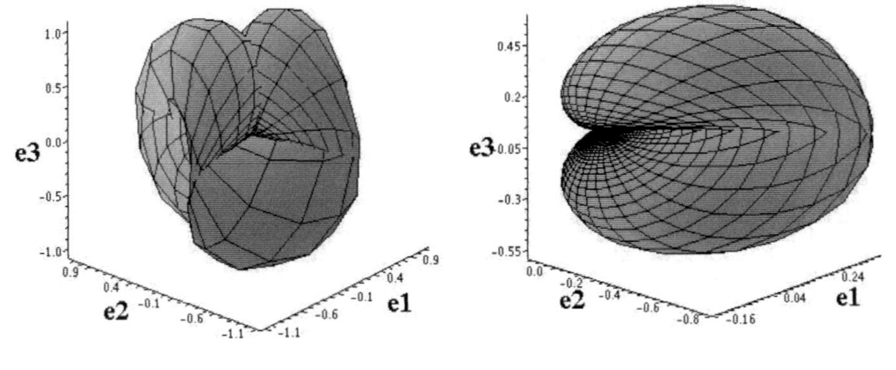

F^G F^C

Figure 2. Surfaces F^G and F^C with $\lambda = i/2$ and $\xi_\pm = x \pm iy$ with x and $y \in [-20, 20]$. The axes indicate the components of the immersion function in the basis (61)

The surface associated with conformal symmetry of the E-L equation (58) with $g(\xi_+) = 1$ is given by

$$\begin{aligned} F^C &= \Phi^{-1}\left(U^1 + U^2\right)\Phi \\ &= \frac{1}{(1-\lambda^2)(1+|\xi|^2)^2}\begin{bmatrix} \xi_- - \xi_+ + \lambda(\xi_+ - \xi_-) & (1+\lambda)\xi_-^2 + 1 - \lambda \\ (\lambda - 1)\xi_-^2 - 1 - \lambda & \xi_+ - \xi_- - \lambda(\xi_+ - \xi_-) \end{bmatrix}. \end{aligned}$$

In both cases the Gaussian and mean curvatures are rational functions though they are too involved to write out in an illustrative fashion. Their graphs are given in figure 2.

6. Future outlook

The approach for constructing soliton surfaces immersed in Lie algebras described in this paper has proven to be very effective in revealing geometric properties of investigated classes of surfaces. The future objective is to extend it to more general sigma models and study surfaces related to the complex Grassmannian models and possibly to models associated with octonion geometry. Since the field theoretic formulation of soliton surfaces by the Euler-Lagrange equation naturally involves the variation principle, it also leads in a straightforward way to geometric objects like geodesics, harmonic maps, minimal surfaces and the like. It also relates closely to the formalism of completely integrable systems, e.g. Lax pairs, conservations laws and Hamiltonian structures. So, it is apparent that the methods bring a strong unifying potential to the prospective research.

Acknowledgments

S Post would like to thank the organizers, Dieter Schuch and Michael Ramek, for their kind invitation and support in attending the conference. The research reported in this paper is supported by NSERC of Canada. S Post acknowledges a postdoctoral fellowship provided by the Laboratory of Mathematical Physics of the Centre de Recherches Mathématiques, Université de Montréal.

References

[1] Bavan H and Marvan M 2010 Classification of integrable Wiengarten surfaces possesing $sl(2)$-valued zero curvature representation *Nonlinearity* **23** 2577–2597

[2] Bobenko A I 1990 Integrable surfaces *Functional Analysis and Its Applications*, vol. 24 pp. 227–228

[3] Bobenko A I 1994 Surfaces in terms of 2 by 2 matrices *Harmonic Maps and Integrable systems*, eds. A Fordy and J C Wood (Braunschweig: Vieweg)

[4] Bobenko A I and Eitner U 1985 Lect. Notes Math. *Painlevé Equations in the Differential Geometry of Surfaces* (Berlin: Springer-Verlag) 1753

[5] Borchers H J and Garber W D 1980 Local theory of solutions for the $O(2k+1)$ σ-model *Comm. Math. Phys.* **72** 77–102

[6] Ceyhan O, Fokas A S and Gürses M 2000 Deformation of surfaces associated with integrable Gauss-Mainardi-Codazzi equations *J. Math. Phys.* **41** 2251–2270

[7] Cieśliński J 1997 A generalized formula for integrable classes of surfaces in Lie algebras *Journal of Mathematical Physics* **38** 4255–4272

[8] Davydov A 1999 *Solitons in Molecular Systems* (NewYork: Kluwer)

[9] Din A M and Zakrzewski W J 1980 General classical solutions in the $\mathbb{C}P^{N-1}$ model *Nucl. Phys. B* **174** 397–406

[10] Din, A.M., Zakrzewski, W.J.: Properties of general classical $\mathbb{C}P^{N-1}$ solutions. *Phys. Lett. B.* 95, 426–30 (1980)

[11] Fokas A S and Gel'fand I M 1996 Surfaces on Lie groups, on Lie algebras, and their integrability *Comm. Math. Phys.* **177** 203–220

[12] Fokas A S, Gel'fand I M, Finkel F and Liu Q M 2000 A formula for constructing infinitely many surfaces on Lie algebras and integrable equations *Sel. Math.* **6** 347–375

[13] Goldstein P P and Grundland A M 2010 Invariant recurrence relations for $\mathbb{C}P^{N-1}$ models. *J. Phys. A.* **43** 265206–24

[14] Goldstein P P and Grundland A M 2011 On surfaces associated with $\mathbb{C}P^{N-1}$ models *J. Phys.: Conf. Ser. GROUP28 Int. Conf. on Group Theoretical Methods in Mathematical Physics***284** 012031.

[15] Gross D G, Piran T and Weinberg S 1992 *Two Dimensional Quantum Gravity and Random Surfaces* (Singapore: World Scientific)

[16] Grundland A M and Post S 2011 Soliton surfaces associated with generalized symmetries of integrable equations *J. Phys. A.:Math. Theor.* **44** 165203

[17] Grundland A M and Post S 2011 Soliton surfaces associated with elliptic equations *J. Phys. A.:Math. Theor.* arXiv:1106.2080

[18] Grundland A M and Post S 2011 Soliton surfaces associated with differential equations written in zero-curvature form arXiv:1111.4162

[19] Grundland A M, Strasburger A and Zakrzewski W 2005 Surfaces immersed in $su(n+1)$ Lie algebras obtained from $\mathbb{C}P^n$ sigma models. *J. Phys. A.* **39** 9187–9213

[20] Grundland A M and Yurdusen I 2009 On analytic discriptions of two-dimensional surfaces associated with the $\mathbb{C}P^{N-1}$ sigma models. *J. Phys. A: Math. Theor.* **42** 172001–172006

[21] Helein F 2001 *Constant Mean Curvature Sufaces, Harmonic Maps and Integrable Systems* (Boston: Birkhäuser)

[22] Konopelchenko B 1996 Induced surfaces and their integrable dynamics. *Stud. Appl. Math.* **96** 9–51

[23] Konopelchenko B and Landolfi G 1999 Induced surfaces and their integrable dynamics: Ii. generalized Weierstrauss representations in 4-d spaces and deformations via DS hierarchy. *Stud. Appl. Math.* **96** 129–169

[24] Konopelchenko B and Taimanov I 1996 Constant mean curvature surfaces via an integrable dynamical system *J. Phys. A: Math. Gen.* **29** 1261–1265

[25] Landolfi G 2003 On the Canham-Helfrich membrane model. *J. Phys. A* **36** 4699–4715

[26] Olver P J 1993 *Applications of Lie Groups to Differential Equations* (New York: Springer-Verlag) second edn.

[27] Post S and Grundland A M 2010 Analysis of $\mathbb{C}P^{N-1}$ sigma models via projective structure *Nonlinearity* (to appear) arXiv:1010.2183

[28] Rajaraman R 2002 $\mathbb{C}P^n$ solitons in quantum Hall systems. *Eur. Phys. B* **28** 157–162

[29] Safram S 1994 *Statistical Thermodynamics of Surfaces Interface and Membranes* (Massachusetts: Addison-Wesley)

[30] Sym A 1985 Lecture notes in physics. *Geometric aspects of the Einstein Equations and Integable systems.*, ed. R Martini (New York, NY: Springer-Verlag) 239 chap. Soliton surfaces and their applications (Soliton Geometry from Spectral Problems)

[31] Tafel J 1995 Surfaces in $\mathbb{R}^3$ with prescribed curvature. *J. Geom. Phys.* **17** 381–390

[32] Zakharov V E and Mikhailov A V 1979 Relativistically invariant two-dimensional models of field theory which are integrable by means of the inverse scatterting method. *Sov. Phys.* **40** 1017–1049
[33] Zakzrewski W J 1989 *Low Dimensional Sigma Models* (Bristol: Adam Hilger)

Spherical Orbifolds for Cosmic Topology

Peter Kramer
Institut für Theoretische Physik der Universität
Tübingen, Germany

E-mail: peter.kramer@uni-tuebingen.de

Abstract. Harmonic analysis is a tool to infer cosmic topology from the measured astrophysical cosmic microwave background CMB radiation. For overall positive curvature, Platonic spherical manifolds are candidates for this analysis. We combine the specific point symmetry of the Platonic manifolds with their deck transformations. This analysis in topology leads from manifolds to orbifolds. We discuss the deck transformations of the orbifolds and give eigenmodes for the harmonic analysis as linear combinations of Wigner polynomials on the 3-sphere. These provide new tools for detecting cosmic topology from the CMB radiation.

1. Motivation and elements of topology

Data observed by the Wilkinson satellite [15] and expected from the Planck satellite display the black-body cosmic microwave background CMB radiation with temperature T= 2.75 K and λ_{max}=19 mm, Fig. 1. This radiation is generally assumed to stem from the big bang. A small amplitude for low CMB multipoles l suggest selection rules. Do they arise from cosmic topology? In this first section we introduce topology and functional analysis for manifolds. Multiply

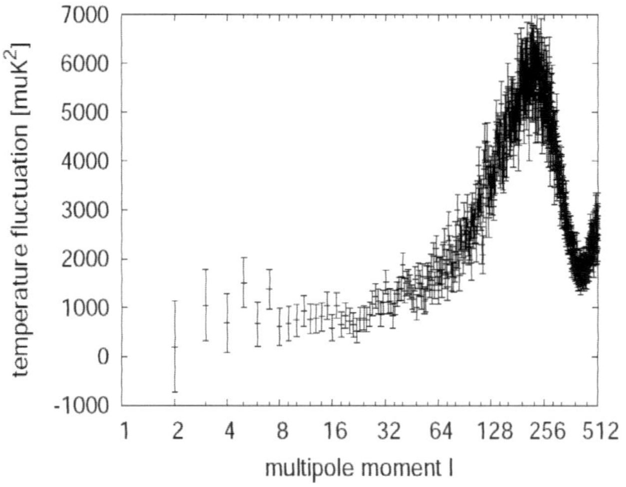

Figure 1. CMB Amplitude from the WMAP team [15] as a function of the multipole order l.

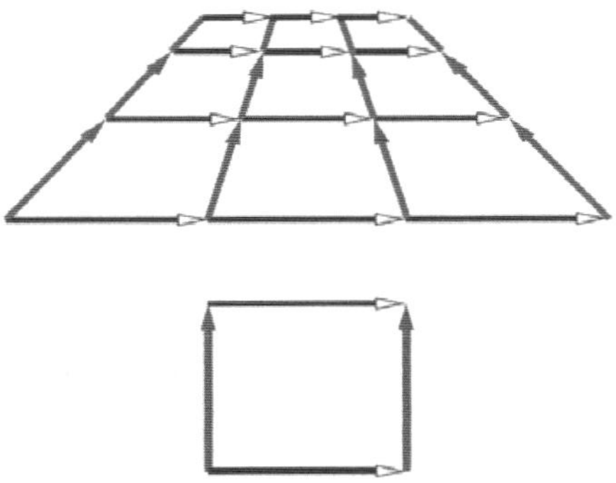

Figure 2. Parallel universes? Lift the square to its universal cover. By matching with horizontal/vertical neighbours we cover the plane by parallel squares.

connected spherical 3-manifolds can model the space part of cosmic space-time. We demonstrate topology in two dimensions from a square. First in Fig. 2 we lift this square to the Euclidean plane E^2, called its universal cover. The square becomes part of a square tiling of E^2. The group that generates the tiling is called the deck group. In a second approach, Fig. 3, we fold and glue the square into the torus. The gluings correspond to a group, termed the homotopy group H of the square. Homotopy and covering are linked by a theorem of Seifert and Threlfall [22] which says that the homotopy and the deck group are isomorphic. Fig. 3 illustrates the gluings, but may be misleading with respect to curvature: The square is Euclidean and shares zero curvature with its cover, the Euclidean plane. So the two views of a topological manifold, in terms of its universal cover or its self-gluing, allow to express different topological aspects and must be taken in conjunction.

1.1. Functions carried on a topological manifold

We pass to functions carried by the square. On the covering plane, periodic boundary conditions on the square tiles yield twofold Fourier analysis. Any Fourier basis function

$$\psi_{n_1,n_2}(x,y) = \exp i(k_1x + k_2y),\ k_1L = 2\pi n_1,\ k_2L = 2\pi n_2,\ \ (n_1,n_2) = 0, \pm 1, \pm 2, ... \tag{1}$$

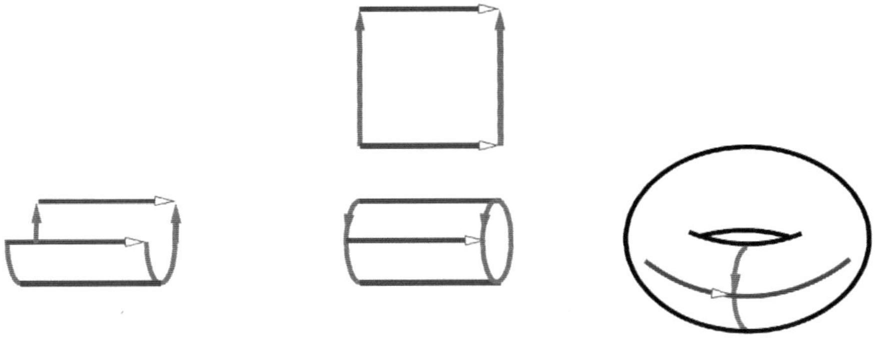

Figure 3. Folded universe? Glue the square by local homotopy. Horizontal edge folding plus gluing yields the cylinder. Successive vertical edge folding plus gluing gives the 2-torus T^2.

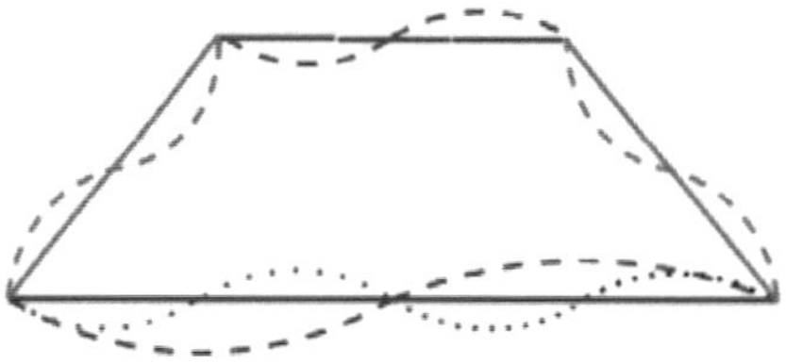

Figure 4. Periodic functions on the square from Fourier analysis.

is invariant under the deck group $H = C_\infty \times C_\infty$. As any deck group, H must act without fixpoints, compare Fig. 2.
A random function on a topological manifold is generated by a random superposition of its characteristic eigenmodes. In cosmic topology one compares such random superpositions with the observed CMB radiation. A similarity of both is thought to indicate the physical topology of the space part from space-time.
Before a comparison can be made, we must pass from the 2-dimensional square to 3-dimensional and more realistic manifolds. In contrast to zero curvature found for the square, positive curvature suggests to study spherical manifolds to which we turn in section 2.
The exploration of point symmetry for topology will lead us to the notion of orbifolds. In contrast to topological manifolds, orbifolds have deck group actions with singular fixpoints, that is, fixpoints of non-zero but finite order.

1.2. Einstein's field equations

Einstein's field equations determine the metric structure of space-time from the energy-momentum tensor.

$$(R_{ik} - \frac{1}{2}g_{ik}) - \lambda\, g_{ik} = -\kappa\, T_{ik}, \tag{2}$$
$$\kappa = \frac{8\pi C}{c^4} = 23.07 \times 10^{-48}\frac{s^2}{g\ cm}.$$

Here, R_{ik} is the Ricci tensor, g_{ik} the metrical tensor, T_{ik} the energy-momentum tensor, C Newton's gravitational, λ the cosmological constant. The Ricci tensor contains second derivatives of the metrical tensor, and so the field equations are second order partial differential equations for the metrical tensor g_{ik} in terms of a given energy-momentum tensor T_{ik}. Note: The cosmic mass distribution is smoothed out into a mass fluid!
Einstein in his first paper [4], as is also explained in [16], applied his equations to space taken as a 3-sphere. This choice assumes positive average curvature and closure, and still is relevant for present-day cosmic topology.

1.3. Selection rules from point symmetry and orbifolds

Angular coordinates with respect to the midpoint of the square are

$$z = \rho\ \exp(i\phi),\ \rho \geq 0,\ 0 \leq \phi < 2\pi. \tag{3}$$

Functions in multipole form may be written as

$$f(z) = R_m(\rho)\exp(im\phi), m = 0, \pm 1, \pm 2, ... \tag{4}$$

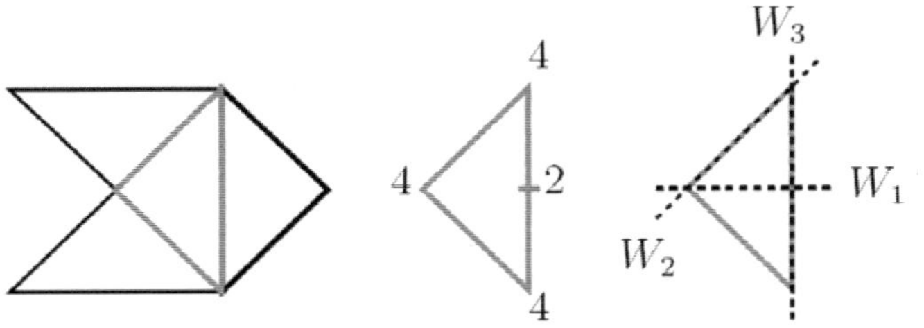

Figure 5. Orbifolds: The square is an orbit under the cyclic group C_4 of the triangular orbifold Δ. Left: Three $\Delta's$ cover edges of Δ. Middle: Singular fixpoints of finite orders $(4, 4, 4, 2)$ on the edges $\partial\Delta$ of the orbifold Δ. Right: Three Weyl reflections W_1, W_2, W_3 in the dashed lines generate a Coxeter group Γ. Its rotational unimodular subgroup $S\Gamma$: $\langle W_1W_2, W_2W_3, W_3W_1\rangle$ of even reflections generates a triangular tiling of the plane into orbifolds, preserves orientation, and yields selection rules. The unimodular subgroup $H = S\Gamma$ is the deck and homotopy group of the orbifold Δ!

Selection rules from point symmetry under ther cyclic group C_4 then imply

$$m \equiv\ 0 \bmod 4 =\ 0, \pm 4, \pm 8, ... \tag{5}$$

The introduction of point symmetry into topology leads to orbifolds rather than manifolds. For the square we explain the connection in Fig. 5. A triangle Δ forms a fundamental domain for the cyclic group C_4 acting on the square. The application of even products from a Coxeter group to a single preimage covers the plane E^2 by copies of the triangle. This action has rotational fixpoints on edges of the triangle, marked in Fig. 5, and so is not fixpoint-free as required for the deck transformation of a topological manifold. Instead, the triangle with its fixpoints forms an orbifold.

2. Spherical spaces

After the introductory section 1 we turn now to realistic models of cosmic 3-space. For these models of cosmic space we must extend all the notions illustrated on the square to appropriate manifolds. The spaces should have dimension 3 and should have positive curvature like spherical spaces. Such spaces result by closing pieces from the 3-sphere as used by Einstein [4] for his initial cosmology. Our exposition now necessarily becomes more technical and uses results given in the references. Harmonic analysis on the manifolds aims at explaining the observed low amplitudes at small multipole order l of the cosmic microwave background. We analyze assumptions of point symmetry and randomness for Platonic spherical spaces. There emerge four new spaces, called orbifolds, with low volume fraction from the 3-sphere and sharp multipole selection rules in their eigenmodes.

3. Eigenmodes for spherical spaces

Spherical spaces or spherical space forms are abstractly described as the quotients of the covering manifold S^3 by the deck group H of order $|H|$ acting fixpoint-free. The topologies are then classified as space forms $\mathcal{M} = S^3/H$. They were characterized with faithful representations of groups H by Wolf in [25] pp. 198-230. This characterization leaves open the geometry of the space, and so gives no access to point symmetry. Therefore we prefer a fully geometric approach [12], starting from homotopies derived by Everitt [5].

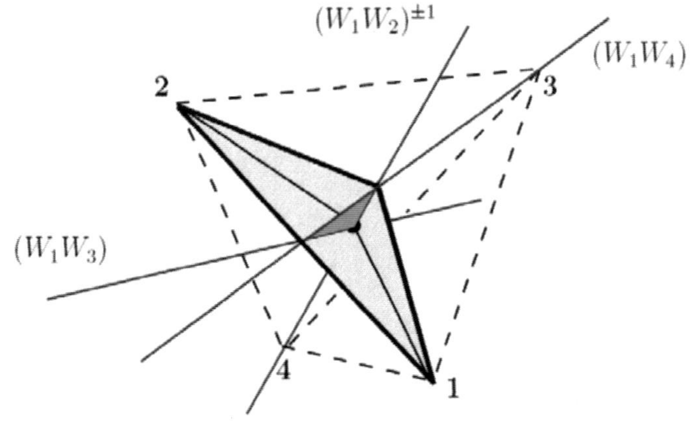

Figure 6. Tetrahedral orbifold. Face gluing deck rotations of the tetrahedral duplex orbifold (yellow) $N8$ in Euclidean version. The vertices of the Platonic tetrahedron are marked by the numbers $(1, 2, 3, 4)$. A small tetrahedron of volume fraction $\frac{1}{12}$ inside the Platonic tetrahedron is the orbifold (yellow colour). The four covering rotations $(W_1W_3), (W_1W_2)^{\pm 1}, (W_1W_4)$ of this orbifold, corresponding to the covering of its four faces, are products of Weyl reflections from the Coxeter group with diagram $\Gamma = \circ - \circ - \circ - \circ$. They are marked by their rotation axes, intersection lines of pairs of Weyl reflection planes. The three intersections of these axes with the orbifold form the edges of the glue triangle (light blue area) of two Coxeter simplices. The inner points of its edges have the orders (2, 3, 2) of their covering rotations.

The 3-sphere S^3 of constant positive curvature $\kappa = 1$, Einstein 1917 [4], covers all spherical space forms. There is a one-to-one correspondence of S^3 and $SU(2, C)$ given by

$$\begin{aligned} & x = (x_0, x_1, x_2, x_3), \\ & \leftrightarrow u(x) = \begin{bmatrix} x_0 - ix_3 & -x_1 - ix_2 \\ x_1 - ix_2 & x_0 + ix_3 \end{bmatrix} \in SU(2, C), \\ & x_0^2 + x_1^2 + x_2^2 + x_3^2 = \det(u) = 1. \end{aligned} \tag{6}$$

The rotation group $SO(4, R)$ acts on the points of the 3-sphere by the correspondence

$$SO(4, R) = (SU^l(2, C) \times SU^r(2, C))/ \pm (e, e), \tag{7}$$

and its elements (g_l, g_r) transform these points by

$$(g_l, g_r) : u \to g_l^{-1} u g_r. \tag{8}$$

Wigner introduced the matrices $D^j_{m_1m_2}(u)$ as the irreducible representation matrices of the group $SU(2, C)$ [3]. Considered as homogeneous polynomials in the four elements of the matrix $u(x)$, we call them Wigner polynomials. Their overall degree is $2j$ and is conserved under the linear transformations eq. 9. The Wigner polynomials are a complete and orthonormal set of functions on the 3-sphere S^3. Moreover they are harmonic [12], that is, vanish under the Laplacian on the Euclidean space E^4 that embeds the 3-sphere.
The action of $SO(4, R)$ on Wigner polynomials has the form

$$\begin{aligned} & (T_{(g_l,g_r)} D^j_{m_1,m_2})(u) = D^j_{m_1m_2}(g_l^{-1} u g_r) \\ & = \sum_{m_1'm_2'} D^j_{m_1'm_2'}(u) D^j_{m_1m_1'}(g_l^{-1}) D^j_{m_2'm_2}(g_r), \end{aligned} \tag{9}$$

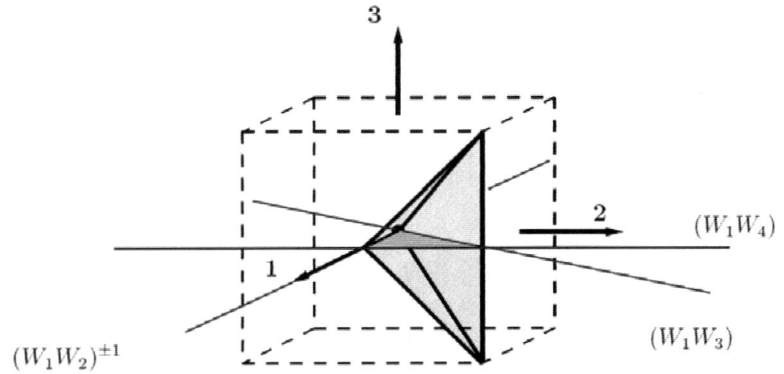

Figure 7. Cubic orbifold. Face gluing deck rotations of the cubic duplex orbifold (yellow) $N9$ in Euclidean version. Its four covering rotations $(W_1W_3), (W_1W_2)^{\pm 1}, (W_1W_4)$, products of pairs of Weyl reflections from the Coxeter group with diagram $\Gamma = \circ \overset{4}{-} \circ - \circ - \circ$, are marked by rotation axes, intersection lines of pairs of Weyl planes. The three intersections of these axes with the orbifold form the edges of the glue triangle (light blue area). Its inner edge points have the orders (2, 4, 2) of their covering rotations.

and so is determined by pairs of representation matrices of the group $SU(2, C)$. By eq. 9 we can take advantage of the elaborate techniques of angular momentum from [3].
Each spherical manifold carries a specific harmonic basis of eigenmodes, invariant under a deck group H, and obeying homotopic boundary conditions.
The eigenmodes of a spherical manifold offer two alternative views on different domains: (i) On a single closed manifold $\mathcal{M}$, they allow to expand square integrable observables. These obey homotopic boundary conditions for their functional values on faces and edges.
(ii) The covering 3-sphere is tiled by copies of $\mathcal{M}$. Any eigenmode now must on each different tile repeat its value, and fulfill the homotopic boundary conditions. It follows that the eigenmodes must display selection rules in comparison to a general polynomial basis on the 3-sphere. This second view allows for a comparison of observables for different topologies on the same domain, Einstein's 3-sphere.
In topology, these two views present (i) the local concept of homotopy of $\mathcal{M}$, and (ii) the concept of deck transformations from the group $H = \text{deck}(\mathcal{M})$ that generates the tiling on the 3-sphere as universal cover. Seifert and Threlfall [22] pp. 195-8 prove the equivalence of the two views: the fundamental or first homotopy group $\pi_1(\mathcal{M})$ that generates all the gluings is isomorphic to the group of deck transformations $\text{deck}(\mathcal{M})$ that generates the tiling on the cover.

4. Spherical Platonic 3-manifolds

We study here the family of Platonic spherical 3-manifolds. For each Platonic polyhedron we construct in [12] on the universal cover, the 3-sphere, a unique group H of fixpoint-free deck transformations acting on the 3-sphere as a subgroup of a Coxeter group Γ . By the theorem from [22], the deck groups H are isomorphic to, and were derived in [12] from, the fundamental or first homotopy groups constructed by Everitt [5].
Coxeter groups Γ [2], [6] will become a main tool of the following analysis. They are generated on Euclidean 4-space with coordinates $x = (x_0, x_1, x_2, x_3)$, and on the 3-sphere, by four involutive Weyl reflections (W_1, W_2, W_3, W_4) in hyperplanes, see Table 2. A reflection hyperplane W_i containing the origin is characterized by a unit vector a_i perpendicular to it. The Coxeter

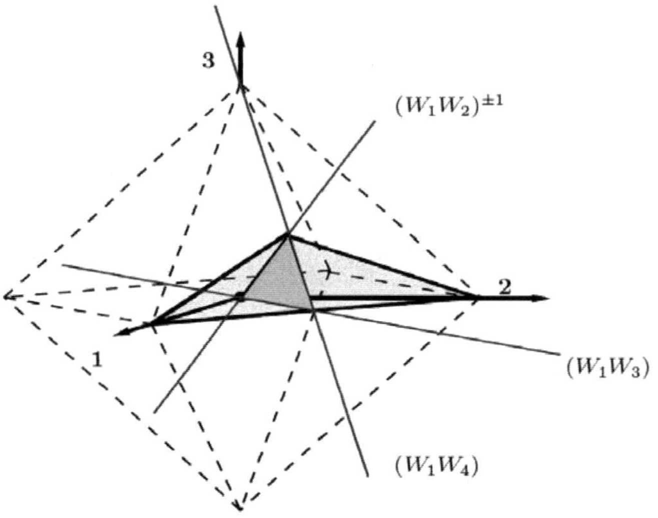

Figure 8. Octahedral orbifold. Face gluing deck rotations of the octahedral duplex orbifold (yellow) $N10$ in Euclidean version. The corresponding four products $(W_1W_3), (W_1W_2)^{\pm 1}, (W_1W_4)$ of pairs of Weyl reflections from $\Gamma = \circ - \circ \overset{4}{-} \circ - \circ$ are marked by rotation axes, intersection lines of pairs of Weyl reflection planes. The three intersections of these axes with the orbifold form the edges of the glue triangle (light blue area). Its inner edge points have the orders (2, 3, 2) of their covering rotations.

Coxeter diagram Γ	$\lvert\Gamma\rvert$	polyhedron $\mathcal{M}$	$H = \mathrm{deck}(\mathcal{M})$	$\lvert H\rvert$	Reference
$\circ - \circ - \circ - \circ$	120	tetrahedron $N1$	C_5	5	[8]
$\circ \overset{4}{-} \circ - \circ - \circ$	384	cube $N2$	C_8	8	[9]
		cube $N3$	Q	8	[12]
$\circ - \circ \overset{4}{-} \circ - \circ$	1152	octahedron $N4$	$C_3 \times Q$	24	[12]
		octahedron $N5$	B	24	[12]
		octahedron $N6$	$\mathcal{T}^*$	24	[12]
$\circ \overset{5}{-} \circ - \circ - \circ$	120^2	dodecahedron$N1'$	$\mathcal{J}^*$	120	[7]

Table 1. 4 Coxeter groups Γ, 4 Platonic polyhedra $\mathcal{M}$, 7 deck groups $H = \mathrm{deck}(\mathcal{M})$ of order $\lvert H\rvert$. C_n denotes a cyclic, Q the quaternion, $\mathcal{T}^*$ the binary tetrahedral, $\mathcal{J}^*$ the binary icosahedral group. The symbols Ni are taken or extended from [5]. The last column gives the reference for the harmonic analysis. The diagram for the icosahedral Coxeter group is corrected compared to [12].

diagram encodes the group relations between the Weyl reflections, associated to its four nodes. A line connecting two nodes implies $(W_iW_{i+1})^3 = e$, a connecting line with integer superscript $k > 3$ implies $(W_iW_{i+1})^k = e$, Weyl reflections for nodes not connected by lines commute with one another. Tables 1, 2 review the relation of the Platonic polyhedra to Coxeter groups. In Table 2 we list the four unit vectors $a_j \in E^4$, perpendicular to the Weyl reflection hyperplanes, for each Coxeter group. The Coxeter group tiles the 3-sphere into $\lvert\Gamma\rvert$ Coxeter simplices. The initial Platonic polyhedron consists of those Coxeter simplices which share a vertex at the center $x = (1, 0, 0, 0)$. In topology we prefer orientable manifolds. In the Coxeter groups this means

Γ	a_1	a_2	a_3	a_4
$\circ - \circ - \circ - \circ$	$(0, 0, 0, 1)$	$(0, 0, \sqrt{\frac{3}{4}}, \frac{1}{2})$	$(0, \sqrt{\frac{2}{3}}, \sqrt{\frac{1}{3}}, 0)$	$(\sqrt{\frac{5}{8}}, \sqrt{\frac{3}{8}}, 0, 0)$
$\circ \overset{4}{-} \circ - \circ - \circ$	$(0, 0, 0, 1)$	$(0, 0, -\sqrt{\frac{1}{2}}, \sqrt{\frac{1}{2}})$	$(0, \sqrt{\frac{1}{2}}, -\sqrt{\frac{1}{2}}, 0)$	$(-\sqrt{\frac{1}{2}}, \sqrt{\frac{1}{2}}, 0, 0)$
$\circ - \circ \overset{4}{-} \circ - \circ$	$(0, \sqrt{\frac{1}{2}}, -\sqrt{\frac{1}{2}}, 0)$	$(0, 0, -\sqrt{\frac{1}{2}}, \sqrt{\frac{1}{2}})$	$(0, 0, 0, 1)$	$(\frac{1}{2}, \frac{1}{2}, \frac{1}{2}, \frac{1}{2})$
$\circ \overset{5}{-} \circ - \circ - \circ$	$(0, 0, 1, 0)$	$(0, -\frac{\sqrt{-\tau+3}}{2}, \frac{\tau}{2}, 0)$	$(0, -\sqrt{\frac{\tau+2}{5}}, 0, -\sqrt{\frac{-\tau+3}{5}})$	$(\frac{\sqrt{2-\tau}}{2}, 0, 0, -\frac{\sqrt{\tau+2}}{2})$

Table 2. The Weyl vectors a_s for the four Coxeter groups Γ from Table 1 with $\tau := \frac{1+\sqrt{5}}{2}$.

Coxeter diagram Γ	$\lvert H\rvert$	polyhedron and orbifold	point group $M, \lvert M\rvert$	deck generators	fixpoint order k
$\circ - \circ - \circ - \circ$	5	tetrah N8	$A(4), 12$	$(W_1W_3), (W_1W_2)^{\pm 1}, (W_1W_4)$	(2, 3, 2)
$\circ \overset{4}{-} \circ - \circ - \circ$	8	cube N9	$O, 24$	$(W_1W_3), (W_1W_2)^{\pm 1}, (W_1W_4)$	(2, 4, 2)
$\circ - \circ \overset{4}{-} \circ - \circ$	24	octah N10	O, 24	$(W_1W_3), (W_1W_2)^{\pm 1}, (W_1W_4)$	(2, 3, 2)
$\circ \overset{5}{-} \circ - \circ - \circ$	120	dodec N11	$\mathcal{J}$, 60	$(W_1W_3), (W_1W_2)^{\pm 1}, (W_1W_4)$	(2, 5, 2)

Table 3.
From four Coxeter groups Γ, $|S\Gamma| = |M| \cdot |H|$ and Platonic polyhedra to orbifolds. Point groups M, deck generators of orbifolds, and selected orders k for fixpoints on edges of the glue triangle of four duplex orbifolds $N8, N9, N10, N11$, see Figs. 6 -9. $A(4)$ is the tetrahedral, O the cubic, $\mathcal{J}$ the icosahedral rotation group. The center of the polyhedron has order $|M|$, inner points of the orbifolds are regular. Note that the Weyl reflections W_l depend on the Coxeter group chosen, see Tables 1, 2.

that we must restrict our attention to the subgroups generated by an even number of Weyl reflections. We call the corresponding subgroups $S\Gamma$, where S stands for unimodularity of the defining representation on the Euclidean space E^4. The order of these subgroups is $|S\Gamma| = |\Gamma|/2$. A set of generators of all unimodular Coxeter groups with four Weyl reflections is given by

$$S\Gamma : \{(W_1W_2), (W_2W_3), (W_3W_4)\}. \tag{10}$$

The first two products of reflection generators generate the point group M of the Platonic polyhedron, the last product enters the group of deck operations of the Platonic tiling of the 3-sphere. Note that any product (W_iW_j) leaves invariant the intersection of the two Weyl reflection hyperplanes perpendicular to the vectors $\{a_i, a_j\}$. On the 3-sphere this intersection is the geodesic rotation axis for the rotation (W_iW_j). The representation of any such product is derived from their Weyl reflection vectors in [8] eq. (60): For two Weyl reflection operators with Weyl unit vectors $\{a_i, a_j\}$, define with eq. 6

$$v_i := u(a_i),\ v_j := u(a_j). \tag{11}$$

Then the rotation operator for the product (W_iW_j) is given by inserting $(g_l, g_r) = (v_iv_j^{-1}, v_i^{-1}v_j)$ into eq. 9,

$$T_{(W_iW_j)} = T_{(v_iv_j^{-1}, v_i^{-1}v_j)}. \tag{12}$$

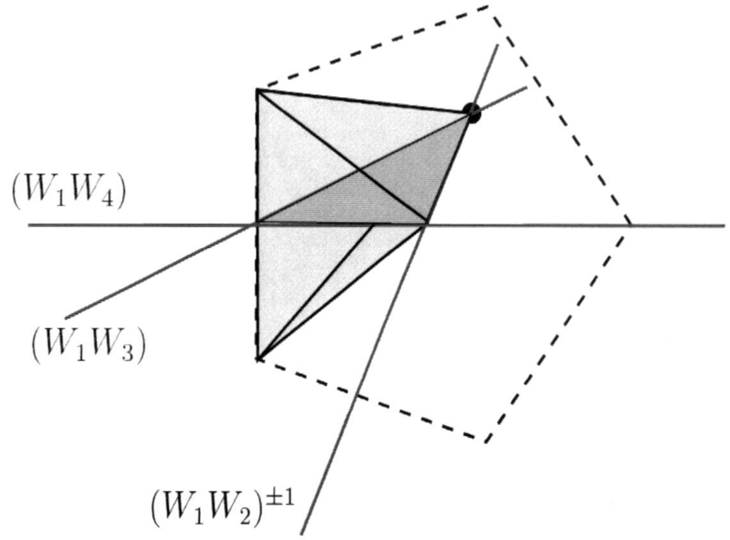

Figure 9. Dodecahedral orbifold. Face gluing deck rotations of the dodecahedral duplex orbifold (yellow) $N11$, attached to a dodecahedral face and the center, in Euclidean version. Only one face and the center are shown. The corresponding four covering products $(W_1W_3), (W_1W_2)^{\pm 1}, (W_1W_4)$ of two Weyl reflections from $\Gamma = \circ \overset{5}{-} \circ - \circ - \circ$ are marked by rotation axes, intersection lines of pairs of Weyl planes. The three intersections of these axes with the orbifold form the edges of the glue triangle (light blue area). Its inner edge points have the orders (2, 5, 2) of their covering rotations.

As an example of an eigenmode we give from [12] the selection rules for the Platonic spherical cube manifold $N3$. The 3-sphere S^3 is tiled into eight spherical cubes. This tiling is the 8-cell. The deck transformations $H = Q$ from [12] form the quaternion group Q generated by the quaternions $\langle \pm e, \pm i, \pm j, \pm k \rangle$. The projector P_Q^0 to the identity representation of $H = Q$ applied to a Wigner polynomial is found to be

$$(P_Q^0 \, D^j_{m_1 m_2})(u) = \frac{1}{8} \left[1 + (-1)^{2j}\right] \left[1 + (-1)^{m_1}\right] \times \left[D^j_{m_1 m_2}(u) + (-1)^{m_1} D_{-m_1 m_2}(u)\right], \tag{13}$$

it selects $j = \text{integer}, m_1 = \text{even}$.

5. Duplices under the point group tile a polyhedron

In [12] we introduce in addition for random functions the notion of random symmetry under the point group M of these manifolds. We argue there that the values of a proper random function on a polyhedral manifold whose shape displays the point symmetry M should be independent of operations from M. Therefore one should explore M for the analysis of the CMB radiation. The point groups are the tetrahedral group $A(4)$ for the tetrahedral 3-manifold, the cubic group O for the cubic and octahedral 3-manifolds, and the icosahedral group $\mathcal{J}$ for the spherical dodecahedron. These point groups are unimodular subgroups w.r.t the 2-sphere

$$x_1^2 + x_2^2 + x_3^2 = 1, \tag{14}$$

and may be characterized on 3-space by unimodularity S and by Coxeter subdiagrams as

$$A(4) = S(\circ - \circ - \circ),\ O = S(\circ \overset{4}{-} \circ - \circ) \sim S(\circ - \circ \overset{4}{-} \circ),\ \mathcal{J} = S(\circ \overset{5}{-} \circ - \circ). \tag{15}$$

The action of the point group M on the Platonic polyhedron with center at $x = (1, 0, 0, 0)$ can now be decomposed into a pre-image and its images under M. The shape of a fundamental domain is not unique, but we can choose it in compact, convex and polyhedral form: Each fundamental domain for M we take as a duplex inside the proto-polyhedron, formed by gluing a Coxeter simplex and its mirror image under reflection in one simplex face. This simplex face we call the glue triangle. It is shown in Figs. 6, 7, 8, 9.

Prop 1: Fundamental domains under point groups: For each Platonic polyhedron, a duplex fundamental domain of their point group M may be chosen as shown in Figures 6, 7, 8, 9. All of them are again spherical simplices with four faces, have one vertex at the center of the polyhedron, and tile the polyhedron.

6. Orbifolds under the group $S\Gamma$ tile the 3-sphere

The Platonic 3-manifolds under their deck group H in turn tile the 3-sphere into $|H|$ copies of the Platonic polyhedron as prototile. These tilings into Platonic polyhedra are the m-cells, $m = |H|$, described in [23] and [12]. By composing the polyhedra, tiled into duplices, as tiles of the $|H|$-cell tilings, it follows that the 3-sphere is tiled into $|S\Gamma| = |M|\cdot|H|$ duplices. With respect to actions on the 3-sphere, the Coxeter duplices, which originated as fundamental domains of the point group acting on the Platonic prototile, now become the fundamental domains of the much bigger groups $S\Gamma$. In the mathematical community they are called orbifolds. In Euclidean crystallography with space groups they are familiar under the name asymmetric units. A planar Euclidean example is provided in Fig. 5.

6.1. The action of deck and point groups.

The subgroup $H < S\Gamma$ of deck transformations for a given 3-manifold acts on the 3-sphere without fixpoints. The point group $M < S\Gamma$ by definition preserves the center of the polyhedron. It follows [14] that the intersection of these two subgroups contains only the identity, $H \cap M = e$. Any image of the orbifold as proto-duplex and pre-image under $S\Gamma$ on the 3-sphere has a unique factorized address, composed of a unique point group element $p \in M$ acting on the initial polyhedron, followed by a unique deck transformation $h \in H$ from the initial polyhedron to an image on the $|H|$-cell tiling. This leads to the following general conclusion on the group structure:

Prop 2: Unimodular Coxeter groups are products of subgroups : Given a prototile duplex, its image tile in the Coxeter duplex tiling of the 3-sphere results from the action of a group element from $S\Gamma$, uniquely factorized into a deck transformation from $h \in S\Gamma$ and a point transformation $p \in M$. This means that the elements $g \in S\Gamma$ obey the unique subgroup product law

$$g \in S\Gamma:\ g = hp,\ h \in H,\ p \in M,\ H \cap M = e,\ |S\Gamma| = |H| \cdot |M|. \tag{16}$$

Proof: The group $S\Gamma$ with subgroups H, M all have faithful (one to one) representations, [25] p. 138, on E^4. This implies that any image under $g \in S\Gamma$ of the orbifold as proto-duplex has a unique factorized address as a product $g = hp,\ h \in H, p \in M$. Similarly one can construct for the same image a unique factorized address $g = p'h'$.

This product structure $S\Gamma = H \cdot M$ differs from a direct or semidirect product. The two subgroups H, M do not commute, and the group H in general is not invariant under conjugation with elements from M, except in case of the cubic Coxeter group, analyzed in [12]. The

Journal of Physics: Conference Series **380** (2012) 012024 doi:10.1088/1742-6596/380/1/012024

icosahedral rotation group for example has no invariant subgroup and so cannot be a semidirect product of two subgroups.
Eq. 16 shows that each of the subgroups generates the cosets for its partner.

6.2. Topology and deck groups for orbifolds.
We turn to the topological significance of the fundamental duplex domains for the group $S\Gamma$. A duplex of a unimodular Coxeter group on the 3-sphere cannot form a topological 3-manifold, because under the point group action it exhibits fixpoints of finite order $k > 1$ on its boundaries. The fixpoints are located on the rotation axis passing through the faces of the duplices, and their order is simply the order of the corresponding rotations.
The order k of a point is defined as the order of its stabilizer. Points of order $k = 1$ in topology are called regular, for order $k > 1$ singular, [21] pp. 664-6. We propose here to move in topology from the standard space forms to orbifolds. This concept is illustrated on the square in Fig. 5. We refer to Montesinos [20] pp. 78-97 and to Ratcliffe [21] pp. 652-714 for the introduction and mathematical terms associated with orbifolds. The notion includes a manifold structure and a covering, but admits singular points of finite order.
We claim:
Prop 3: Spherical Platonic 3-manifolds with point symmetry are appropriately described as orbifolds.
We shall demonstrate orbifold coverings by deck transformations from $S\Gamma$, and derive the harmonic analysis for their use in cosmic topology.
In figures 6, 7, 8, 9 we reproduce, with minor changes, based on [12], [13], a set of fundamental duplex domains for the point groups. Now we interprete them as 3-orbifolds under $S\Gamma$ acting on the 3-sphere, based on the tetrahedron, the cube, the octahedron or the dodecahedron, drawn in their Euclidean version. The edges of the Platonic polyhedra are given in dashed lines, the orbifolds are marked by yellow color. Note that the polyhedra in the figures are drawn in Euclidean fashion, but stand for spherical polyhedra on S^3.
For topological 3-manifolds, the deck group on the 3-sphere is generated by the operations which map the pre-image of the manifold to all its face neighbours. We present a corresponding analysis for the deck group of an orbifold on the 3-sphere.
We use the tetrahedral manifold $N2$ from Table 3 for the demonstration of orbifold covering. Here the Coxeter group is $S(5)$, the symmetric group on 5 objects, with unimodular subgroup $S\Gamma = A(5)$, the group of even permutations. In Fig. 6 we mark the four vertices of the tetrahedron by the numbers $1, 2, 3, 4$. A different enumeration is used in [12], Fig. 4. The Weyl reflection operators are in one-to-one correspondence to the vertex transpositions

$$W_1 = (1,2),\ W_2 = (2,3),\ W_3 = (3,4),\ W_4 = (4,5). \tag{17}$$

From eq. 17 and Fig. 6 we find for the four covering deck rotations of this orbifold the simple expressions

$$\begin{aligned}
&(W_1W_3) = (1,2)(3,4), \\
&(W_1W_2) = (1,2)(2,3) = (1,2,3),\ (W_1W_2)^{-1} = (2,3)(1,2) = (3,2,1), \\
&(W_1W_4) = (1,2)(4,5),
\end{aligned} \tag{18}$$

compare Fig. 6. The three even permutations in eq. 18 cover the orbifold and generate $A(5)$.
We return to all four orbifolds associated with Platonic polyhedra and denote them by $N8, N9, N10$ as in [13] and by $N11$ for the Platonic dodecahedron. Each face of the orbifold as proto-tile of the duplex tiling is covered by a face-neighbour, a copy of the orbifold. Instead of drawing the neighbours, we give in the Figures 6-9 the deck rotations from $S\Gamma$ which map the preimage into

its face neighbours. These deck rotations have axes which intersect with faces of the orbifold. In the figures we denote the rotation axes by blue lines and give in each case the rotation as an even product of Weyl reflections. The first three deck rotations all have axes passing through the center of the polyhedron and generate the point group M. The last even product always involves a Weyl reflection that passes through an outer face of the Platonic polyhedron. The rotation containing this Weyl reflection has order $k = 2$. It transforms the orbifold into a face neighbour inside a new Platonic neighbour polyhedron. By examining the deck operations from the figures and comparing them with the Coxeter group relations, it can be verified in each case that the four deck operations that cover the faces of the orbifold generate the full unimodular Coxeter group $S\Gamma$. This is in full analogy to the role of deck groups H for spherical 3-manifolds [12].

The intersections of the covering rotation axes with the 3-orbifold determine singular points and their orders. The center point of the initial spherical polyhedron, chosen as $x = (1, 0, 0, 0)$ and marked in the figures in black, is a fixpoint of maximal order $k = |M|$. For the orbifolds under inspection, more fixpoints appear on the inner points of edges of the glue triangle of each duplex, with area marked in the figures in light blue. The order k of these fixpoints agrees with the order of the covering rotation. Note that, in contrast to the Euclidean drawings, we always infer the order k of the rotations from the Coxeter group relations and their spherical settings on the 3-sphere. We give the covering rotations and orders of these fixpoints in Table 3.

The obvious rotation axes from the point group M determine additional singular points on inner positions of edges of the orbifolds. For all four orbifolds we find by inspection:

Prop 4: Deck transformations of orbifolds generate unimodular Coxeter groups: Deck rotations cover the four faces of the four Platonic 3-orbifolds, they generate the duplex tiling of S^3 by all elements of the corresponding unimodular Coxeter deck groups $S\Gamma$. It follows that the orbifolds can abstractly be characterized as quotient spaces $S^3/S\Gamma$.

7. Harmonic analysis on orbifolds

The harmonic analysis can follow the two views described in section 3. Clearly the homotopic boundary conditions imply selection rules compared to the full basis on the 3-sphere. The CMB radiation on an assumed topology usually is modelled by random coefficients in the specific basis. Here we analyze the basis construction, but postpone any numerical modelling.

On the 3-sphere, after replacing the Cartesian coordinates x by matrix coordinates $u = u(x) \in SU(2, C)$ [12], the harmonic basis may be spanned by Wigner polynomials $D^j_{m_1 m_2}(u)$ [3]. More convenient for pure three-dimensional rotations of 3-space (x_1, x_2, x_3) as they appear in the point groups are the spherical polynomials $\psi(jlm)(u)$, linearly related to the Wigner polynomials by

$$\psi(jlm)(u) = \sum_{m_1 m_2} D^j_{m_1 m_2}(u)\langle j - m_1 j m_2 | lm\rangle(-1)^{(j-m_1)}, \tag{19}$$

$$D^j_{m_1 m_2}(u) = \sum_{lm} \psi(jlm)(u)\langle j - m_1 j m_2 | lm\rangle(-1)^{(j-m_1)},\ 0 \le l \le 2j,$$

and with summations restricted by the Wigner coefficients [3]. The spherical polynomials transform, under rotations of the last three coordinates (x_1, x_2, x_3) while preserving x_0, like the spherical harmonics Y^l_m [12],

$$(T_{(g_l, g_l)}\psi)(jlm)(u) = \psi(jlm)(g_l^{-1} u g_l) = \sum_{m'=-l}^{m'=l} \psi(jlm')(u) D^l_{m',m}(g_l). \tag{20}$$

They are adapted to the application of rotations from the point group. Point symmetry selects a lowest multipole order l_0, compare [12] Table 3, [13]. The Wigner polynomial basis is suited

for the projection of the subbasis invariant under the deck group H. In the cubic case, the deck group of the 3-manifold $N3$ is the quaternion group $H = Q$. Invariant under conjugation with the cubic point group O, it forms a semidirect product $S\Gamma = (Q \times_s O)$.
The harmonic analysis on the four spherical orbifolds $N8, N9, N10, N11$ is given by polynomials invariant under their unimodular Coxeter group $S\Gamma$. These and only these polynomials repeat their values on any Coxeter duplex from the tiling.
Prop 5: Harmonic analysis on spherical 3-orbifolds: The harmonic analysis on a spherical orbifold is spanned by harmonic polynomials, invariant under its group of deck transformations $S\Gamma$.
We have seen in Prop. 2 that any element of $S\Gamma$ admits a unique factorization. For the projectors to the identity representation of $S\Gamma$ we claim by use of eq. 16:
Prop 6: Factorization of projectors: The projector to the identity representation, denoted by Γ_1, for the unimodular Coxeter group $S\Gamma$ factorizes into the product of the projectors to the identity representations Γ_1 for the two subgroups H, M,

$$P_{S\Gamma}^{\Gamma_1} = P_H^{\Gamma_1} P_M^{\Gamma_1} = P_M^{\Gamma_1} P_H^{\Gamma_1}. \tag{21}$$

Proof: In the group operator algebra of the unimodular Coxeter group $S\Gamma$ we have from eq. 16 the factorization

$$\begin{aligned} P_{S\Gamma}^{\Gamma_1} &= \sum_g T_g = \sum_{h,\,p} T_{hp} = \sum_{p,\,h} T_{ph}, \;\; h \in H, p \in M, \\ &= \sum_{h,\,p} T_h T_p = (\sum_h T_h)(\sum_p T_p) = (\sum_p T_p)(\sum_h T_h) = P_H^{\Gamma_1}\, P_M^{\Gamma_1} = P_M^{\Gamma_1} P_H^{\Gamma_1}. \end{aligned} \tag{22}$$

This result greatly simplifies the projection to the identity representation:
The projectors $P_H^{\Gamma_0}$ for the group H of deck transformations are given in [12] in the Wigner polynomial basis, see eq. 19.
Prop 7: Orbifolds give sharp multipole selection rules: $S\Gamma$-invariant polynomials by their point group M-invariance select a lowest non-zero multipole order l_0, see [12] Table 3. This projection is carried out in [13] for the spherical cubic 3-orbifold and multipole order $0 \le l \le 8$. The results are reproduced, and in part corrected compared to [13], in Tables 4 and 5 in explicit multipole order.
We note that the results depend on a convenient choice of the coordinates in the spherical harmonics with respect to the symmetry axes of the orbifolds. Under a general rotation of the system of these coordinates one would obtain many additional terms of the same multipole order l. The function $R_{2j+1,l}(\chi)$ is a Gegenbauer polynomial given in [12] eq. (55). The analysis can be extended by the method described in section 9.
A similar analysis applies to the Platonic spherical tetrahedral, octahedral and dodecahedral 3-orbifolds.

8. Homotopic boundary conditions from orbifolds

From [12] we know that a fixed geometric shape of a Platonic 3-manifold can have different and inequivalent topologies, characterized by different groups H of homotopies and deck transformations. These differences give rise to different homotopic boundary conditions. We now examine the boundary conditions for orbifolds.
We have seen that the orbifold is covered face-to-face by its rotational images. It follows, as in the case of 3-manifolds, that the topology on 3-manifolds implies homotopic boundary conditions on the faces of the 3-orbifolds. Since with the orbifolds we introduce point symmetry in addition to deck transformations, we find from the arguments given in [12] for random point symmetry,

l	$Y^{\Gamma_1,l} = \sum_m a_{lm} Y^l_m(\theta,\phi)$
0	Y^0_0
4	$\frac{1}{2}\sqrt{\frac{7}{3}}Y^4_0 + \frac{1}{2}\sqrt{\frac{5}{6}}(Y^4_4 + Y^4_{-4})$
6	$\frac{1}{2\sqrt{2}}Y^6_0 - \frac{\sqrt{7}}{4}(Y^6_4 + Y^6_{-4})$
8	$\frac{\sqrt{33}}{8}Y^8_0 + \frac{1}{4}\sqrt{\frac{7}{6}}(Y^8_4 + Y^8_{-4}) + \frac{1}{8}\sqrt{\frac{65}{6}}(Y^8_8 + Y^8_{-8})$

Table 4. The first cubic invariant spherical harmonics $Y^{\Gamma_1,l}$, expressed by spherical harmonics $Y^l_m(\theta,\phi)$. Γ_1 denotes the identity representation of the point group H.

$2j$	l	$\psi^{0,\Gamma_1,2j} = \sum_l b_l R_{2j+1\,l}(\chi) Y^{\Gamma_1,l}(\theta,\phi)$
0	0	$R_{10}Y^{\Gamma_1,0}$
4	0, 4	$\sqrt{\frac{2}{5}}R_{50}Y^{\Gamma_1,0} + \sqrt{\frac{3}{5}}R_{54}Y^{\Gamma_1,4}$
6	0, 4, 6	$\sqrt{\frac{1}{7}}R_{70}Y^{\Gamma_1,0} + \sqrt{\frac{6}{11}}R_{74}Y^{\Gamma_1,4} + 2\sqrt{\frac{6}{77}}R_{76}Y^{\Gamma_1,6}$
8	0, 4, 6, 8	$\sqrt{\frac{1}{57}}R_{90}Y^{\Gamma_1,0} + 9\sqrt{\frac{2}{2717}}R_{94}Y^{\Gamma_1,4}$
		$+2\sqrt{\frac{38}{165}}R_{96}Y^{\Gamma_1,6} + \sqrt{\frac{2}{1235}}R_{98}Y^{\Gamma_1,8}$

Table 5. Harmonic analysis on the cubic Platonic orbifold $N3$: The first ($S\Gamma = (Q \times_s O)$)-invariant polynomials $\psi^{0,\Gamma_1,2j}$ of degree $2j = 0,4,6,8$ on the 3-sphere, expressed by Gegenbauer polynomials $R_{nl}(\chi)$ and cubic invariant spherical harmonics $Y^{\Gamma_1,l}$ from Table 4. $(Q \times_s O)$-invariance enforces superpositions of several cubic invariant spherical harmonics. Note the suppression of odd and $l = 2$ multipole orders in the eigenmodes.

Prop 8: Topological universality from point symmetry: If we demand, for a function on a given Platonic 3-manifold, point symmetry under M in addition to the boundary conditions set by homotopy on faces and edges, then new boundary conditions apply universally, that is, dependent on the point group M but independent of the specific group H of deck transformations chosen.

Prop 9: Universal homotopic boundary conditions from 3-orbifolds: If in addition to homotopy we demand on the manifold symmetry under the rotational point group M, the homotopic boundary conditions for different deck and homotopy groups on the same Platonic geometrical shape coincide with one another and reduce to the homotopic boundary conditions for the orbifolds. Their boundary conditions are determined by the generators in Table 1 of the covering rotations.

9. Recursive computation of the bases invariant under the orbifold deck groups $S\Gamma = H \cdot M$

We describe the recursive construction of $S\Gamma$-invariant bases by use of the factorization given in Prop. 4. This recursive construction is used to obtain the results of Tables 4, 5. The multiplicity $m(l,\Gamma_1)$ of the identity representation Γ_1 of M for given multipole order l is given in [17] pp. 436-8.

Our basic tool are the relations [12] between Wigner and spherical harmonic polynomials on the 3-sphere given in eq. 19.

(i) We start from a linear combination of spherical harmonics of multipole order l, invariant under the point group M, and construct with a Gegenbauer polynomial $R_{nl}(\chi)$ and its coefficients a_{lm} a M-invariant linear combination of spherical polynomials $\psi_{jlm}(u)$, $l \leq 2j$ eq. 22, on the 3-sphere

[12] with lowest degree $l = l_0 = 2j$,

$$\psi^0_{j,l}(u) = \sum_m^l a_{lm'} \psi_{j,l,m'}(u). \tag{23}$$

By the upper index 0 we denote invariance under the point group M. The starting linear combinations of spherical harmonics for lowest multipole order $l_0 > 0$ for the point groups involved can be found in the literature. For tetrahedral point symmetry the lowest multipole order from [12] Table 3 is $l_0 = 3$. For cubic point symmetry $M = O$ this function has $l_0 = 4$ and is given in Table 4. Since the point group M acts by rotations eq. 20 only, the coefficients $a_{lm'}$ in eq. 23 are independent of the degree $n = 2j$ of the polynomial.
(ii) Next we transform with eq. 19 from the spherical to the Wigner basis and apply the projector $P^{\Gamma_1}_H$ for H-invariance, for example the one with $H = Q$ for the cubic manifold $N3$ given in eq. 13. If the resulting function does not vanish, we transform with eq. 19 back to the spherical basis. In this way we find from the starting function a new one, invariant from Prop. 4 under both M and H and hence under $S\Gamma$, given by

$$\begin{aligned} \psi_{j,\Gamma_1}(u) &= \sum_i \left[\sum_m a^{\Gamma_1}_{l+i,m} \psi_{j,l+i,m}(u) \right], \\ a^{\Gamma_1}_{l+i,m} &= \sum_{m',m'_1,m'_2,m_1,m_2} a_{lm'} \langle j - m'_1 j m'_2 | l m' \rangle (-1)^{(j-m'_1)} \\ &\langle j m'_1 m'_2 | P^{\Gamma_1}_H | j m_1 m_2 \rangle \langle j - m_1 j m_2 | l + i\, m \rangle (-1)^{(j-m_1)}. \end{aligned} \tag{24}$$

The polynomial eq. 24, if non-vanishing, can be normalized.
The Wigner coefficients are given in [3], and the matrix elements of the projector $P^{\Gamma_1}_H$ for the identity representation of the group H are given in the Wigner polynomial basis by

$$\langle j m'_1 m'_2 | P^{\Gamma_1}_H | j m_1 m_2 \rangle = \frac{1}{|H|} \sum_{h=(h_l,h_r) \in H} D^j_{m'_1 m_1}(h_l^{-1}) D^j_{m'_2 m_2}(h_r), \tag{25}$$

and specified in [12] for each Platonic 3-manifold. Due to universality Prop. 5, we can choose the most convenient deck group H for a fixed geometric shape. The recursion relation eq. 24 involves Wigner coefficients, the elements of the group H given as pairs $h = (h_l, h_r)$ in [12], and Wigner D^j-representations for the group $SU(2, C)$. The $S\Gamma$-invariant bases appear as linear combinations of M-invariant spherical functions with fixed multipole order $l + i$.
(iii) Moreover, since the point group action cannot change the multipole order, each new partial sum of eq. 24 in square brackets for fixed $l + i$ must generate a (new) invariant under the point group M. This allows to restart the computation with $l \to l + i$ by going again from spherical to Wigner polynomials, followed by projection of an invariant under H. In this way we can increase the polynomial degree $2j$. In Tables 4 and 5 we show the first steps in the recursive computation of eigenmodes for the group $Q \times_s O$. By character technique, see [8], [9], we can control the number of invariants for given degree $2j$ of the polynomials.

10. Applications

Before applying the eigenmodes as found in Tables 4, 5 to the analysis of CMB radiation, one has to consider several questions from astrophysics. In principle the CMB radiation spectrum reflects the density fluctuations of the primordial plasma at the time of last scattering. In the cosmological process of creation and propagation of the CMB radiation one has to consider the Sachs-Wolfe and the Doppler effect. An obstacle is a strong foreground contamination of the observed radiation due to CMB sources within our own galaxy, see [24], which the WMAP team tried to eliminate. The Ulm group of physicists has experience in incorporating all these astrophysical data. The standard procedure then is to choose a topological manifold, to consider random superpositions of its eigenmodes up to a maximal polynomial degree and multipole order, to incorporate the astrophysical data, and compute spectra similar to Fig. 1. An application of the cubic Platonic 3-manifold $N3$ that includes all astrophysical data is given in cooperation with the Ulm group in [1].

11. Conclusion

In the present analysis, we follow J Levin in [18] p. 305, who in the context of hyperbolic 3-manifolds sees no convincing reason to omit the study of orbifolds in relation to the CMB. We illuminate the topological background of cosmic topology and then develop new methods of group representations for the analysis of eigenmodes for spherical orbifolds. We see the future interpretation and use of spherical 3-orbifolds from Table 1 in cosmic topology along two lines:
(I) We can retain the strict original notion and analysis of topological 3-manifolds. The new harmonic basis characterized by point symmetry forms a subbasis of the harmonic analysis based on the deck group H. If this subbasis fails to model the CMB fluctuations, one can return to the larger basis of [12] for the 3-manifold without point symmetry.
(II) We formulate the notion of a topological 3-manifolds with (random) point symmetry M in terms of spherical 3-orbifolds. The relevant groups of deck transformations are shown to be four unimodular Coxeter groups $S\Gamma$. They generate, from a Coxeter duplex orbifold as prototile, a tiling of the 3-sphere into $|S\Gamma|$ copies. We factorize these groups as $S\Gamma = H \cdot M$ into products of a group H of polyhedral deck transformations with the point group M of the polyhedron. The Platonic 3-manifolds under the assumption of random point symmetry are replaced by 3-orbifolds, and their harmonic bases are supported on a fraction $1/|S\Gamma|$ of the 3-sphere.
For the deck group $S\Gamma = H \cdot M$ we construct a new harmonic analysis which can model the CMB radiation. Point symmetry implies topological universality: all fundamental groups for the same geometrical polyhedral shape produce the same boundary conditions. In contrast to general spherical space forms, point symmetry enforces, depending on the chosen point symmetry, a sharp non-zero onset of the multipole expansion of CMB radiation with harmonic polynomials of minimal non-trivial degree $2j_0 = l_0$. This phenomenon is in line with the physics of molecules with point symmetry [17]. The multipole selection rules, if present, must show up already in the lowest multipole orders of the CMB radiation. Moreover, topology correlates polynomials of increasing multipole order with one another. These phenomena are exemplified in Tables 4, 5 for orbifolds from spherical cubes.
Acknowledgement: The author is grateful to Tobias Kramer for the recursive algebraic computation of the results given in Tables 4 and 5.

References

[1] Aurich R, Kramer P and Lustig S 2011 *Physica Scripta* **84** 055901 (arXiv:1107.5214)
[2] Coxeter H S M 1973 *Regular polytopes* (New York: Dover)
[3] Edmonds A R 1957 *Angular momentum in quantum mechanics* (Princeton: Princeton University Press)
[4] Einstein A 1917 *Preuss. Akad. Wiss. Berlin Sitzber.* 142
[5] Everitt B 2004 *Topol. Appl.* **138** 253
[6] Humphreys J E 1990 *Reflection groups and Coxeter groups* (Cambridge: Cambridge University Press)

[7] Kramer P 2005 *J. Phys A: Math. Gen.* **38** 3517
[8] Kramer P 2009 *Physica Scripta* **79** 045008 (arXiv:0810.3403)
[9] Kramer P 2009 *Physica Scripta* **80** 025902 (arXiv:0901.0511)
[10] Kramer P 2010 *Physica Scripta* **81** 025005 (arXiv:0908:.1000)
[11] Kramer P 2010 *Physica Scripta* **82** 019802
[12] Kramer P 2010 *Class. Quantum Grav.* **27** 095013(arXiv:0909.2758v5)
[13] Kramer P 2011 in: *Symmetries in Nature, Symposium in Memoriam Marcos Moshinsky, AIP Conference Proceedings 1323* ed K B Wolf, L Benet, J M Torres and P O Hess (Melville, NY: American Inst. of Physics)
[14] Kramer P 2010 in: *Latin-American School of Physics XL ELAF: Symmetries in Physics, AIP Conference Proceedings 1334* ed O Castaños, R Bijker, R Jáuregui, and O Rosas-Ortiz (Melville, NY: American Inst. of Physics)
[15] Larson D, Dunkley J, Hinshaw G, Komatsu E, Nolta M R, Bennett C L, Gold B, Halpern M, Hill R S, Jarosik N, Kogut A, Limon M, MeyerS S , Odegard N, Page L, Smith K M, Spergel D N, Tucker G S, Weiland J L, Wollack E, and Wright E L 2011 *Astrophysical J Supplement Series* **192** 16
[16] Laue M v 1956 *Die Relativitätstheorie II* (Braunschweig: Vieweg und Sohn)
[17] Lax M 1974 *Symmetry principles in solid state and molecular physics* (New York: Wiley)
[18] Levin J 2002 *Phys. Rep.* **365** 251
[19] Misner Ch W, Thorne K S, and Wheeler J A 1970 *Gravitation* (San Francisco: W H Freeman and Co)
[20] Montesinos J M 1987 *Classical tesselations and three-manifolds* (Berlin: Springer)
[21] Ratcliffe J G 1994 *Foundations of hyperbolic manifolds* (Berlin: Springer)
[22] Seifert H and Threlfall W 1934 *Lehrbuch der Topologie* (Leipzig: Teubner; Providence, RI: AMS Chelsea Publishing, 1980 reprint)
[23] Sommerville D M Y 1958 *An introduction to the geometry of N dimensions* (New York: Dover)
[24] Weeks J 2004 *Notices of the AMS* **51** 610
[25] Wolf J A 1984 *Spaces of constant curvature* (Washington: Publish or Perish)

Quantum Particles From Quantum Information

T Görnitz
Fachbereich Physik, J. W. Goethe-Universität Frankfurt/Main, Germany
Mail: Karl-Mangold-Str. 13, D-81245 München, Germany

E-mail: goernitz@em.uni-frankfurt.de

U Schomäcker,
Institut für Mathematische Physik, TU Braunschweig
Mendelssohnstraße 3, D-38106 Braunschweig, Germany

E-mail: u.schomaecker@tu-bs.de

Abstract. Many problems in modern physics demonstrate that for a fundamental entity a more general conception than quantum particles or quantum fields are necessary. These concepts cannot explain the phenomena of dark energy or the mind-body-interaction. Instead of any kind of "small elementary building bricks", the Protyposis, an abstract and absolute quantum information, free of special denotation and open for some purport, gives the solution in the search for a fundamental substance. However, as long as at least relativistic particles are not constructed from the Protyposis, such an idea would remain in the range of natural philosophy. Therefore, the construction of relativistic particles without and with rest mass from quantum information is shown.

1. Introduction

Modern physics has to deal with problems which refer to a principal understanding of nature. On the one end of the line of the actual problems, we find the so-called dark energy. It seems to be responsible for an essential portion of the cosmic expansion and it is often declared as an unknown entity. At the other end, science is confronted with the mind-body-interaction.

At first view, it seems to be odd that these problems may have something in common, but it is not. The material objects from visible and dark matter seem to constitute only a small part of reality. For the dark energy it seems to be evident that a particle-like basis does not exist. Then again thoughts and feelings are real but the elementary objects with or without a rest mass – the photons or atoms – cannot explain the reality of the human psyche.

The solution of both issues and for some others is related to a new and fundamental understanding of matter because quantum particles and even quantum fields, which can be reduced to quantum particles, are not enough to explain the phenomena.

Therefore it can be often read in the last years that there is a need for a more general idea. The millennium-old idea must be transcended that matter ultimately consists of some types of small building bricks. Such "elementary bricks" were envisaged primarily as the atoms, later on as the hadrons, then as the leptons and quarks and in the last years as the strings. However, such an imagination of small elementary objects prevents a solution of the problems and it can also be defeated by modern quantum theory. Quantum theory knows already structures which do not appear as free particles in space. Structural quanta like phonons, quarks, and so on, cannot appear as free relativistic particles, nevertheless, they can achieve effects.

The key for a solution is abstract and absolute quantum information. It must be introduced as having primarily no special denotation, but it has to be open for some purport. To avoid the obvious identification of information with destination or meaning, a new term is introduced: Protyposis. In the 1950^{ies} C. F. v. Weizsäcker had started to think about quantum information as the basis for all of the physical objects.[1] However, the conceptions of particles and fields are too successful to abandon them. Not only chemistry is incomprehensible without the particle conception. Therefore, to connect such an idea with the established parts of physics, it is inevitable to show in which way particles can be constructed from qubits. Wigner [2] gave the mathematical construction for the description of relativistic particles. The group of motions in Minkowski-space consists of the Lorentz-group and the translations. Therefore, Wigner spoke about the "inhomogeneous Lorentz-group". Today, the name Poincaré-group became popular for it.

If the protyposis should be connected with the established parts of physics, then, beside other requirements, the construction of relativistic particles from qubits has to be given. This means that irreducible representations of the Poincaré-group must be constructed. Today the related physics can be found in many textbooks[3]. So, let us first describe the qubits.

The states of the qubits are normalized vectors in a complex two-dimensional space $\boldsymbol{C}^2$:

$$e^{i\varphi}\,(x+iy,\; z+iw)$$

$$\text{with } x^2+y^2+z^2+w^2=1$$

The invariance group for the expectation values consists of the complex conjugation, a U(1) for the phase, and a SU(2). The SU(2) is locally isomorphic to O(3), the rotational group in three dimensions. Therefore v. Weizsäcker's conjecture was that the three-dimensional cosmic position space is a consequence of quantum information and he had the hope that the real existing quantum particles can be constructed from quantum bits.

In these early times, one of the problems was that the models were in conflict with general relativity and that the intended numbers for the qubits of a particle seems to be too large for the imagination of the physicists. In this time the idea seemed to be unacceptable for most of them that a proton is 10^{40} qubits. In the 1980^{ies}, Görnitz had shown [4] that such numbers can be connected to the existing parts of physics by using the Bekenstein-Hawking-entropy and a rational cosmology.

The states of a qubit can be understood as forming a two-dimensional sub-representation of the regular representation of SU(2). These states are represented by functions having only one knot surface on the SU(2), a three-dimensional sphere $\boldsymbol{S}^3$ which will be identified with the cosmic position space. Such functions will divide the $\boldsymbol{S}^3$ into only two parts. Many qubits are represented by the tensor product of the two-dimensional representations of the SU(2). These product representations can be reduced to irreducible representations. In such a reduction scheme, more localized functions appear. Therefore, something bounded can be localized by many qubits on the $\boldsymbol{S}^3$. We will illustrate this by a trivial example.

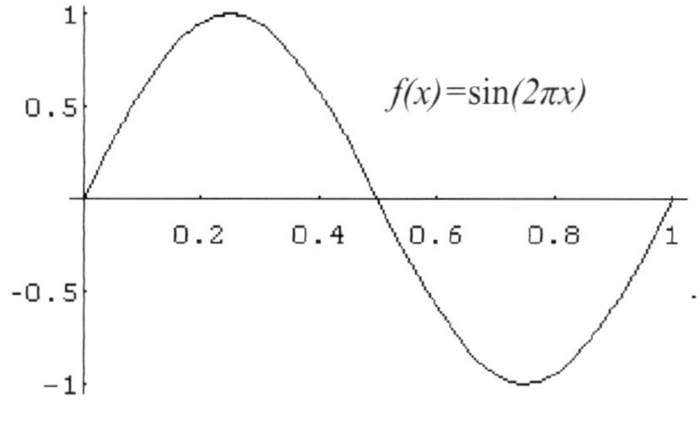

Fig. 1: The sinus divides the interval [0,1] into two parts.

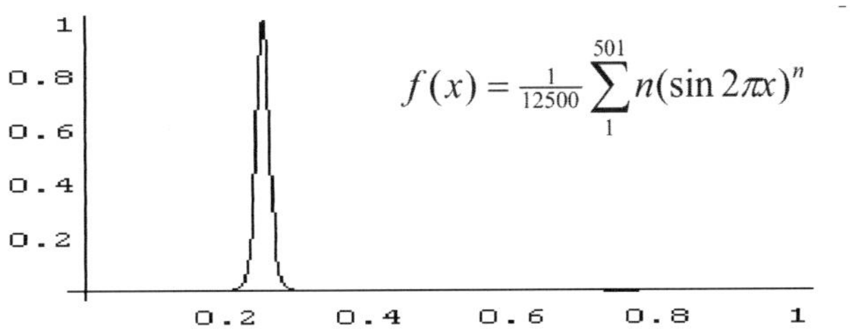

Fig. 2: By means of higher powers of the sinus a better localization can be achieved.

If N qubits are present in the cosmic position space, the tensor product of the two-dimensional representations ${}^2D_{1/2}$ can be reduced to irreducible representations. The resulting Clebsch-Gordan-series results as

$$({}^2D_{1/2})^{\otimes N} = \bigoplus_{j=0}^{|N/2|} \frac{N!(N+1-2j)}{(N+1-j)!\,j!}\left({}^{2|N/2|-2j+1}D_{|N/2|-j}\right)$$

with the factor of multiplicity

$$f(j) = \frac{N!(N+1-2j)}{(N+1-j)!\,j!}$$

Set $k=N/2-j$. The representations ${}^{2k+1}D_k$ with $0 \leq k \leq N/2$ have the factors for the multiplicity

$$f(k) = N!(2k+1)/(N/2+1+k)!(N/2-k)!$$

has its maximum for ${}^{2k+1}D_k$ at the value $k \approx \frac{1}{2}\sqrt{N}$. Then an exponential decrease follows and functions with shorter wavelengths disappear more and more quickly. As an illustration, a plot of $f(k)$ for $N=900$ is given.

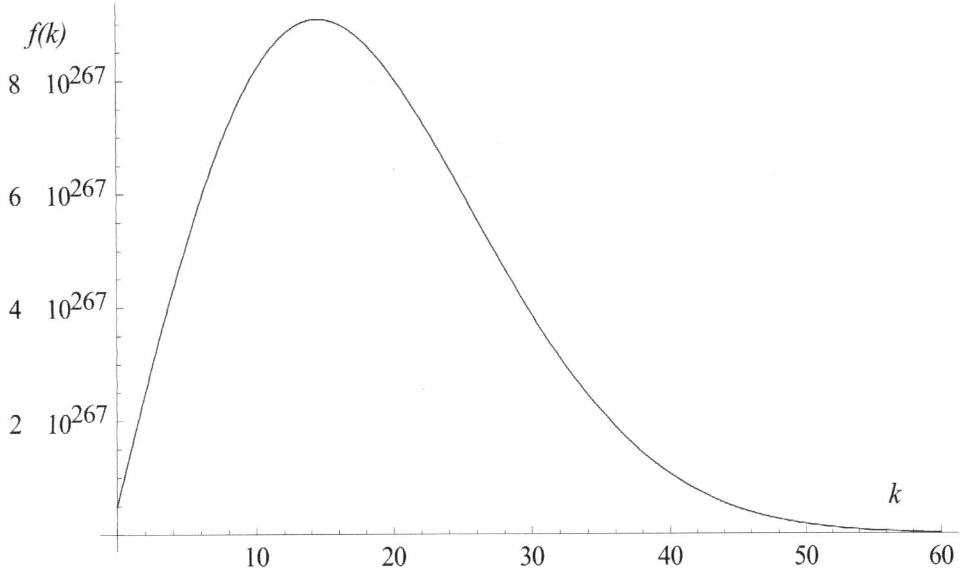

Fig. 3: $f(k)$ for $N=900$, $\sqrt{N}=30$, $k_{maximum}=14.514$,
$f(0)=4.98\times10^{266}$, $f(14.514)=9.082\times10^{267}$, $f(30)=3.86\times10^{267}$, $f(60)=1.76\times10^{265}$

Let R be the radius of the $\boldsymbol{S^3}$. With respect to the exponential decrease, it seems rational to introduce a smallest physically accessible length between $R/(1/2)\sqrt{N}$ and $R/2\sqrt{N}$ and to ignore the states with a shorter wavelength.

For a relation between the number of qubits and the evolution of cosmic space, this means for a cosmological model, the basic assumptions of the three fundamental theories of physics , i.e. special relativity, quantum theory, and thermodynamics, were demanded.

1. There exists a universal and distinguished velocity.

This assumption introduces c, the velocity of light.

Recently, Riess et. al. [5] published the most accurate determination of H_0. With an uncertainty of only 3%, they found the following value: $H_0 = 73.8 \pm 2.4$ km s^{-1} Mpc^{-1}. Assuming the age of the universe as 1.325×10^{10} yr, then $H_0 t_0=1$. A value $H_0 t_0=1$ means that the universe is expanding with velocity c. This supports the earlier work of Tonry et al [6] which, among other results, has shown that for the value of the Hubble parameter H_0 and the age of the universe t_0, the relation $H_0\ t_0 = 0.96 \pm 0.04$ holds true. Set $c=1$ then the age of the universe t_{cosmos} and the curvature Radius R become equal and the number of qubits N increases with $t_{cosmos}{}^2$.

2. The energy of a quantum system is inversely proportional to its characteristic wavelength.

This second assumption is the familiar Planck relation, $E=h\nu=hc/\lambda$. A qubit has a wavelength of order $2R$. Set h = 1. Then a qubit has an amount of energy of order $1/2R$.

3. The first law of thermodynamics is valid.

This allows to define a cosmological pressure p according to $dU+pdV=0$. By the second assumption U increases proportional to R and therefore the energy density decreases with $1/R^2$. According to the first law of thermodynamics, the state equation for the cosmic substrate, the protyposis, follows as $\mu=-p/3$.

If the Planck-length is introduced as $l_{Planck} = R/\sqrt{(3/2N)} = R/1.225\ \sqrt{N}$, then Einstein's Equations follow as the description for the local disturbances of a homogeneous and isotropic universe. [7]

This cosmological model has no horizon problem, needs no unphysical ad-hoc-inflation or ghost dimensions. It has no problem with a cosmological constant or with flatness and has also no problem of empiricism. It does not violate the strong energy condition: $\mu+3p \geq 0$ and induces a photon-to-baryon-ratio of 10^9 [8].

In Planck's units, the age of the universe is $10^{61}\ t_{Planck}$ and the number of qubits is of order 10^{122}. A proton has the mass of $7.68 * 10^{-20}\ m_{Planck}$. This corresponds to 10^{41} qubits and is of the order of v. Weizsäcker's conjecture long time ago.[9].

2. From cosmos to particles

The meaning-open absolute quantum information, the protyposis, is primarily a cosmological entity. It is not localized in time and space, but it is acting on the expansion of the universe.

Nevertheless, special forms of it can appear as either energetic or material objects. Energetic objects can be localized in time, material objects also in space. Both types of objects can act as carrier of localized and meaningful information. For instance, mind is quantum information with a living brain as its carrier. Therefore an interaction of mind – being quantum information that is able to know itself – and matter – being quantum information in the special form that has rest mass – will no longer be counted among the outside of scientific descriptions.

Such a philosophical idea will only become part of science if one can show in which way the particles can be reconstructed from quantum information. It is well known in physics from the effects of the Hawking radiation that fields of acceleration are able to produce particles. Therefore, the idea of the vacuum as an area without any particle becomes questionable outside Minkowski space. For a lucid definition of a relativistic particle, we have to leave the real cosmic space with its inhomogenities and accelerations and have to work in the idealization of the cotangential space, i.e. in the Minkowski space. Here an "elementary object" is clearly defined as an entity that can be moved in space and time, thereby changing only its state but not the object itself. Mathematics shows that any such object can be represented by an irreducible representation of the Poincaré group. The parameters that characterize such an irreducible representation are mass and spin.

By the transition from the physical space to the flat Minkowski space, we loose the naturally given measure that can be introduced by the relation between curvature radius of cosmic space and Planck's length. Therefore, the connection between qubits and space, respectively between qubits and energy is lost and the mass of a particle is a parameter which must be prescribed from the outset.

3. Second quantization of qubits

To construct a particle from quantum bits, the process of "second quantization" can give hints. Einstein's idea that an object with a very large or infinite number of degrees of freedom, for instance the electromagnetic field, can sometimes be better described by a very large or infinite number of objects with few degrees of freedom, like photons. In an analogous way, the formation of a relativistic particle can be achieved from an indefinite number of qubits. If all quantum bits of the Protyposis are supposed to be different, such a condition of a "Boltzmann quantization" appears as being too strong because no commutation relations are possible and no group representation can be constructed. If all bits are supposed to be equal, then particles can be constructed with such Bose commutation relations , but under this condition only massless objects. Para-Bose-statistic appeared as a fruitful conjecture.

The Parabose commutation relations are three-linear.

$$\tfrac{1}{2}[\{a_r, a_s^\dagger\}, a_t] = -\delta_{st} a_r$$
$$[\{a_r, a_s\}, a_t] = [\{a_r^\dagger, a_s^\dagger\}, a_t^\dagger] = 0$$

In the consequence we get also

$$\tfrac{1}{2}[a_t, \{a_r^\dagger, a_s^\dagger\}] = -\delta_{rt} a_s^\dagger - \delta_{st} a_r^\dagger$$
$$\tfrac{1}{2}[a_t^\dagger, \{a_r, a_s\}] = -\delta_{rt} a_s - \delta_{st} a_r$$
$$\tfrac{1}{2}[a_t^\dagger, \{a_r^\dagger, a_s\}] = -\delta_{st} a_r^\dagger$$

By the Green decomposition[10] it becomes evident that Para-Bose statistics can be interpreted as bosons which are anticommuting in case they are of different types.

$$a_r = \sum_{\alpha=1}^{p} b_r^\alpha \qquad a_s^\dagger = \sum_{\alpha=1}^{p} b_s^{\dagger\alpha}$$

with

$$[b_r^\alpha, b_s^{\dagger\alpha}] = \delta_{rs}$$
$$[b_r^\alpha, b_s^\alpha] = [b_r^{\dagger\alpha}, b_s^{\dagger\alpha}] = 0$$
$$\{b_r^\alpha, b_s^{\dagger\beta}\} = \{b_r^\alpha, b_s^\beta\} = \{b_r^{\dagger\alpha}, b_s^{\dagger\beta}\} = 0 \qquad \text{for } \alpha \neq \beta$$

Finally

$$a_s a_r^\dagger |\Omega\rangle = \delta_{rs} p |\Omega\rangle$$

$|\Omega>$ is the vacuum of the Protyposis, p is the Para-Bose-order, $p=1$ restores Bose statistics.
The vacuum in Minkowski space, the Lorentz vacuum $|0\rangle$, is an eigenstate of the Poincaré-group with mass, energy and spin zero. It can be constructed over the vacuum of qubits $|\Omega\rangle$ as

$$|0\rangle = \sum_{n_1=0}^{\infty} \sum_{n_2=0}^{\infty} \frac{(-1)^{n_1+n_2}}{n_1! \cdot n_2!} \left(\frac{a_1^\dagger a_3^\dagger + a_3^\dagger a_1^\dagger}{2} \right)^{n_1} \left(\frac{a_2^\dagger a_4^\dagger + a_4^\dagger a_2^\dagger}{2} \right)^{n_2} |\Omega\rangle$$

It shows that the statement "at every point in Minkowski-space there is no particle" corresponds to a potentially infinite amount of information. Besides the qubits (index 1 and 2), also anti-qubits (index 3 and 4) are introduced. This is a possibility to linearly represent the complex conjugation which is a subgroup of the symmetry of the qubits. For the vacuum of the Minkowski-space, the annihilation of a qubit is equivalent to the creation of its anti-qubit:

$$a_1|0\rangle = -a_3^\dagger|0\rangle \qquad a_2|0\rangle = -a_4^\dagger|0\rangle \qquad a_3|0\rangle = -a_1^\dagger|0\rangle \qquad a_4|0\rangle = -a_2^\dagger|0\rangle$$

4. The Poincaré group

The Poincaré group is the ten-parameter group of motions in the Minkowski space R(3,1). It is a semidirect product of the Lorentz group SO(3,1) with the four translations. The SO(3,1) is locally equivalent with SL(2,C). This special linear group in two dimensions would be a general symmetry group for the qubits if the preservation of the norm was no longer stipulated. It was an early conjecture of v. Weizsäcker that this could be the deeper reason for the role of this group in special relativity.[11]

The calculations of particle states and of the Casimir-operators quickly become very complicated. Therefore, the aid of the computer is very useful. For the calculations with *mathematica®,* we use a form for the creation- and destruction-operators that is more valuable at the computer. We set

$$a_r^\dagger \Rightarrow e[r] \qquad \text{(Erzeuger)}$$
$$a_r \Rightarrow v[r] \qquad \text{(Vernichter)}$$
$$\{a_r^\dagger, a_s^\dagger\} \Rightarrow 2f[r,s] \qquad \{a_r, a_s\} \Rightarrow 2w[r,s]$$
$$\{a_r^\dagger, a_s\} \Rightarrow 2d[r,s] \qquad |0\rangle \Rightarrow lvac$$

Then the generators of the Poincaré group get the following form:

Translations:
P1 = (-w[2,3]-f[3,2]-w[1,4]-f[4,1] -d[1,2]-d[2,1]-d[4,3]-d[3,4])/2
P2 = I*(-w[2,3] +f[3,2] +w[1,4]-f[4,1] -d[1,2] +d[2,1]-d[4,3] +d[3,4])/2
P3 = (-w[1,3]-f[3,1] +w[2,4] +f[4,2] -d[1,1] +d[2,2]-d[3,3] +d[4,4])/2
P0 = (-w[1,3]-f[3,1]-w[2,4]-f[4,2] -d[1,1]-d[2,2]-d[3,3]-d[4,4])/2

Boosts:
M10 = I*(w[1,4]-f[4,1] +w[2,3]-f[3,2])/2
M20 = (w[1,4] +f[4,1]-w[2,3]-f[3,2])/2
M30 = I*(w[1,3]-f[3,1]-w[2,4] +f[4,2])/2

Rotations:
M32 = (d[2,1] +d[1,2]-d[3,4]-d[4,3])/2
M21 = (d[1,1]-d[2,2]-d[3,3] +d[4,4])/2
M31 = I*(d[2,1]-d[1,2]-d[3,4] +d[4,3])/2

By ** non-commutative multiplication is denoted and the non-commutative product of n times f[r,s], i.e. f[r,s]**f[r,s]** …**f[r,s] is abbreviated by f[r,s,n].
The two Casimir-Operators of the Poincaré group are the square of the mass

$m^2 = P0^2 - P1^2 - P2^2 - P3^2 =$
-(d[1,1] + d[3,3] + f[3,1] + f[4,2] - d[1,1]**w[2,4] + d[1,2]**w[1,4] + d[2,1]**d[1,2] + d[2,1]**w[2,3] - d[2,2]**d[1,1] - d[2,2]**w[1,3] - d[3,3]**d[2,2] - d[3,3]**w[2,4] + d[3,4]**d[1,2] + d[3,4]**w[2,3] + d[4,3]**d[2,1] + d[4,3]**d[3,4] + d[4,3]**w[1,4] - d[4,4]**d[1,1] - d[4,4]**d[3,3] - d[4,4]**w[1,3] - f[3,1]**d[2,2] - f[3,1]**d[4,4] - f[3,1]**w[2,4] + f[3,2]**d[1,2] + f[3,2]**d[4,3] + f[3,2]**w[2,3] + f[4,1]**d[2,1] + f[4,1]**d[3,4] + f[4,1]**f[3,2] + f[4,1]**w[1,4] - f[4,2]**d[1,1] - f[4,2]**d[3,3] - f[4,2]**f[3,1] - f[4,2]**w[1,3] - w[1,3]**w[2,4] + w[1,4]**w[2,3] + w[1,3] + w[2,4])

and the Pauli-Lubanski-operator $W^2 = W_a W^a$ with $W_d = (1/2)\varepsilon_{abcd} M^{ab} P^c$.
In the appendix we give the explicit form of W^2. Its ugly expression shows that it appears evident that without computer-aid, such expressions seem to be hardly manageable.

5. Relativistic particles

Now examples of states for massless particles with different helicities and for massive particles at rest with spin 0 and 1/2 will be presented.

We denote the Para-Bose-order by p[0]. Let the momentum be P0=P3=m, P1=P2=0 and the spin equal to zero, then the state has the form

$$\sum_{p[1]=0}^{\infty} \frac{(p[0]-1)!\ (-m)^{p[1]}}{p[1]!\ (p[0]-1+p[1])!} f[3,1,p[1]]**\mathrm{lvac}$$

It is also possible to define states with helicity p[2]=1,2,3,…, which get the structure

$$\sum_{p[1]=0}^{\infty} \frac{(2\,p[2]+p[0]-1)!\ (-m)^{p[1]}}{p[1]!\ (p[0]-1+p[1]+2\,p[2])!} f[3,1,p[1]]**f[1,1,p[2]]**\mathrm{lvac}$$

For the state of a massless fermion with helicity: p[2]+(1/2) we find

$$\sum_{p[1]=0}^{\infty} \frac{(2\,p[2]+1+p[0]-1)!\ (-m)^{p[1]}}{p[1]!\ (p[0]-1+p[1]+2\,p[2]+1)!} e[1]**f[3,1,p[1]]**f[1,1,p[2]]**\mathrm{lvac}$$

The state of a massive spinless boson at rest is more complicated. Here the Para-Bose-order has to be greater than one: p[0] > 1. The momentum at rest is P0=m, P1=P2=P3=0 and the spin is 0.

$$\sum_{p[3]=0}^{\infty}\sum_{p[2]=0}^{\infty}\sum_{p[1]=0}^{\infty} \frac{(-1)^{(p[1]+p[2]+p[3])}\ (m)^{(2\,p[1]+p[2]+p[3])}\ (p[1]+p[2]+p[3]+p[0]-2)!}{(2\,p[1]+p[2]+p[3]+p[0]-1)!\,(p[3]-1+p[0]-1+p[1])!\,(p[2]-1+p[0]-1+p[1])!\,p[1]!\,p[2]!\,p[3]!} *$$

$$f[4,2,p[3]]**f[4,1,p[1]]**f[3,2,p[1]]**f[3,1,p[2]]**\mathrm{lvac}$$

The state of a massive fermion at rest is still more complicated. Again the Para-Bose-order has to be greater than one: p[0] > 1. The momentum at rest is P0=m, P1=P2=P3=0, and the spin in z-direction is s_z=1/2. The expression has two parts:

$$\sum_{p[3]=0}^{\infty}\sum_{p[2]=0}^{\infty}\sum_{p[1]=0}^{\infty}\frac{(-1)^{(p[1]+p[2]+p[3])}\,(m)^{(2p[1]+p[2]+p[3])}}{p[1]!\,p[2]!\,p[3]!\,(2p[1]+p[2]+p[3]+p[0])!}*$$
$$\frac{(p[1]+p[2]+p[3]+p[0]-1)!}{(p[1]+p[3]+p[0]-2)!\,(p[1]+p[2]+p[0]-1)!}*$$
$$e[1]**f[4,2,p[3]]**f[4,1,p[1]]**f[3,2,p[1]]**f[3,1,p[2]]**lvac$$

$$+\sum_{p[3]=0}^{\infty}\sum_{p[2]=0}^{\infty}\sum_{p[1]=0}^{\infty}\frac{(-1)^{(p[1]+p[2]+p[3])}\,(m)^{(1+2p[1]+p[2]+p[3])}}{p[1]!\,p[2]!\,p[3]!\,(2p[1]+p[2]+p[3]+p[0]+1)!}*$$
$$\frac{(p[1]+p[2]+p[3]+p[0]-1)!}{(p[1]+p[3]+p[0]-1)!\,(p[1]+p[2]+p[0]-1)!}*$$
$$e[2]**f[4,2,p[3]]**f[4,1,p[1]+1]**f[3,2,p[1]]**f[3,1,p[2]]**lvac$$

There is a possibility for another state of a massive fermion at rest with the same mass, momentum, spin and spin direction. It is not far to seek this second state as the state of the antiparticle of the former one.

$$\sum_{p[3]=0}^{\infty}\sum_{p[2]=0}^{\infty}\sum_{p[1]=0}^{\infty}\frac{(-1)^{(p[1]+p[2]+p[3])}\,(m)^{(2p[1]+p[2]+p[3])}}{p[1]!\,p[2]!\,p[3]!\,(2p[1]+p[2]+p[3]+p[0])!}*$$
$$\frac{(p[1]+p[2]+p[3]+p[0]-1)!}{(p[1]+p[3]+p[0]-2)!\,(p[1]+p[2]+p[0]-1)!}*$$
$$e[4]**f[4,2,p[3]]**f[4,1,p[1]]**f[3,2,p[1]]**f[3,1,p[2]]**lvac$$

$$+\sum_{p[3]=0}^{\infty}\sum_{p[2]=0}^{\infty}\sum_{p[1]=0}^{\infty}\frac{(-1)^{(p[1]+p[2]+p[3])}\,(m)^{(1+2p[1]+p[2]+p[3])}}{p[1]!\,p[2]!\,p[3]!\,(2p[1]+p[2]+p[3]+p[0]+1)!}*$$
$$\frac{(p[1]+p[2]+p[3]+p[0]-1)!}{(p[1]+p[3]+p[0]-1)!\,(p[1]+p[2]+p[0]-1)!}*$$
$$e[3]**f[4,2,p[3]]**f[4,1,p[1]+1]**f[3,2,p[1]]**f[3,1,p[2]]**lvac$$

With the Para-Bose-operators beside the Casimir-operators, also other operators can be constructed that do commutate with all the generators of the Poincaré group. Among them, two operators are interesting which can be interpreted as operators transforming a charge.

In consequence of the calculation process, these two operators are denoted as P832 and P906.

P832 = - 2d[1,4] + 3d[2,3] + d[1,1]**d[1,4] - d[1,3]**w[1,4] + d[1,4]**w[1,3] - d[2,1]**d[1,3] - d[2,2]**d[2,3] - d[2,3]**w[2,4] + d[2,4]**d[1,2] + d[2,4]**w[2,3] - d[3,3]**d[2,3] - d[3,3]**w[3,4] + d[3,4]**d[1,3] + d[3,4]**w[3,3] - d[4,3]**d[2,4] - d[4,3]**w[4,4] + d[4,4]**d[1,4] + d[4,4]**w[3,4] - f[1,1]**d[2,1] - f[1,1]**w[1,4] + f[2,1]**d[1,1] - f[2,1]**d[2,2] + f[2,1]**w[1,3] - f[2,1]**w[2,4] + f[2,2]**d[1,2] + f[2,2]**w[2,3] - f[3,1]**d[2,3] - f[3,1]**w[3,4] + f[3,2]**d[1,3] + f[3,2]**w[3,3] - f[4,1]**d[2,4] - f[4,1]**w[4,4] + f[4,2]**d[1,4] + f[4,2]**w[3,4]

P906 = - 3d[3,2] + 4d[4,1]- d[1,1]**w[1,2] + d[1,2]**w[1,1]- d[2,1]**w[2,2] + d[2,2]**w[1,2] + d[3,1]**d[1,2]- d[3,1]**w[2,3] + d[3,2]**d[2,2] + d[3,2]**d[3,3] + d[3,2]**w[1,3]- d[4,1]**d[1,1]- d[4,1]**d[4,4]- d[4,1]**w[2,4]- d[4,2]**d[2,1] + d[4,2]**d[3,4] + d[4,2]**w[1,4]- d[4,3]**d[3,1]- f[3,1]**d[4,1]- f[3,1]**w[1,2]- f[3,2]**d[4,2]- f[3,2]**w[2,2]- f[3,3]**d[4,3]- f[3,3]**w[2,3] + f[4,1]**d[3,1] + f[4,1]**w[1,1] + f[4,2]**d[3,2] + f[4,2]**w[1,2] + f[4,3]**d[3,3]- f[4,3]**d[4,4] + f[4,3]**w[1,3]- f[4,3]**w[2,4] + f[4,4]**d[3,4] + f[4,4]**w[1,4]

For any generator M of the Poincaré group it holds [M, P832] = [M, P906] = 0.

Further on, P832 as well as P906 annihilates massless particle states and massive spinless states. However, (1/m)*P906 transforms the massive spin-(1/2)-particle state into the corresponding antiparticle state, and (-1/m)*P832 transforms the massive spin-(1/2)-antiparticle state into the corresponding particle state.

6. Conclusions

By the construction of states of relativistic particles, a necessary step is done for a connection of the conception of abstract and absolute quantum information, protyposis, with the existing areas of physics. This has some crucial consequences. There is no longer a need for looking into smaller and smaller areas in space to find "the last objects" for an explanation of nature. For a quantum description of the universe, it seems no longer necessary to quantize Einstein's equations. Also a scientific treatment of consciousness becomes possible with this new model for matter. As long as only the cells and atoms in the brain are grasped as real, consciousness cannot be seen as an acting reality. However, with matter as a special form of quantum information that is able to gather and to emit not only energy but also information, an interaction of matter, energy and information came into the realm of physics. Consciousness is quantum information with the living brain as its carrier, so an interaction between mind and matter can be scientifically described. [12]

7. Appendix

The explicit form of the Pauli-Lubanski-operator $W^2 = W_a W^a$ with $W_d = (1/2)\varepsilon_{abcd} M^{ab} P^c$.

WW= -(1/2d[1,1]**d[1,1] +1/2d[1,1]**w[1,3] +1/2d[1,1]**w[2,4] -1/2d[2,2]**d[1,1] -1/2d[2,2]**w[1,3] - 1/2d[2,2]**w[2,4] +d[3,3]**d[1,1] -1/2d[3,3]**d[2,2] +1/2d[3,3]**d[3,3] +1/2d[3,3]**w[1,3] +1/2d[3,3]**w[2,4] +2d[3,4]**d[1,2] +2d[4,3]**d[2,1] -1/2d[4,4]**d[1,1] -1/2d[4,4]**d[3,3] -1/2d[4,4]**w[1,3] -1/2d[4,4]**w[2,4] +1/2f[3,1]**d[1,1] -1/2f[3,1]**d[2,2] +1/2f[3,1]**d[3,3] -1/2f[3,1]**d[4,4] +2f[3,1]**w[1,3] +f[3,1]**w[2,4] +f[3,2]**w[2,3] +f[4,1]**w[1,4] +1/2f[4,2]**d[1,1] -1/2f[4,2]**d[2,2] +1/2f[4,2]**d[3,3] -1/2f[4,2]**d[4,4] +f[4,2]**w[1,3] +2f[4,2]**w[2,4] +1/4d[1,1]**d[1,1]**d[1,1] +1/4d[1,1]**d[1,1]**w[1,3] - 1/4d[1,1]**d[1,1]**w[2,4] +1/2d[1,1]**d[1,2]**w[1,4] -1/2d[1,1]**w[1,3]**w[2,4] +1/2d[1,1]**w[1,4]**w[2,3] +1/2d[2,1]**d[1,1]**d[1,2] +1/2d[2,1]**d[1,1]**w[2,3] +2d[2,1]**d[1,2]**w[2,4] +1/2d[2,1]**d[2,2]**d[1,2] - 1/2d[2,1]**d[2,2]**w[2,3] -d[2,2]**d[1,1]**d[1,1] -d[2,2]**d[1,1]**w[1,3] -1/2d[2,2]**d[1,2]**w[1,4] +3/4d[2,2]**d[2,2]**d[1,1] +3/4d[2,2]**d[2,2]**w[1,3] +1/4d[2,2]**d[2,2]**w[2,4] +1/2d[2,2]**w[1,3]**w[2,4] - 1/2d[2,2]**w[1,4]**w[2,3] -5/4d[3,3]**d[1,1]**d[1,1] -3/2d[3,3]**d[1,1]**w[1,3] -1/2d[3,3]**d[1,1]**w[2,4] +1/2d[3,3]**d[1,2]**w[1,4] -1/2d[3,3]**d[2,1]**d[1,2] -3/2d[3,3]**d[2,1]**w[2,3] +3/4d[3,3]**d[2,2]**d[2,2] +d[3,3]**d[2,2]**w[1,3] -5/4d[3,3]**d[3,3]**d[1,1] +d[3,3]**d[3,3]**d[2,2] +1/4d[3,3]**d[3,3]**d[3,3] +1/4d[3,3]**d[3,3]**w[1,3] -1/4d[3,3]**d[3,3]**w[2,4] -3/2d[3,3]**d[3,4]**d[1,2] +1/2d[3,3]**d[3,4]**w[2,3] - 1/2d[3,3]**w[1,3]**w[2,4] +1/2d[3,3]**w[1,4]**w[2,3] -3/2d[3,4]**d[1,1]**d[1,2] +1/2d[3,4]**d[1,1]**w[2,3] - 2d[3,4]**d[1,2]**w[1,3] -3/2d[3,4]**d[2,2]**d[1,2] -1/2d[3,4]**d[2,2]**w[2,3] -3/2d[4,3]**d[1,1]**w[1,4] - 3/2d[4,3]**d[2,1]**d[1,1] -3/2d[4,3]**d[2,1]**d[2,2] -2d[4,3]**d[2,1]**w[2,4] +3/2d[4,3]**d[2,2]**w[1,4] - 3/2d[4,3]**d[3,3]**d[2,1] +1/2d[4,3]**d[3,3]**d[3,4] +1/2d[4,3]**d[3,3]**w[1,4] -1/2d[4,3]**d[3,4]**d[1,1] +3/2d[4,3]**d[3,4]**d[2,2] +2d[4,3]**d[3,4]**w[2,4] -3/2d[4,3]**d[4,4]**d[2,1] +1/2d[4,3]**d[4,4]**d[3,4] - 1/2d[4,3]**d[4,4]**w[1,4] +d[4,4]**d[1,1]**d[1,1] +d[4,4]**d[1,1]**w[1,3] -1/2d[4,4]**d[1,2]**w[1,4] +3/2d[4,4]**d[2,1]**d[1,2] +3/2d[4,4]**d[2,1]**w[2,3] -1/2d[4,4]**d[2,2]**d[1,1] -1/2d[4,4]**d[2,2]**w[1,3] +1/2d[4,4]**d[2,2]**w[2,4] -1/2d[4,4]**d[3,3]**d[2,2] -d[4,4]**d[3,3]**d[3,3] -d[4,4]**d[3,3]**w[1,3] - 3/2d[4,4]**d[3,4]**d[1,2] -1/2d[4,4]**d[3,4]**w[2,3] +3/4d[4,4]**d[4,4]**d[1,1] +3/4d[4,4]**d[4,4]**d[3,3] +3/4d[4,4]**d[4,4]**w[1,3] +1/4d[4,4]**d[4,4]**w[2,4] +1/2d[4,4]**w[1,3]**w[2,4] -1/2d[4,4]**w[1,4]**w[2,3]

+1/4f[3,1]**d[1,1]**d[1,1] +f[3,1]**d[1,1]**w[1,3] -1/2f[3,1]**d[1,1]**w[2,4] +f[3,1]**d[1,2]**w[1,4] -
f[3,1]**d[2,2]**d[1,1] +3/4f[3,1]**d[2,2]**d[2,2] -1/2f[3,1]**d[2,2]**w[2,4] -3/2f[3,1]**d[3,3]**d[1,1]
+f[3,1]**d[3,3]**d[2,2] +1/4f[3,1]**d[3,3]**d[3,3] +f[3,1]**d[3,3]**w[1,3] -1/2f[3,1]**d[3,3]**w[2,4]
+f[3,1]**d[3,4]**w[2,3] -2f[3,1]**d[4,3]**d[2,1] +f[3,1]**d[4,4]**d[1,1] -1/2f[3,1]**d[4,4]**d[2,2] -
f[3,1]**d[4,4]**d[3,3] +3/4f[3,1]**d[4,4]**d[4,4] -1/2f[3,1]**d[4,4]**w[2,4] +f[3,1]**f[3,1]**w[1,3]
+f[3,1]**w[1,3]**w[1,3] -f[3,1]**w[1,3]**w[2,4] +2f[3,1]**w[1,4]**w[2,3] +1/2f[3,2]**d[1,1]**d[1,2]
+3/2f[3,2]**d[1,1]**w[2,3] +2f[3,2]**d[1,2]**w[2,4] -1/2f[3,2]**d[2,2]**d[1,2] +1/2f[3,2]**d[2,2]**w[2,3] -
3/2f[3,2]**d[3,3]**d[1,2] +1/2f[3,2]**d[3,3]**w[2,3] +1/2f[3,2]**d[4,3]**d[1,1] -1/2f[3,2]**d[4,3]**d[2,2]
+1/2f[3,2]**d[4,3]**d[3,3] -1/2f[3,2]**d[4,3]**d[4,4] +f[3,2]**d[4,3]**w[1,3] +f[3,2]**d[4,3]**w[2,4]
+3/2f[3,2]**d[4,4]**d[1,2] +3/2f[3,2]**d[4,4]**w[2,3] +f[3,2]**f[3,1]**w[2,3] +f[3,2]**w[1,3]**w[2,3]
+f[3,2]**w[2,3]**w[2,4] +1/2f[4,1]**d[1,1]**w[1,4] +1/2f[4,1]**d[2,1]**d[1,1] -1/2f[4,1]**d[2,1]**d[2,2]
+f[4,1]**d[2,1]**w[1,3] +f[4,1]**d[2,1]**w[2,4] +3/2f[4,1]**d[2,2]**w[1,4] +1/2f[4,1]**d[3,3]**d[2,1]
+1/2f[4,1]**d[3,3]**d[3,4] +3/2f[4,1]**d[3,3]**w[1,4] -3/2f[4,1]**d[3,4]**d[1,1] +3/2f[4,1]**d[3,4]**d[2,2]
+2f[4,1]**d[3,4]**w[2,4] -1/2f[4,1]**d[4,4]**d[2,1] -1/2f[4,1]**d[4,4]**d[3,4] +1/2f[4,1]**d[4,4]**w[1,4]
+f[4,1]**f[3,1]**w[1,4] +1/2f[4,1]**f[3,2]**d[1,1] -1/2f[4,1]**f[3,2]**d[2,2] +1/2f[4,1]**f[3,2]**d[3,3] -
1/2f[4,1]**f[3,2]**d[4,4] +2f[4,1]**f[3,2]**w[1,3] +2f[4,1]**f[3,2]**w[2,4] +f[4,1]**w[1,3]**w[1,4]
+f[4,1]**w[1,4]**w[2,4] -1/4f[4,2]**d[1,1]**d[1,1] -1/2f[4,2]**d[1,1]**w[1,3] -f[4,2]**d[1,1]**w[2,4]
+f[4,2]**d[1,2]**w[1,4] +2f[4,2]**d[2,1]**d[1,2] +2f[4,2]**d[2,1]**w[2,3] +1/4f[4,2]**d[2,2]**d[2,2] -
1/2f[4,2]**d[2,2]**w[1,3] +2f[4,2]**d[2,2]**w[2,4] -1/2f[4,2]**d[3,3]**d[1,1] -1/4f[4,2]**d[3,3]**d[3,3] -
1/2f[4,2]**d[3,3]**w[1,3] -f[4,2]**d[3,3]**w[2,4] -2f[4,2]**d[3,4]**d[1,2] +f[4,2]**d[3,4]**w[2,3]
+2f[4,2]**d[4,3]**d[3,4] +2f[4,2]**d[4,3]**w[1,4] +1/2f[4,2]**d[4,4]**d[2,2] +1/4f[4,2]**d[4,4]**d[4,4] -
1/2f[4,2]**d[4,4]**w[1,3] +2f[4,2]**d[4,4]**w[2,4] -1/2f[4,2]**f[3,1]**d[1,1] +1/2f[4,2]**f[3,1]**d[2,2] -
1/2f[4,2]**f[3,1]**d[3,3] +1/2f[4,2]**f[3,1]**d[4,4] -f[4,2]**f[3,1]**w[1,3] -f[4,2]**f[3,1]**w[2,4]
+f[4,2]**f[3,2]**w[2,3] +f[4,2]**f[4,1]**w[1,4] +f[4,2]**f[4,2]**w[2,4] -f[4,2]**w[1,3]**w[2,4]
+2f[4,2]**w[1,4]**w[2,3] +f[4,2]**w[2,4]**w[2,4] -1/4d[1,1]**d[1,1]**d[1,1]**w[2,4]
+1/4d[1,1]**d[1,1]**d[1,2]**w[1,4] -1/4d[1,1]**d[1,1]**w[1,3]**w[2,4] +1/4d[1,1]**d[1,1]**w[1,4]**w[2,3]
+1/4d[2,1]**d[1,1]**d[1,1]**d[1,2] +1/4d[2,1]**d[1,1]**d[1,1]**w[2,3] -d[2,1]**d[1,1]**d[1,2]**w[2,4]
+d[2,1]**d[1,2]**d[1,2]**w[1,4] -d[2,1]**d[1,2]**w[1,3]**w[2,4] +d[2,1]**d[1,2]**w[1,4]**w[2,3]
+d[2,1]**d[2,1]**d[1,2]**d[1,2] +d[2,1]**d[2,1]**d[1,2]**w[2,3] -3/2d[2,1]**d[2,2]**d[1,1]**d[1,2] -
1/2d[2,1]**d[2,2]**d[1,1]**w[2,3] -d[2,1]**d[2,2]**d[1,2]**w[1,3] +1/4d[2,1]**d[2,2]**d[2,2]**d[1,2]
+1/4d[2,1]**d[2,2]**d[2,2]**w[2,3] -1/4d[2,2]**d[1,1]**d[1,1]**d[1,1] -1/4d[2,2]**d[1,1]**d[1,1]**w[1,3]
+1/2d[2,2]**d[1,1]**d[1,1]**w[2,4] -1/2d[2,2]**d[1,1]**d[1,2]**w[1,4] +1/2d[2,2]**d[1,1]**w[1,3]**w[2,4] -
1/2d[2,2]**d[1,1]**w[1,4]**w[2,3] +1/2d[2,2]**d[2,2]**d[1,1]**d[1,1] +1/2d[2,2]**d[2,2]**d[1,1]**w[1,3] -
1/4d[2,2]**d[2,2]**d[1,1]**w[2,4] +1/4d[2,2]**d[2,2]**d[1,2]**w[1,4] -1/4d[2,2]**d[2,2]**d[2,2]**d[1,1] -
1/4d[2,2]**d[2,2]**d[2,2]**w[1,3] -1/4d[2,2]**d[2,2]**w[1,3]**w[2,4] +1/4d[2,2]**d[2,2]**w[1,4]**w[2,3]
+1/4d[3,3]**d[1,1]**d[1,1]**w[2,4] -1/2d[3,3]**d[1,1]**d[1,2]**w[1,4] +1/2d[3,3]**d[1,1]**w[1,3]**w[2,4]
+1/2d[3,3]**d[1,1]**w[1,4]**w[2,3] -d[3,3]**d[1,2]**w[1,3]**w[1,4] -1/2d[3,3]**d[2,1]**d[1,1]**d[1,2]
+1/2d[3,3]**d[2,1]**d[1,1]**w[2,3] -d[3,3]**d[2,1]**d[1,2]**w[1,3] -1/2d[3,3]**d[2,1]**d[2,2]**d[1,2]
+1/2d[3,3]**d[2,1]**d[2,2]**w[2,3] +d[3,3]**d[2,1]**w[2,3]**w[2,4] +1/4d[3,3]**d[2,2]**d[1,1]**d[1,1]
+1/2d[3,3]**d[2,2]**d[1,1]**w[1,3] -1/2d[3,3]**d[2,2]**d[1,2]**w[1,4] -1/4d[3,3]**d[2,2]**d[2,2]**d[2,2] -
1/2d[3,3]**d[2,2]**d[2,2]**w[1,3] -1/4d[3,3]**d[2,2]**d[2,2]**w[2,4] -1/2d[3,3]**d[2,2]**w[1,3]**w[2,4] -
1/2d[3,3]**d[2,2]**w[1,4]**w[2,3] +1/4d[3,3]**d[3,3]**d[1,1]**w[2,4] -3/4d[3,3]**d[3,3]**d[1,2]**w[1,4] -
3/4d[3,3]**d[3,3]**d[2,1]**d[1,2] +1/4d[3,3]**d[3,3]**d[2,1]**w[2,3] +1/4d[3,3]**d[3,3]**d[2,2]**d[1,1] -
1/2d[3,3]**d[3,3]**d[2,2]**d[2,2] -1/4d[3,3]**d[3,3]**d[2,2]**w[1,3] -1/2d[3,3]**d[3,3]**d[2,2]**w[2,4] -
1/4d[3,3]**d[3,3]**d[3,3]**d[2,2] -1/4d[3,3]**d[3,3]**d[3,3]**w[2,4] +1/4d[3,3]**d[3,3]**d[3,4]**d[1,2]
+1/4d[3,3]**d[3,3]**d[3,4]**w[2,3] -1/4d[3,3]**d[3,3]**w[1,3]**w[2,4] +1/4d[3,3]**d[3,3]**w[1,4]**w[2,3]
+1/2d[3,3]**d[3,4]**d[1,1]**d[1,2] -1/2d[3,3]**d[3,4]**d[1,1]**w[2,3] +d[3,3]**d[3,4]**d[1,2]**w[1,3]
+1/2d[3,3]**d[3,4]**d[2,2]**d[1,2] +1/2d[3,3]**d[3,4]**d[2,2]**w[2,3] +1/4d[3,4]**d[1,1]**d[1,1]**d[1,2] -
3/4d[3,4]**d[1,1]**d[1,1]**w[2,3] +d[3,4]**d[1,1]**d[1,2]**w[1,3] -d[3,4]**d[1,1]**w[1,3]**w[2,3]
+d[3,4]**d[1,2]**w[1,3]**w[1,3] -d[3,4]**d[2,1]**d[1,2]**w[2,3] -d[3,4]**d[2,1]**w[2,3]**w[2,3]
+1/2d[3,4]**d[2,2]**d[1,1]**d[1,2] +1/2d[3,4]**d[2,2]**d[1,1]**w[2,3] +d[3,4]**d[2,2]**d[1,2]**w[1,3]
+1/4d[3,4]**d[2,2]**d[2,2]**d[1,2] +1/4d[3,4]**d[2,2]**d[2,2]**w[2,3] +d[3,4]**d[2,2]**w[1,3]**w[2,3]
+1/4d[4,3]**d[1,1]**d[1,1]**w[1,4] +d[4,3]**d[1,1]**w[1,4]**w[2,4] -d[4,3]**d[1,2]**w[1,4]**w[1,4]
+1/4d[4,3]**d[2,1]**d[1,1]**d[1,1] +d[4,3]**d[2,1]**d[1,1]**w[2,4] -d[4,3]**d[2,1]**d[1,2]**w[1,4]
+1/2d[4,3]**d[2,1]**d[2,2]**d[1,1] +1/4d[4,3]**d[2,1]**d[2,2]**d[2,2] +d[4,3]**d[2,1]**d[2,2]**w[2,4]
+d[4,3]**d[2,1]**w[2,4]**w[2,4] +1/2d[4,3]**d[2,2]**d[1,1]**w[1,4] -3/4d[4,3]**d[2,2]**d[2,2]**w[1,4] -
d[4,3]**d[2,2]**w[1,4]**w[2,4] +1/2d[4,3]**d[3,3]**d[1,1]**w[1,4] +1/2d[4,3]**d[3,3]**d[2,1]**d[1,1]

+1/2d[4,3]**d[3,3]**d[2,1]**d[2,2] +d[4,3]**d[3,3]**d[2,1]**w[2,4] -1/2d[4,3]**d[3,3]**d[2,2]**w[1,4]
+1/4d[4,3]**d[3,3]**d[3,3]**d[2,1] +1/4d[4,3]**d[3,3]**d[3,3]**d[3,4] +1/4d[4,3]**d[3,3]**d[3,3]**w[1,4] -
1/2d[4,3]**d[3,3]**d[3,4]**d[1,1] -1/2d[4,3]**d[3,3]**d[3,4]**d[2,2] -d[4,3]**d[3,3]**d[3,4]**w[2,4] -
3/4d[4,3]**d[3,4]**d[1,1]**d[1,1] -d[4,3]**d[3,4]**d[1,1]**w[1,3] -d[4,3]**d[3,4]**d[1,2]**w[1,4] -
2d[4,3]**d[3,4]**d[2,1]**d[1,2] -d[4,3]**d[3,4]**d[2,1]**w[2,3] +1/2d[4,3]**d[3,4]**d[2,2]**d[1,1] -
3/4d[4,3]**d[3,4]**d[2,2]**d[2,2] -d[4,3]**d[3,4]**d[2,2]**w[2,4] +d[4,3]**d[3,4]**d[3,4]**w[2,3] -
d[4,3]**d[3,4]**w[1,3]**w[2,4] +d[4,3]**d[3,4]**w[1,4]**w[2,3] +d[4,3]**d[4,3]**d[3,4]**d[3,4]
+d[4,3]**d[4,3]**d[3,4]**w[1,4] +1/2d[4,3]**d[4,4]**d[1,1]**w[1,4] +1/2d[4,3]**d[4,4]**d[2,1]**d[1,1]
+1/2d[4,3]**d[4,4]**d[2,1]**d[2,2] +d[4,3]**d[4,4]**d[2,1]**w[2,4] -1/2d[4,3]**d[4,4]**d[2,2]**w[1,4]
+1/2d[4,3]**d[4,4]**d[3,3]**d[2,1] -3/2d[4,3]**d[4,4]**d[3,3]**d[3,4] -1/2d[4,3]**d[4,4]**d[3,3]**w[1,4] -
1/2d[4,3]**d[4,4]**d[3,4]**d[1,1] -1/2d[4,3]**d[4,4]**d[3,4]**d[2,2] -d[4,3]**d[4,4]**d[3,4]**w[1,3]
+1/4d[4,3]**d[4,4]**d[4,4]**d[2,1] +1/4d[4,3]**d[4,4]**d[4,4]**d[3,4] +1/4d[4,3]**d[4,4]**d[4,4]**w[1,4] -
1/4d[4,4]**d[1,1]**d[1,1]**d[1,1] -1/4d[4,4]**d[1,1]**d[1,1]**w[1,3] -1/2d[4,4]**d[1,1]**d[1,1]**w[2,4]
+1/2d[4,4]**d[1,1]**d[1,2]**w[1,4] -1/2d[4,4]**d[1,1]**w[1,3]**w[2,4] -1/2d[4,4]**d[1,1]**w[1,4]**w[2,3]
+d[4,4]**d[1,2]**w[1,3]**w[1,4] -1/2d[4,4]**d[2,1]**d[1,1]**d[1,2] -1/2d[4,4]**d[2,1]**d[1,1]**w[2,3] -
d[4,4]**d[2,1]**d[1,2]**w[2,4] -1/2d[4,4]**d[2,1]**d[2,2]**d[1,2] -1/2d[4,4]**d[2,1]**d[2,2]**w[2,3] -
d[4,4]**d[2,1]**w[2,3]**w[2,4] +1/2d[4,4]**d[2,2]**d[1,1]**w[2,4] +1/2d[4,4]**d[2,2]**d[1,2]**w[1,4]
+1/4d[4,4]**d[2,2]**d[2,2]**d[1,1] +1/4d[4,4]**d[2,2]**d[2,2]**w[1,3] +1/2d[4,4]**d[2,2]**w[1,3]**w[2,4]
+1/2d[4,4]**d[2,2]**w[1,4]**w[2,3] +1/4d[4,4]**d[3,3]**d[1,1]**d[1,1] +1/2d[4,4]**d[3,3]**d[1,1]**w[1,3]
+1/2d[4,4]**d[3,3]**d[1,2]**w[1,4] +1/2d[4,4]**d[3,3]**d[2,1]**d[1,2] +1/2d[4,4]**d[3,3]**d[2,1]**w[2,3]
+1/4d[4,4]**d[3,3]**d[2,2]**d[2,2] +1/2d[4,4]**d[3,3]**d[2,2]**w[2,4] +1/4d[4,4]**d[3,3]**d[3,3]**d[1,1] -
1/4d[4,4]**d[3,3]**d[3,3]**d[3,3] -1/4d[4,4]**d[3,3]**d[3,3]**w[1,3] +1/2d[4,4]**d[3,3]**d[3,3]**w[2,4]
+1/2d[4,4]**d[3,3]**d[3,4]**d[1,2] -1/2d[4,4]**d[3,3]**d[3,4]**w[2,3] +1/2d[4,4]**d[3,3]**w[1,3]**w[2,4] -
1/2d[4,4]**d[3,3]**w[1,4]**w[2,3] +1/2d[4,4]**d[3,4]**d[1,1]**d[1,2] -1/2d[4,4]**d[3,4]**d[1,1]**w[2,3]
+d[4,4]**d[3,4]**d[1,2]**w[1,3] +1/2d[4,4]**d[3,4]**d[2,2]**d[1,2] +1/2d[4,4]**d[3,4]**d[2,2]**w[2,3] -
1/2d[4,4]**d[4,4]**d[1,1]**d[1,1] -1/2d[4,4]**d[4,4]**d[1,1]**w[1,3] -1/4d[4,4]**d[4,4]**d[1,1]**w[2,4]
+1/4d[4,4]**d[4,4]**d[1,2]**w[1,4] -3/4d[4,4]**d[4,4]**d[2,1]**d[1,2] -3/4d[4,4]**d[4,4]**d[2,1]**w[2,3]
+1/4d[4,4]**d[4,4]**d[2,2]**d[1,1] +1/4d[4,4]**d[4,4]**d[2,2]**w[1,3] +1/4d[4,4]**d[4,4]**d[3,3]**d[2,2]
+1/2d[4,4]**d[4,4]**d[3,3]**d[3,3] +1/2d[4,4]**d[4,4]**d[3,3]**w[1,3] -1/4d[4,4]**d[4,4]**d[3,3]**w[2,4]
+1/4d[4,4]**d[4,4]**d[3,4]**d[1,2] +1/4d[4,4]**d[4,4]**d[3,4]**w[2,3] -1/4d[4,4]**d[4,4]**d[4,4]**d[1,1] -
1/4d[4,4]**d[4,4]**d[4,4]**d[3,3] -1/4d[4,4]**d[4,4]**d[4,4]**w[1,3] -1/4d[4,4]**d[4,4]**w[1,3]**w[2,4]
+1/4d[4,4]**d[4,4]**w[1,4]**w[2,3] -1/4f[3,1]**d[1,1]**d[1,1]**w[2,4] +f[3,1]**d[1,1]**w[1,4]**w[2,3] -
f[3,1]**d[1,2]**w[1,3]**w[1,4] +f[3,1]**d[2,1]**d[1,1]**w[2,3] -f[3,1]**d[2,1]**d[1,2]**w[1,3] -
f[3,1]**d[2,1]**d[2,2]**d[1,2] +f[3,1]**d[2,1]**w[2,3]**w[2,4] -1/4f[3,1]**d[2,2]**d[1,1]**d[1,1]
+1/2f[3,1]**d[2,2]**d[1,1]**w[2,4] -f[3,1]**d[2,2]**d[1,2]**w[1,4] +1/2f[3,1]**d[2,2]**d[2,2]**d[1,1] -
1/4f[3,1]**d[2,2]**d[2,2]**d[2,2] -1/4f[3,1]**d[2,2]**d[2,2]**w[2,4] -f[3,1]**d[2,2]**w[1,4]**w[2,3]
+1/2f[3,1]**d[3,3]**d[1,1]**w[2,4] -f[3,1]**d[3,3]**d[1,2]**w[1,4] -f[3,1]**d[3,3]**d[2,1]**d[1,2]
+f[3,1]**d[3,3]**d[2,1]**w[2,3] +1/2f[3,1]**d[3,3]**d[2,2]**d[1,1] -1/2f[3,1]**d[3,3]**d[2,2]**d[2,2] -
1/2f[3,1]**d[3,3]**d[2,2]**w[2,4] -1/4f[3,1]**d[3,3]**d[3,3]**d[2,2] -1/4f[3,1]**d[3,3]**d[3,3]**w[2,4]
+f[3,1]**d[3,3]**w[1,4]**w[2,3] -f[3,1]**d[3,4]**d[1,1]**w[2,3] -f[3,1]**d[3,4]**w[1,3]**w[2,3]
+f[3,1]**d[4,3]**d[1,1]**w[1,4] +f[3,1]**d[4,3]**d[2,1]**d[1,1] +f[3,1]**d[4,3]**d[2,1]**d[2,2]
+2f[3,1]**d[4,3]**d[2,1]**w[2,4] +f[3,1]**d[4,3]**d[3,3]**d[2,1] +f[3,1]**d[4,3]**d[3,3]**w[1,4] -
f[3,1]**d[4,3]**d[3,4]**d[1,1] -f[3,1]**d[4,3]**d[3,4]**w[1,3] +f[3,1]**d[4,3]**d[4,4]**d[2,1] -
f[3,1]**d[4,3]**d[4,4]**d[3,4] +f[3,1]**d[4,3]**w[1,4]**w[2,4] -1/4f[3,1]**d[4,4]**d[1,1]**d[1,1] -
1/2f[3,1]**d[4,4]**d[1,1]**w[2,4] +1/4f[3,1]**d[4,4]**d[2,2]**d[2,2] +1/2f[3,1]**d[4,4]**d[2,2]**w[2,4]
+1/2f[3,1]**d[4,4]**d[3,3]**d[1,1] -1/4f[3,1]**d[4,4]**d[3,3]**d[3,3] +1/2f[3,1]**d[4,4]**d[3,3]**w[2,4] -
f[3,1]**d[4,4]**d[3,4]**w[2,3] -1/2f[3,1]**d[4,4]**d[4,4]**d[1,1] +1/4f[3,1]**d[4,4]**d[4,4]**d[2,2]
+1/2f[3,1]**d[4,4]**d[4,4]**d[3,3] -1/4f[3,1]**d[4,4]**d[4,4]**d[4,4] -1/4f[3,1]**d[4,4]**d[4,4]**w[2,4] -
f[3,1]**d[4,4]**w[1,4]**w[2,3] +f[3,1]**f[3,1]**d[2,1]**w[2,3] +f[3,1]**f[3,1]**d[4,3]**d[2,1]
+f[3,1]**f[3,1]**d[4,3]**w[1,4] +f[3,1]**f[3,1]**w[1,4]**w[2,3] +1/4f[3,2]**d[1,1]**d[1,1]**d[1,2] -
3/4f[3,2]**d[1,1]**d[1,1]**w[2,3] +f[3,2]**d[1,1]**d[1,2]**w[1,3] -f[3,2]**d[1,1]**d[1,2]**w[2,4] -
f[3,2]**d[1,1]**w[1,3]**w[2,3] +f[3,2]**d[1,2]**d[1,2]**w[1,4] +f[3,2]**d[1,2]**w[1,3]**w[1,3] -
f[3,2]**d[1,2]**w[1,3]**w[2,4] +f[3,2]**d[1,2]**w[1,4]**w[2,3] +f[3,2]**d[2,1]**d[1,2]**d[1,2] -
f[3,2]**d[2,1]**w[2,3]**w[2,3] -1/2f[3,2]**d[2,2]**d[1,1]**d[1,2] +1/2f[3,2]**d[2,2]**d[1,1]**w[2,3]
+1/4f[3,2]**d[2,2]**d[2,2]**d[1,2] +1/4f[3,2]**d[2,2]**d[2,2]**w[2,3] +f[3,2]**d[2,2]**w[1,3]**w[2,3]
+1/2f[3,2]**d[3,3]**d[1,1]**d[1,2] -1/2f[3,2]**d[3,3]**d[1,1]**w[2,3] +f[3,2]**d[3,3]**d[1,2]**w[1,3]
+1/2f[3,2]**d[3,3]**d[2,2]**d[1,2] +3/2f[3,2]**d[3,3]**d[2,2]**w[2,3] +1/4f[3,2]**d[3,3]**d[3,3]**d[1,2]

+1/4f[3,2]**d[3,3]**d[3,3]**w[2,3] +f[3,2]**d[3,3]**w[2,3]**w[2,4] -f[3,2]**d[3,4]**d[1,2]**w[2,3] -
f[3,2]**d[3,4]**w[2,3]**w[2,3] -3/4f[3,2]**d[4,3]**d[1,1]**d[1,1] -f[3,2]**d[4,3]**d[1,1]**w[1,3] -
f[3,2]**d[4,3]**d[2,1]**d[1,2] -f[3,2]**d[4,3]**d[2,1]**w[2,3] +1/2f[3,2]**d[4,3]**d[2,2]**d[1,1]
+1/4f[3,2]**d[4,3]**d[2,2]**d[2,2] +f[3,2]**d[4,3]**d[2,2]**w[2,4] -1/2f[3,2]**d[4,3]**d[3,3]**d[1,1]
+1/2f[3,2]**d[4,3]**d[3,3]**d[2,2] +1/4f[3,2]**d[4,3]**d[3,3]**d[3,3] -f[3,2]**d[4,3]**d[3,4]**d[1,2]
+f[3,2]**d[4,3]**d[4,3]**d[3,4] +f[3,2]**d[4,3]**d[4,3]**w[1,4] -1/2f[3,2]**d[4,3]**d[4,4]**d[1,1]
+1/2f[3,2]**d[4,3]**d[4,4]**d[2,2] -1/2f[3,2]**d[4,3]**d[4,4]**d[3,3] +1/4f[3,2]**d[4,3]**d[4,4]**d[4,4] -
f[3,2]**d[4,3]**d[4,4]**w[1,3] +f[3,2]**d[4,3]**d[4,4]**w[2,4] -f[3,2]**d[4,3]**w[1,3]**w[2,4]
+f[3,2]**d[4,3]**w[1,4]**w[2,3] +f[3,2]**d[4,3]**w[2,4]**w[2,4] -1/2f[3,2]**d[4,4]**d[1,1]**d[1,2] -
1/2f[3,2]**d[4,4]**d[1,1]**w[2,3] -f[3,2]**d[4,4]**d[1,2]**w[2,4] -1/2f[3,2]**d[4,4]**d[2,2]**d[1,2] -
1/2f[3,2]**d[4,4]**d[2,2]**w[2,3] +1/2f[3,2]**d[4,4]**d[3,3]**d[1,2] +1/2f[3,2]**d[4,4]**d[3,3]**w[2,3] -
3/4f[3,2]**d[4,4]**d[4,4]**d[1,2] -3/4f[3,2]**d[4,4]**d[4,4]**w[2,3] -f[3,2]**d[4,4]**w[2,3]**w[2,4] -
f[3,2]**f[3,1]**d[1,1]**w[2,3] +f[3,2]**f[3,1]**d[2,2]**w[2,3] -f[3,2]**f[3,1]**d[4,3]**d[1,1]
+f[3,2]**f[3,1]**d[4,3]**d[2,2] -f[3,2]**f[3,1]**d[4,3]**w[1,3] +f[3,2]**f[3,1]**d[4,3]**w[2,4] -
f[3,2]**f[3,1]**w[1,3]**w[2,3] +f[3,2]**f[3,1]**w[2,3]**w[2,4] -f[3,2]**f[3,2]**d[1,2]**w[2,3] -
f[3,2]**f[3,2]**d[4,3]**d[1,2] -f[3,2]**f[3,2]**d[4,3]**w[2,3] -f[3,2]**f[3,2]**w[2,3]**w[2,3]
+1/4f[4,1]**d[1,1]**d[1,1]**w[1,4] +f[4,1]**d[1,1]**w[1,4]**w[2,4] -f[4,1]**d[1,2]**w[1,4]**w[1,4]
+1/4f[4,1]**d[2,1]**d[1,1]**d[1,1] +f[4,1]**d[2,1]**d[2,1]**d[1,2] +f[4,1]**d[2,1]**d[2,1]**w[2,3] -
1/2f[4,1]**d[2,1]**d[2,2]**d[1,1] +1/4f[4,1]**d[2,1]**d[2,2]**d[2,2] -f[4,1]**d[2,1]**d[2,2]**w[1,3]
+f[4,1]**d[2,1]**d[2,2]**w[2,4] -f[4,1]**d[2,1]**w[1,3]**w[2,4] +f[4,1]**d[2,1]**w[1,4]**w[2,3]
+f[4,1]**d[2,1]**w[2,4]**w[2,4] +1/2f[4,1]**d[2,2]**d[1,1]**w[1,4] -3/4f[4,1]**d[2,2]**d[2,2]**w[1,4] -
f[4,1]**d[2,2]**w[1,4]**w[2,4] -1/2f[4,1]**d[3,3]**d[1,1]**w[1,4] -1/2f[4,1]**d[3,3]**d[2,1]**d[1,1] -
1/2f[4,1]**d[3,3]**d[2,1]**d[2,2] -f[4,1]**d[3,3]**d[2,1]**w[1,3] -1/2f[4,1]**d[3,3]**d[2,2]**w[1,4] -
3/4f[4,1]**d[3,3]**d[3,3]**d[2,1] +1/4f[4,1]**d[3,3]**d[3,3]**d[3,4] -3/4f[4,1]**d[3,3]**d[3,3]**w[1,4]
+1/2f[4,1]**d[3,3]**d[3,4]**d[1,1] -1/2f[4,1]**d[3,3]**d[3,4]**d[2,2] +f[4,1]**d[3,3]**d[3,4]**w[1,3] -
f[4,1]**d[3,3]**d[3,4]**w[2,4] -f[4,1]**d[3,3]**w[1,3]**w[1,4] +1/4f[4,1]**d[3,4]**d[1,1]**d[1,1]
+f[4,1]**d[3,4]**d[1,1]**w[1,3] -f[4,1]**d[3,4]**d[1,2]**w[1,4] -f[4,1]**d[3,4]**d[2,1]**d[1,2]
+1/2f[4,1]**d[3,4]**d[2,2]**d[1,1] -3/4f[4,1]**d[3,4]**d[2,2]**d[2,2] -f[4,1]**d[3,4]**d[2,2]**w[2,4]
+f[4,1]**d[3,4]**d[3,4]**w[2,3] +f[4,1]**d[3,4]**w[1,3]**w[1,3] -f[4,1]**d[3,4]**w[1,3]**w[2,4]
+f[4,1]**d[3,4]**w[1,4]**w[2,3] -f[4,1]**d[4,3]**d[2,1]**w[1,4] -f[4,1]**d[4,3]**d[3,4]**d[2,1]
+f[4,1]**d[4,3]**d[3,4]**d[3,4] -f[4,1]**d[4,3]**w[1,4]**w[1,4] +3/2f[4,1]**d[4,4]**d[1,1]**w[1,4]
+1/2f[4,1]**d[4,4]**d[2,1]**d[1,1] +1/2f[4,1]**d[4,4]**d[2,1]**d[2,2] +f[4,1]**d[4,4]**d[2,1]**w[2,4] -
1/2f[4,1]**d[4,4]**d[2,2]**w[1,4] +1/2f[4,1]**d[4,4]**d[3,3]**d[2,1] -1/2f[4,1]**d[4,4]**d[3,3]**d[3,4]
+1/2f[4,1]**d[4,4]**d[3,3]**w[1,4] +1/2f[4,1]**d[4,4]**d[3,4]**d[1,1] -1/2f[4,1]**d[4,4]**d[3,4]**d[2,2]
+1/4f[4,1]**d[4,4]**d[4,4]**d[2,1] +1/4f[4,1]**d[4,4]**d[4,4]**d[3,4] +1/4f[4,1]**d[4,4]**d[4,4]**w[1,4]
+f[4,1]**d[4,4]**w[1,3]**w[1,4] -f[4,1]**f[3,1]**d[2,1]**w[1,3] +f[4,1]**f[3,1]**d[2,1]**w[2,4] -
f[4,1]**f[3,1]**d[3,3]**d[2,1] -f[4,1]**f[3,1]**d[3,3]**w[1,4] +f[4,1]**f[3,1]**d[4,4]**d[2,1]
+f[4,1]**f[3,1]**d[4,4]**w[1,4] -f[4,1]**f[3,1]**w[1,3]**w[1,4] +f[4,1]**f[3,1]**w[1,4]**w[2,4]
+1/4f[4,1]**f[3,2]**d[1,1]**d[1,1] +f[4,1]**f[3,2]**d[1,1]**w[1,3] -f[4,1]**f[3,2]**d[1,1]**w[2,4]
+f[4,1]**f[3,2]**d[1,2]**w[1,4] +f[4,1]**f[3,2]**d[2,1]**d[1,2] +f[4,1]**f[3,2]**d[2,1]**w[2,3] -
1/2f[4,1]**f[3,2]**d[2,2]**d[1,1] +1/4f[4,1]**f[3,2]**d[2,2]**d[2,2] -f[4,1]**f[3,2]**d[2,2]**w[1,3]
+f[4,1]**f[3,2]**d[2,2]**w[2,4] +1/2f[4,1]**f[3,2]**d[3,3]**d[1,1] -1/2f[4,1]**f[3,2]**d[3,3]**d[2,2]
+1/4f[4,1]**f[3,2]**d[3,3]**d[3,3] +f[4,1]**f[3,2]**d[3,3]**w[1,3] -f[4,1]**f[3,2]**d[3,3]**w[2,4]
+f[4,1]**f[3,2]**d[3,4]**w[2,3] +f[4,1]**f[3,2]**d[4,3]**d[3,4] +f[4,1]**f[3,2]**d[4,3]**w[1,4] -
1/2f[4,1]**f[3,2]**d[4,4]**d[1,1] +1/2f[4,1]**f[3,2]**d[4,4]**d[2,2] -1/2f[4,1]**f[3,2]**d[4,4]**d[3,3]
+1/4f[4,1]**f[3,2]**d[4,4]**d[4,4] -f[4,1]**f[3,2]**d[4,4]**w[1,3] +f[4,1]**f[3,2]**d[4,4]**w[2,4]
+f[4,1]**f[3,2]**w[1,3]**w[1,3] -2f[4,1]**f[3,2]**w[1,3]**w[2,4] +2f[4,1]**f[3,2]**w[1,4]**w[2,3]
+f[4,1]**f[3,2]**w[2,4]**w[2,4] -f[4,1]**f[4,1]**d[2,1]**w[1,4] -f[4,1]**f[4,1]**d[3,4]**d[2,1] -
f[4,1]**f[4,1]**d[3,4]**w[1,4] -f[4,1]**f[4,1]**w[1,4]**w[1,4] -1/4f[4,2]**d[1,1]**d[1,1]**d[1,1] -
1/4f[4,2]**d[1,1]**d[1,1]**w[1,3] -f[4,2]**d[1,1]**w[1,4]**w[2,3] +f[4,2]**d[1,2]**w[1,3]**w[1,4] -
f[4,2]**d[2,1]**d[1,1]**d[1,2] -f[4,2]**d[2,1]**d[1,1]**w[2,3] -f[4,2]**d[2,1]**d[1,2]**w[2,4] -
f[4,2]**d[2,1]**w[2,3]**w[2,4] +1/2f[4,2]**d[2,2]**d[1,1]**d[1,1] +1/2f[4,2]**d[2,2]**d[1,1]**w[1,3]
+f[4,2]**d[2,2]**d[1,2]**w[1,4] -1/4f[4,2]**d[2,2]**d[2,2]**d[1,1] -1/4f[4,2]**d[2,2]**d[2,2]**w[1,3]
+f[4,2]**d[2,2]**w[1,4]**w[2,3] +1/4f[4,2]**d[3,3]**d[1,1]**d[1,1] +1/2f[4,2]**d[3,3]**d[1,1]**w[1,3] -
1/4f[4,2]**d[3,3]**d[2,2]**d[2,2] -1/2f[4,2]**d[3,3]**d[2,2]**w[1,3] +1/4f[4,2]**d[3,3]**d[3,3]**d[1,1] -
1/2f[4,2]**d[3,3]**d[3,3]**d[2,2] -1/4f[4,2]**d[3,3]**d[3,3]**d[3,3] -1/4f[4,2]**d[3,3]**d[3,3]**w[1,3]
+f[4,2]**d[3,3]**d[3,4]**d[1,2] -f[4,2]**d[3,3]**w[1,4]**w[2,3] +f[4,2]**d[3,4]**d[1,1]**d[1,2]

+2f[4,2]**d[3,4]**d[1,2]**w[1,3] +f[4,2]**d[3,4]**d[2,2]**d[1,2] +f[4,2]**d[3,4]**d[2,2]**w[2,3]
+f[4,2]**d[3,4]**w[1,3]**w[2,3] -f[4,2]**d[4,3]**d[2,2]**w[1,4] -f[4,2]**d[4,3]**d[3,3]**d[3,4] -
f[4,2]**d[4,3]**d[3,3]**w[1,4] -f[4,2]**d[4,3]**d[3,4]**d[2,2] -f[4,2]**d[4,3]**d[3,4]**w[2,4] -
f[4,2]**d[4,3]**w[1,4]**w[2,4] -1/2f[4,2]**d[4,4]**d[1,1]**d[1,1] -1/2f[4,2]**d[4,4]**d[1,1]**w[1,3]
+f[4,2]**d[4,4]**d[1,2]**w[1,4] -f[4,2]**d[4,4]**d[2,1]**d[1,2] -f[4,2]**d[4,4]**d[2,1]**w[2,3]
+1/2f[4,2]**d[4,4]**d[2,2]**d[1,1] +1/2f[4,2]**d[4,4]**d[2,2]**w[1,3] +1/2f[4,2]**d[4,4]**d[3,3]**d[2,2]
+1/2f[4,2]**d[4,4]**d[3,3]**d[3,3] +1/2f[4,2]**d[4,4]**d[3,3]**w[1,3] +f[4,2]**d[4,4]**d[3,4]**d[1,2]
+f[4,2]**d[4,4]**d[3,4]**w[2,3] -1/4f[4,2]**d[4,4]**d[4,4]**d[1,1] -1/4f[4,2]**d[4,4]**d[4,4]**d[3,3] -
1/4f[4,2]**d[4,4]**d[4,4]**w[1,3] +f[4,2]**d[4,4]**w[1,4]**w[2,3] -1/4f[4,2]**f[3,1]**d[1,1]**d[1,1] -
f[4,2]**f[3,1]**d[1,2]**w[1,4] -f[4,2]**f[3,1]**d[2,1]**d[1,2] -f[4,2]**f[3,1]**d[2,1]**w[2,3]
+1/2f[4,2]**f[3,1]**d[2,2]**d[1,1] -1/4f[4,2]**f[3,1]**d[2,2]**d[2,2] +1/2f[4,2]**f[3,1]**d[3,3]**d[1,1] -
1/2f[4,2]**f[3,1]**d[3,3]**d[2,2] -1/4f[4,2]**f[3,1]**d[3,3]**d[3,3] -f[4,2]**f[3,1]**d[3,4]**w[2,3] -
f[4,2]**f[3,1]**d[4,3]**d[3,4] -f[4,2]**f[3,1]**d[4,3]**w[1,4] -1/2f[4,2]**f[3,1]**d[4,4]**d[1,1]
+1/2f[4,2]**f[3,1]**d[4,4]**d[2,2] +1/2f[4,2]**f[3,1]**d[4,4]**d[3,3] -1/4f[4,2]**f[3,1]**d[4,4]**d[4,4] -
2f[4,2]**f[3,1]**w[1,4]**w[2,3] +f[4,2]**f[3,2]**d[1,2]**w[1,3] -f[4,2]**f[3,2]**d[1,2]**w[2,4]
+f[4,2]**f[3,2]**d[3,3]**d[1,2] +f[4,2]**f[3,2]**d[3,3]**w[2,3] -f[4,2]**f[3,2]**d[4,4]**d[1,2] -
f[4,2]**f[3,2]**d[4,4]**w[2,3] +f[4,2]**f[3,2]**w[1,3]**w[2,3] -f[4,2]**f[3,2]**w[2,3]**w[2,4]
+f[4,2]**f[4,1]**d[1,1]**w[1,4] -f[4,2]**f[4,1]**d[2,2]**w[1,4] +f[4,2]**f[4,1]**d[3,4]**d[1,1] -
f[4,2]**f[4,1]**d[3,4]**d[2,2] +f[4,2]**f[4,1]**d[3,4]**w[1,3] -f[4,2]**f[4,1]**d[3,4]**w[2,4]
+f[4,2]**f[4,1]**w[1,3]**w[1,4] -f[4,2]**f[4,1]**w[1,4]**w[2,4] +f[4,2]**f[4,2]**d[1,2]**w[1,4]
+f[4,2]**f[4,2]**d[3,4]**d[1,2] +f[4,2]**f[4,2]**d[3,4]**w[2,3] +f[4,2]**f[4,2]**w[1,4]**w[2,3])

8. References

[1] Weizsäcker C F v 1958 *Weltbild der Physik*, (Stuttgart: Hirzel) pp 281, 1982 *Aufbau der Physik*, (München: Hanser)

[2] Wigner E P 1939 *Annals of Mathematics* **40** 149

[3] see e.g. Barut A O, Rączka R 1986 (Singapore: World Scientific) *Theory of group representations and applications*; Tung W-K 1985 (Singapore: World Scientific) *Group Theory in Physics*; Scheck F 2007 (Springer: Berlin) *Theoretische Physik 4*

[4] Görnitz T 1988 *Intern. J. of Theoret. Phys.* **27** 527, 1988 *Intern. J. of Theoret. Phys.* **27** 659

[5] Riess A G et al 2011 A 3% Solution: Determination of the Hubble Constant with the Hubble Space Telescope and Wide Field Camera 3, *Astrophysical J.*, **730** 119

[6] Tonry J L et al 2003 Cosmological Results from High-z Supernovae, *Astrophysical J.* **594** 1

[7] Görnitz T 2011 *Adv. Sci. Lett.* **4** 577-585

[8] Görnitz T 1988 *Intern. J. of Theor. Phys.* **27** 527

[9] Weizsäcker C F v 1972 *Nova Acta Leopoldina*, **206** 515

[10] Green H S 1953 *Phys. Rev.* **90** 270

[11] Scheibe E, Süßmann G, Weizsäcker C F v 1958 *Z. Naturforschung* **13a** 705

[12] Görnitz T, Görnitz B 2008 *Die Evolution des Geistigen* (Göttingen: Vandenhoeck&Ruprecht), 2002 *Der kreative Kosmos - Geist und Materie aus Information* (Heidelberg: Spektrum Akadem. Verl.)